Developn
Wooı

ACS SYMPOSIUM SERIES **982**

Development of Commercial Wood Preservatives

Efficacy, Environmental, and Health Issues

Tor P. Schultz, Editor
Mississippi State University

Holger Militz, Editor
Georg-August University

Michael H. Freeman, Editor
Consultant

Barry Goodell, Editor
University of Maine

Darrel D. Nicholas, Editor
Mississippi State University

Sponsored by the
Division of Cellulose and Renewable Materials

American Chemical Society, Washington, DC

Library of Congress Cataloging-in-Publication Data

Development of commercial wood preservatives : efficacy, environmental, and health issues / Tor P. Schultz, editor...[et al.] ; sponsored by the Division of Cellulose and Renewable Materials.

p. cm.—(ACS symposium series ; 982)

Includes bibliographical references and index.

ISBN 978–0–8412–3951–7 (alk. paper)

1. Wood—Preservation.

I. Schultz, Tor P., 1953- II. American Chemical Society. Division of Cellulose and Renewable Materials.

TA424.D48 2008
674′.386—dc22

2007060563

The photographs on the cover (top 3) are coutesy of Tor P. Schultz (Mississippi State University), the bottom left is coutesy of Terry Amburgey (Mississippi State University), and the bottom left is courtesy of woodbug.com (photograph by Scott Bauer of the Agricultural Research Service, U.S. Depratent of Agriculture.)

The paper used in this publication meets the minimum requirements of American National Standard for Information Sciences—Permanence of Paper for Printed Library Materials, ANSI Z39.48–1984.

Distributed by Oxford University Press

PRINTED IN THE UNITED STATES OF AMERICA

Foreword

The ACS Symposium Series was first published in 1974 to provide a mechanism for publishing symposia quickly in book form. The purpose of the series is to publish timely, comprehensive books developed from ACS sponsored symposia based on current scientific research. Occasionally, books are developed from symposia sponsored by other organizations when the topic is of keen interest to the chemistry audience.

Before agreeing to publish a book, the proposed table of contents is reviewed for appropriate and comprehensive coverage and for interest to the audience. Some papers may be excluded to better focus the book; others may be added to provide comprehensiveness. When appropriate, overview or introductory chapters are added. Drafts of chapters are peer-reviewed prior to final acceptance or rejection, and manuscripts are prepared in camera-ready format.

As a rule, only original research papers and original review papers are included in the volumes. Verbatim reproductions of previously published papers are not accepted.

ACS Books Department

Contents

Molds and Mold Health Issues

Chemical Biocides, Termite Control Methods, and Biocide Depletion

Non-Biocidal Chemicals and Processes to Protect Wood

Chemical Wood Preservative Systems

Registration and Approval, and Environmental Disposal Issues of Wood Preservatives

Worldwide Trends in Wood Preservation

Indexes

Preface

Wood is one of the oldest materials employed to benefit mankind. Recently, however, there is a perception among the public that alternative materials are more technically advanced than wood, and a growing segment of the public is willing to pay a premium for new and modern non-wood construction products. Further, starting in the 1990s concerns arose over arsenic, chromium, pentachlorophenol and creosote exposure, and disposal issues of wood treated with these preservatives. While such concerns have been debated at considerable length, the result was that wood preservation underwent a rapid and dramatic change worldwide.

The view of wood as an "inferior" non-technically advanced material has changed. We now see forest products as a renewable resource to make sustainable products that require much less energy to manufac-ture then alternative non-renewable materials such as concrete blocks, steel studs, and plastic lumber. Furthermore, as trees grow they sequester carbon dioxide, providing additional ecological benefits.

The field of wood preservation will continue to evolve to address society's changing environmental perceptions and needs, including developing cost-effective, desirable, and dependable construction material with good weathering and bioresistant properties that can be safely and easily disposed of at the end of its useful service life.

The editors appreciate and enjoy the privilege of working in the profession of wood science and protection, and the challenge of developing environmentally sound and sustainable wood products that will benefit mankind in the coming decades. We thank all our professional colleagues who have provided much assistance in this endeavor, especially the authors and many reviewers. We also thank the professional staff of the American Chemical Society Books Department and Oxford University Press for their kind and patient assistance, especially Dara Moore and Margaret Brown. Some of the photos were kindly provided by Professors Terry Amburgey and H. Mike Barnes, Mississippi State University, and Mitteramskogler GmbH, of Mirako brand heat-treated lumber.

Tor Schultz, Darrel Nicholas, and Barry Goodell also gratefully recognize the financial support for basic, long-term research provided by the U.S. Department of Agriculture/Cooperative State Research, Education, and Extension Service through the Wood Utilization Research and National Research Initiative programs.

The following corporations and organizations provided financial support which made the symposium and this resulting book possible:

- ACS Division of Cellulose and Renewable Materials
- Akzo Nobel Functional Chemicals
- Arch Wood Protection Inc.
- Ciba Specialty Chemicals
- Eastman Chemical Company
- ISK Biocides, Inc.
- Janssen Pharmaceutica
- Lonza Inc.
- Merichem Company
- Mississippi State University, FWRC
- Nisus Corporation
- Osmose Holding, Inc.
- Phibro-Tech Inc.
- Rio Tinto Minerals/U.S. Borax
- Sostram Corporation
- Viance LLC
- Weyerhaeuser Company

We five editors have enjoyed working together and with our many professional colleagues, and hope that this book proves useful to researchers, managers and students working in the challenging yet rewarding field of wood preservation.

Tor P. Schultz
Forest Products Laboratory/FWRC
Box 9820
Mississippi State University
Mississippi State, MS 39762–9820

Holger Militz
Institute of Wood Biology and Wood Technology
University of Gottingen
Busgenweg 4
D–37077 Gottingen, Germany

Mike H. Freeman
Independent Wood Scientist
7421 Hunters Tree Cove
Memphis, TN 38125

Barry Goodell
Wood Science and Technology
University of Maine
5755 Nutting Hall
Orono, ME 04469–5755

Darrel D. Nicholas
Forest Products Laboratory/FWRC
Box 9820
Mississippi State University
Mississippi State, MS 39762–9820

Development of Commercial Wood Preservatives

Introduction and Wood Deterioration Overview

These photographs are courtesy of Darrel D. Nicholas of Mississippi State University. (See page 1 of color inserts.)

Chapter 1

Introduction to Developing Wood Preservative Systems and Molds in Homes

Tor P. Schultz and Darrel D. Nicholas

Forest Products Laboratory/FWRC, Box 9820, Mississippi State University, Mississippi State, MS 39762–9820

Treating wood with biocides greatly lengthens its service life, saving consumers billions of dollars annually and extending the world's forests. Despite these positive attributes the public has a negative perception of treated wood. Recent health and environmental concerns with the older 1st-generation preservative systems has led to rapid and profound changes on a worldwide basis as we moved to the 2nd-generation copper-rich preservatives. This rapid change is continuing, with some countries now requiring 3rd-generation totally organic systems. Furthermore, 4th-generation processes that protect wood by non-biocidal means are already commercially used in Europe (heat-treated lumber) or will shortly be (acetylated lumber). Recently, molds in homes and other structures have also had much negative publicity - and resulting litigation. The purpose of this book is to describe, chapter-by-chapter, the many steps involved in developing a new wood preservative system. Also included are overview chapters that cover molds in homes, review different aspects of wood deterioration, and discuss the current and expected future status of wood preservation in Europe, North America, and Asia/Oceania.

The Past

Wood has long been employed for mankind's benefit. However, wood used to build homes, bridges, fences, ships and many other structures is degraded by decay fungi, insects/termites, bacteria, and marine borers. Non-wood degrading organisms, such as molds and stains, can also colonize wood. Finally, wood is also damaged by abiotic sunlight and weathering/rain mechanisms. This can be a curse. For example, wood deterioration is estimated to cost U.S. homeowners over $5 billion annually, and about 10% of the annual production of forests is required to merely replace degraded products. Wood degradation is also a blessing, as it is nature's way to recycle dead trees and limbs and decay fungi and their enzymes benefit man in many different ways.

Historically, some treatments have long been known to protect wood such as heating wood in an anoxic environment or applying copper salts or cedar oil. In addition physical barriers, such as copper plating the bottom of ships for protection against marine borers, were also employed. However, these historical wood protection treatments were only practiced on a small scale and man mainly relied on forests to provide new lumber to replace deteriorated wood. In the process, however, civilizations cut down many trees.

About 200 years ago the first major wood preservative, creosote, was developed. Pressure treating wood with creosote greatly increased the service life of ships and other wooden structures. Creosote is a viscous black coal-tar by-product composed of a complex mixture of polyaromatic hydrocarbons (PAHs), polynuclear aromatics (PNAs), and other organic compounds. In 2000 in North America, creosote was used primarily to treat railroad ties, utility poles and marine pilings, accounting for about 10% by volume of all treated wood. In the 1930's chlorinated phenolics were evaluated as wood preservatives and by the early 1950's oilborne pentachlorophenol, or penta, was first employed to treat utility poles and crossarms. Despite some initial resistance by creosote providers penta sales increased, and by 2000 in North America penta was used in about 10% of all treated wood. These two organic systems, creosote and penta, were the main wood preservatives up to the early 1960's, with industrial applications being the major market.

A metallic system, based on metallic salts/oxides of arsenic, chromium and copper, was also developed in the 1930's. Known first as Erdalith and later as chromated copper arsenate, or CCA, it was initially promoted by Bell Laboratories as being friendlier to their linemen. Being waterborne, CCA-treated wood also had no unpleasant odor, unlike creosote or penta. With the suburban building boom of the 1960's homeowners desired decks and patios to enjoy their backyard, and fences to block the neighbor's view, CCA-treated lumber experienced a rapidly increasing market. In 2000 in North America, CCA-treated wood accounted for about 80% by volume of all treated wood. While CCA had some industrial uses, the major application for CCA-treated wood was the residential market, where CCA accounted for over 95% of all treated wood. The large residential market, about 70% of the total volume of treated wood in 2000 in North America, was a total change from 40 years earlier when industrial uses represented the major market for treated wood products.

Some minor wood preservative systems were also utilized during this time, and lumber from naturally durable heartwood, such as redwood and cypress, was employed in relatively small amounts. In addition, some early research on wood modification, principally acetylation and heating wood in a reducing environment, was conducted. However, the primary purpose of the wood modification research was not to develop an environmentally-benign process to protect wood but to improve the dimensional stability of lumber.

The above three 1st-generation preservatives, creosote, penta and CCA, were the major systems due to their long term effectiveness, low cost, and robust formulations that were easy to use in a treating facility. They also had a long and proven track record. Thus, the early 1970's were good years for the wood preservation industry.

Starting in the late 1970's, however, health and environmental concerns with CCA, penta and creosote arose in Europe, Asia, and North America. At this time the U.S. Environmental Protection Agency conducted the Rebutable Presumption Against Re-registration study, RPAR, on the 1st-generation systems. This stimulated research to reexamine some minor systems, such as copper naphthenate and oxine copper, and to identify new wood preservatives from commercial organic agrochemicals used in non-wood applications. Much of the initial research was conducted by various wood preservative companies and the New Zealand Forest Research Institute and Michigan Technological University. RPAR only resulted in minor restrictions in the U.S., but concerns over possible restrictions on the 1st-generation systems in other countries accelerated the development of non-arsenical waterborne systems for residential applications and penta and creosote alternatives for industrial uses.

The Present

Overnight, it seems, the wood preservation industry underwent rapid and dramatic changes. Basically, wood preservatives were no longer selected based on economics and efficacy but on governmental regulations and environmental, health and disposal issues. Public perceptions, whether based on facts or misconceptions, also became important. In addition, due to rising energy costs homes were built "tighter" which increased the potential for water to become trapped within walls. This greatly increased the likelihood for mold growth and, with some negative but unsupported publicity, molds quickly became an important public health issue.

In the U.S., penta and creosote have had some additional but relatively minor use restrictions imposed in the past few years. The big change was with

CCA, due to public concerns on health and disposal issues with lumber that contained arsenic and chromium. Starting in 2004, CCA was no longer labeled [permitted to be used] for almost all residential applications. This reduced CCA usage by about 68%.

The 2nd-generation preservatives that replaced CCA for residential applications were copper-rich waterborne systems with an organic co-biocide to control copper-tolerant fungi. These new systems required a bit more attention in the treating plants than CCA, and the absence of chromium led to increased metal corrosion problems. Also, wet lumber treated with these systems often had obvious mold growth on the surface which resulted in consumer concerns. [CCA-treated lumber can also have surface mold, but it is usually mistaken for dirt. Consequently, mold growth on CCA-treated lumber was not a serious public issue.] The new systems also leach more copper than CCA-treated wood which can negatively impact aquatic systems, and disposal of metallic-treated lumber from residential structures – including CCA-treated lumber – is a growing concern.

As mentioned earlier, creosote and penta have so far faced only relatively minor restrictions in the U.S., but concern over possible restrictions has led to increased research and the development of new systems. Some of these systems were perhaps rushed into service without being fully tested and/or adequate education and technical support for users, with the result that some wood products treated with a few new systems failed. These failures and the resulting negative publicity, along with public concern with perceived health issues, has led to an increasing market share for non-wood construction alternatives such as plastic lumber and steel studs. Further increasing the public demand for non-wood alternatives is the desire by many homeowners for premium products that require little maintenance. The poor dimensional stability of wood planks, which results in lumber warping, bowing, and checking, increased the market share for dimensionally-stable plastic lumber in the large decking market. Health issues, and the public misconception that naturally-durable lumber is safe, have led to increased demand for redwood and other durable lumber. However, restrictions on harvesting trees with naturally durable heartwood have limited the availability of this lumber.

The use of wood composites is increasing due to their good dimensional stability and other desirable properties. However, wood composites that are exposed to the weather will degrade and/or have mold growth. In applications where this may occur, the composite must be protected with a biocide. Zinc borate, which has low water leachability, is used to protect many wood composites.

Waterborne borates are low-cost and safe, and have long been used in residential construction in New Zealand to protect lumber against insects. Borates are increasingly used in the U.S. for non-exposed structural lumber and other wood products in regions with drywood or Formosan termites, such as the

Gulf Coast or Hawaii. However, borates are quickly leached from wood exposed to moisture and, despite much research and some claims, no borate preservative system has yet been shown effective in long-term outdoor exposure tests.

The rapid changes in the past few years in North America occurred a decade earlier in Europe and Japan, where CCA, penta and creosote have been greatly restricted or banned outright. Copper-rich waterborne systems have been commercially available in these countries for over a decade. However, disposal and copper leaching concerns with these 2nd-generation systems recently caused three European countries to require totally-organic systems, and other countries may also shortly mandate 3rd-generation preservatives.

Regulatory pressures and public demand in Europe have brought increased attention to the 4th-generation systems, the non-biocidal modification of wood to prevent biodeterioration. Lumber protected by various heat-treatment processes has been commercially available for several years in Europe, and an European plant to produce acetylated lumber is being constructed.

In North America it is expected that 3rd-generation systems may be required for residential applications in the near future, perhaps as early as 2010. Furthermore, at least one company is considering commercial acetylation of lumber.

The Future

Likely future North American 3rd-generation organic wood preservatives are already commercially available in Europe and Asia. Totally-organic systems have several disadvantages compared to the metallics, however, including the relatively high cost of organic biocides compared to copper(II) and that organic biocides are depleted by various biological and chemical pathways that do not affect metallic biocides. Furthermore, residential wood preservatives will be waterborne, but most organic biocides are not soluble in water. Thus, emulsion formulations, or alternative technologies to the traditional pressure treating processes, need to be developed. Use of wood composites will increase and these products must also be protected with approved preservatives.

Development of effective and economical 3rd-generation organic systems for areas with high or severe deterioration hazards, such as exist in the southeastern U.S., will be "interesting". As noted by Alan Preston at CSI, the wood preservation industry will face conflicting issues – durability versus cost versus consumer expectations. Some systems may be commercialized without the lengthy testing previously employed and, as a result, some unanticipated problems may arise. In addition, the desire to minimize costs has resulted in some preservative proposals being submitted to regulatory organizations with recommended biocide levels that many professionals feel are too low. Rushing systems into commercial production without long-term testing or a comprehensive review, or having systems with too low a biocide level, may result in failures that will further increase consumer demand for non-wood alternatives. In addition, the poor weathering properties of the current 2nd-generation treated lumber will increase demand for the relatively expensive plastic decking.

Many North American consumers are interested in non-biocidal treatments or naturally-durable lumber. As mentioned earlier, however, governmental restrictions are limiting the harvesting of trees that contain naturally durable heartwood. Also, the heat-treatment process would likely not be effective in outdoor exposure in high or severe deterioration hazard areas such as the southeastern U.S. What may prove successful are chemical-modification non-biocidal processes, such as acetylation. Lumber treated with non-biocidal silicates, with borates sometimes co-added, is being advertised as suitable for outdoor above-ground or ground-contact applications. However, to the best of our knowledge no long-term outdoor exposure tests with publicly reported results have shown that a silicate and/or borate treatment is effective.

The recent dramatic increase in energy costs will cause structures to be built even more "tightly". Unfortunately, these structures are susceptible to trapping water within walls which will lead to mold growth. While much scientific knowledge has been developed by professionals that could greatly lower the potential for mold growth and reduce wood deterioration, for various reasons this knowledge is not available or utilized by many home builders.

Basically, the near future looks pretty grim for the wood preservation industry – ever increasing governmental regulations, a product that the public views negatively, increased competition from non-wood alternative construction materials, a high chance for litigation, and continuing low prices for treated wood. Furthermore, the mature nature of the treated wood industry has resulted in limited R&D funds to develop future products.

The long-term future prospects, however – at least to our eyes – looks very bright! How can we be so optimistic? Trees provide an organic material that is easily and with minimal energy converted into a wide variety of useful products, is renewable and can be easily and safely disposed (and, if burned, provides energy when disposed of), and traps carbon as the raw material is grown. However, the wood preservation profession needs to develop high-value products with desirable and dependable properties, with products that have a profit margin sufficient to encourage companies to undertake the long-term and expensive research necessary to develop additional wood-based products for the future.

Book Objective

Our first ACS symposium and subsequent book on wood deterioration emphasized the science of wood biodegradation. For this second symposium and book, we five co-organizers addressed the practical side of wood preservation with presentations by world-recognized experts that covered the many steps involved in developing a commercial wood preservative system. This includes developing new biocides or identifying commercial biocides used in non-wood applications, formulation, testing the new system, registration and approval of all active components by appropriate governmental agencies, submitting the proposed system for standardization [approval to use in clearly specified products and applications] by regulatory organizations, and environmental and disposal issues. Additionally, for readers unfamiliar with wood preservation we have provided overview chapters on different aspects of wood deterioration and that summarize the current and future status of wood preservation in Europe, Asia/Oceania, and North America. In addition to the above, we also organized a mini-symposium on Molds in Structures, with these overview chapters also presented.

We hope that this book will prove useful to our profession, and sincerely thank our many colleagues and friends who spent considerable time writing the many chapters and the industries/organizations that provided the necessary financial support. Many chapters cover areas that are extremely complex, but the authors had a limited number of pages. In addition, the subject matter is rapidly changing. Thus, readers should feel free to contact any particular author, or one of the five co-editors, for further information.

We five co-editors have enjoyed working with everyone involved and each other. We especially thank the professionals with the American Chemical Society, particularly Dara Moore of the ACS Books Deptartment and our many collegues and friends in the ACS Cellulose Division. One of us, Tor, also thanks Jake, Red, Irv, Tom and Darrel for all that you have taught me.

Chapter 2

Fungal Decay of Wood: Soft Rot—Brown Rot—White Rot

Barry Goodell[1], Yuhui Qian[1], and Jody Jellison[2]

[1]Wood Science and Technology and [2]Biological Sciences, [1]Nutting Hall/ AEWC, and [2]Hitchner Hall, University of Maine, Orono, ME 04469–5755

Wood decay by fungi is typically classified into three types: soft rot, brown rot and white rot. Brown rot fungi are the most prevalent with regard to attack on coniferous, structural wood products in North America. The wood decayed by brown rot fungi is typically brown and crumbly and it is degraded via both non-enzymatic and enzymatic systems. A series of celluloytic enzymes are employed in the degradation process by brown rot fungi, but no lignin degrading enzymes are typically involved. White rot fungi are typically associated with hardwood decay and their wood decay patterns can take on different forms. White rotted wood normally has a bleached appearance and this may either occur uniformly, leaving the wood a spongy or stringy mass, or it may appear as a selective decay or a pocket rot. White rot fungi possess both cellulolytic and lignin degrading enzymes and these fungi therefore have the potential to degrade the entirety of the wood structure under the correct environmental conditions. Soft rot fungi typically attack higher moisture, and lower lignin content wood and can create unique cavities in the wood cell wall. Less is known about the soft rot degradative enzyme systems, but their degradative mechanisms are reviewed along with the degradative enzymatic and non-enzymatic systems known to exist in the brown rot and white rot fungi. As we learn more about the non-enzymatic systems involved in both brown and white rot degradative systems, it changes our perspective on the role of enzymes in the decay process. This in turn is affecting the way we think about controlling decay in wood preservation and wood protection schemes, as well as how we may apply fungal decay mechanisms in bioindustrial processes.

Introduction

Wood decay fungi have long caused problems for the durability of structures, but these fungi have also served an important ecological function as a beneficial agent in breaking down waste materials in the environment. Since wood remains the most abundant natural resource used in construction (*1*), methods to protect wood remain of paramount importance in the built environment. Although insects, marine organisms, weathering and other agents of deterioration cause losses in durability of wood and wood products, more destruction of wood is caused worldwide by decay fungi than by any of these other agents. Selected environments may favor other destructive organisms in some locations however.

Typically the term "decay" refers to the degradation of wood by fungal action, ultimately resulting in wood strength loss. Wood staining fungi and the colonization of wood by mold fungi are not considered to be decay organisms even though these fungi can increase moisture uptake in wood and minor strength losses in the wood are observed in some instances in temperate zones because of the action of these fungi. For example, it is well known that some staining fungi can reduce the toughness of stain-colonized wood although other mechanical properties remain unaffected in this wood. Interestingly, significant strength loss can occur in wood attacked by stain fungi in certain tropical environments.

Wood decay fungi may be divided into three categories. Classically, the action of white rot and brown rot fungi have been described because of the way that the wood appears after colonization by these fungi and advanced degradation of the wood has occurred. However, in the last 60 years, another category of decay known as soft rot has been discovered and this fungal group has also been observed to cause significant problems in wood products.

Both brown rot and white rot fungi are grouped in the Eumycota (true filamentous fungi) and more specifically in the sub-division of Basidiomycetes (Basidiomycotina) fungi. The soft rot fungi typically have been found in the Ascomycetes (Ascomycotina) sub-division or they have been classed as Fungi Imperfecti because their sexual state has not clearly been identified. In most cases, these "imperfect" fungi key out to Ascomycetes once their sexual state is revealed or when molecular techniques are utilized for fungal analysis. The Basidiomycete decay fungi produce filamentous hyphae that are 20 to 30 times smaller (1 – 5 μm) than the wood cell walls they invade. Typically, brown and white rot fungi attack the wood ray cells initially and most hyphae will pass from one cell to the next by growing through the pits of the wood cells. As the growth and invasion process continues beyond the initial stages, the fungi extend lengthwise through the lumens of the wood cells secreting non-enzymatic and enzymatic metabolites in an effort to degrade the wood cell wall to obtain nutrients from these walls. The soft rot fungi also initially invade the wood in a similar manner. For decay to occur, all decay fungi must have an adequate micro-environment to foster the growth of the fungus. In addition this micro-environment must permit the secretion and diffusion of fungal metabolites permitting attack of the wood cell wall. A favorable micro-environment would include wood at the appropriate temperature, moisture content and pH. Heartwood containing natural fungitoxic agents, or wood treated with fungicidal preservative chemicals, can act in several ways to prevent the growth or function of the decay fungi. Typically wood must have a moisture content (MC) at or above the fiber saturation point for wood degradation by fungi to progress. Optimal moisture contents will range from 50% to beyond 150% MC depending on the specific gravity of the wood and the species of decay fungi colonizing the wood. Many soft rot fungi are known to prefer high MC environments however, and can attack wood at levels close to the wood-lumen saturation level.

Some thermophilic fungi will grow in, and degrade, wood cell walls at high temperatures in excess of 80 °C; however, most decay fungi prefer temperatures in the range of 5-42 °C. Optimal temperature ranges will vary considerably dependent upon fungal species and isolate. Above and below the optimal range, the fungi typically may not be killed, but if the temperatures are not extreme the fungi will form "resting" structures such as spores or resistant mycelial fragments. These resting structures can then germinate to form new fungal hyphae when conditions are suitable for fungal growth and metabolism. Some resting structures/spores are capable of surviving for years and can therefore be a source of fungal colonization if wood is rewetted.

Detecting the Presence of Decay

Basidiomycetous fungi often produce large fruiting bodies, commonly known as conks, punks or bracket/shelf mushrooms in the later stages of wood decay processes. In some cases however, at the macroscopic level the fruiting bodies can be indistinguishable from a mycelial mat growing on the wood surface, typically on the underside. Both white and brown rot fungi can grow through and decay wood even in the homokaryotic (haploid) state. Because a dikaryotic state from the fusion of two homokaryon hyphae is required to produce sexual fruiting bodies, and decay fungi in the homokaryon genetic state have been observed to produce active decay of wood, decay fungi do not necessarily produce sexual fruiting bodies even in advanced decay stages. Production of fruiting bodies can also be dependent upon environmental conditions. Even among dikaryons, many decay fungi do not produce sexual fruiting bodies even in advanced decay stages. The purpose of the fruiting bodies is to produce sexual spores which can infect new wood and to provide a mechanism for sexual recombination. However, asexual spores and mycelial fragments have been found to be a primary source of fungal decay colonization in some cases (*2*).

Determining if wood decay is occurring in its early stages can be difficult and has been examined from a number of perspectives. Isolation via culturing of the fungi from the wood can be used for determining if decay fungi are present, but this method can pose problems in interpretation. First, only living fungi and spores capable of germinating can be isolated in culture from wood undergoing decay. If decay occurred in wood and caused a loss of strength, but the fungus has become inactive over time, it may not be possible to culture the fungus from the wood. Conversely, recent colonization of the wood may not have produced significant decay, yet the fungus can still be cultured from the wood. Therefore, fungal culturing and allied methods using immunological or molecular techniques to detect the presence of decay fungi or fungal metabolites require interpretation or they can be misleading. Even if fungi can be cultured from wood, since many fungi and bacteria often inhabit wood the fungi isolated must be carefully sub-cultured, and techniques which favor the isolation of basidiomycetes are often employed (*3*). Once pure, isolates of the decay fungus must be keyed out, a time consuming and sometimes tedious process. For positive identification of some isolates, mating of the unknown with a known candidate culture, maintained in a collection, must be performed. Methods based on immunogenic response to components of the fungi or their metabolites have been developed for detection of the presence of decay fungi in wood (*4-7*). The specificity of these assays can be good, and a field test detection system has been developed (*8-10*); however, like microscopy and culturing analysis they test only for the presence of the fungus, not for the presence of decay of the wood. Newer methods for detection of decay fungi have also been developed based on DNA analysis and these methods have the ability to test for either specific or broad classes of fungi (*4*). Polymerase Chain Reaction (PCR) methods exist allowing such assays to be used reliably in the laboratory (*11*). With regard to the assessment of degraded wood or the presence of active or inactive decay however, the same concerns exist with this method as with immunological methods and culturing methods.

Other methods of decay detection include examining the decayed wood microscopically. This is a good method for determining if decay fungi are present; however, it depends upon the observation of key features indicative of decay. Microscopic analysis of wood necessarily requires that only a small portion of a sample be observed at any one time, and large samples would require extensive analysis in a number of locations to obtain an overall assessment of the condition of sample as a whole. The presence of fungal hyphae in wood cells is not an accurate method of assessing the presence or absence of decay because non-decay fungi may also be present in the wood. Some basidiomycetous fungi will produce "clamp connections" as a method for transferring genetic information in the cell division process. Clamp connections appear as septate nodes along the sides of the fungal hyphae but, as discussed above, both brown and white rot fungi can decay wood in the homokaryon state or under other conditions where the fungi would not have clamp connections. Some species of wood decay basidiomycetes however, do not produce clamp connections even in the dikaryotic state. So, whereas the presence of clamp connections in a wood-inhabiting fungus positively indicates the presence of a decay fungus, the absence of clamp connections is not proof of a lack of decay.

Other methods of decay detection are continually being studied. These methods range from chemical assays, electrical resistance analysis, mechanical tests such as "resistance drilling", to sonic and acoustical methods (*4,12-14*). These methods all measure different wood properties that can be affected by the action of decay fungi. To date, however, no method has proven to be fully accurate in the detection of early stage decay in field samples where a great number of variables typically exist. The difficulties arise because the variability in the wood itself is typically greater than any specific factor contributed by the decay process, so a baseline control value is critical for accurate assessment, particularly if early stages of decay are to be detected. Several of these systems do work well for laboratory monitoring of decay, where careful controls and baseline measurements can be taken.

Fungal Decay Mechanisms

The decay processes employed by the brown-, white- and soft-rot fungi are incompletely understood and hence are still being examined by a number of groups. Although enzymatic degradation has been studied for many years and a number of cellulolytic and lignolytic enzymes have been isolated, research over the last 15 years has also determined that non-enzymatic systems are involved in decay processes. These processes will be detailed in the specific discussions of the fungi below.

Soft Rot Decay

Soft rotted wood was initially described as a surface attack of wood caused by a variety of Fungi Imperfecti and Ascomycetes (*15,16*). The name soft rot was suggested by Savory (*17*) to describe surface wood, typically in a waterlogged condition, that was degraded by these fungi. The surface of this moist wood was softer than sound wood. Softening of wood is typical of all types of decay, so the name may be considered a misnomer, but the term is currently accepted by experts in the field. Daniel and Nilsson (*18*) broadened the scope of fungi included as soft rot species to include all Ascomycetes and Fungi Imperfecti that cause decay, but suggested that a more definitive classification scheme should be explored before alternate names for these fungi are developed. Soft rot fungi typically attack the outer surface of wood in relatively wet environments (*19*). However they are known to cause extensive and deep degradation (extending several centimeters deep into the wood) of utility poles in Scandanavia and northern Europe with other reports of deep penetrating soft rot from other areas of the world appearing periodically (*20*). Deeply penetrating soft rot occurs when the wood is at a high moisture content, but not saturated. It is unknown whether this type of deep-penetrating degradation occurs more extensively in other parts of the world, or if it simply has been less frequently reported because of the limited number of studies on this type of fungal attack.

As soft rotted wood dries it develops surface checks across the grain as the wood shrinks. The wood becomes brown in color. Although the decay may be superficial, the surface appearance may be similar to brown rot decay. In advanced stages of decay the wood will fail in a brash or brittle manner when a surface sliver is lifted. Typically the wood is described as having a weathered appearance like unpainted "barn board".

In North America, decay by soft rot fungi became more widely recognized when severe damage to utility poles treated using a new type of preservative treatment occurred. The process, known as the Cellon process (*21,22*), used liquified propane (LP) gas or other volatile solvents as the carrier for pentacholophenol which was then pressure infused into the wood. The solvent carrier was volatilized from the wood leaving a clean, residue-free surface. Unfortunately, even though bulk chemical analysis indicated that adequate retention of pentachlorophenol was present, the chemical often only passed into the cell lumens without extensive penetration into the wood cell walls, or the treatment was localized because of the volatilization of the solvent producing a variable treatment that the fungi could "grow around". Washing the surface of some poles with caustic to reduce pentachlorophenol blooming also resulted in the surface leaching of pentachlorophenol resulting in even greater treatment variability. Since soft rot fungi have the capability to penetrate and extend through the wood within the S2 layer and also typically have greater resistance to pentachlorophenol treatments, they could invade and degrade the wood and were less affected by the preservative treatment present in the cell lumens. Over 800,000 utility poles were treated using the Cellon process (*23*) starting in the 1960s, and many were degraded in service by soft rot fungi over a number of years following installation. Prior to this experience with soft rot, this type of fungus was known primarily to attack wood in very wet locations such as the wooden slats commonly used in cooling towers in the previous century, which are still widely used. Soft rot in the US, although still less common than brown rot, is now better understood to occur in wood treated with a variety of different preservative systems

Soft rot fungi produce two different types of attack on the wood cell wall. The first of these is known as cavity formation or Type I attack of the S2 layers of the wood cell wall (*24*). While Type I cavity formation is always occurs in the S2 layers of the wood cell wall, these cavities are also formed in some cases in the S1 layers, often in the same cell (G. Daniel, personal communication, 2006). The second type (Type II) is a general erosion of the wood cell wall layers starting from the S3-lumen interface and working outward. Type II soft rot is similar to white rot in this regard. Often, particularly in hardwoods, both Type I and Type II attack can be produced by the same fungus in the same sample.

Not all soft rot fungi produce cavities in the S2 layer of the wood cell wall; however, Type I cavity formation (*24*) has been studied more because of its uniqueness. Typically, soft rot fungi initiate Type I attack by penetrating from the lumen, using a fine microhyphae, and boring perpendicularly into the wood cell wall (*25,26*). The fungal micro-hyphae re-orient when they reach the S2 layer to align the hyphae in the direction of the S2 cellulose microfibrils. Alternately, the hyphae can branch, in a characteristic manner known as T-branching, to allow the hyphae to grow in opposite directions paralleling the wood cell wall S2 microfibrils. The microfine hyphae will extend for a short distance and then secretion of enzymes and other degradative metabolites for attack of cell wall material in the immediate environment around the fungal hyphae is thought to occur. This secretion of metabolites results in the formation of cavities within the cell wall. A complete complement of cellulosic enzymes, including endo-1,4-glucanase, exo-1,4-ß-glucanase and 1,4-ß-glucosidase have been reported to be produced by soft rot fungi (*27-29,26*). Other researchers *(18)* have hypothesized that a low molecular weight free radical generating system also functions with phenol oxidase enzymes to oxidize lignin, and earlier work (*28*) provides evidence that laccase is produced by these fungi to degrade the wood cell wall lignin. The degradative action typically produces a diamond shaped cavity with conically pointed ends, presumably because of the way the enzymes interact with the cellulose structure of the wood cell wall. Following production of a cavity, the microhyphae extends further through the S2 cell wall and at this point it is known as a proboscis hyphae. The process of enzymatic secretion and the cavity formation then repeats to produce a chain of diamond-shaped cavities. It has been reported that soft rot cavity forming fungi lose the ability to produce cavities when invading wood cells that have been delignified, suggesting that the orientation of cellulose microfibrils or macrofibrils as well as the encapsulating lignin matrix may play a role in cavity formation (*18*). Multiple hyphae often invade a wood cell wall and can completely riddle the wall in advanced stages of degradation.

Daniel and Nilsson (*18,26*) have reported on the chemical nature and the amount of lignin in wood, and how these appear to affect the type and severity of attack by soft rot fungi. The authors were able to correlate high lignin levels in wood with a reduced severity of attack. Hardwoods posses a combination of syringyl- and guiaicyl-type lignins and generally have lesser amounts of total lignin in the cell walls as compared to softwoods. Although both hardwoods and softwoods are attacked by soft rot fungi, Type II erosion attack occurs in a more limited fashion in softwoods. The reduced capacity for attack of the S3 layer of softwoods appears to be because of the highly lignified nature of this layer in softwoods and its guiaicyl lignin composition. Although the macroscopic appearance of the wood may appear similar to brown rot, the erosion patterns occurring at the microscopic level appear similar to those produced by white rot fungi. Worrall et al. (*30*) have shown however that white rot fungi produce erosion troughs with rounded ends whereas the soft rot fungi observed produce angular erosion troughs.

Brown Rot Decay

Brown rot fungi primarily metabolize the holocellulose component of wood and cause the wood to rapidly lose strength in comparison to the rate of wood metabolism. In later stages of decay, brown rotted wood becomes brown, crumbly and checked across the grain, similar to the appearance of soft rot decay but with a coarser checking pattern. In early stages of decay however, the wood may appear little changed other than being “water-stained” or wet. In these early stages, little cell wall material is metabolized by the fungus; however, extensive depolymerization of the cellulose within the wood cell wall occurs. In practical application, the lack of noticeable signs of decay in early decay stages presents a dangerous situation in load-bearing applications of the wood since the rapid depolymerization of cellulose in the wood cells results in early strength losses during the initial decay stages. Because the wood does not visually appear to be degraded at this stage, unexpected failures of wood structures can occur when loads are applied. This points to the need for regular inspection and maintenance of wooden structures.

In initial stages of colonization, brown rot fungi selectively attack the simple sugars in the rays and the hemicellulose in the wood cell wall (*31*). The fungi ramify through the wood, often using the pits to pass from one cell to another. Once established, pits may still be used to move through the wood; however, the brown rot hyphae also produce bore holes directly through the wood cell walls. The hyphae will secrete both enzymatic and non-enzymatic degradative metabolites within a hyphal sheath (an extracellular glucan matrix). This hyphal sheath surrounds the fungal hyphae tips, protecting the hyphae and connecting the fungus to the wood cell wall to provide a watery environment channeling the diffusion of the metabolites into the wood cell wall (*32,33*). The hyphal sheath has been observed to penetrate within degraded wood cell walls (*34*), permitting passage of degradative enzymes into the wall in advanced stages of degradation.

Enzymatic Brown Rot Degradation Mechanisms

Both an enzymatic and non-enzymatic system are required for attack of the cellulose in the wood cell wall. Prior to the realization that a non-enzymatic system was required, cellulolytic enzyme systems were studied extensively. For this reason the enzymatic systems of the brown rots are well understood compared to non-enzymatic systems. The proposed "Goodell non-enzymatic mechanism" is diagramed here for clarification (Figure 1). Brown rot fungi are known to produce endoglucanases to cleave the ß-1,4-glucosidic linkages, and ß-glucosidases to hydrolyse cellobiose or other short oligosaccharides (*35,36*). Only one brown rot fungus, *Coniophora puteana*, is known to also produce a cellobiohydrolase (*37,38*). Brown rot fungi also produce a number of endo-xylanases and ß-xylosidases important in the breakdown of hemicellulose (*39-41*).

Non-Enzymatic Brown Rot Mechanisms

Cowling (*42*) first recognized that all known enzymes were too large a molecular size to allow penetration into the pore structure of the intact wood cell wall to produce the extent of cellulose depolymerization that had been observed in early stages of fungal attack. Cowling postulated at the time that an undiscovered, very small mass "enzyme" may be involved (*43*) because at that time, non-enzymatic catalytic systems had not yet been studied in microbial systems. With further study by others --see earlier brown rot review by Goodell (*43*) for reference citations-- it became apparent that a low molecular weight system was necessary to penetrate the wood cell wall permitting cellulose depolymerization in advance of enzymatic penetration of the wood cell wall.

Low molecular weight catechols, quinones and hydroxybenzene derivatives produced by brown rot fungi were first proposed to act as low molecular weight agents in non-enzymatic degradation of the wood cell by Goodell et al. (*44*). Prior research by this group had shown that low molecular weight phenolate compounds produced by the brown rot fungus *Gloeophyllum trabeum (45,46)* could catalyze the production of hydroxyl radicals in a mediated Fenton reaction (*47*). Continuation of this research identified specific compounds produced by *G. trabeum* involved in this mechanism (*48*). Other groups have confirmed this mechanism and have identified other compounds produced by the fungi that also function to mediate the reaction, advancing research in this area and confirming the validity of this mechanism (*49-52*). Research on an alternate non-enzymatic mechanism involving a glycopeptide isolated from wood degrading fungi has also been explored for several years (*29*). This mechanism provides a potential

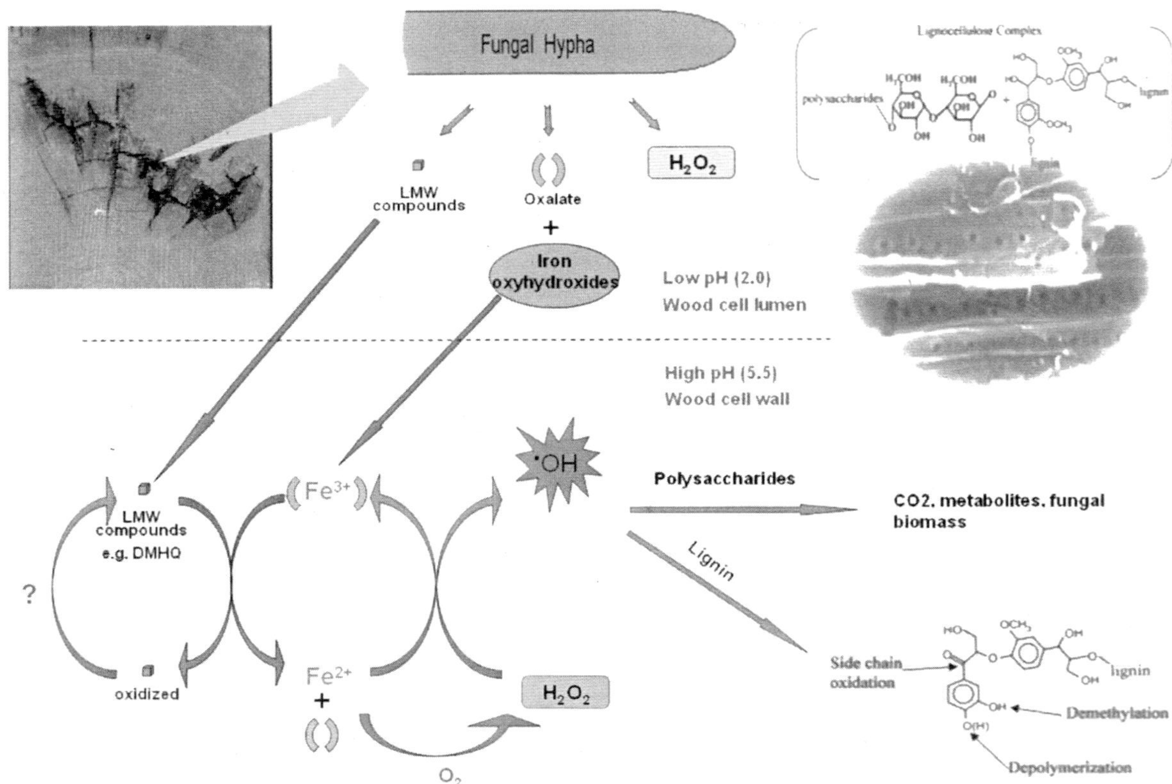

Figure 1. Simplified overview of the Goodell non-enzymatic brown-rot decay mechanism.

alternate pathway for free radical production required for non-enzymatic degradation of wood cell wall components. However, the glycopeptide(s) remain only partially purified and their chemical mechanism therefore remains unclear. See Goodell et al. (*43*) for a more complete review of brown rot degradation mechanisms.

White Rot Decay

White rot fungi attack wood and, in advanced stages of decay, cause the wood to appear bleached in appearance. In some common types of white rot decay, the wood will become soft and develop a "stringy" character where the softened wood fibers can be easily separated allowing the wood to be peeled apart. In other types of white rot, pockets of softened, deteriorated zones of wood may appear.

The unifying feature of the white rot fungi is that they produce a complete enzymatic system capable of directly or indirectly oxidizing and mineralizing lignin (*53,54*). The enzymes involved in this ligninolytic activity include lignin peroxidase, manganese peroxidase, and laccase (Figure 2). Individual white rot fungal species will possess one or more of these enzymes.

Beyond the physical appearance, two types of white rot are known based on the manner in which the wood cell wall components are oxidized. "Simultaneous white rot" is the most common type of white rot of wood and wood products in nature and typically, cellulose, hemicellulose and lignin are all degraded at some relatively uniform rate with this type of white rot. "Selective white rot" occurs when hemicellulose and lignin are attacked preferentially to cellulose, in some cases and with some fungal species, allowing the cellulose to remain relatively undegraded. Fungi that produce this later type of degradation have been studied for several years in selective delignification systems for biotechnological applications (*55-57*). In nature there is often an intergradation of white rot types and both simultaneous and selective white rot decays have been reported to be produced by the same fungus (*58,59*).

Historically, white rotted wood has been used for a variety of applications ranging from its use as insulated paneling in Russian refrigerator trucks/vehicles, and white rotted wood and fiber has been used for cattle feed in isolated regions of the world. In the later application, the selective white rot fungi free the cellulose from the lignin fraction of the wood or fiber, improving the digestibility in ruminant animals. More recently, but for several decades now, uses for white rot fungi in the field of biotechnology have been studied (*60-64*). The white rot fungus *Ceriporiopsis subvermispora* and *Phlebia subserialis* have been used in commercial bio-pulping applications to reduce the energy required in refining

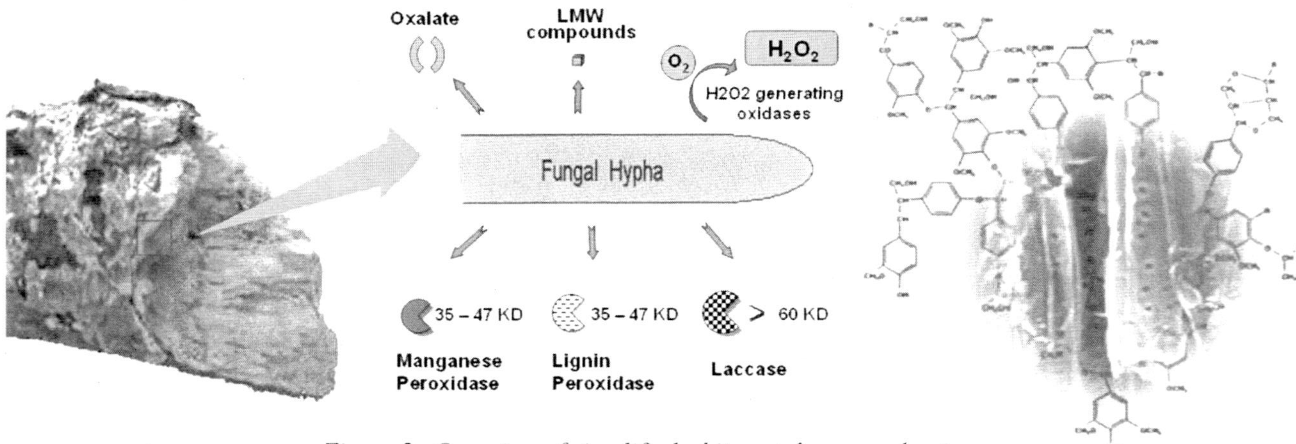

Figure 2. Overview of simplified white-rot decay mechanisms.

wood chips in mechanical pulp production, and to improve certain strength and brightness properties of paper produced from the inoculated chips. Other white rot fungi such as *Phanerochaete chrysosporium*, *Phanerochaete crassa*, and *Pleurotus pulmonarius* have also been studied as bio-bleaching agents for both kraft pulp and sulphite pulp (*65*).

Cellulose and hemicellulose degradation in the white rot fungi is known to occur via enzymatic processes and the white rot fungi produce both endo-glucanases and exo-glucanases which can act synergistically on the crystalline cellulose. The non-enzymatic processes which are known to be involved (*26,59*) particularly for hemicellulose depolymerization and selective white rot attack are not well understood. Enzymatic systems for break down of holocellulose include endoglucanases, ß-glucosidases and cellobiohydrolase enzymes (*66,67*) as well as xylosidase, xylanase, acetyl xylan esterase, glucuronidase and arabinofuranosidase; these later enzymes being necessary for complete depolymerization and oxidation of hemicellulose (*68,69*).

Because of the historic interest in white rot fungi for biotechnological applications in the latter half of the 20th century, white rot enzyme biochemistry for lignin oxidation was studied more intensely than the degradative mechanisms in the brown rot fungi. For this reason, more is known about white rot lignolytic enzyme systems; however, additional work remains to fully elucidate the complete degradation schemes. Work revealing the non-enzymatic mechanisms involved in brown rot decay has to some extent stimulated research to explore the role of non-enzymatic degradative systems in the white rot fungi (*70,71*). However, non-enzymatic mediators involved in some white rot enzyme systems, such as laccase, have been known for more than a decade. Further research on these systems may lead to an improved understanding, not just of the basic fungal biochemistry, but on the improved use of white rot fungi in industrial applications. It is well known that non-enzymatic systems penetrate deeply into the wood cell wall in advance of, or preferentially to penetration of the enzymes (*26*). Particularly in selective white rot, low molecular weight agents have been observed to penetrate completely through wood cell walls and the middle lamella regions (*26,59*). Although metals and radical ions generated from enzymatic action have been identified using in vitro biochemical assays, and the penetration of these low molecular weight components into the wood cell wall has been postulated, a conclusive study demonstrating the action of specific low molecular weight compounds on the wood cell wall is lacking.

Lignin degrading enzymes and the biochemical mechanisms employed by these enzymes to oxidize lignin have been described in a number of reviews (*72-74*). The discussion below briefly summarizes these mechanisms and also introduces information on how some enzymatic systems employ the use of naturally produced substrates, or "mediators", where this has been reported in the literature.

Lignin Peroxidase

Lignin peroxidase (LiP) has a relatively high redox potential (*75*) and is known to oxidize both phenolic and non-phenolic lignin groups; the latter composing approximately 90% of the lignin fraction of wood. The substrate typically used for this enzyme in laboratory studies has been veratryl alcohol and the mechanism for production of the veratryl alcohol radical in the presence of hydrogen peroxide is shown in Figure 3. As with all enzymes, lignin peroxidase is incapable of penetrating the intact wood cell wall because of its size, and it is capable only of attacking the exposed components of the wood cell wall at the lumen surface. The veratryl alcohol radical however, would have the capacity to penetrate into the wood cell wall. This radical has a relatively short half-life however (*76*); and if it participates in oxidizing lignin in the wood cell wall, then its short half life would help to explain the slow erosion of the wood cell which occurs in simultaneous white rot. It would not however, explain the observed degradation of the cell wall in selective white rot decay because this type of attack requires cell wall penetration. With this enzyme's ability to oxidize non-phenolic lignin, lignin peroxidase was one of the early enzymes explored for use in bioindustrial applications, but it has proven to be less suited for these applications than desired and research on its application, for example in pulp bleaching, has declined.

Manganese Peroxidase and Versatile Peroxidase

The primary role of manganese peroxidase (MnP) in white rot wood degrading systems is to oxidize Mn^{2+} to Mn^{3+} (*73,77*) as overviewed in Figure 4. It has been demonstrated that certain aliphatic organic acids such as oxalate, malonate and lactate produced as de novo metabolites by white-rot fungi (*78,79*) function as metal chelating agents. These organic acids increase the oxidation rate of Mn^{2+} (*80*) by binding with Mn^{3+}, resulting in a relatively stable complex, and effectively pulling the oxidation equilibrium by the complexation of Mn^{3+} by these chelators. The more abundant Mn^{2+} is then oxidized by MnP to Mn^{3+} (*59,72,81*). The complexed Mn^{3+} will then diffuse into the wood cell wall and oxidize phenolic lignin compounds.

The mechanism for this oxidation is not completely elucidated. It is likely that it includes the redox of the manganese metal although other organic radicals such as phenoxy radical may also be involved. MnP has a lower redox potential than LiP and can therefore only oxidize phenolic lignin structures under physiological conditions. Comparatively recently a new enzyme, Versatile Peroxidase (VP), has been discovered which can be considered a hybrid between MnP and LiP (*82-85*). VP can directly oxidize non-phenolic substrates such as veratryl alcohol as well as phenolic substrates. It is currently unknown how widespread VPs are in among the white rot fungi although they have been

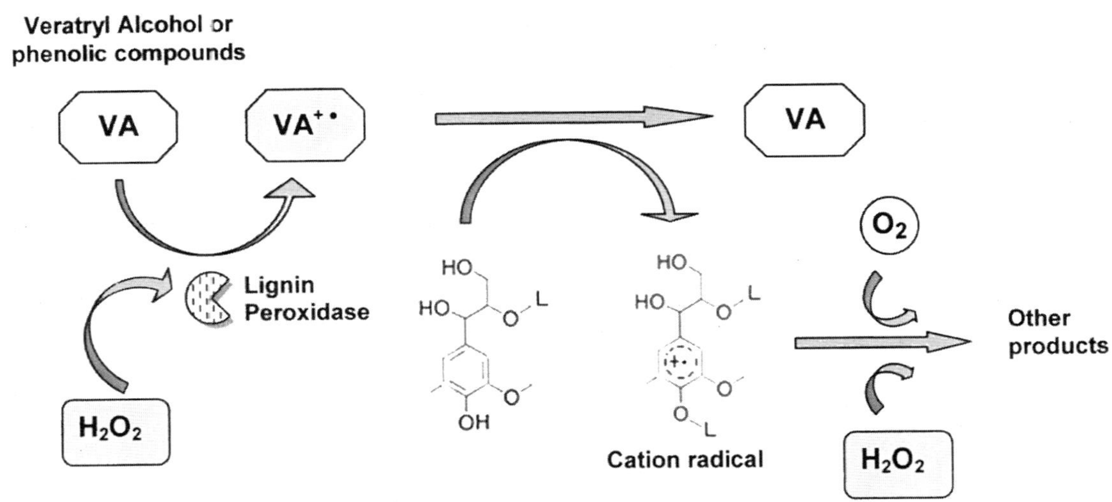

Figure 3. Lignin peroxidase activity involved in white-rot decay.

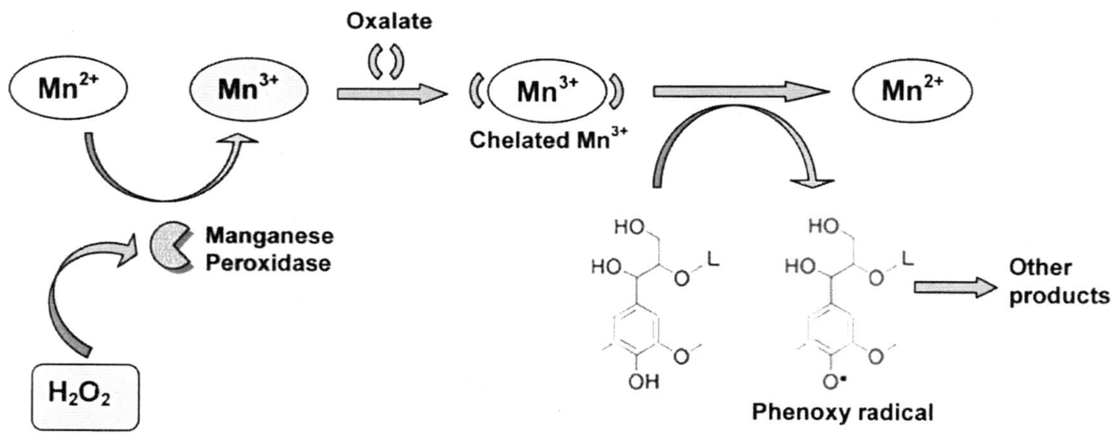

Figure 4. Manganese peroxidase activity involved in white-rot decay.

discovered in fungi such as *Pleurotus eryngii* (*86*) and *Bjerkandera sp* (*87*). Timofeevski et al. (*88)* reports that a mutant of MnP from *P. chrysosporium*, in which only one amino acid position had been changed, was able to oxidize both Mn^{2+} and typical LiP substrates such as veratryl alcohol. This points out the close similarity between MnP and LiP enzymes. More information about the action and distribution of this enzyme is expected in the future.

Laccase

Laccase differs from LiP, MnP and VP in that it can use oxygen as an oxidant to degrade lignin and ultimately produce water with a four electron reduction of oxygen (Figure 5). Laccase is produced by many microbial organisms, both degradative and non-degradative and may play multiple metabolic roles. Laccase alone is unable to oxidize non-phenolic lignin compounds directly. However, it was discovered (*89*) that synthetic substrates known as "mediator" compounds can be oxidized to their radical forms, and in the presence of laccase these compounds then catalyze the oxidation of non-phenolic lignin. Since laccase has better properties for industrial use, greater exploration of this enzyme for use in bioindustrial applications has been seen (*90*). Mediators such as 2,2'-azinobis(3-ethylbenzthiazoline-6-sulfonate) (ABTS) and 1-hydroxy benzotriazole (HBT) have been used effectively in trial applications but issues such as cost and toxicity remain. More importantly from the standpoint of understanding the fungal mechanisms, researchers in Finland (*91*) have demonstrated that several fungal hydroxamate siderophores can be successfully used as natural mediators in laccase-aided delignification processes. It was shown that iron-binding structures and free hydroxyl groups in these siderophores are the key targets for laccase. Iron chelating compounds with phenolate functional groups similar to other types of microbial siderophores have also been isolated from brown rot fungi (*44-48*), which raises the intriguing suggestion of phylogenetically-related mechanisms for lignin and cellulose degradation by white rot and brown rot fungi, respectively.

Summary - Conclusions

Soft rot, brown rot and white rot fungi are the most destructive microorganisms of wood products. Ecologically they are vitally important, but it is their destructive nature in the built environment that has driven the interest in understanding how these fungi function at the physical and biochemical level. In the first part of this chapter the progressive affects of the three types of fungi on wood are detailed, the similarities and differences in the modes of attack are reviewed, and the advantages and disadvantages of different methods for

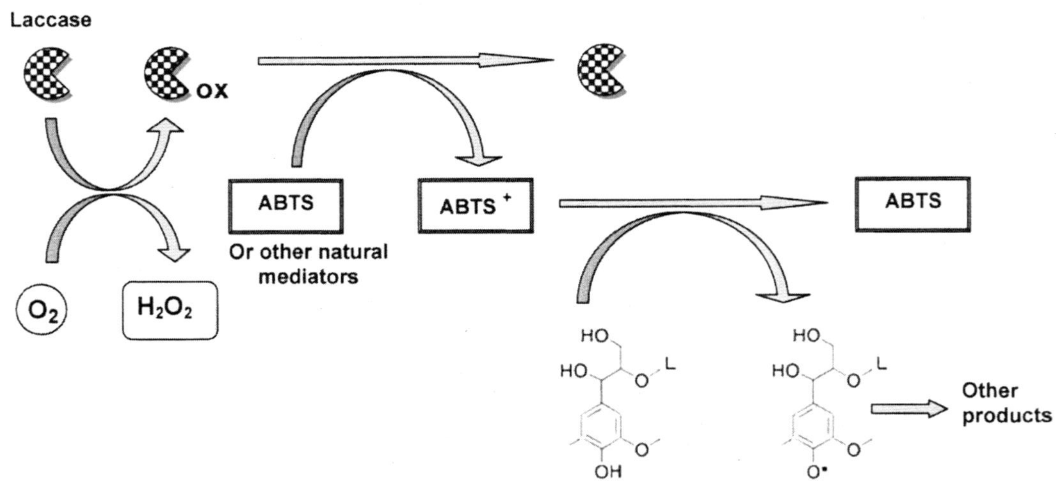

Figure 5. Laccase activity involved in white-rot decay.

detecting decay are discussed. The biochemical mechanisms and spatial relationships in the decay process are then discussed. In later chapters of this book, discussions of new methods to protect wood are presented, as are discussions of the fate of chemical protection methods using chemical preservatives that have been withdrawn from use in several counties around the world. As older broad-spectrum chemical treatments are abandoned it will become increasingly important to understand the biochemical mechanisms and microspatial relationships between wood and the fungi involved in the different types of decay processes. This chapter reviews the known mechanisms and spatial relationships required for enzymes and non-enzymatic metabolites to penetrate and function as cell wall depolymerization agents. New non-biocidal processes for protecting wood (*92-94*) are already being developed based on biochemical studies of fungal decay functions that were uncovered only in the last decade. Because information presented in this review chapter indicates the mechanisms of the different decay organisms vary, combined specific treatments may be needed to prevent attack by all types of decay fungi. Wood modification methods under development should also consider the variety of mechanism used by the decay fungi to avoid the failures of past treatments that were effective against some, but not all fungi.

Acknowledgements

The authors would like to thank our many collaborators worldwide who offered images for the ACS talk on this subject in March 2005. These include: Geoffrey Daniel, Jaime Rodriguez, Tor Schultz, Darrel Nicholas, Jonathan Schilling, Holger Militz, Kurt Messner and Tim Filley. Professor Goodell would like to thank his co-organizers for the ACS meeting which spawned this book. Tor Schultz as always has done an excellent job of organizing us all. In addition, both Professors Jellison and Goodell would like to thank the many students, current and past, including Yuhui Qian an author on this paper, for their insight into decay biochemistry and the value of discussions with them in developing new insights for research.

Funding for the work research on this topic was by the Wood Utilization Research program and by the CSREES McIntire-Stennis program. This is paper 2976 of the Maine Agricultural and Forest Research Station.

References

1. Bowyer, J.L.; Shmulsky, R.; Haygreen, J.G. *Forest Products and Wood Science: An Introduction (4th edition)*. Iowa State Press: Ames, IA, **2003**, 554 pp.

2. Przybylowicz, P.R.; Kropp, B.R.; Corden, M.E.; Graham, R.D. *Forest Products J.* **1987**, *37*, 17-23.
3. Nobles, M.K. *Can. J. Bot.* **1965**, *43*, 1097-1139.
4. Jellison, J.; Jasalavich, C. *Internat. Biodeter. Biodegrad.* **2000**, *46*, 241-244.
5. Goodell, B.; Jellison, J. In *Biodeterioration Research: Volume 3*. (Eds. Llewellyn, G.C.; O'Rear, C.E.). Plenum Publishing, **1991**, pp. 361-375.
6. Jellison, J.; Goodell, B. *Wood Sci. Technol.* **1988**, *22*, 293-297.
7. Jellison, J.; Goodell. B. *J. Biomass.* **1988**, *15*, 109-116.
8. Green, F.; Clausen, C.A.; Larsen, M.J.; Highley, T.L. *Canadian J. Microbiology.* **1992**, *38*, 898-904.
9. Clausen, C.A. *Internat. Biodeter. Biodegrad.* **1997**, *39*, 133-143.
10. Palfreyman, J.W.; Bruce, A.; Button, D.; Glancy, H.; Vigrow, A.; King, B. *Internat. Biodeter. Biodegrad.* **2001**, *48*, 74-78.
11. Jellison, J.; Jasalavich, C.; and A. Ostrofsky, A. In *Wood Deterioration and Preservation: Advances in Our Changing World.* (Eds. Goodell, B.; Nicholas, D.; Schultz, T.), American Chemical Society Series, Oxford University Press, **2003**, pp. 346-357.
12. Tornberg, K.; Olsson, S. *FEMS Microbio. Ecology.* **2002**, *40*, 13-20.
13. Zaremski, A.; Ducousso, M.; Domergue, O.; Fardoux, J.; Rangin, C.; Fouquet, D.; Joly, H.; Sales, C.; Dreyfus, B. *Canadian J. Forest Res.* **2005**, *35*, 1256-1260.
14. Craighead, I.A; Thackery, S.; Redstall, M.; Thomas, M.R. *J. Mechanical Engineering Sci.* **2001**, *215*, 905-917.
15. Bailey, I.W.; Vestal, M.R. *J. Arnold Arbor.* **1937**, *18*, 196-205.
16. Barghoorn, E.S.; Linder, D.H. *Farlowia.* **1944**, *1*, 395-467.
17. Savory, J.G. *Ann. Appl. Biol.* **1954**, *41*, 336-347.
18. Daniel, G.; Nilsson, T. In *Forest Products Biotechnology.* (Eds. A. Bruce; J.W. Palfreyman), Taylor and Francis, London, **1998**, pp. 37-62.
19. Zabel, R.A.; Lombard, F.F.; Wang, C.J.K.; Terracina, F. *Wood Fiber Sci.* **1985**, *17*, 75-91.
20. Greaves, H. *Holzforschung.* **1977**, *31*, 71-79.
21. Resch, H.; Arganbright, D.G. *Forest Products J.* **1971**, *21*, 38-43.
22. Arsenault, R.D. *Proc., Am. Wood-Preservers' Assoc.* **1966**, *62*, 132-147.
23. Davies, D.L. *Proc., Am. Wood-Preservers' Assoc.* **1971**, *67*, 37-42.
24. Corbett, N.H. *J. Inst. Wood Sci.* **1965**, *14*, 18-29.
25. Hale, M.; Eaton, R. *Mycologia.* **1985**, *77*, 277-288.
26. Daniel, G. In *Wood Deterioration and Preservation: Advances in Our Changing World.* (Eds. Goodell, B.; Nicholas, D.; Schultz, T.), American Chemical Society Series, Oxford University Press, **2003**, pp. 34-72.
27. Tanaka, H.; Enoki, A.; Fuse, G. *Gakkaishi.* **1985**, *31*, 935-945.
28. Tanaka, H.; Itakura, S.; Enoki, A. *Holzforschung.* **2000**, *54*, 463-468.
29. Enoki, A.; Tanaka, H.; Itakura, S. In *Wood Deterioration and Preservation: Advances in Our Changing World.* (Eds. Goodell, B.; Nicholas, D.; Schultz,

T.), American Chemical Society Series, Oxford University Press, **2003**, pp. 140-153.
30. Worrall, J.J.; Anagnost, S.E.; Zabel, R.A. *Mycologia.* **1997**, *89*, 199-219.
31. Winandy, J.E.; Morrell, J.J. *Wood Fiber Sci.* **1993**, *25*, 278-288.
32. Illman, B.; Highley, T.L. In *Biodeterioration Research: Vol 2.* (Ed. Llewellyn, G.C.), Perseus Books, Basic Books, NY, **1989**.
33. Jellison, J.; Chen, Y.; Fekete, F. *Holzforschung.* **1997**, *51*, 503-510.
34. Green, F.; Larsen, M.; Highley, T. In *3rd Meeting of the Pan American Biodeterioration Society.* (Eds. Llewellyn, G.C.; O'Rear, C.E.), Plenum Press, Washington, D.C., **1990**, pp. 311-325.
35. Herr, D.; Baumer, F.; Dellweg, H. *Arch. Microbiol.* **1978**, *117*, 287-292.
36. Mansfield, S.D.; Saddler, J.N.; Gübitz, G.M. *Enzyme Microb. Technol.* **1998**, *23*, 133-140.
37. Shimidhalter, D.R.; Canevescini, G. *Arch. Biochem. Biophys.* **1993**, *300*, 551-558.
38. Hyde, S.M.; Wood, P.M. *Microbiology.* **1997**, *143*, 259-266.
39. Highley, T.L. *Mat. u. Org.* **1976**, *11*, 33-46.
40. Green, F.; Clausen, C.A.; Micales, J.A.; Highley, T.L.; Wolter, K.E. *Holzforschung.* **1989**, *43*, 25-31.
41. Kvesitadze, E.G.; Lomitashvili, T.B.; Khutsishvili, M.P.; Lamed, R.; Bayer, E.A. *Appl. Biochem. Biotechnol.* **1995**, *50*, 137-143.
42. Cowling, E.B. *USDA Technical Bulletin No. 1258.* **1961**, pp. 79.
43. Goodell, B. In *Wood Deterioration and Preservation: Advances in Our Changing World.* (Eds. Goodell, B.; Nicholas, D.; Schultz, T.), American Chemical Society Series, Oxford University Press, **2003**, pp. 97-118.
44. Goodell, B.; Jellison, J.; Liu, J.; Daniel, G.; Paszczynski, A.; Fekete, F.; Krishnamurthy, S.; Jun, L.; Xu, G. *J. Biotechnology.* **1997**, *53*, 133-162.
45. Jellison, J.; Chandhoke, V.; Goodell, B.; Fekete, F. *Appl. Microbiology Biotechnol.* **1991**, *35*, 805-809.
46. Fekete, F.; Chandhoke, V.; Jellison, J. *Appl. Environmental Microbiology.* **1989**, *55*, 2720-2722.
47. Chandhoke, V.; Goodell, B.; Jellison, J.; Fekete, F. *Federation of European Microbiology (FEMS).* **1991**, *90*, 263-266.
48. Paszczynski, A.; Crawford, R.; Funk, D.; Goodell, B. *Appl. Environmental Microbiology.* **1999**, *65*, 674-679.
49. Kerem, Z.; Jensen, K.A.; Hammel, K.E. *FEBS Letters.* **1999**, *446*, 49-54.
50. Wang, W.; Gao, P.J. *J. Biotechnol.* **2003**, *101*, 119-130.
51. Hammel, K.E.; Kapich, A.N.; Jensen, K.A.; Ryan, Z. *Enzyme and Microbial Technol.* **2002**, *30*, 445-453.
52. Shimokawa, T.; Nakamura, M.; Hayashi, N.; Ishihara, M. *Holzforschung.* **2004**, *58*, 305-310.
53. Hatakka, A. *FEMS Microbiology Reviews.* **1994**, *13*, 125-135.

54. Hatakka, A.; Lundell, T.; Jeffries, T. *Enzyme Microbial Technol.* **2002**, *30*, 423-424.
55. Eriksson, K.-E.; Kirk, T.K. *Pract of Biotech, Spec Prod and Serv Act.* **1985**, *4*, 271-294
56. Akhtar, M.; Blanchette, R.A.; Myers, G.; Kirk, T.K. In: *Environmentally Friendly Technologies for the Pulp and Paper Industry* (Ed by Young, R.; Akhtar, M.), John Wiley & Sons, New York, **1998**, pp. 309-340.
57. Kirk, T.K.; Tien, M.; Faison, B.D. *Biotechnology Advan.* **1984**, *2*, 183-199.
58. Otjen, L,; Blanchette, R.A. *Can. J. Bot.* **1986**, *64*, 905–11.
59. Messner, K.; Fackler, K.; Lamaipis, P.; Gindl, W.; Srebotnik, E.; Watanabe, T. In *Wood Deterioration and Preservation: Advances in Our Changing World.* (Eds. Goodell, B.; Nicholas, D.; Schultz, T.), American Chemical Society Series, Oxford University Press, **2003**, pp. 73-96.
60. Goodell, B. In *Concise Encyclopedia of Wood & Wood-Based Materials.* (Ed. Schniewind, A.P.), Pergamon Press, NY, **1989**, 374 pp.
61. Blanchette, R.A.; Burnes, T.A.; Leatham, G.F.; Effland, M.J. *Biomass.* **1988**, *15*, 93-101.
62. Chae, K.S.; Selvam, K.; Swaminathan, K. *Bioresource Technol.* **2003**, *88*, 115-119.
63. Bajpai, P. *Critical Reviews in Biotechnology.* **2004**, *24*, 1-58.
64. Kamei, I.; Kondo, R. *Appl. Microbiology Biotechnol.* **2005**, *68*, 560-566.
65. Hossain, M.; Das, M.; Ibrahim, S.H. *J. Institution Engins.* **2003**, 83, 33-38.
66. Eriksson, K.E.; Hamp, S.G. *Eur. J. Biochem.* **1978**, *90*, 183-190.
67. Uzcategui, E.; Ruiz, A.; Montesino, R.; Johansson, G.; Pettersson, G. *J. Biotechnol.* **1991**, *19*, 271-286.
68. Kirk, T.K.; Cowling, E.B. In *Chemistry of Solid Wood.* (Ed. Rowell, R.M.), American Chemical Society Press, Washington, DC, **1984**, pp. 455-487.
69. Blanchette, R.A.; Ahad, A.R.; Farrell, R.L.; Leathers, T.D. *Appl. Environ. Microbiol.* **1989**, *55*, 1457-1465.
70. Enoki, A.; Itakura, S., Tanaka, H. *J. Biotechnology.* **1997**, *53*, 265-272.
71. Tanaka, H.; Itakura, S.; Enoki, A. *J. Biotechnology.* **1999**, *75*, 57-70.
72. Have, R.T.; Teunissen, J.M. *Chem. Rev.* **2001**, *101*, 3397-3413.
73. Hofrichter, M. *Enzyme and Microbial Technol.* **2002**, *30*, 454-466.
74. Li, K. In *Wood Deterioration and Preservation: Advances in Our Changing World.* (Eds. Goodell, B.; Nicholas, D.; Schultz, T.), American Chemical Society Series, Oxford University Press, **2003**, pp. 196-209.
75. Kersten, P.J.; Kalyanaraman, B.; Hammel, K.E.; Reinhammar, B.; Kirk, T.K. *Biochemical J.* **1990**, *268*, 475-480.
76. Candeias, L.P.; Harvey, P.J. *J. Biol. Chem.* **1995**, *270*, 16745-16748.
77. Hatakka, A.; Lundell, T.; Hofrichter, M. In *Applications of Enzymes to Lignocellulosics.* (Eds. Mansfield, S.; Saddler, J.), ACS Symp. Series 855, Oxford University Press, **2003**, pp. 230-243.
78. Shimada, M.; Akamtsu, Y.; Tokimatsu, T.; Mii, K.; Hattori, T. *J. Biotechnol.* **1997**, *53*, 103.

79. Dutton, M.V.; Evans, C.S. *Can. J. Microbiol.* **1996**, *42*, 881.
80. Kuan, I. C.; Tien, M. *Proc. Natl. Acad. Sci.* **1993**, *90*, 1242.
81. Banci, L.; Bertini, I.; Pozzo, L.D.; Conte, R.D.; Tien, M. *Biochemistry.* **1998**, *37*, 9009-9015.
82. Camarero, S.; Sarkar, S.; Ruiz-Dueñas, F.J.; Martínez, M.J.; Martínez, A.T. *J. Biol. Chem.* **1999**, *274*, 10324-10330.
83. Ruiz-Dueñas, F.J.; Martínez, M.J.; Martínez, A.T. *Mol. Microbiol.* **1999**, *31*, 223-235.
84. Hofrichter, M. *Enzyme Microbial Technol.* **2002**, *30*, 454-466.
85. Martínez, Á.T.; Speranza, M.; Ruiz-Dueñas, F.J.; Ferreira, P.; Camarero, S.; Guillén, F. *Int. Microbiol.* **2005**, *8*, 195-204.
86. Martínez, M.J.; Ruiz-Dueñas, F.J.; Guillén, F.; Martínez, A.T. *Eur J. Biochem.* **1996**, *237*, 424-432.
87. Mester, T.; Field, J.A. *J. Biol. Chem.* **1998**, *273*, 15412-15417.
88. Timofeevski, S.L.; Nie, G.; Reading, N.S.; Aust, S.D. *Arch. Biochem. Biophys.* **2000,** *373*, 147-153.
89. Bourbonnais, R.; Paice, M.G. *FEBS Lett.* **1990**, *267*, 99-102.
90. Call, H-P.; Mucke, I. *J. Biotechnol.* **1997**, *53*, 163-202.
91. Niku-Paavola, M.-L.; Anke, H.; Poppius-Levlin, K.; Viikari, L. In Applications of Enzymes to Lignocellulosics, ACS Symposium Series 855 (Eds. Mansfield, S.D; Saddler, J.N.), Oxford University Press, **2003**, pp. 176-191.
92. Goodell, B.; Jellison, J.; Liu, J.; Krishnamurthy, S. **2000**, *U.S. Patent 6,046,375.*
93. Schultz, T.P.; Nicholas, D.D. **2001**, *U.S. Patent 6,231,651.*
94. Schultz, T.P.; Nicholas, D.D.; Henry, W.P.; Pittman, C.U.; Wipf, D.U.; Goodell, B. *Wood Fiber Sci.* **2005**, *37*, 175-184.

Chapter 3

Insects That Infest Seasoned Wood in Structures

Terry L. Amburgey

Department of Forest Products, Mississippi State University, Mississippi State, MS 39762

Insects causing damage to wood products or building components comprise two groups: (1) those that infest non-seasoned wood (eg, logs, freshly-cut materials), or (2) those that can infest (or re-infest) seasoned wood. Many species of insects can infest wood before it has been dried, but relatively few species can infest or re-infest wood that has been dried to below approximately 20% moisture content. Since most building components used in the United States are both heated and dried in kilns or presses to well below approximately 20% moisture content, insects that have infested the wood prior to seasoning are killed and no additional damage will occur. However, wood salvaged from structures (eg, barn wood) and re-used often contains insects and, unless fumigated or kiln-dried prior to re-use, the insects can continue to infest wood in the new structure. Insects capable of infesting and feeding on seasoned wood include beetles such as anobiids, lyctids, old-house borers, and bostrichids as well as subterranean, damp-wood and dry-wood termites. Insects that excavate nest sites in seasoned wood but do not consume wood as a food source include carpenter bees and carpenter ants. An understanding of the basic biology of the insects and the types of wood that they infest will enable one to identify the insect causing the damage and to formulate a management program.

Introduction

Insects form the largest class in the Animal Kingdom; more than three-fourths of all animals are insects. The Class Insecta is subdivided into more than 30 Orders by grouping those with similar characteristics. Fortunately, relatively few insects can infest and re-infest seasoned wood and use it as a source of food or cause deterioration by excavating nest sites within seasoned wood. This chapter, therefore, will deal with a limited number of members of only three Orders: Isoptera (termites), Coleoptera (beetles), and Hymenoptera (bees, ants, and wasps). Some insects that only infest non-seasoned wood (trees or logs with bark intact) will be discussed briefly since the damage which they cause as their larvae bore through wood will be seen as a defect in lumber sawn from infested material. Several texts and manuals discuss insects in general (eg, refs *1* and *2*) but relatively few discuss those insects that infest seasoned wood and are problems in structures (eg, refs *3* and *4*).

Identification

Identification of the insects causing wood deterioration is important to determine whether or not management measures must be implemented to prevent additional damage, or no management procedures are necessary (eg, the damage observed was caused by insects such as Ambrosia beetles that infest non-seasoned logs and cannot re-infest seasoned wood). Usually, it is sufficient to identify a wood-deteriorating insect to genus, but sometimes the identity of the infesting species is necessary (*5*). Key features to use when identifying these organisms is the appearance and size of adults; the type of wood infested (Angiosperms –hardwoods vs Conifers – softwoods); the age of the wood (new wood is less than 10 years since the lumber was cut); the size, shape, or color of galleries formed in the wood during infestation; the size and shape of the exit holes made by emerging adults (if present); or the color and texture of frass (digested wood particles) found in galleries or adjacent to exit holes (ranging from very fine powder to fibrous strands). Another key to the identification of insects causing wood deterioration is the appearance of the insect form actually feeding on wood. (eg, adult or larva). Most insects undergo complete metamorphosis, that is, the immature insect stages look very different from the adult (eg, the immature caterpillar looks very different from the adult butterfly). Members of the orders Coleoptera (beetles) and Hymenoptera (bees, ants, wasps) undergo complete metamorphosis. Other insects (eg, Isoptera or termites) undergo incomplete metamorphosis. That is, the immature termites look very similar to the adults (*6*).

Insects That Nest in Seasoned Wood

Of the insects that nest in seasoned wood, those causing the greatest damage are carpenter bees and carpenter ants. These insects do not feed on wood. Rather, they damage wood by excavating nest sites within it.

Carpenter bees

Carpenter bees (*Xylocopa* spp.) resemble bumble bees, but their abdomen, unlike bumble bees, is black in the most common species and has no hair. This feature makes their abdomen shine as they fly about on sunny days. Carpenter bees prefer to nest in non-finished wood in shaded exposure. The adult bee chews into the wood surface, leaving a hole approximately 0.5 to 0.7 inch (12 to 17 mm) in diameter. The bee then turns and continues to excavate a 0.5 to 0.7-inch gallery lengthwise, parallel to the wood grain, in the interior of the piece. This becomes the site where eggs are laid. Beginning at the end of the gallery farthest from the entrance hole, the adult forms a rough ball-shaped larval food source comprised of pollen and regurgitated nectar, lays an egg on the food ball, and walls off the "egg chamber" with wood fibers held together with saliva. Then a second egg is laid and the "egg chamber" formed as with the first egg. This procedure continues until the entrance hole is reached. The presence of these galleries is evidenced by piles of fibrous "sawdust" on the soil, concrete, etc. below the entrance hole.

As the eggs hatch, each developing larva feeds on the pollen ball in its "egg chamber". The larva in the "egg chamber" closest to the entrance hole (last egg laid) matures into an adult first, chews its way through the "egg chamber" wall, and exits through the entrance/exit hole that was made by the adult who laid the eggs. This process continues until the first-laid egg hatches and matures into an adult, with all new adults emerging at approximately the same time. The time of development from eggs to adults is 7 weeks or longer, depending on temperatures during development. The presence of such a gallery rarely causes the colonized wood to fail, but these bees frequently return to the same wood component and, eventually, the multiple galleries can reduce the cross-section of the wood sufficiently to cause failure.

Carpenter bee damage can be largely prevented (or greatly reduced) by painting all surfaces of wood in exterior exposure, especially those in shaded areas. Since carpenter bees can form galleries in wood treated with water-borne preservatives (eg, CCA), even treated wood must be painted to prevent their damage. This is because the adults do not digest the treated wood that they have excavated, so the preservative is non-toxic to adult bees. These bees can be controlled by spraying volatile insecticides or insecticidal dusts labeled for this use into exit holes and plugging the entrance/exit holes with caulk. A mild fumigant such as moth balls or crystals also can be packed into the entrance/exit holes and the holes caulked to control these bees. Additional information on carpenter bees may be found in a text by Kard (*7*).

Carpenter Ants

Carpenter ants (*Camponotus* spp.), unlike carpenter bees, are social insects that occur in large colonies. Depending on the species and colony maturity, colony size can vary from a few hundred to more than 100,000 individuals. The adults enter the wood through checks, splits, or other natural openings or they can chew small entrance holes. Also unlike carpenter bees, carpenter ants primarily excavate the softer springwood (early wood) in successive growth rings, giving the colonized wood a laminar appearance (*8*). While the areas of wood excavated by carpenter ants resemble those of subterranean termites, carpenter ant galleries conatin no soil or debris and their walls are very smoothe. Relatively large numbers of eggs then are laid in the nest where they are cared for by colony members. The ants are very large (0.16 – 0.47 in. or 4 – 12 mm) and, depending on the species, they may be black, brown, red, or yellow. They do not sting but can bite. As with the nest of carpenter bees, their nest site often can be located by the piles of fibrous "sawdust", often containing pieces of insects that they have fed on, that accumulate below the entrance. They often nest in decayed wood (eg, interior decay in log-home logs or power poles), but nests frequently occur in sound wood, even in wood that has been pressure-treated with a water-borne preservative such as CCA. The treated wood that is excavated is not digested, so the preservative is non-toxic to ant adults. Hollow-core doors are a favored nest site. Carpenter ants tend to be major pests in the Pacific Northwest and the northeastern United States. However, species are found throughout the United States.

Carpenter ants can be managed by spraying liquid or powder insecticidal dusts labeled for this use directly into the nest site. Non-insecticide measures include repairing plumbing or roof leaks and decayed wood associated with them, caulking areas where plumbing or electric lines enter a structure, and cutting tree limbs that contact a structure (*9*).

Insects That Feed on Seasoned Wood

Insects that feed on (consume as a food source) seasoned wood in structures primarily belong to either the Coleoptera (beetles) or Isoptera (termites).

Coleoptera (Beetles) – Biology

Wood-infesting beetles are non-social insects that often are very specific in the types of wood that they colonize (*10-12*). Many of these beetles are very small (less than 0.25 inch or 6 mm long), and these are collectively often termed powderpost beetles. However, the various genera and species involved have different growth requirements. Some wood-infesting beetles approach one-inch (25mm) in length. In general, Coleoptera have two pairs of wings with the outer wings (elytra) being thickened and leathery and the hind wings being membranous and are folded under the front wings when the beetles are at rest.

The damage to wood is caused by beetle larvae that are yellowish–white with dark mandibles. As they feed through infested wood, the larvae form galleries that usually are packed with frass (wood fragments and fecal material) that varies from the consistency of talcum powder to long, fibrous wood particles. As they approach maturity, the larvae form pupal chambers near the wood surface where the pupae form. As the pupae develop into adults, the adults chew their way to the surface and make exit holes conforming to their size and shape. To emphasize the different groups of beetles that infest seasoned wood, Williams (*11*) has coined the acronym ALBOW that is short for **A**nobiids, **L**yctids, **B**ostrichids, **O**ld house borer (a cerambycid), and **W**eevils (Curculionidae - a minor problem). Characteristics of the damage caused by wood-boring beetles in structures are summarized in Table 1. Additional authoritative information on wood-infesting beetles is presented in a text and table comparing the characteristics of damage caused by beetles that infest seasoned and/or non-seasoned wood (*13*).

Anobiidae

Members of the family Anobiidae often are commonly called powderpost beetles in the United States but are usually referred to as furniture or deathwatch beetles in other countries. Adult anobiids that cause deterioration of wood in structures range from 1/8 to ¼ inch (3 mm to 7 mm) in length. The head of adults is hidden when viewed from above because the pronotum (the segment just behind the head) projects over it. The hardened, outer wings of most of most anobiid species have conspicuous grooves and/or rows of pits. The larvae are C-shaped and white, except for the brownish head and mouthparts. The frass of anobiids feels gritty, and this is because, under magnification, it can be seen to contain lemon-shaped fecal pellets.

Anobiids infest older (greater than 10 years old) unfinished wood of both conifers and hardwoods, often in the damp crawlspaces and framing of older structures with a minimum wood moisture content of 13 to 20 percent (*10*). If not managed, and the wood moisture content remains high, the anobiids will move into the wall framing and will infest any older softwoods and hardwoods with an unfinished surface. They usually lay their eggs on wood surfaces, but sometimes also in small splits and checks or existing exit holes in the wood. The life cycle (time from egg to adult) may range from one to five years, but usually ranges from two to three years. The length of the life cycle, as with other beetle species, is dependent on wood moisture content, the type of wood infested, and the temperature. Warmer temperatures and higher wood moisture contents are associated with a shorter life cycle. Anobiid infestations usually can be managed by spraying insecticide on the exposed surfaces of all wood in the crawlspace and instituting measures such as installing a soil cover, foundation vents, or fans to decrease the wood moisture content. Anobiids sometimes are found in new homes, but, when this occurs, the infestation usually is found to have originated from recycled wood from an old structure that was used in the new construction. When using wood from older structures in new construction, it is prudent to have the materials heated in a dry kiln or fumigated by a certified pesticide applicator prior to use.

Bostrichids

Another group of wood-boring beetles referred to as false powderpost beetles are members of the family Bostrichidae. In nature, the bostrichids typically infest the dying and dead branches of hardwoods, but they may also infest softwoods. The adult Bostrichids that typically are found in structures are reddish-brown to black, are elongate and cylindrical with heads directed downward which, when viewed from above, are hidden by the pronotum (as with the Anobiids) which often is rough or rasp-like on the front edge. The adults of most species that infest structures are 1/8 to ¼ inch (3 to 6 mm) long with wing covers that are concave at the posterior end, with spines projecting from these posterior ends in some species. The larvae are curved and wrinkled, have six well-developed legs, and their front half is larger than their rear half. Although the larvae vary in size in different species, most are 3/16 to 5/16 inch (5 to 8 mm) long. The larval galleries are round and tightly packed with frass that is meal-like (no pellets as with Anobiid frass). One wood-infesting species that occurs in the western U. S. and in the Rocky Mountains, the black polycaon, is larger (1/2 to 1.0 inch or 12 to 25 mm), is coal-black, has a prominent head which extends forward , and has a smooth, oval pronotum that does not project over the head. Unlike the other "powderpost" beetles, adult bostrichid females bore into the sapwood and prepare "egg tunnels" for laying eggs.

Bostrichids infest the outer sapwood (high starch content) of new (less than 10 years old) hardwood and, to a lesser extent, softwood and frequently are introduced into structures in hardwood firewood or infested bamboo, rattan, or

Table 1. Characteristics of damage caused by common wood-boring beetles in houses[1]

WOOD ATTACKED			RECOGNITION OF DAMAGE			
Type of borer	**Part and type**	**Condition**	**Exit holes**	**Galleries (tunnels)**	**Frass**	**Reinfes-tation**
Anobiid powderpost beetles	Sapwood of hard- woods and soft- woods: rarely in heartwood	Seasoned	Circular, 1/16 to 1/ 8 in (1.6 to 3mm) diameter	Circular, up to 1/8 in (3 mm) diameter; numerous; random	Fine powder with elongate pellets conspicuous; loosely packed[2]	Yes
Bostrichid powderpost beetles	Sapwood of hardwoods primarily: minor in soft- woods	Seasoning and newly seasoned	Circular, 3/32 to 9/32 in (2.5 to 7mm) diameter	Circular, 1/16 to 3/8 in (1.6 to 10mm) diameter; numerous; random	Fine to coarse powder; tightly packed, tends to stick together	Rarely
Lyctid powderpost beetles	Sapwood of ring- and diffuse-porous hardwoods only	Newly seasoned, with high starch content	Circular, 1/32 to 1/16 in (0.8 to 1.6mm) diameter	Circular, 1/16 in (1.6 mm) diameter; numerous; random	Fine, flour-like, loose in tunnels	Yes
Round-headed borers (general)	Sapwood of softwoods and hardwoods; some in heartwood	Unseasoned, logs and lumber	Oval to circular, 1/8 to 3/8 in (3 to 10 mm) long diameter	Oval. up to 1/2 in (13 mm) long diameter, size varies with species	Coarse to fibrous; may be mostly absent	No
Old house borer	Sapwood of softwoods, primarily pine	Seasoning to seasoned	Oval, 1/4 to 3/8 in (6 to 10 mm) long diameter	Oval, up to 3/8 in (10 mm) long diameter; numerous in outer sapwood, ripple marks on walls	Very fine powder and tiny pellets; tightly packed in tunnels	Yes
Flat oak borer	Sapwood and heartwood of hardwoods, primarily oak	Seasoning and newly seasoned	Slightly oval; 1/16 to 1/12 in (1.6 to 2 mm)	Oval, up to 1/12 in (2 mm) long diameter	Fine granules	No

Table 1. *Continued.*

WOOD ATTACKED			RECOGNITION OF DAMAGE			
Type of borer	Part and type	Condition	Exit holes	Galleries (tunnels)	Frass	Reinfestation
Flat headed borers	Sapwood and heartwood of softwoods and hardwoods	Seasoning	Oval, 1/8 to 1/2 in (3 to 13 mm) long diameter;	Flat oval, up to 3/8 in (10 mm) long diameter; winding	Sawdust-like, may contain light and dark portions if under bark: tightly packed	No
Bark beetles	Inner bark and surface of sap- wood only	Unseasoned, under bark only	Circular, 1/16 to 3/32 in (1.6 to 2.5 mm) diameter	Circular, up to 3 32 in (2.5 mm) diameter; random	Coarse to fine powder, bark-colored, tightly packed in some tunnels	No
Ambrosia beetles	Sapwood and heartwood of hardwoods and softwoods	Unseasoned, logs and lumber	Circular, 1/50 to 1/8 in (0.5 to 3 mm) diameter; stained walls	Circular, same diameter as holes; across grain; walls stained	None present	No
Wood-boring weevils	Sapwood and heartwood of hardwoods and softwoods	Slightly damp, decayed	Raggedly round or elongate, 1/16 to 1/12 in (1.6 to 2 mm) diameter	Circular, up to 1/16 in (1.6 mm) diameter	Very fine powder and very tiny pellets, tightly packed	Yes

[1] From: (*3*) p. 111.

[2] Pellets may be absent and frass tightly packed in hardwoods.

cane furniture or furniture stock, baskets, etc. Bamboo, rattan and cane imported from the Pacific rim now is fumigated or heat-treated to avoid importing Bostrichid-infested material. Many species require bark for egg laying. Depending on the wood moisture content, the life cycle of bostrichids may be from one to five years. There is little danger of re-infestation of seasoned wood after the first generation has completed its life cycle.

Lyctidae

Members of the family Lyctidae were the first to be called powderpost beetles, and many refer to them as true powderpost beetles. Unlike those of the anobiids and bostrichids, lyctid adults, when viewed from above, have the head clearly visible. They are about the same size as anobiid adults (1/8 to ¼ inch or 3 to 7 mm) but are slender, somewhat flattened, elongate, and reddish brown to black in color. The mature larvae are ¼ inch (6 mm) long and are enlarged at the thorax and have six distinct legs. Also unlike the anobiids and bostrichids, the frass of lyctids contains no pellets, has the consistency of talcum powder or flour, and is easily dislodged from the frass-filled larval galleries.

Lyctids, in contrast to the anobiids, only infest the sapwood of unfinished new hardwoods (material less than 10 years old). They favor large-pored hardwoods such as oak, hickory and ash as well as many species of imported tropical hardwoods (*11*, *14*, *15*). The majority of instances of lyctid beetle infestations reported to my laboratory are in newly-installed ash cabinets. Adult lyctids usually lay their eggs in the large pores (vessel elements) of these species, but some species will lay their eggs in small cracks and crevices. While they primarily infest wood whose moisture content is between 10 and 20 percent, they can infest wood with moisture contents ranging from 6 to 32 percent. Their preference for new sapwood is because their major source of food is starch as well as sugars and protein found in sapwood cells. The starch content of wood tends to decrease with age, both in sawn material and position in the tree. Heartwood has little starch. Lyctids tend to prefer new kiln-dried sapwood since it retains a higher starch content than air-dried sapwood. Frequently, mold fungi growing on air-dried sapwood utilize the starch and other storage products as they colonize sapwood.

Adult lyctid activity is greatest in the late winter and early spring, and their life cycle for most species is 9 to 12 months but can be as short as 4 months if the environmental factors are ideal (high-starch wood, moisture, and temperature). As the larvae feed (an adult can lay between 20 and 50 eggs), they literally can convert the interior of a piece of sapwood to fine powder, with no external indication of beetle activity until emerging adults bore their exit holes (1/32 to 1/16 inch or 0.8 to 1.6 mm) in the wood surface and piles of very fine frass accumulate on or below the infested material. The probability of infestations by lyctids can be lessened by kiln-drying sapwood materials and keeping them dry and by finishing all surfaces of hardwood products. Warehouses containing inventories of hardwood materials should be carefully and regularly inspected for signs of lyctid infestation. Inventories should be subjected to "first-in, first-out" rotation and wood scraps should be disposed of on a regular basis. If signs of lyctids are found (eg, piles of fine frass below stacks of lumber), either the entire facility should be fumigated or all of the wood removed and heated in a dry kiln while the structure is cleaned and sprayed with an insecticide labeled for that use (Table 2).

Lyctid activity in small areas of infested flooring, for example, can usually be managed by removing the finish and flooding the area with insecticide (labeled for that use) so that it will enter the exit holes and be distributed in the larval galleries. Small items (eg, kitchen cabinets) can be removed and replaced, fumigated individually (eg, infested pieces of furniture), heated in an oven (eg, picture frames), or placed in a freezer for a period of time (eg, 1 to 2 weeks).

Cerambycidae/Old House Borer

The Old House Borer (*Hylotrupes bajulus* L.), is the only species among the estimated 1200 species of Cerambycidae (long-horned borers) in the U. S.,that is a serious pest of structures. Most Cerambycids infest living hardwood and softwood trees or recently–felled trees and hardwood and softwood logs but cannot infest seasoned wood. Most species infest only sapwood, but some species can infest both sapwood and heartwood. Some species are large (up to 2 inches or 50 mm) and all are characterized by long antennae which, in some species, are much longer than their bodies. The larval galleries of these species are large (nearly ½ inch or 13 mm) across and may be tightly packed with coarse frass or be relatively free of frass. These galleries or "worm holes" cause degrade in hardwood lumber but are rarely a problem in softwood framing material. Since these galleries are in the logs prior to processing, their "worm holes" may appear as circular when the saw cut was directly across a gallery or elliptical when sawn at an angle. Occasionally, an adult will emerge from lumber that was not kiln-dried prior to use in a structure.

The old house borer is unusual among the Cerambycidae in that it can infest seasoned wood and cause significant damage to the sapwood of softwood framing or trim of relatively new (less than 10 years old) structures. Modern log homes located in the southern and eastern U. S.. and constructed with small-diameter, non-seasoned, untreated pine logs may be infested by both old house borers and buprestid beetles (discussed later in the text) during the first few years after construction (*16-20*). Old house borers do not infest hardwood materials.

Table 2. Schedule for treating wood to stop damage by powderpost beetles

Relative humidity	Lethal temperature required		Thickness of timber		Time Required to overcome lag after kiln has attained lethal temperature	Additional margin of safety	Time then held at lethal temp.	Total period of exposure after kiln has attained required conditions
Percent	*°F*	*°C*	*Inches*	*(cm)*	*Hours*	*Hours*	*Hours*	*Hours*
			1	(2.5)	1/2	1/2	1-1/2	2-1/2
100	130	54	2	(5.1)	2	1/2	1-1/2	4
			2-1/2	(6.3)	3-1/4	1/2	1-1/2	5-1/4
			3	(7.6)	4-1/2	1/2	1-1/2	6-1/2
			1	(2.5)	1/2	1/2	2	3
	125	52	2	(5.1)	2	1/2	2	4-1/2
			2-1/2	(6.3)	3-1/4	1/2	2	5-3/4
			3	(7.6)	4-1/2	1/2	2	7
			1	(2.5)	1/2	1-1/2	6	8
80	120	49	2	(5.1)	2	1-1/2	6	9-1/2
			2-1/2	(6.3)	3-1/4	1-1/2	6	10-3/4
			3	(7.6)	4-1/2	1-1/2	6	12
			1	(2.5)	1/2	7-1/2	30	38
	115	46	2	(5.1)	2	7-1/2	30	39-1/2
			2-1/2	(6.3)	3-1/4	7-1/2	30	40-3/4
			3	(7.6)	4-1/2	7-1/2	30	42-1/2
			1	(2.5)	1/2	1	4	5-1/2
	125	52	2	(5.1)	2	1	4	7
			2-1/2	(6.3)	3-1/4	1	4	8-1/4
			3	(7.6)	4-1/2	1	4	9-1/2
			1	(2.5)	1/2	2	7	9-1/2
60	120	49	2	(5.1)	2	2	7	11
			2-1/2	(6.3)	3-1/4	2	7	12-1/4
			3	(7.6)	4-1/2	2	7	13-1/2
			1	(2.5)	1/2	9	36	45-1/2
	115	46	2	(5.1)	2	9	36	47
			2-1/2	(6.3)	3-1/4	9	36	48-1/4
			3	(7.6)	4-1/2	9	36	49-1/2

[1] From: Ref. *3*, p. 107.

Old house borer adults are brownish-black in color, with several gray hairs on their heads and the fore-part of their bodies which are 5/8 to 1 inch (16 to 25 mm) long and slightly flattened. The wing covers frequently have two distinct gray cross bands or spots formed by patches of gray hairs. The larvae are up to 1.25 inch (31 mm) long and have three black eyespots in a row on each side of their heads. Old house borer larvae require wood with a high protein content (larvae require a minimum of 0.2% protein for their development), so the outer sapwood of pine is a preferred food. Larval development occurs most rapidly in wood with a moisture content of 15 to 25%, with the larval period as short as two to three years in the southern part of its range (a mean August temperature of about 73 F (23 C) favors larval development). The life cycle commonly is three to five years, but it can be as long as 12 to 15 years in very dry wood. The old house borer range in the U. S. is the eastern half of the country, with the heaviest infestations occurring in the mid-Atlantic states. According to Moore (*3*), evidence indicates that there is little probability of re-infestation in heated, well-ventilated, occupied structures. However, fumigation of homes with small to moderate old house borer infestations continues to occur. Infestations occur in sapwood of coniferous (especially pine) framing in crawlspaces, basements, and storage areas (southern portion of its range) or attic framing (northern part of its range). Larval galleries are oval in cross-section and are up to 3/8 inch (9 mm) across. The surfaces of most galleries have a rippled pattern. Since this does not occur in the galleries of most wood borers, this is a very distinctive feature. The galleries are loosely filled with fine frass that is readily dislodged from galleries that are disturbed.

Wood-boring weevils (Curculionidae)

Wood-boring weevils (Curculionidae) differ in appearance from other beetles in that their head is elongated into a "snout". This is why some refer to them as "snout beetles". Wood-boring Curculionid adults are black to reddish–brown with heavily pitted wing covers, cylindrical shape, and are 1/8 to 1/5 inch (3 to 5 mm) long when mature. Their larvae are whitish, legless, and about 1/8 inch (3 mm) long when mature.

Curculionids occur in situations similar to those where anobiids are found. That is, older hardwoods and softwoods with a relatively high moisture content, except that they prefer to infest wood with an even higher moisture content than the anobiids. However, both anobiids and weevils may be found infesting the same wood. Some weevil species infest older, moist hardwoods whereas other species infest older, moist softwoods. Their exit holes are smaller than those of the anobiids and their frass, unlike that of anobiids, consists of fine powder with tiny pellets and is tightly packed into the galleries. Also unlike the anobiids, it is common to find both weevil adults and larvae feeding in infested wood. Many species infest only non-seasoned wood. Curculionids are a minor problem in wood structures.

Coleoptera (Beetles) – Management

Before attempting to manage wood-boring beetles, one must be certain that the infestation still is active. Since exit holes often are the first sign that wood is infested, active infestations can be identified by answering the following questions. Is fresh frass the color of the wood infested seen around the exit holes, or is it discolored or clumped together by spider webs? Is the wood visible in the outer portions of the exit holes clean and bright or discolored? Can larvae be heard feeding within the wood when other noises are absent (eg, old house borer larvae at night) or can active larvae be found by probing the galleries? Are dead adults accumulating on window sills or around lights?

Non-Insecticide Control

In general, beetles infest wood more readily in structures with **moisture problems** that result in wood moisture contents near or above 20%. Anobiids and weevils require wood moisture contents above about 13%. Infestations by other beetles develop more rapidly, and the beetle life cycles are shortened, at higher wood moisture contents. Since the wood in structures would not be expected to have a moisture content above about 8 to 12%, the moisture contents discussed above indicate that a moisture problem exists. Moisture content control practices, depending on the structure involved, may include site drainage; soil covers in crawlspaces; increased ventilation in crawlspaces and/or attics; venting stoves, showers, and/or clothes dryers to the exterior; increasing air movement within structures with fans; installation of humidifiers; etc.

With large infestations, whole houses may be fumigated. However, after fumigation, no residual insecticide remains. For this reason, fumigation usually should be coupled with moisture control measures. Infested items (eg, furniture, picture frames, etc) can be placed in a freezer for a week or two, heated in an oven or microwave, or placed in a sealed container and fumigated. Fumigations should be done by a certified pesticide applicator. Alternatively, small infestations (eg, in furniture or cabinets) can be managed by removing the infested items from the structure and replacing them, or pieces of damaged framing (eg, floor joists) can be re-enforced with pressure-treated wood while moisture control procedures are instituted.

Residual Insecticides

Most structural damage resulting from beetle infestations can be prevented by using pressure-treated framing (eg, ACQ-ammoniacal copper quat, copper azole, or borates). However, this will not prevent beetles from infesting other building components.

Unlike subterranean termites, which will be discussed next, preventive procedures for beetles rarely are practiced. Typically, management procedures for beetles are instituted after an infestation has been discovered. Application of residual insecticides labeled to manage beetle infestations should be applied by certified pesticide applicators. However, residual insecticides should be coupled with the moisture control procedures discussed above. With unfinished wood (eg, floor joists infested with anobiids), all exposed surfaces should be treated, thereby eliminating egg-laying sites and flooding existing exit holes. Beetles infesting finished wood (eg, lyctids in hardwood furniture or flooring) cannot re-infest the wood except by laying eggs in existing exit holes, cracks through the finish, or unfinished surfaces (most furniture and flooring is only finished on the front and top surfaces). Infestations in furniture can be controlled by the non-insecticide procedures discussed above followed by applying a finish to all surfaces. With infested flooring, it may be necessary to remove the finish prior to applying insecticide.

Isoptera (Termites) – Biology

Members of the order Isoptera (termites) are characterized by having four membranous wings of equal size and shape that extend past the end of their bodies.Termites that infest wooden components of structures have been divided into groups, based on feeding requirements, and are referred to as **subterranean termites, drywood termites, and dampwood termites**. Information on termite classification and species distribution within the United States is presented by Jones (21). The subterranean termites are the most destructive species in the U. S. Termites can ingest wood, but they must rely on the symbiotic protozoa and/or bacteria (depending on termite species) to digest the wood and make it available as a termite food source. Termites, in contrast to beetles, are social insects with incomplete metamorphosis. That is, different castes with different functions co-exist in colonies, and the young that hatch from eggs resemble the adults. The social structure of the colonies is comprised of reproductives, soldiers, immature nymphs, and workers. The "balance" between the numbers of each caste in a mature colony, and the size of a mature colony, vary by species. For instance, *Reticulitermes* spp., the subterranean termites native to the U. S. that cause the most damage to structures, have approximately 2% soldiers per colony. A mature colony of *Reticulitermes flavipes* Kollar (the Eastern U. S.

Subterranean Termite) has approximately 51,000 to 5 million individuals, depending on environmental factors (*22*, *23*). By contrast, the introduced (to the U. S.) subterranean termite *Coptotermes formosanus* Shiraki has approximately 10% soldiers per colony, which is significantly higher than other termite species. A mature colony of *C. formosanus* averages 2 million individuals, with large colonies exceeding 10 million individuals (*23*, *24*). In general, mature colonies of *C. formosanus* are two to three times larger than those of *R. flavipes*. Differences between subterranean termites and wood-infesting beetles are summarized in Table 3.

Subterranean Termites (Rhinotermitidae)

Subterranean termite workers forage at random to find sources of food and, when a suitable food source is found, they lay a chemical pheromone trail back to the nest. Other workers then will follow the trail to the food source and re-enforce the trail by laying down their own pheromone. This is done by compressing a gland on their abdomen, thus expelling pheromone. The workers feed the reproductives, immature nymphs, and soldiers and are the caste that actually feeds on wood and causes damage to structures. Studies suggest that the activities of termites are influenced by several pheromones (*25-27*). Termites cannot digest the wood that they have ingested. Depending on the termite species involved, wood ingested is digested and converted into nutrients used by termites by symbiotic protozoa and/or bacteria in their gut. As the colony size continues to grow, the level of juvenile hormones regulated within the colony stimulates the transformation of some nymphs into secondary reproductives. These individuals form wings and develop coloration that differs with species. For instance, *R. flavipes* alates (mature secondary reproductives) become black to dark brown in color, whereas *C. formosanus* alates are yellowish-brown in color. Dispersal flights of alates give rise to new colonies as the alates fly short distances (termites are poor flyers), lose their wings, and males pair with females and follow them to seek new nest sites. These flights normally are in the spring of the year but still differ by species. *R. flavipes* alates have dispersal flights in the spring during the daylight hours. By contrast, alates of *C. formosanus* have dispersal flights primarily in the spring in the early evening and are attracted to lights. A single mature colony of *C. formosanus* can produce more that 60,000 alates. This is why light traps can be used to estimate populations of *C. formosanus* in different areas. Formosan termite colonies begin to produce alates after 5 or 6 years. Since both ants and termites have dispersal flights in the spring, flights of ants often are thought to be termites. Physical differences between termites and ants are given in Table 4.

Characteristics to distinguish between our native (U.S.) species of *Reticulitermes* and *C. formosanus* are given in Table 5.

Table 3. Differences between subterranean termites and wood-infesting beetles.

Subterranean Termites	Beetles
1. Social insects	1. Non-social insects
2. Incomplete metamorphosis	2. Complete metamorphosis
3. Four (4) wings of equal size and shape	3. Two (2) outer hard wings and Two (2) inner membranous wings
4. Exist in organized colonies	4. Solitary
5. Feed on broad range of wood species	5. Tend to feed on specific kinds of wood
6. Travel between wood and soil (usually)	6. Spend most of life feeding within wood
7. Colonize wood found in random	7. Colonize wood by adults laying eggs on or in it
8. Dispersal flights outdoors or sometimes within structures	8. Dispersal flights usually within structures (lyctids)
9. Wood damage usually concentrated and limited to a few boards or small areas	9. Wood damage often widespread in many boards
10. Usually apply treatments or baits to prevent termites	10. Usually apply treatment to control existing infestations
11. Can infest wood with a wide range of moisture contents	11. Different species have a relatively specific minimum and maximum wood moisture content requirements
12. Can infest wood of many ages (new to very old)	12. Different species have relatively specific ages of wood that they can infest (nutrient depletion with wood age)

Table 4. Differences in winged forms of termites and ants.

Termites	Ants
1. Four (4) membranous wings of equal size and shape	1. Four (4) membranous wings with two (2) wings much larger than the remaining two (2)
2. Body shape cylindrical	2. Body hour-glass-shaped with a constriction between the thorax and abdomen
3. Antennae straight and composed of bead-like segments	3. Antennae elbowed and all segments are not the same size and shape (some club-shaped)

Table 5. Differences between species of *Reticulitermes* and *Coptotermes formosanus*.

Reticulitermes	*C. formosanus*
1. Native to U.S.	1. Introduced to U.S.
2. Dispersal flights during day	2. Dispersal flights during early evening and are attracted to lights
3. Swarmers (winged forms) are dark brown to black	3. Swarmers (alates) are light yellowish-brown
4. Soldiers have rectangular head with large mandibles	4. Soldiers have oval-shaped head with both mandibles and a frontal pore which exudes a whitish sticky fluid
5. Usually construct nests in soil	5. Often construct nests within walls of structures. Nests are composed of carton that resembles a sponge
6. Wings have no hair	6. Wings are hairy
7. Length of swarmers 1/3 to ½ inch (8 to 12 mm)	7. Length of swarmers up to 5/8 inch (15 mm)
8. Few soldiers per colony (2%)	8. Noiceably more soldiers per colony (10%) that are more aggressive when colonies are disturbed

Subterranean termites usually nest in the soil, but the nests can occur in structures. Workers can feed on wood in structures that do not have a moisture problem, but they must return to the soil to feed their nest members and to obtain water. However, if a source of moisture occurs within structures, termites can nest within them. Although interior nests can occur with species of *Reticulitermes*, interior nests are common with *C. formosanus* infestations. Interior nests of *C. formosanus* can encompass large wall sections, completely filling wall voids with their carton nesting material. Carton is solid and resembles a sponge. Carton is composed of digested wood particles and fecal material. These large nests often can be located by the heat generated by respiration of the large numbers of colony members using infra-red or thermal detectors. When located, carton nests should be removed from structures after applying localized injections of labeled insecticides into the carton. *C. formosanus* also can form carton nests in the center of hollow trees.

When subterranean termites must cross exposed areas (eg, from the soil, across foundation materials, to house framing), they form shelter tubes to protect them from both desiccation and predators such as ants. Shelter tubes are formed from soil and termite body fluids and are a visible indication of a termite infestation.

Non-decayed wood infested by subterranean termites has a number of interior galleries that occur within the softer springwood of adjacent annual growth rings that superficially appear like those in carpenter ant nest sites. Unlike carpenter ant galleries that are clean and with smooth walls that appear to be sanded, those of *Reticulitermes* spp. often have soil in them and the walls are plastered with their fecal material. Galleries formed *by C. formosanus* are much cleaner than those of species of *Reticultitermes* and more closely resemble those of carpenter ants. However, carpenter ant nest sites are much more limited in size and often occur in decayed wood.

Studies have shown that species of *Reticulitermes* are attracted to wood being decayed by several species of brown-rot decay fungi (*25-27*). Galleries in decayed wood often cut across both springwood and summerwood, but they do not appear as clean as carpenter ant nest sites in decayed wood.

The geographic range of subterranean termites also varies by species. Species of *Reticulitermes* are more prevalent in the warm, humid southeastern U. S. and California, and their activity decreases to relatively low levels in the northern U. S. and Canada. However, the numbers of *Reticulitermes* in the colder areas of North America have increased due to the increased use of central heating in structures, especially those that occur in clusters in cities. Since *C. formosanus* is an introduced tropical species, its spread northward from the southern states and southern California is estimated to be as far north as approximately 35th North Latitude (border between Mississippi and Tennessee) and possibly farther up the eastern U. S. coastal areas due to the influence of the Gulf Stream and central heating in structures. *C. formosanus* is common in the Hawaiian Islands (*28*).

Drywood Termites (Kalotermitidae)

In general, drywood termite colonies are much smaller than those of subterranean termites. As their name implies, drywood termites can infest relatively dry wood without bringing moisture to it from other sources such as the soil. In contrast to subterranean termites, areas of wood fed upon are relatively small and are not confined to galleries in springwood. Rather, drywood termites cut across both the springwood and the summerwood of adjacent annual growth rings to form irregular cavities of relatively small size. Drywood termites form fecal pellets as the moisture is extracted from the ingested wood and the size, shape, and markings of these pellets tend to be characteristic for each species. There are 18 species of drywood termites in 8 genera, with several species being of economic importance (*2*). Drywood termite cavities are very clean, except for some fecal pellets. Most fecal pellets are expelled from their workings through holes cut to the wood surface. Piles of fecal pellets tend to accumulate near drywood termite workings and are used as a sign that drywood activity is present.

The drywood termites are tropical and are common in the Caribbean Islands, Mexico, and Pacific islands such as Hawaii, where they can cause extensive damage to structures. In the contiguous 48 states of the U.S., however, the drywood termite range is confined to Florida and coastal areas of North Carolinia, South Carolina, Georgia, Alabama, Mississippi, Louisiana, and Texas as well as southern New Mexico, Arizona and California. Occasionally, drywood termite damage occurs north of this range but it usually is confined to infested articles (eg, furniture) that were shipped from their natural range to northern regions. The accumulation of fecal pellets under infested articles usually is how the infestations are discovered. Normally, such infested articles are fumigated, placed in a freezer for a time, or simply discarded to manage the infestation.

Within the drywood termite range, dispersal flights (as with subterranean termites) occur as a colony matures, and the swarmers (alates) often are carried by winds and are attracted to lights. Dispersal flights indicate that a colony is 5 or more years old. Drywood termite alates can enter the attics or crawlspaces of structures through vents (blocking vents with 18 to 20 mesh screen usually is not effective), in cracks between siding and trim, and in spaces between adjacent pieces of siding.

Dampwood Termites

Dampwood termites are larger (up to one inch or 25 mm long) than subterranean or drywood termites and, as with drywood termites, their colonies are relatively small. Dampwood termites, as their name implies, usually do not infest dry wood and then transport water to it. Rather, dampwood termites infest damp wood. Since both decay fungi and dampwood termites require moist wood, both pests often are found in the same area. Since moist wood often occurs in the lower portions of wood-frame walls as the result of poor site drainage or rain splash, these are the areas where dampwood termite damage occurs. The ranges of the principal dampwood termites causing damage to structures is in the Pacific coastal areas from British Columbia to Baja, California. (*Zootermopsis angusticollis* Harger, the Pacific Dampwood Termite) and southern Florida (*Prorhinotermes simplex* Hagen, the Florida Dampwood Termite). Dampwood termite biology is discussed in a text by Nalepa (*29*). Moisture control procedures used to stop the growth of decay fungi will manage dampwood termites.

Tree-Nesting Termites

Although tree-nesting termites can infest wood in structures in other regions, these termites are not a problem in the U.S.

Isoptera (Termites) – Management

Subterranean termites

Establishing a termiticide chemical barrier in the soil under concrete slabs or foundation walls and piers, coupled with good building practices, has been the primary (and often sole) method of protecting U.S. structures from infestation by subterranean termites for more than 50 years (*30-33*). This method initially relied on the long-term persistence of chlorinated hydrocarbons such as chlordane and, if a **complete** soil barrier is established and maintained, the protection afforded structures is very good. The shortcoming of this method of control is the difficulty in establishing and maintaining a **complete** termiticidal soil barrier around structures. Placing mulch around house perimeters, permitting vines and other vegetation to grow up the sides of structures, altering the house perimeter by adding earth-filled appurtenances (eg, porches, patios, flower planters) and failure to maintain site drainage away from the structure are among homeowner practices that disrupt soil barriers and permit termites to gain access to structures (*34*). The removal of chlorinated hydrocarbons for this use meant that less-persistent alternative chemicals that require more frequent retreatments had to be used for termiticidal soil treatments. Currently, approximately 400,000,000 gallons of termiticide are used annually under and around living spaces in the continental U. S. to provide a toxic barrier to termites (*35*).

These and other factors stimulated the development of **termite bait stations** that are placed around structures at regular spacings (eg, 10 feet) as either a stand alone termite management procedure or use in conjunction with soil barriers. This procedure relies on random foraging of termite workers to find untreated wood in the bait stations. As stations are infested, the untreated wood in them is replaced with wood or other substrates treated with a slow-acting stomach poison, insect growth regulators, or termiticides to manage termites by disrupting colonies near structures as the workers pass the chemicals to other nest-mates through mutual grooming and feeding (trophallaxis).

Antibiotics also have been developed to control termites by killing the symbiotic protozoa and/or bacteria in the termite gut, thereby making them unable to digest wood that they have ingested and obtain nourishment from it. Presently, no antibiotic baits are being used commercially.

The use of preservative-treated building components, used either alone or in combination with soil treatments or bait stations, also prevents termite infestation of structures or appurtenances such as decks or porches. The durability of solid wood, engineered wood (wood-based composite beams and panel products),and non-wood (eg, foam panels) building products all can be protected from insects and/or fungi by adding insecticides and/or fungicides during processing or as surface-applied treatments before or during construction (*36*, *37*). Treated sill plates and framing will prevent structural damage by termites if they breach the termiticidal soil barriers and/or bait stations (*38*). Borate-treated framing is specified in Hawaii and is being marketed on the U.S. mainland to protect structures from *C. formosanus* and other termite species. Likewise, spray-treatment of sill plates, the lower two feet (610 mm) of wall studs, and other framing has been shown to be an effective termite prevention method (*39*).

Recent studies have demonstrated that the movements of subterranean termite workers can be directed toward or away from given areas by low-intensity electronic fields (*40-42*). This technology likely will be incorporated into termite management procedures after experimental parameters have been fully evaluated.. Other recent studies have indicated that stainless-steel mesh (0.66 mm x 0.45 mm mesh) will serve as a barrier to termites (eg, TermiMesh), provided that it completely covers potential termite entry points into structures and all seams are fused (usually welded leaving no voids). Other barriers now in use to prevent access of subterranean termites into structures include non-biocidal soil aggregate (eg, Basaltic Termite Barrier, with basalt aggregate 4.75 mm to 1.18 mm... to pass #4 to #16 square mesh openings), and plastic sheeting impregnated with termiticide (eg, Impasse Termite Blocker) for use where pipes, electrical conduit, etc extend through concrete slabs (*43*, *44*)

Current research to reduce the reliance on soil treatments as the primary (or only) procedure for termite prevention is testing the hypothesis that barriers established with non-biocidal soil amendments, landscaping plants whose roots and surrounding soil have termite-repellent activity, and mulches containing termite-repellent plant tissues, used either alone or in combination, will discourage termite foraging toward structures and form natural termite "shields" around the perimeter and under structures (*45*). Plants such as Vetiver (Vetiveria zizanioides) have been evaluated for their termite repellency, but there has been very little, if any, focus on utilizing the plants in landscaping schemes or incorporating the plant tissues into mulches (*46*). Many professionals prefer not to rely on a single procedure for preventing access of subterranean termites into structures. Rather, the "belts and suspenders" approach is favored, whereby multiple procedures for termite prevention are used to complement one another.

Drywood Termites

Since drywood termites do not nest in the soil, termiticidal soil barriers cannot protect structures from them. With limited drywood termite infestations, location of the colony or colonies by the accumulation of fecal pellets nearby, followed by drilling into the colony site and injecting insecticide, is quite effective. This is known as the **drill and treat** method. In some instances, infestations are confined to discrete building components such as window frames and sills that can be removed and discarded. Before replacement components are installed, the surrounding framing should be spray-treated with an insecticide such as borate. Extensive infestations can be controlled by fumigating entire structures, but it must be remembered that no residual insecticide remains after the fumigant dissipates. Within the United States, the greatest economic losses caused by drywood termites occur in Florida, California and Hawaii. Additional information on drywood termites can be found in Scheffrahn and Su (*47*).

Dampwood Termites

Dampwood termite infestations can be controlled by altering the conditions that are permitting the wood to stay damp (eg, wood-soil contact, water drainage toward a structure) followed by treating the foundation and framing in the area with an approved insecticide such as borate. The soil in the area of the infestation should be treated with an insecticide labeled for use as a soil treatment. Pressure-treated wood should be used to replace infested wood.

Damage by Insects that Infest Non-Seasoned Wood

Beetles (Coleoptera) that infest trees and/or non-seasoned logs and lumber cause damage that often is found in seasoned wood products and is confused with that caused by insects infesting seasoned wood. It is important to recognize the cause of this damage and to understand that no insect control procedures are required (Table 1). One key feature to look for when damage is suspected to have occurred prior to lumber being sawn from logs is that the shapes of the galleries may range from circular to elliptical to oblong within a short distance, depending on the orientation of the galleries in the log to the saw blade (*48*). In addition, unlike those made by insects infesting seasoned wood, the gallery walls may not be clean and the color of the wood infested. Rather, the walls of galleries made by ambrosia beetles infesting non-seasoned wood often are discolored by the growth of mold and/or stain fungi growing on them and the galleries usually are free of frass. Since mold fungi cannot colonize seasoned wood, their presence on gallery walls indicates that the galleries were made prior to the wood being seasoned. The galleries of many species of buprestids and cerambycids are not discolored and contain frass.

Buprestids

The Buprestids, or metallic wood borers, have caused a great deal of damage to the logs of modern log homes fabricated from small-diameter, unseasoned logs, primarily of softwoods (*16-18*, *20*). Such logs contain both a large percentage of sapwood and a high moisture content. Buprestids lay their eggs in cracks and crevices in bark, and the larvae hatch and feed on the sapwood. Therefore, their presence indicates inadequate log handling practices whereby the logs are not debarked in a timely fashion (*19*).

Buprestid infestations in these structures can be eliminated by debarking the logs rapidly and treating them with a diffusible biocide such as borate or kiln-drying the debarked logs and treating the exteriors of the log structures with biocides and water repellents (*49*, *50*) .

Alternatively, log home manufacturers could use large–diameter logs that are squared prior to use in structures. Squaring the logs will remove most of the sapwood and a large portion of the water in the log because softwood heartwood has a much lower moisture content than sapwood.

Bark Beetles (Scolytidae)

Bark beetles, as the name implies, infest the bark and cambium areas of hardwood and softwood logs. If all bark is not removed while cutting materials, bark beetle larvae can continue feeding until the wood becomes seasoned. Since the larvae feed in the area of the cambium, some of the frass produced by them is the color of the bark and some is the color of the sapwood. Since the larvae do not penetrate the sapwood, bark beetles cause no structural damage.

Ambrosia Beetles (Scolytidae and Platypodidae)

Unlike most beetles, the adults rather than the larvae of ambrosia beetles cause the damage seen in various wood products. Adult ambrosia beetles bore circular galleries (between 1/50 and 1/8 inch, 0.5 to 3mm diameter depending on the species) through the bark of stored hardwood and softwood logs deeply into the sapwood (*19*, *51*). Eggs are laid in small "egg cradles" (short galleries) formed on either side of the entrance gallery. The developing larvae do not feed on wood. Rather, they feed on the mycelium of the "ambrosia" fungus introduced into the gallery by the adult. This fungus discolors the galleries, egg cradles, and often the surrounding sapwood to a black color. As the adult bores through the sapwood, the frass is expelled through the entrance hole, so logs infested with ambrosia beetles will have piles of fine frass below them or in bark crevices. The presence of dark-stained galleries with side egg cradles that are deep within the sapwood, free of frass, and cut at different angles along their lengths (appear as circular to oblong in diameter) is characteristic of ambrosia beetle damage.

Ambrosia beetle damage has been seen in the lower 2 to 4 feet of pine utility poles that have been treated with a water-borne preservative such as chromated copper arsenate (CCA) and placed in swampy areas where drying is retarded for several weeks. In such instances, the galleries do not contain the "ambrosia" fungus because its growth is controlled by the CCA. Adult ambrosia beetles are not harmed by the CCA because they excavate wood to form the galleries but do not ingest it.

References

1. Baker. W. L. USDA Forest Service,1972. Misc. Publ. No. 1175.
2. Arnett, R. H., Jr. American Insects. A Handbook of the Insects of America North of Mexica, 2nd Ed. CRC Press, Boca Raton, FL. 2000.
3. Moore, H. B. Wood-Inhabiting Insects in Houses: Their Identification, Biology, Prevention and Control. USDA Forest Service and Dept. Housing and Urban Develop. 1979.
4. Gold, R. E.; S. C. Jones (eds). Handbook of Household and Structural Insect Pests. Entomol. Soc. Amer., Lanham, MD. 2000.
5. Natl. Pest Control Assn. 1962. NPCA Tech. Release 26-62.
6. Hatfield, I. *Pest Control* **1958**, *264*, 50-60.
7. Kard, B. M. Carpenter Bees. In: Gold, R. E. and S. C. Jones (eds). Entomol. Soc. Amer. 2000, pp. 110-112.
8. Hansen. L. D. Amer. Wood-Preserv. Assn. Newsline Oct/Nov. 1999.
9. Cook, J. L. Wood-destroying insects. Carpenter Ants. In: Gold, R. E. and S. C. Jones (eds). Entomol. Soc. Amer. 2000, pp. 108-110.
10. Williams, L. H. *Pest Control*, **1973**, *41*(2), 24-28.
11. Williams, L. H. *Pest Control*, **1973**, *41*(5), 30-40.
12. Williams, L. H. *Pest Control*, **1973**, *41*(6), 18-44.
13. Williams, L. H. Wood-destroying Insects. Beetles. In: Gold, R. E. and S. C. Jones (eds). Entomol. Soc. Amer. 2000, pp. 99-108.
14. LaFage, J. P. and L. H. Williams. LA State Univ., Agr. Exp. Sta. Circul. No. 106. 1979.
15. Natl. Pest Control Assn. NPCA Tech. Release 19-621. 1961.
16. Amburgey, T. L. and L. H. Williams. Forest Prod. Lab., MS State Univ. Inform. Series No. 22. 1982.
17. Amburgey, T.; L.; L. H. Williams. *Log Home Guide for Builders and Buyers* **1983**, *6*(1), 24-38.

18. Amburgey, T. L.; L. H. Williams. *Log Home Guide for Builders and Buyers* **1984**, *7*(3), 30-75.
19. Amburgey, T. L. In: Goyer, R. A.; J. P. Jones (eds). LA Agr. Expt. Sta. and Coop. Ext. Serv. 1985.
20. Price, T. S., L. H. Williams; T. L. Amburgey. GA Forest. Commiss. Bull. Dec 1987/235-LH.
21. Jones, S. C. Wood-destroying Insects. Subterranean Termites. In: Gold, R. E.; S. C. Jones (eds). Entomol. Soc. Amer. 2000, pp. 119-124.
22. Wells, J. D.; G. Henderson. *Ecolog. Entomol.* **1993**, *18*, 270-274.
23. Jarratt, J. Miss. State Univ., Coop. Ext. Serv. Publ. No. 2247. 1994.
24. Yates III, J. R. Univ. Hawaii, Manoa,. Ext. Publ. www.exteto.hawaii.edu. 1992.
25. Amburgey, T. L.; R. V. Smythe. *Sociobiology* **1977**, *3*(1), 3-12.
26. Amburgey, T. L.; R. V. Smythe. *Sociobiology* **1977**, *3*(1), 13-25.
27. Amburgey, T. L.; R. V. Smythe. *Sociobiology* **1977**, *3*(1), 27-34.
28. Jones, S. C.; H. N. Howell. Wood-destroying Insects. Formosan Subterranean Termite. In: Gold, R. E.; S. C. Jones (eds). Entomol. Soc. Amer. 2000, pp. 119-124.
29. Nalepa, C. A. Wood-destroying Insects: Dampwood Termites. In: Gold, R. E.; S. C. Jones (eds). Entomol. Soc. Amer. 2000, pp. 113-116.
30. Beal, R. H., J. K. Mauldin, and S. C. Jones. USDA, Forest Serv., Home and Garden Bull. No. 64. 1989.
31. Jarratt, J., J. Mauldin; J. Haskins. Miss. State Univ, Coop. Ext. Serv. Publ. No. 531. 1989.
32. Amburgey, T. L. *Pest Management* **1983**, *2*(5), 32-37.
33. Amburgey, T. L. Effect of Building Design and Construciton Techniques on the Colonizaiton of structures by Subterranean Termites and Fungi. In: Forest Prod. Soc. Proc. 2002, pp. 291-294.
34. Amburgey, T. L. *Pest Control* **1971**, *39*(8), 34-36.
35. Nisus Corp. 2003. www.nisuscorp.com/pdfspr toxic soil.pdf.
36. Gardner, D. J., C. Tascioglu; M. E. Walinder. In: Goodell, B., D. D. Nicholas; T. P. Schultz (eds). Amer. Chem. Soc. Symp. Ser. 845, 2003, pp. 399-419.
37. Kirkpatrick, J. W. MS Thesis, Miss. State Univ. 2005.
38. Williams, L. H. and T. L. Amburgey. Enhancing Durability by Building with Treated Wood: Termies Damage and Boron Contents in 12-year-old Houses in Mississipppi. In: Forest Prod. Soc. Proc. 2002, pp. 295-303.
39. Williams, L. H. and T. L. Amburgey. (In Preparation).
40. Amburgey, T. L., M. G. Sanders, and C. Bell. U.S. Patent No. 6,837,001. 2005.
41. Ragon, K. Ph. D. Dissertation (manuscript in progress) . 2005.
42. Ragon, K., T. L. Amburgey, M. G. Sanders, and P. Donohoe. U.S. Patent Office Docket No. 2343-215-07. 2005.

43. Univ. Toronto. Newsletter of the Urban Entomol. Program, No. 7. 1991.
44. Termite control – Alt. Meas. Physical Barriers. www.toolbase.org/techinv/techDetails.aspx?technologyID=257. 2006.
45. Kitchens, S. C. Ph D Dissertation (work in progress), Miss. State. Univ. 2005.
46. Sanaa, A. I., G. Henderson, B. C. R. Zhu, H. Fei; R. A. Laine. *J. Econ. Entomol.* **2004**, *97*(1), 102-111.
47. Scheffrahn, R. H.; N. Y. Su. Wood-destroying Insects. Drywood Termites. In: Gold, R. E.; S. C. Jones (eds). Entomol. Soc. Amer. 2000, pp. 116-119.
48. Amburgey, T. L. Protecting Timbers from Fungi and Insects During Seasoning, Storage, Transit, and Use. In: Wang, S. Y.; M. C. Yeh (eds). Forest Prod. Assn. of R.O.C., Bull. No. 16. 1999, pp. 538-540.
49. Williams, L. H.; T. L. Amburgey. *Forest Prod. J.* **1987**, 37(2), 10-17.
50. Amburgey, T. L. Univ. Idaho WUD Publ. No. 93-1. 1993.
51. Amburgey, T. L. *Southern Lumberman* **1979**, *238*(2945), 13-15.

Chapter 4

Molds and Stain Fungi

Jeffrey J. Morrell

Department of Wood Science and Engineering, Oregon State University, Corvallis, OR 97331

The presence of fungi on the surfaces of both freshly sawn lumber as well as wood surfaces in buildings has become an increasing public concern. Formerly, most wood users were concerned about decay fungi that could weaken the wood, creating unsafe structural conditions; however, surface molds and stain fungi have become an increasing concern because of their potential as allergens. This paper describes the types of fungi that cause molds and stains and outlines the conditions required for growth on wood.

Introduction

While concerns about the risks of mold in indoor environments have become a major public issue in recent years, these fungi have long been a component in the built environment wherever conditions are suitable for their growth (*1*). Fungal spores are nearly always present in the air, although the levels present and species involved can vary (*2-9*). As a result, creating conditions that are conducive to fungal growth will inevitably lead to fungal colonization. Molds, which have adapted to use simple sugars as a food source and are capable of rapid growth, will tend to colonize the wood first followed by decay fungi.

There are a number of different conceptions of what constitutes a mold; but for the purposes of this chapter, molds are defined as those fungi that discolor the surface of wood through the production of pigmented spores. These spores can typically be brushed from the surface leaving the wood appearance unmarred. Stain fungi, which permanently discolor the interior of the wood, also produce dark pigmented structures on the wood surface and may be confused with molds. Similarly, decay fungi can produce color changes in wood that may be confused with mold.

All of the fungi involved in these changes in appearance have two common traits. First, the fungi involved are all growing into the wood to obtain nutrients; principally from sugars, starches and proteins stored in the ray cells. In addition, all of the fungi require that some free water be present. Free water is usually present when the wood moisture content is 30% (by dry weight) or greater, although there is considerable debate about the ability of molds to grow at wood MCs as low as 20 percent and this value is easily achieved in most properly designed, constructed and maintained buildings. It is also important to note that moisture in wood can change markedly in a short time, particularly near the surface. Thus, wood can become wet enough to support mold growth but this moisture can be gone by the time the damage is detected, leading to the conclusion that the fungus somehow sorbed moisture from the air.

Molds are all fungi which are heterotrophic, typically filamentous organisms that obtain nutrients from a variety of materials including wood. Estimates of the number of fungal species range up to 300,000 or more species (*10*). While many of these species are highly adapted for special niches such as leaves of a single tree species, molds are generally cosmopolitan in nature. As a result, many species grow on a broad range of materials under a variety of environmental conditions.

Nearly all mold species are members of the Fungi Imperfecti, a group of fungi characterized by asexual spore production. This allows these fungi to produce abundant quantities of spores in a relatively short time. A number of mold fungi are also capable of sexual spore production. These fungi are typically members of the Ascomycetes. In many cases, the anamorph-telomorph connections between sexual and asexual stages remain unknown.

Although there are innumerable mold species, wood-based materials tend to be dominated by a few genera, including members of the *Trichoderma*, *Penicillium*, *Aspergillus*, *Fusarium*, *Alternaria*, and *Aureobasidium*. In addition, some species of Ascomycetes, notably *Chaetomium globosum*, have been implicated in indoor air issues. Finally, a few mold species have drawn considerable attention, although they are not among the more prevalent species. The most notorious among these is *Stachybotrys atra* (*chartarum*). This fungus is prevalent on cellulose containing materials, particularly paper, including that used in drywall. *S. chartarum* has gained considerable attention because it produces a potent mycotoxin as a byproduct of its metabolic processes. Mycotoxins can induce a variety of human health responses (*11-24*). They are a particular problem for people with compromised immune systems, but there remains considerable debate about the real health risks associated with casual exposure to this fungus. While *S. chartarum* is found in wood structures it is more commonly found on drywall and other paper products, and its presence is an indicator of serious moisture issues in a structure.

The species present in a given wood-based material depends on a number of factors including moisture levels, temperature, and time of year. A number of early studies found that a diverse array of fungi colonized logs or freshly sawn lumber (*25-29*). For example, Kang and Morrell (*30*) studied fungal colonization of freshly sawn Douglas-fir sapwood and found that species diversity and isolation frequency increased steadily over time. *Trichoderma* spp. and other fast growing molds tended to predominate in this material. The type of wood based material will also influence colonization as well as fungal growth rates (*31-35*). The variations in building materials, environmental conditions, and capabilities of fungal species make it difficult to make global statements concerning the importance of these fungi in structures. Formerly, the presence of mold was viewed as an indicator of excessive moisture, which would eventually lead to more serious structural damage by decay fungi. This approach has clearly shifted and the presence of molds is, itself, now the primary concern and is often the cause of extensive remediation efforts.

Effects of Mold on Wood Properties

While many mold species have cellulase systems (i.e. they can degrade cellulose), most are unable to attack the lignocellulose matrix, making them relatively innocuous to wood from a structural aspect. The growth of most mold fungi is limited to the sapwood, where they scavenge proteins, lipids, sugars and other readily assimilated compounds (Figure 1). The hyphae move through the cells via the pits, degrading the pectin in the pit membranes, primarily at the surface of the wood. As a result, moldy wood tends to have greater permeability than non-colonized wood. This increased permeability can affect coatings and finishes, preservative treatment, drying rates and susceptibility to rewetting once in service (*36*).

Colonization by mold fungi has also been reported to condition wood to colonization by wood decay fungi, either by degrading toxic extractives or enhancing water uptake. Conversely, some mold fungi produce an array of antibiotic compounds that have been shown to inhibit colonization by decay fungi. The roles of these compounds in situ are poorly understood, although they are clearly toxic under laboratory conditions. Some of these fungi are also mycoparasitic and will attack and degrade the hyphae of competing organisms. A number of members of the genus *Trichoderma* are capable of mycoparasitism and antibiosis and have been used for inhibiting the growth of other fungi. Stain

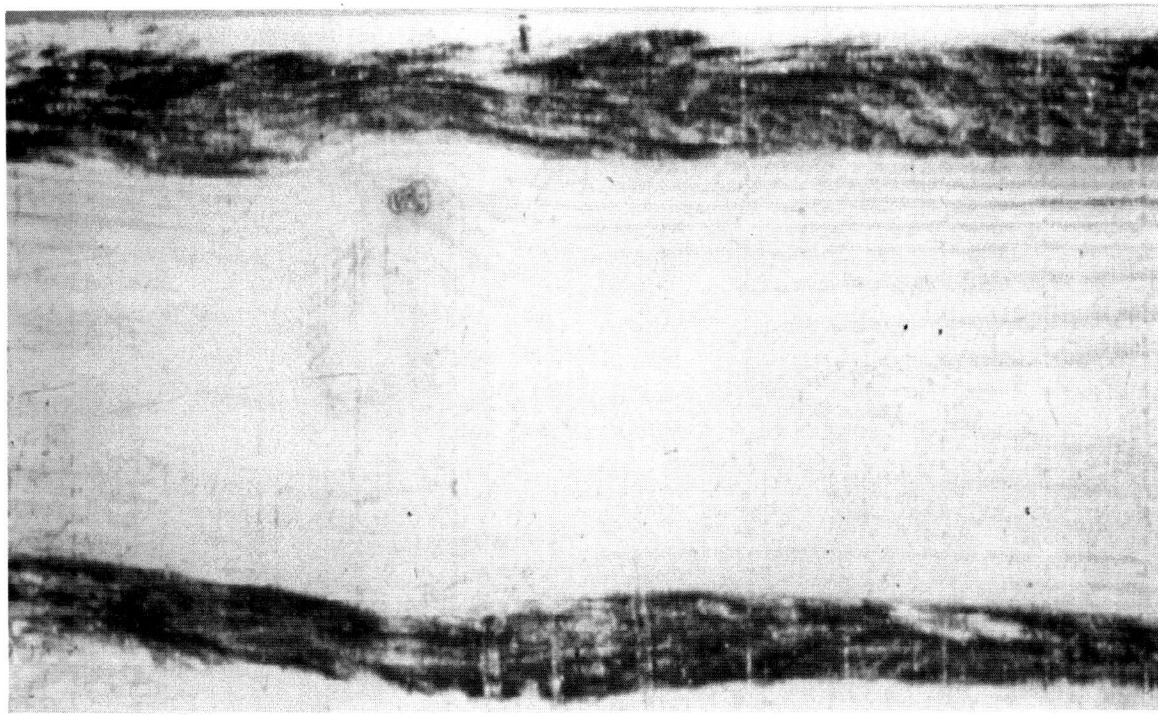

Figure 1. Mold and stain fungi tend to preferentially colonize sapwood in most species, with the heartwood(center of the wood piece above) not typically colonized.

fungi also preferentially colonize the sapwood, where they degrade pit membranes as well as ray parenchyma. As with molds, stain fungi increase wood permeability. Unlike molds, stain fungi can reduce the impact strength of the wood and, with prolonged exposures, some stain fungi can produce soft rot attack.

The most important effects of molds and sapstain fungi on wood-based materials are cosmetic, as a result of pigmented spores and hyphae on the wood surface by molds and the pigmented hyphae of stain fungi deeper in the wood (*37*). While most mold damage can be eliminated by brushing or power washing, its presence can raise concerns among wood users.

Fungal Life Cycle

The life cycle of a mold fungus is relatively simple (Figure 2). It generally begins with a spore landing on a wood surface. The spore imbibes water from the wood or the surrounding surface, and swells. Eventually, the spore produces a hyphal initial that begins to penetrate the substrate. The hyphae secrete digestive enzymes as they grow and these enzymes degrade and solubilize the simple carbon compounds present on or in the wood that then diffuse back to the fungi. Although hyphae will be found throughout the wood, they tend to preferentially colonize parenchyma cells because these cells contain the stored nutrients. Once the fungus has obtained a sufficient amount of nutrients, it will begin to produce spores, usually asexually (Figure 3). In some cases, continued growth will lead to sexual reproduction, although this does not occur with all fungi and many mold fungi are known primarily by their asexual states. The total time from germination to sporulation can be as little as 24 hours, although the time varies with fungus, substrate and environmental conditions.

Fungal Requirements for Growth

Like most living agents, fungi require nutrients, adequate temperature, oxygen and water to survive. They may also have specific requirements for pH, vitamins or other compounds, but the former four requirements are the most critical for successful colonization.

The stored nutrients in the ray cells provide an abundant resource for fungi. These cells contain proteins, lipids and carbohydrates and most wood inhabiting fungi preferentially colonize these cells. The process of lumber manufacturing ruptures many of these cells, making access to the nutrients even easier. Pollen and other organic debris falling on the wood surfaces further enhance this nutrient mix and possibly increase the level of the essential nutrient nitrogen.

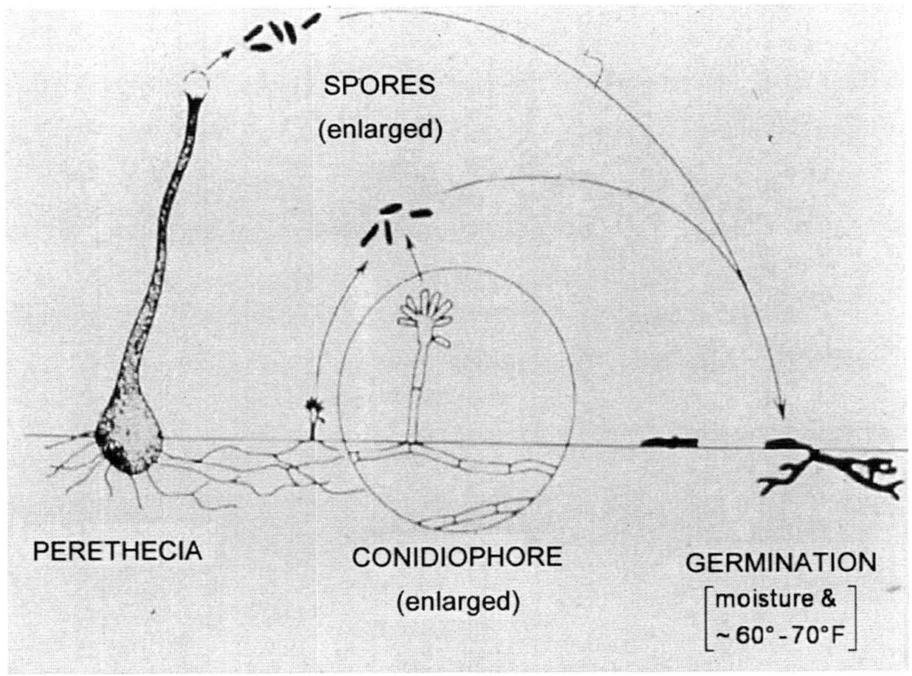

Figure 2. Example of a typical life cycle of a fungus with both asexual and sexual reproduction.

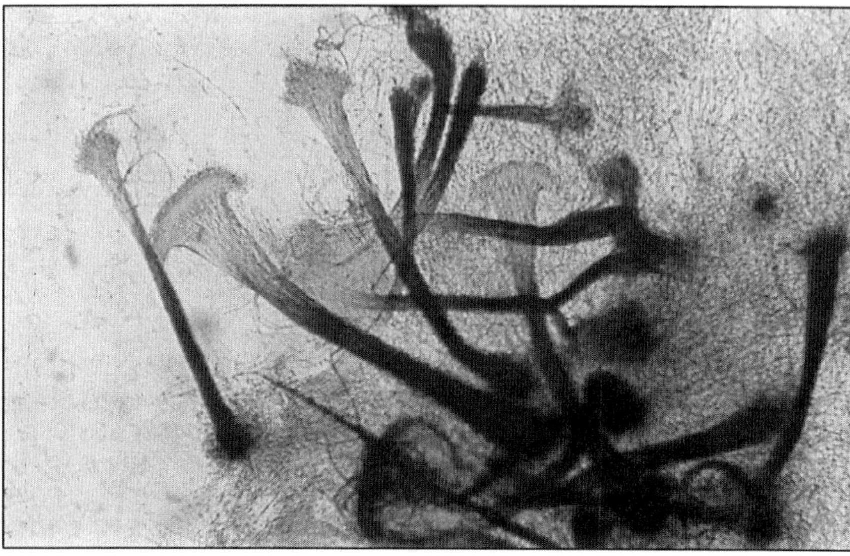

Figure 3. Example of fungal hyphae and spores of the Graphium stage of a blue stain fungus Ophiostoma picea in ponderosa pine sapwood.

As a result, the area colonized by a given fungus before it produces spores can be quite limited. For example, Xiao (*38*) noted that the *Graphium* stage of *Ophiostoma picea* extended only a few cells inward from the wood surface, but was still capable of producing a fructification. Other species clearly more extensively colonize this substrate.

Mold fungi tend to be primarily present in the sapwood of most species, although limited growth also occurs in the heartwood. Limited fungal growth in the heartwood reflects the tendency for stored compounds in this region to be converted to phenolic compounds that provide a less nutritious substrate and are, in some cases, toxic to fungi.

Temperatures for fungal growth can be quite extreme, with fungal colonization occurring at or near freezing temperatures (5°C) and well as up to 40°C and beyond. Most fungi, however, grow best between 20 and 28°C, coincidentally, the temperatures found in most homes. Temperature is rarely a limiting factor for mold attack for long periods of time (i.e. eventually it will warm).

Oxygen is essential for aerobic respiration, however, most fungi tolerate very low oxygen tensions. For example, soft rot fungi and many heart rotting basidiomycetes continue to growth at 2 to 5 ppm oxygen. By comparison, the air we breathe contains approximately 20% oxygen (200,000 ppm). Oxygen can become limiting in very waterlogged wood, as moisture fills the wood cell lumens and excludes air. This phenomenon is often observed in very wet wall cavities. For example, wet insulation pressed up against either a stud or drywall paper can create wet pockets that limit fungal growth. Growth does occur at the edge of the pockets because oxygen can diffuse inward to support fungal growth. For practical purposes, however, oxygen is rarely a limiting factor in mold growth in a structure.

The most important factor in mold related issues is water. Wood is considered to be immune from fungal attack at moisture levels below 20% (wt/wt). By its nature, wood is hygroscopic and will tend to sorb moisture. This moisture is initially sorbed by the hydroxyl groups in the wood polymers (primarily cellulose and hemicellulose) and is termed bound water. As these sites are occupied, the water eventually accumulates in the interstitial spaces and cell lumens and is termed free water. The point where free water begins to accumulate in the wood is termed the fiber saturation point (fsp). The fsp for most wood species lies between 25 and 32% (*39*).

Fungi generally require free water to colonize wood, but there are numerous reports of mold growth at lower moisture levels. The reasons for these anomalies are probably several fold. First, biological organisms are inherently variable and it is very possible that some fungi have evolved more efficient strategies for sorbing moisture. These fungi, however, appear to be the exception rather than the rule. The more likely explanations for apparent fungal growth in seemingly dry wood reflects the relationship between water and wood. Wood in humid environments will sorb moisture to specific levels at a given temperature and relative humidity (*39*). For practical purposes, wood under nearly saturated relative humidity conditions will reach moisture levels between 17 and 21% MC. These levels still do not allow free water to accumulate in the wood and would therefore be marginal for fungal growth. However, if the temperature of the wood were to drop slightly, while the amount of moisture in the air remained the same, then liquid condensation would occur on the wood surface and any fungal spores present (there are always spores present) could sorb moisture and initiate growth. This process could continue, resulting in wet wood with associated fungal attack. However, one can imagine another scenario where wetting occurs, the fungi grow and sporulate, and no further wetting occurs. The moisture present in the wood continues to diffuse into the wood which equilibrates to moisture levels below 20 %. Examination of the now moldy wood at a later time reveals fungal growth on wood at or below 20% MC leading to the conclusion that the fungus grew on very dry wood. This scenario commonly occurs on kiln dried lumber in containers. The container heats up, evaporating water from the wood (which is still at 17% MC). Condensation occurs as the container cools, allowing for fungal growth. Some exporters add moisture sorbent material to containers to preclude this possibility.

As noted earlier, however, growth of fungi in residential structures is rarely the result of short term condensation, but rather occurs because of plumbing leaks, failure to caulk around windows or other building elements or design flaws that allow moisture to build up to the point where condensation or water intrusion is inevitable. In all these cases, it is important to remember that water can enter wood as a liquid or gas, but it can only exit as a gas. Thus, wood will invariably dry far more slowly than it wets, allowing some organisms to grow. Although a large industry has evolved to deal with moisture intrusion and related mold issues, prevention is far less expensive and more effective.

Significance of Molds

The presence of mold and stain fungi on the surfaces of building materials is clearly disconcerting to the homeowner. The responses to mold can vary, but generally small patches can be cleaned with dilute bleach or soap. These treatments will not completely kill the fungi, but will help reduce the spore load and brighten the surface (*40*). Fungal growth is also common on framing lumber, but this colonization generally ceases when the wood is seasoned and is sealed in by the sheathing and drywall. Thus, any spores present on these surfaces are unlikely to invade the inhabited space to impact the occupants. Larger areas of mold must be approached more carefully since careless cleaning can disperse spores, leading to more widespread contamination. The use of a professional remediation specialist may be advisable under these circumstances. In all cases, the first step in remediation is to remove the source of moisture. Failure to remove moisture will allow fungi to reinvade the substrate, negating the value of the treatment.

Conclusions

Concerns about molds have become a prominent building issue and are unlikely to go away. In the end, molds problems are intimately associated with moisture problems and careful p planning is required to limit moisture peneration and accumulation in a structure. Practices that include designs to avoid wetting, ventilation to limit moisture accumulation and proper construction can all limit the potential for both moisture intrusion and mold.

References

1. *Wood microbiology: decay and its prevention*; Zabel R.A.; Morrell, J.J., Eds.; Academic Press, San Diego, CA, 1992; 474 pages.
2. Daggett, D.A.; Chamberlain, M.; Smith, W. 1999. *Proceedings of the 2nd Annual Conference on Durability and Disaster Mitigation* **1999**, Madison, WI: Wisconsin Department of Health and Family Services.
3. Horner, W.E.; Worthan, A.G.; Morey. P.R. *Appl. Environ. Microbio.* **2004**, *70*, 6394.
4. Kozak, P.P. Jr.; Gallup, J.; Cummins, L.H.; Gillman, S.A. 1979. *Ann. Allergy* **1979**, 88.
5. Li, D.W.; Yang, C.S. *Mycotaxon* **2004**, *89*:473.
6. Lumpkins, E.D.; Corbit, S.L.; Tiedeman, G.M. *Ann. Allergy* **1973**, *31*:361.
7. Peltola, J.; Andersson, M.A.; Haahtela, T.; Mussalo-Rauhamaa, H.; Rainey, F.A.; Kroppenstedt, R.M.; Samson, R.A.; Salkinoja-Salonen, M.S. *App. Environ. Microbio.* **2001**, *67*, 3269.
8. Shelton, B.G.; Kirkland, K.H.; Flanders, W.D.; Morris, *App. Environ. Microbio.* **2002**, *68*, 1743.
9. Solomon, W.R. *J. Allergy Clinical Immunology* **1975**, *56*(3), 235.
10. Hawksworth, D.L. *Mycological Res.* **1991**, *95*(6):641.
11. Ceigler, A.; Bennett, J.W. *Bio-Sci.* **1980**, *30*, 512.
12. CDC. Acute pulmonary hemorrhage/hemosiderosis among infants–Cleveland, January 1993-November 1994. *Morbidity and Mortality Weekly Report (MMWR)* **1994**, *43*, 881.
13. CDC. Update: Pulmonary hemorrhage/hemosiderosis among infants–Cleveland, Ohio, 1993-1996. *MMWR* **1997**, *46*, 33.
14. CDC. Update: Pulmonary hemorrhage/hemosiderois among infants–Cheveland, Ohio, 1993-1996. *MMWR* **2000**, *49*(09), 180.
15. CDC. Questions and Answers on *Stachybotrys chartarum* and other molds. March 9, 2000b. http://www.cdc.gov/nceh/asthma/factsheets/molds/default.htm.
16. Dobrotko, V.G. *Amer. Review Soviet Medicine* **1945**, *2*, 238.
17. Fung, F.; Clark, R.; Williams, S. *Clinical Toxicology* **1998**, *36*(1&2), 79.
18. Hodgson, M.J.; Morey, P.; Leung, W-Y.; Morrow, L.; Miller, D.; Jarvis, B.B.; Robbins, H.; Halsey, J.F.; Storey, E. J. *Occupational Environ. Med.* **1998**, *40*(3), 241.
19. Jarvis, B.B.; Lee, Y-W.; Comezoglu, S.N.; Yatawara, C.S. *App. Environ. Microbio.* **1986**, *51*, 915.
20. Johanning, E.; Biagini, R.; Hull, D.; Morey, P.; Jarvis, B.; Landsbergis, P. *Internat. Archives Occupational Environ. Health* **1996** *68*, 207.
21. Kelman, B.J.; Robbins, C.A.; Swenson, L.J.; Hardin, B.D. *Internat. J. Toxicology* **2004**, *23*, 3.

22. Pasanen, A-L.; Korip, A.; Kasanen, J.P.; Pasanen, P. *Environ. Internat.* **1996**, *24*(7), 703.
23. Robbins, C.A.; Swenson, L.J.; Nealley, M.N.; Gots, R.E.; Kelman, B.J. 2000. *App. Occupational Environ. Hygiene* **2000**, *15*(10), 773.
24. Schiefer, H. *Proceedings of the Fiftieth Internat. Conference on Indoor Air and Climate.* Toronto, Canada, 1990; p 167.
25. Davidson, R.W. *J. Agric. Res.* **1935**, *50*, 789.
26. Dowding, P. *Transactions British Mycological Soc.* **1970**, *55*(3), 399.
27. Kaarik, A. *Internat. Res. Group on Wood Preservation Document* IRG/WP./199. Stockholm, Sweden. **1980**, 112 pages.
28. Scheffer, T.C. In: *Wood deterioration and its prevention by preservative treatments*, Nicholas, D.D. Ed. Syracuse University Press, Syracuse, NY, **1973**.
29. Scheffer, T.C.; Lindgren, R.M. *Stains of sapwood products and their control.* USDA Technical Bulletin 714, 1940, Washington, D.C.
30. Kang. S.M.; Morrell, J.J. *Mycologia* **2000**, *92*(4), 609.
31. Anderson, B.; Nielsen, K.F.; Jarvis, B.B. *Mycologia* **2002**, *94*, 392.
32. Laks, P.E.; Richter, D.L.; Larkin, G.M. *Forest Prod. J.* **2002**, *52*(5), 41.
33. Murtoniemi, T.; Nevalainen, A.; Hirvonen, M.R. *App. Environ. Microbio.* **2003**, *69*, 3751.
34. Nieminen, S.M.; Karki, R.; Auriolà, S.; Toivola, M.; Laatsch, H.; Laatikainen, R.; Hyarinen, A.; Von Wright, A. *App. Environ. Microbio.* **2002**, *68*, 4871.
35. Ren, P.; Ahearn, D.G.; Crow, S.A. *J. Indus. Microbio.* **1999**, 209.
36. Lindgren, R.M. *Proceedings Amer. Wood Preservers' Assoc.* **1952**, *48*, 158.
37. Zink, P.; Fengel, D. *Holzforschung* **1998**, *42*(4), 217.u
38. Xiao, Y. *Application of green fluorescent protein (GFP) for studying interactions between Ophiostoma piceae and Trichoderma harzianum in freshly sawn Douglas-fir sapwood.* Ph.D. Dissertation, Oregon State University, Corvallis, Oregon, **2004**.
39. U.S. Department of Agriculture. *Wood Handbook: Wood as an engineering material.* USDA Forest Service Forest Products Laboratory General Tech. Report FPL-GTR-113. Madison, WI, **1996**, 463 p.
40. Taylor, A.M.; Freitag, C.M.; Morrell, J.J. *Forest Prod. J.* **2004**, *54*(4), 45.

Chapter 5

Weathering and Photoprotection of Wood

Philip D. Evans

Centre for Advanced Wood Processing, University of British Columbia, Vancouver V6T 1Z4, Canada (phil.evans@ubc.ca)

This Chapter describes the weathering of wood exposed outdoors, the environmental factors and mechanisms responsible for such deterioration, and the ability of chemical treatments to prevent the weathering of wood. Also included are the prospects for the development of more effective protection systems that restrict surface deterioration and protect the structural properties of wood.

Preservative treated wood is used in a wide variety of industrial (poles, cross-arms, ties, highway posts), agricultural (fences, vineyard and orchard posts, barns) and residential (decks, windows, framing) applications. The market for treated wood has increased significantly in the last 30 years due, in part, to the increased use of wood in residential applications, particularly decks. The decking market alone consumes 2 billion board feet of Southern pine (*Pinus* sp.) lumber per annum, and 38 percent of all of the treated Southern pine in the USA is used for outdoor decking (*1*).

The performance requirements of treated decking differ from those needed for wood in industrial and agricultural applications. In particular, the appearance of the wood assumes greater importance because consumers regard decks as extensions of household living areas and expect decks to maintain their visual appeal over time. Evidence to support this assertion is the finding that surface deterioration of treated wood is a significant reason for the premature replacement of chromated-copper-arsenate (CCA) treated timber decks in the USA (*2*). The inability of conventional wood preservatives to prevent the surface deterioration of wood is well known, and there has been longstanding interest in developing hydrophobic emulsion additives to increase the water repellency of treated wood and reduce surface checking (*3, 4*). In general, however, most of the wood preservatives currently on the market fall short of meeting the desire of consumers for a treatment that will maintain the visual appeal of wood exposed outdoors as well as protect its structural integrity. This has provided an opportunity for alternative decking products, for example wood-plastic and plastic lumber, that are more resistant to surface deterioration, to capture an increasing share of the market for treated timber decking (*5*). These lumber substitutes are two to five times the cost of treated wood and their success indicates the price premium that consumers are willing to pay for an aesthetically pleasing, maintenance-free, deck (*5*). This has led to renewed interest in surface treatments and wood protection systems that are more effective at restricting the weathering of wood.

This Chapter describes the weathering of wood exposed outdoors, the environmental factors and mechanisms responsible for such deterioration, and the ability of chemical treatments to prevent the weathering of wood. Also included are the prospects for the development of more effective protection systems that restrict surface deterioration and protect the structural properties of wood.

Weathering of Wood

The surface degradation that occurs when wood is used outdoors and above ground is termed weathering (*6*). Weathering should not be confused with decay caused by basidiomycete fungi, which can extend deeply into wood and significantly reduce the strength of structural timber (*6*). In contrast, the effects of weathering on the mechanical properties of structural lumber are small because they only affect the surface of wood. Accordingly, there are examples of wooden buildings, such as the stave (pole) churches in Norway, that are still structurally sound despite having been exposed to weathering for over 1000 years (*7*). The most obvious features of weathered wood are its gray coloration, rough surface texture and presence of checks (Figure 1). Microorganisms and lichens colonize weathered wood, but conditions at exposed wood surfaces generally do not favor decay. Hence, the defining features of weathering are its superficial nature and the minor role of microorganisms compared to environmental factors in degradative processes (*6*).

Environmental Factors Affecting the Weathering of Wood

The main environmental factors involved in the weathering of wood are solar electromagnetic radiation (ultraviolet and visible light), molecular oxygen (O_2), water, heat, wind-blown particulate matter, atmospheric pollutants and certain specialized micro-organisms (*6*).

Solar radiation

The maximum amount of solar radiation available at the earth's surface on a clear day is normally 1000 W/m^2. Such radiation consists of approximately 5% ultraviolet (UV) (286-380 nm), 45% visible (380-780 nm) and 50% infrared (780-3000 nm). Solar radiation is the main environmental factor responsible for the surface weathering of wood. UV light, particularly UVB (280 to 320 nm), is more energetic than visible light and capable of cleaving the carbon-carbon, carbon-oxygen, and carbon-hydrogen bonds that connect the polymeric components of wood; i.e. cellulose, hemicelluloses, and lignin [Equation 1]. Nevertheless, wood exposed to visible light degrades at about half the rate of material exposed to the full solar spectrum (*8*).

$$E = \frac{2.86 \times 10^4}{\lambda} \qquad [1]$$

where, E = energy of a photon (kcal/mol) and λ wavelength (nm)

The surface photodegradation of wood is more rapid when levels of solar radiation are elevated. Table 1 shows the seasonal variation in weight losses of thin radiata pine (*Pinus radiata* D. Don) wood veneers inclined at three angles

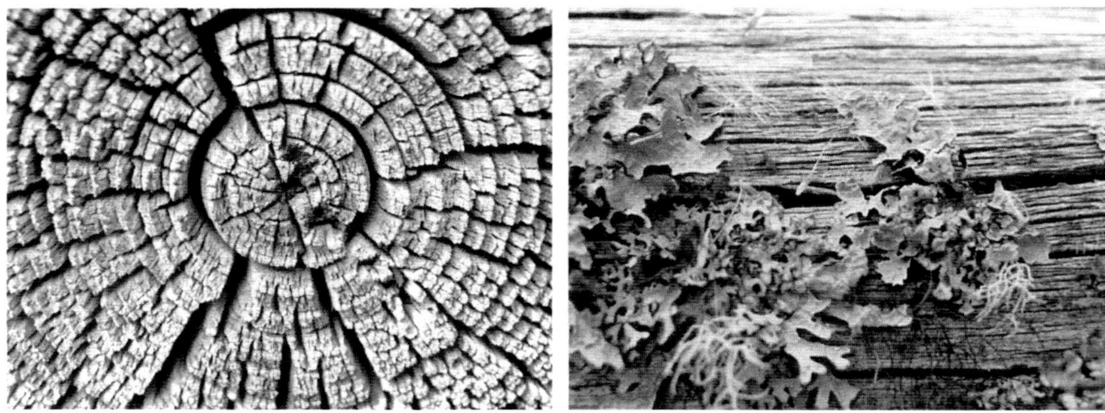

Figure 1. Appearance of weathered wood; end grain of a coniferous log used in the construction of a log cabin showing severe checking and erosion of wood (left); longitudinal surface of a log from the same building showing gray coloration, checking and colonization of the surface by lichens (Reproduced with permission from reference 157. Copyright 2003 For. Prod. J.)

(0° (horizontal), 45°, and 90° (vertical)) and exposed to the weather in Canberra, Australia (*9*). Weight losses were greatest in the summer (December to February in the Southern hemisphere) when levels of solar radiation were high (*9*). Degradation was also highest for samples exposed horizontally to the sun (*9*, *10*). This effect is related to the fact that low angles of exposure maximize the levels of ultraviolet radiation at exposed surfaces (*11*). Increased levels of solar radiation at high altitudes are also thought to be responsible for more pronounced bleaching of wood exposed in mountainous regions compared to wood exposed at sea level (*12*).

Table 1. Effect of season and angle of exposure on the weight losses of thin wood veneers exposed to natural weathering in Canberra, Australia

Angle °	% Weight losses of veneers during different exposure periods					
	April/May	June/Jul	Aug/Sept	Oct/Nov	Dec/Jan	Feb/ Mar
0 Horiz	44.2	27.9	41.4	40.9	51.1	42.0
45	40.9	31.5	38.7	37.8	47.6	39.7
90 Vert	24.5	19.0	25.4	23.3	23.8	24.2

Losses in tensile strength of thin wood veneers exposed outdoors can be related to the level of photochemically active radiation received by the veneers using Equation [2]

$$S = So \exp. (-kD) \quad [2]$$

This equation has been used to derive photodegradation rates for wood exposed to natural or artificial weathering, and different wood species exposed to natural weathering (*13*, *14*). These studies found small differences in the rate of photodegradation of six softwood species, and concluded that artificial weathering was a valid alternative to natural weathering for studies of the effect of temperature and wood moisture content on rate of photodegradation (*13*, *14*).

Water

Water plays an important role in the weathering of wood. Dimensional change as a result of the wetting and drying of wood generates surface tensile stresses that cause checking and warping of timber (*6*). Water leaches degraded lignin fragments and hemicellulose-derived sugars from photodegraded wood surfaces, and hence under natural weathering conditions losses in weight of wood veneers are correlated with rainfall events (*15*, Figure 2). The leaching of colored unsaturated (quinoid) lignin breakdown products from weathered wood

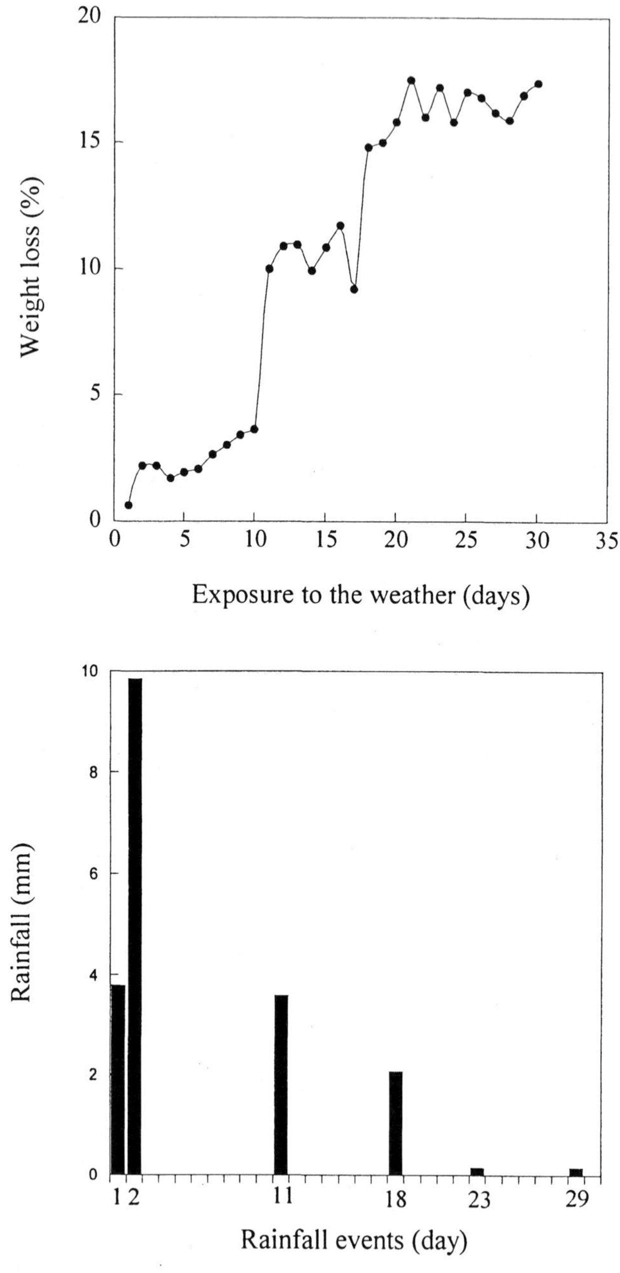

Figure 2. Loss in weight of thin wood veneers exposed over a 30 day exposure trial in Canberra, Australia (top) and rainfall during the same period (below); note the correlation between weight loss of veneers and rainfall events

surfaces by rain explains why weathered wood, which yellows initially, eventually becomes gray and is mainly composed of cellulose. In the absence of such leaching, photodegraded lignin fragments accumulate within the wood and the wood becomes darker (*16*). Water can hydrolyze the non-cellulosic components of wood and prolonged exposure of wood to water at mild temperatures (50-65°C) *in vitro* results in degradation of hemicelluloses and lignin with little degradation of cellulose (*17, 18*). Moisture in the presence of superficial solar heating (see below) may thus degrade the non-cellulosic components of wood, although measurements in the field to confirm this are lacking.

Heat

The maximum surface temperature that has been recorded at wood surfaces exposed outdoors in the USA is 52°C (*19*). Higher surface temperatures presumably occur for wood exposed in tropical climates, however, the temperatures attained are unlikely to approach the glass transition temperature of lignin (130-150°C) or those at which significant structural degradation of wood's chemical components occurs (200°C). Therefore, it is unlikely that heat directly causes surface degradation of wood in temperate climates, but rather accelerates photo-oxidation and hydrolysis of wood caused by light and water, respectively. Evidence to support this suggestion is the finding that loss in tensile strength of thin softwood veneers subjected to artificial weathering is temperature dependent, and increases with rising temperature within the range 25 to 55°C (*20*). Heat may also accelerate the surface drying of wood thereby generating stresses that cause checking of wood. Exposure of wood to low temperatures and repeated freezing and thawing may also cause checking of wood (*7*), and wood in cold climates is susceptible to abrasion by wind-blown particles of ice. For example, the iconic Australian explorer Mawson (*21*) wrote in 1915 following his pioneering expedition to the Antarctic: "the abrasion-effects produced by the impact of the snow particles were astonishing ... A deal (*Pinus* sp.) box, facing the wind, lost all its painted bands and in a fortnight was handsomely marked; the hard knotty fibers (latewood) being only slightly attacked, whilst the softer, pithy laminae (earlywood) were corroded to a depth of one-eighth of an inch (approximately 3 mm)." Windblown sand and salt can also cause similar, if less spectacular, abrasion of wood exposed outdoors (*6*).

Atmospheric pollutants

Observations in the field and laboratory experiments all tend to suggest that the weathering of wood is more rapid in polluted than unpolluted environments.

The main atmospheric pollutants are dust and smoke particles, and volatile pollutants including sulfur compounds, ammonia, nitrogen oxides, carbon monoxide, and saturated / unsaturated aliphatic and aromatic hydrocarbons and their derivatives. Nitric dioxide is an initiator of photo-oxidation and sulfur dioxide in the atmosphere may be converted to sulfuric acid, which can degrade wood surfaces (*22*). Elevated levels of airborne sulfur dioxide arising from the burning of coal were thought to be responsible for significant losses in tensile strength of Norway spruce (*Picea abies* (L) Karsten) wood veneers exposed to the weather during the winter in Poland (*23*). Pre-soaking wood samples in dilute (pH 3.0-4.0) sulfuric or nitric acid increases the degradation of wood during artificial weathering (*24, 25*).

Organisms

Bacteria and certain microfungi are frequently observed colonizing weathered wood, even though conditions at wood surfaces do not favor decay. The dimorphic black yeast, *Aureobasidium pullulans* (de Bary) Arnaud, in particular, is frequently isolated from weathered wood in both temperate and tropical climates where it contributes to the gray coloration of the wood. *A. pullulans* is capable of withstanding temperatures of 80°C, growing over a pH range of 1.9 to 10.1, surviving for long periods without moisture and metabolizing lignin breakdown product. Hence, it is particularly well suited to the micro-environment of weathered wood (*26, 27*). Other species isolated from weathered wood surfaces include *Cladosporium* spp., *Alternaria* spp., *Penicillium* spp., *Aspergillus* spp., *Sclerophoma* spp., *Macrosporium* spp. and *Tetracoccosporium* spp. (*28, 29*). In humid tropical climates, where conditions at wood surfaces are much more favorable for microbial activity the weathering of wood may encourage subsequent colonization by decay fungi, particularly in wood species that lack natural durability (*30*).

A. pullulans is capable of forming symbiotic lichenized associations with green algae (*26*), and lichens are frequently observed colonizing weathered wood surfaces, particularly in unpolluted environments (Figure 1). The role that lichens play in the weathering of wood has received little attention, however, it is possible that they alter the micro-environment of weathered wood through the production of lichen acids, such as usnic, atranorin and evernic acid, which are known to show antimicrobial and antioxidant activity (*31*). Contact dermatitis of forestry workers (Woodcutters disease) has been linked to exposure to lichen acids through handling weathered firewood colonized by lichens (*31*).

A number of insect species, particularly paper wasps (Polistinae) and yellow jackets and hornets (Vespinae) mine weathered wood surfaces to obtain delignified wood fragments (Figure 3). The weathered wood is intensively chewed and mixed with proteins in the insect's saliva to form a water-repellent, fibrous, paper-like, composite that the insects use to construct their elaborate

Figure 3. A wasp (Vespula vulgaris L.) removing wood from the surface of weathered pine; note the lighter areas where insects have removed weathered wood (See page 2 of color inserts.)

nests. These insects show a preference for weathered wood and rarely utilize fresh unweathered wood (*32*).

Interactive effects

The individual factors involved in the weathering of wood rarely act in isolation, but tend to operate together and sometimes synergistically. The most obvious example of such synergy is the role that heat plays in accelerating photodegradation (*20*) and hydrolysis (*17, 18*). Photodegradation weakens the wood surface and degrades its micro-structure, which makes wood more susceptible to checking (*33*). Changes in the surface chemistry of wood influences subsequent colonization of wood by micro-organisms (*27*). The photo-depolymerization of lignin and holocellulose and the production of low molecular weight phenolic compounds and sugars, respectively, provide a nutritional source for microrganisms and explain, in part, why *A. pullulans* is frequently isolated from weathered wood (*27*). Exposure of wood to artificial UV radiation also encourages colonization and degradation of wood by mold fungi (*34*). Ultraviolet light has also been shown to enhance the rate of discoloration of wood caused by sulfur dioxide and nitrogen dioxide (*35*). As mentioned above, water leaches lignin photodegradation products from wood surfaces, which reduces the nutrients available for organisms such as *A. pullulans*. Hence, fully exposed wood surfaces may show less colonization by *A. pullulans* than surfaces, such as those beneath clear finishes, which retain lignin photodegradation products. Photodegradation of wood is greater in the presence of moisture than under dry conditions (*36, 37*), possibly because water molecules swell wood thereby opening up inaccessible regions of the cell wall to photodegradation (*38*). The full extent of the interaction of the factors involved in the weathering of wood has yet to be elucidated.

Mechanisms

The range of environmental factors involved in the weathering of wood suggests that it may be difficult to define a single mechanism responsible for the weathering of wood. The superficial nature of weathering, however, is one of its defining features, and this can be linked to the limited penetration of light into wood (*39*). The other factors involved in the weathering of wood (water, heat, micro-organisms etc), have the ability, in principle, to degrade wood to a much greater depth, yet beyond the zone affected by light, the chemical and physical properties of wood exposed to the weather are largely unchanged. Clearly, light plays the major role in the weathering of wood. An exception to the rule that weathering is only superficial in nature is the surface checking caused by moisture-induced dimensional changes, which may extend several millimeters into wood (*40*). Knowledge of both of the mechanisms responsible for the photodegradation and checking of wood is essential to understand the weathering of wood, and develop more effective protection systems.

Photodegradation of wood

In order to act upon wood, solar radiation must be absorbed by one of wood's chemical constituents, cellulose, hemicelluloses, lignin or low molecular weight extractives. Experimentation has shown that lignin strongly absorbs UV light with a distinct maximum at 280 nm, and decreasing absorption extending beyond 380 nm into the visible region of the spectrum (*41*). The chromophoric groups in lignin that absorb UV light are double bonds, phenolic and carbonyl groups, quinones, quinonemethides and biphenyls (*42*). Cellulose and hemicelluloses absorb UV light between 200 and 300 nm (*43*, *44*), and show some absorption of visible light, but the absorption of light by lignin is thought to be the first step leading to the photodegradation of wood. In fact, lignin is so effective at absorbing light that very little (<10%) photoactive light (<420 μm) penetrates wood beyond a depth of 220 μm (*39*). The penetration of light into wood is inversely proportional to the logarithm of wood density, but it is also affected by the chemical composition of the wood (*45*). Species containing high concentrations of phenolic extractives, which absorb UV light, show less penetration of light than predicted by their density (*45*). The accumulation of lignin photodegradation products at wood surfaces can also restrict the penetration of light into wood (Figure 4).

Energy absorbed from solar radiation can be dissipated in wood (and polymers in general) through the cleavage of molecular bonds resulting in the formation of a free radical, a molecular species that is highly reactive because it has an unpaired valence electron. The critical wavelengths to cleave individual carbon-carbon, carbon-oxygen, and carbon-hydrogen bonds are 346, 334, and 289 nm, respectively (*46*). Aromatic radicals can be readily detected using electron spin resonance techniques when wood is irradiated with UV light, indicating the photolysis of lignin (*47-49*). It has been suggested that these aromatic radicals include gaiacoxyl, cetyl and phenacyl radicals (*50*) (Figure 5).

Generalized schemes for the photodegradation of polymers describe the reaction of free radicals with atmospheric oxygen to form peroxy radicals. Such radicals typically attack polymer chains via hydrogen abstraction to form a hydroperoxide and another free radical. Hydroperoxides are susceptible to UV radiation and undergo photolysis to form additional free radicals. Evidence for the photo-oxidation of wood and the participation of oxygen in photodegradation is the finding that free radicals formed in wood subjected to irradiation are stable in a vacuum or inert atmosphere, but readily disappear in the presence of oxygen (*38*).

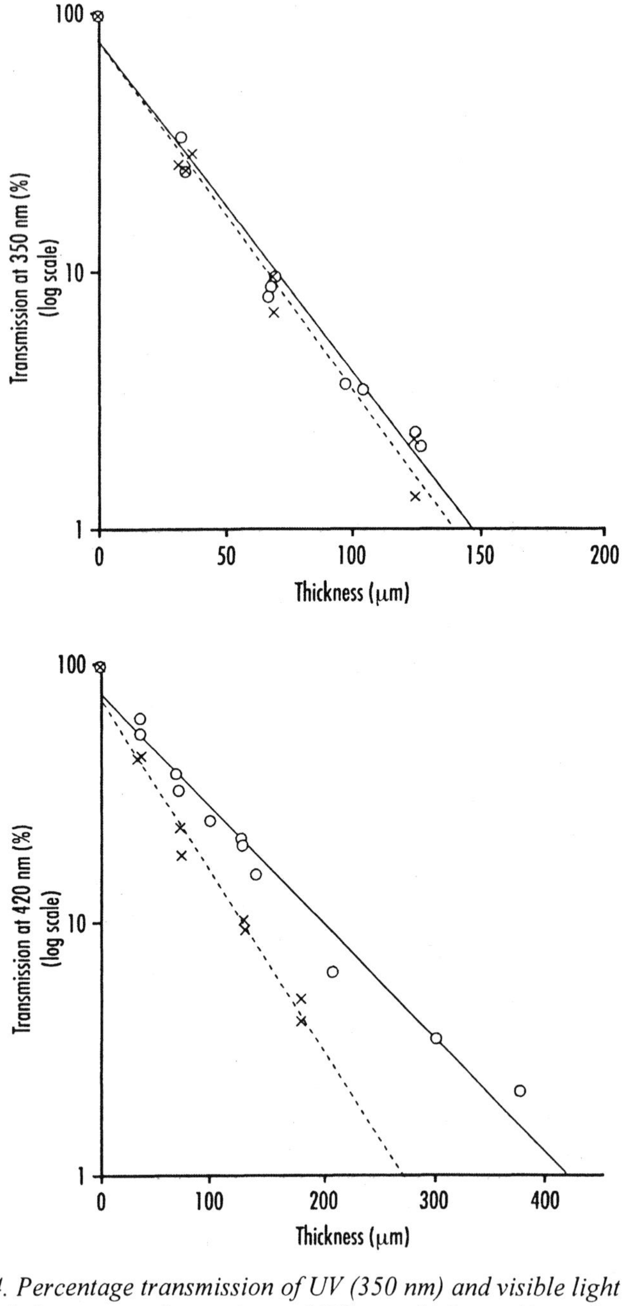

Figure 4. Percentage transmission of UV (350 nm) and visible light (420 nm) through Japanese cedar sections of different thickness (Reproduced with permission from reference 39. Copyright 2004.)

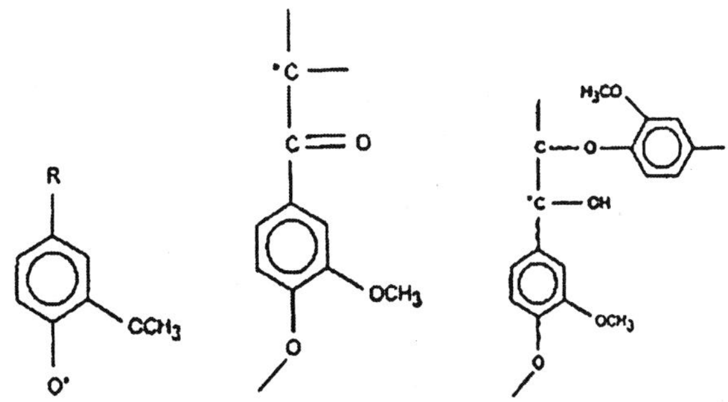

Figure 5. Suggested structures of different lignin-derived radicals; (a), Gaiacoxyl radical; (b), Phenacyl radical; (c), Cetyl radical (50)

The literature indicates that free radical formation and degradation of wood fits some, but not all, of the aspects of the general scheme for the photodegradation of synthetic polymers. The photodegradation of wood is undoubtedly more complicated than that of synthetic homopolymers because it consists of a blend of polymers (lignin, cellulose, and hemicellulose) and low molecular weight extractives that differ in their susceptibility to solar radiation. Furthermore, it is clear that the precise mechanisms and reaction pathways involved in the photodegradation of each of these components have not been fully elucidated. However, the key step involved in the photodegradation of wood appears to be photolysis and fragmentation of lignin resulting in the formation of aromatic radicals (*38*). These free radicals are clearly capable of reacting with oxygen to cause photo-oxidation of cellulose and hemicelluloses. Free radicals may be terminated by reaction of radicals to form colored unsaturated carbonyl compounds, which explains why wood initially yellows when exposed to light (*38*, *51*). Vanillin has consistently been identified as one of the end products of the photodegradation of lignin in wood (*52*). A scheme for the photodegradation of lignin that is consistent with the formation of vanillin is shown below (Figure 6). This scheme involves cleavage of the α-β bond in a typical aryl-ether lignin subunit. Fragmentation of the same bond is also important in lignin breakdown during chemical oxidation (*53*) and fungal decay (*54*).

Checking of wood

Checks represent the macroscopic response of wood to surface stresses generated by anisotropic shrinkage caused by changes in the bound water

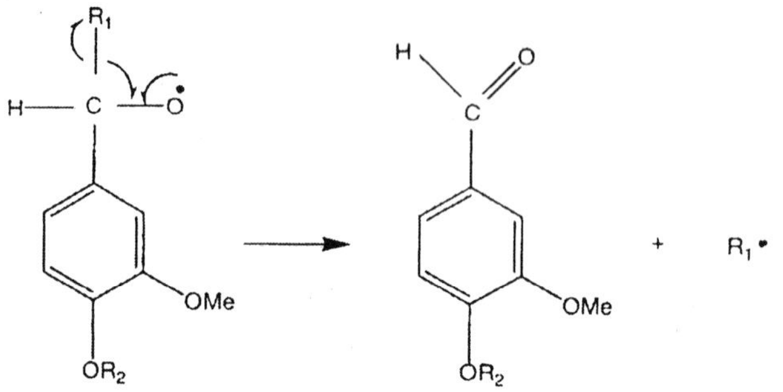

Figure 6. Proposed key fragmentation mechanism for lignin photolysis

content of wood (*55*). An alternative response to such stresses is for wood to warp. Both forms of degrade occur when wood is exposed to the weather, although warping is often restrained by the presence of fixings (nails or screws). Shrinkage anisotropy of wood exposed outdoors results from more rapid drying of surfaces exposed to the weather, and the well known anisotropy caused by the higher shrinkage of wood tangentially than radially (*56*). A scheme to account for the development of surface checks in weathered wood is outlined below (Figure 7). This scheme assumes that fixings reduce the release of stresses by preventing warping, and drying of wood is restricted on the underside of the board.

Wetting and absorption of water by wooden decking boards exposed outdoors will result in the swelling of surface layers. Such swelling will be restrained by fixing of boards, and the underlying wood, and surface layers will be placed under compression. Checks will not form at this stage, and there is little microscopic evidence in weathered wood of damage caused by exposure to high compression stresses. Water absorbed by wood surfaces will penetrate sub-surface layers by capillarity (in permeable species) and diffusion processes (Figure 7a). Subsequent exposure of wood to solar (infra-red) radiation will cause surface wood layers to dry (Figure 7c). Shrinkage of these surface layers will occur as their moisture content decreases below the fiber saturation point, but will be restrained by sub-surface layers, which are still above the fiber saturation point, and by boards being fixed in place by nails or screws (Figure 7c). Such restraint will place the surface layers under a tension stress and cause checks to form (Figure 7d-e). The magnitude of this stress will depend on moisture loss and shrinkage of the surface layers, and will be greatest for wood surfaces whose growth rings are oriented parallel (tangential) rather than perpendicular (radial) to the surface, because shrinkage of wood tangentially is approximately twice that occurring in the radial direction (*56*). Accordingly, checking of quarter-sawn decking boards exposed outdoors is significantly less than that in flat sawn boards (*57*). Less checking has also been observed in flat sawn boards whose growth rings are more perpendicular to the exposed surface

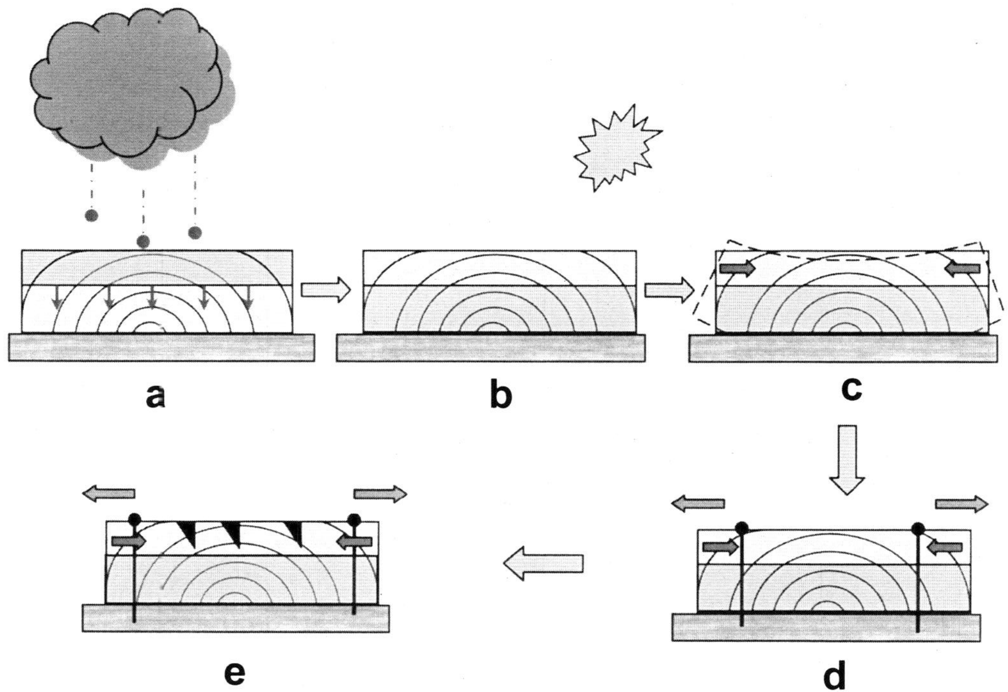

Figure 7. Simplified schematic diagram showing how wetting and drying generates surface stresses that cause restrained wooden decking boards to check during exterior exposure *(See page 2 of color inserts.)*

(ie, pith side up) than in boards whose growth rings are parallel to the surface (ie, bark side up) (*58*).

Checking of wood during kiln drying occurs when stresses cause surface strains of 0.2. Checking at weathered wood surfaces may require lower stresses and strains than unweathered wood due to the weakening of wood surfaces caused by photodegradation, and the creation of voids or microchecks in the wood, which act as focal points for the development of visible checks (*33*, *59*).

Effects of Weathering

The effects of weathering can be observed at different length scales, molecular, microscopic and macroscopic. Changes at the molecular level occur rapidly, and our understanding of the rate of change occurring at wood surfaces exposed outdoors has advanced considerably in recent years due to the widespread availability of spectroscopic techniques that can probe the chemical composition of surfaces (e.g., Fourier transform infrared spectroscopy (FTIR), X-ray photoelectron spectroscopy (XPS), and nuclear magnetic resonance (NMR) spectroscopy).

Molecular

All of the macromolecular components of wood are degraded during weathering. Over 130 years ago it was reported that weathered wood surfaces (the gray layer at the surface) consisted of pure or nearly pure cellulose with a very low lignin content (*6*, *38*). Recent studies have confirmed this finding and have found that lignin at wood surfaces exposed outdoors is rapidly degraded. FTIR spectroscopy of radiata pine veneers weathered outdoors in the summer in Australia showed a remarkably rapid decrease in the peak at 1505 cm^{-1}, which corresponds to aromatic C=C bond stretching in lignin (*60*). Spectra suggested perceptible surface (1-2 μm) delignification after only 4 hours exposure, substantial delignification after 3 days and almost complete surface delignification after 6 days (Figure 8). Changes in peaks at 1601 cm^{-1} (C=C bond stretching), 1263 cm^{-1} (C-O bond stretching vibration in lignin and hemicelluloses), and 870 cm^{-1} (CH out-of-plane bending vibration in lignin) also suggested substantial and rapid degradation of lignin (*60, 61*). These results for naturally weathered wood surfaces confirmed earlier FTIR studies of wood subjected to artificial accelerated weathering, which also showed that lignin was very susceptible to photodegradation (*6*, *38*, *62*). X-ray photoelectron spectroscopy of weathered and UV-irradiated wood surfaces have shown increases in the intensities of carbon-oxygen and oxygen-carbon-oxygen bond signals and decreases in the intensities of carbon-carbon and carbon-hydrogen

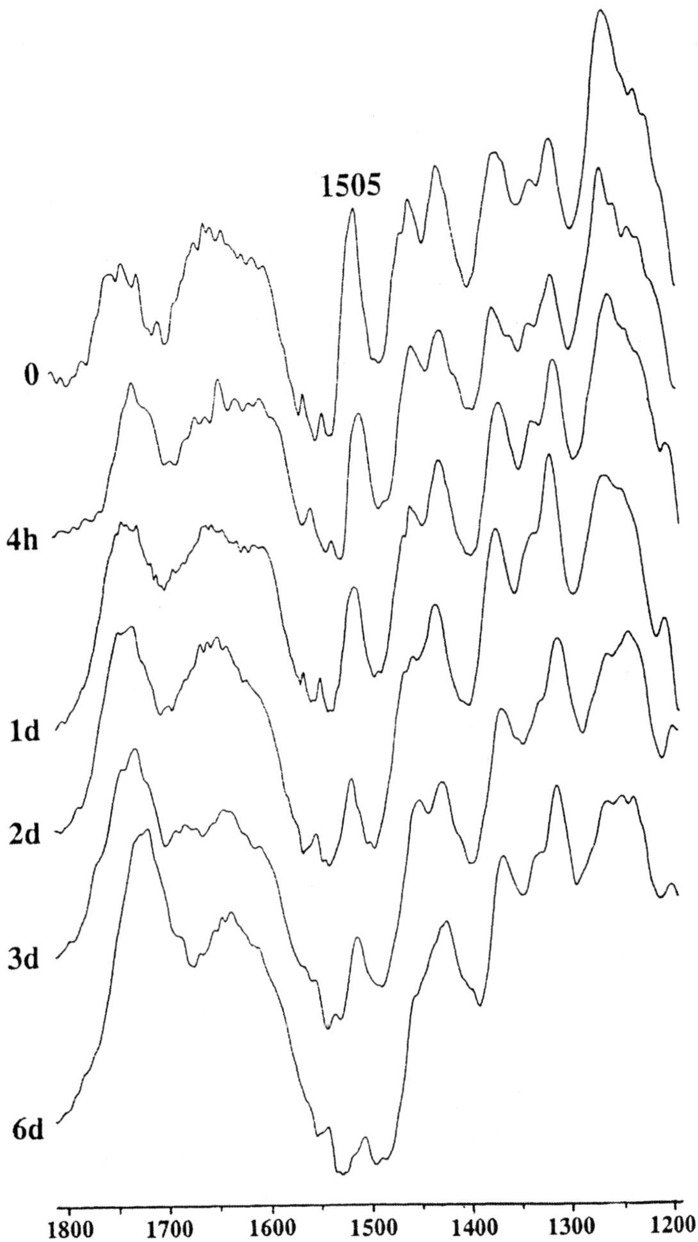

Figure. 8. Fourier transform infra red spectra of radiata pine veneers weathered for short periods of time in Canberra, Australia (Reproduced with permission from reference 60. Copyright 1996 Springer-Verlag.)

bond signals, suggestive of degradation of lignin and enrichment of the surface with cellulose (*63*). Lignin degradation at wood surfaces is accompanied by reduction in the methoxyl content at the surface and an increase in its acidity and carbonyl content (*64*, *65*). These changes occur as a result of the depletion of lignin at the wood surface and accumulation of lignin degradation products in the wood including vanillin, syringaldehyde and organic (carboxylic) acids. Formaldehyde, carbon monoxide, carbon dioxide, hydrogen, water and methanol have also been identified as degradation products of wood during weathering (*41*).

Hemicelluloses are degraded during weathering and are leached from weathered wood surfaces (*41*). Chromatography of leachates from weathered softwood surfaces has shown a high proportion of mannose and xylose, suggestive of degradation of the hemicelluloses galactoglucomannan and arabinoglucuronoxylan, respectively (*61*). FTIR spectroscopy of weathered softwood surfaces has shown reductions in the peak at 1728 cm^{-1} (C=O stretching vibration in acetyl and carboxyl in hemicelluloses) and 809 cm^{-1} (mainly vibration of mannan in hemicelluloses and CH out-of-plane bending vibration in lignin), confirming that hemicelluloses in wood are degraded during weathering (*61*).

Weathered wood surfaces are rich in cellulose, as mentioned above, and this led to suggestions that cellulose was resistant to the effects of weathering. This is clearly not the case, as cellulose at wood surfaces exposed under natural conditions is rapidly depolymerised (*8, 60*). This has been confirmed by both viscometry and high performance size exclusion chromatography of solutions of holocellulose from weathered wood (*8, 60, 66*). Reductions in the degree of polymerization of cellulose account for the rapid and large losses in tensile strength of thin wood veneers exposed to natural and artificial accelerated weathering (*8, 13, 14, 20*).

Extractives are degraded by UV light, which accounts for why many distinctly colored woods fade over time, even when exposed indoors (*67*).

Chemical changes in weathered wood are confined to the surface layers because light does not penetrate deeply into the wood, as mentioned above (*47*, *68*) (Figure 9). Degradation of lignin has been detected at depths of approximately 400 to 700 μm in low density wood species exposed to artificial UV light (*39*, *45*, *69*, *70*). The depth to which photodegradation extends into wood depends on the spectral characteristics of the light source, duration of exposure, wood density and orientation of the wood elements and their chemical composition (*39*, *45*). It has been suggested that radiation can penetrate more deeply through the open pores of transverse surfaces resulting in more complete degradation of such surfaces compared to radial surfaces (*71*). The suggestion that the depth of weathering degradation sometimes extends more than 2540 μm in "well-weathered wood" (*68*) may be accounted for by redistribution of water soluble lignin degradation products from the surface to sub-surface layers.

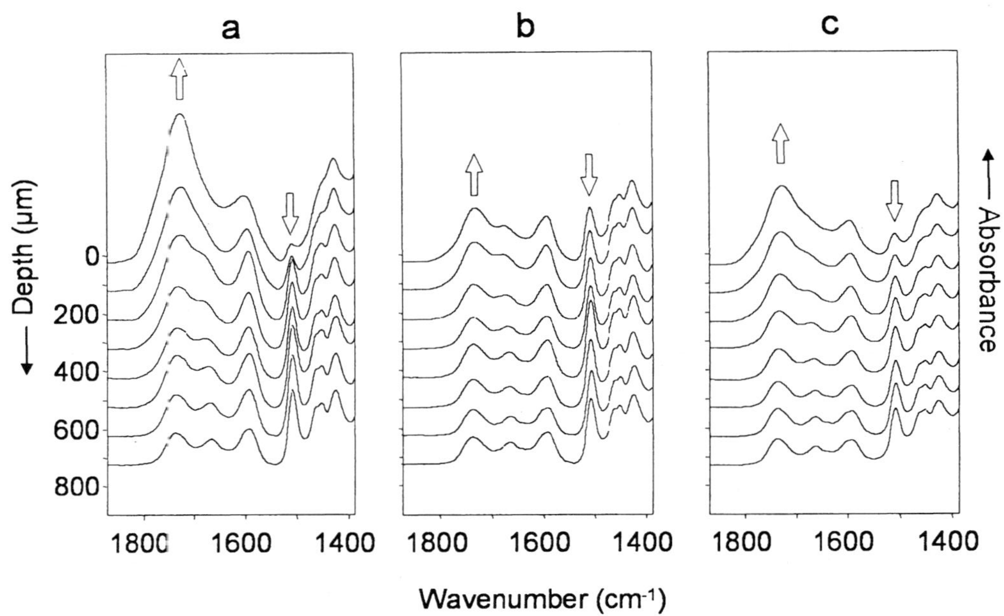

Figure 9. Depth profile of photodegradation in (a) Japanese cedar earlywood and (b) latewood and (c) Japanese cypress (Chamaecyparis obtusa Endl.) earlywood after 1500 hours exposure to UV light (Reproduced with permission from reference 45. Copyright 2005 Springer.)

Microscopic

The microscopic structure of wood within the surface layer affected by weathering is radically altered during exterior exposure. Microscopic changes are pronounced in regions of cell walls, such as the middle lamella, that contain high concentrations of lignin, or are perforated by pits. Thin walled cells are very susceptible to degradation, particularly where they are found adjacent to thicker walled cells.

Erosion of the middle lamella occurs rapidly during exterior exposure and can be clearly seen in transverse surfaces of radiata pine exposed outdoors for 5 days (*15*). The warty layer that covers the S_3 layer of the secondary wall is rich in polyphenolic materials and is also rapidly eroded during exposure to UV light (*72*). Erosion of the middle lamella is more pronounced in the cell corners where the concentration of lignin is highest and between latewood tracheids (Figure 10c, e). Eventually, complete erosion of the middle lamella and partial separation of individual tracheids along their length occurs (Figure 10h). Primary and secondary wall layers delaminate during exposure to the weather, and as a result cells become progressively thinner and more distorted (Figure 10c-h)).

Degradation of the middle lamella can also be seen in longitudinal surfaces where double cell walls are exposed to the weather (*74*, *75*). In this case erosion of the middle lamella again causes cells to separate, but whole tracheids are generally not detached from the surface because microchecks that develop in pits cause tracheids and fibers to fragment.

Microchecking of the cell wall in softwoods first develops in bordered pits (Figure 11a, b) and appears as two small notches in the aperture orientated at an angle to the long axis of the tracheids (Figure 11c, d). Subsequently, these notches propagate on either side of the pit aperture creating a diamond-shaped micro-check (Figure 11e, f) (*76*). The enlargement of these micro-checks in areas of the cell wall where several pits are found close together creates thin bridges of cell wall material between the microchecks (Figure 11f). Microchecking within these bridges creates small cell wall fragments that easily erode from the cell wall. Microchecking of half bordered pits in rays also occurs during weathering and creates fragile areas of cell wall that easily erode from the exposed surface (Figure 11e). Numerous voids are created in wood cell walls as a result of microchecking causing progressive loss in integrity of the exposed surface (Figure 11g).

The aforementioned description of the microchecking of bordered pits in radial cell walls of radiata pine is a generalized scheme and considerable variation exists in the form that the checking takes, depending on wood species and exposure. For example, in Californian redwood (*Sequoia sempervirens* (D. Don.) End.) exposed to artificial UV light, microchecking and erosion of adjacent biseriate pits created large voids that were separated by cell wall material, crassulae, that appeared to be resistant to degradation (*75*). Where pitting was uniseriate the pit annulus was eventually lost with increasing exposure, again creating voids in cell walls (*75*). It has also been reported that microchecking of bordered pits did not develop in a range of hardwoods including European beech (*Fagus sylvatica* L), opepe (*Nauclea diderrichii* Merrill) and English oak (*Quercus robur* L.) exposed to artificial weathering (*77*). Microchecks can also develop in the absence of pits in tangential softwood cell walls (Figure 11h). These generally follow the microfibril angle of the S_2 layer of the secondary wall (*8*) or the direction of orientation of tracheids (*75*). Larger checks often develop at the interfaces between different cell types, for example at growth ring boundaries in softwoods (Figure 12c-e), and at the interfaces between rays and tracheids (Figure 12b, c). In each case the check develops within the thinner walled tissues (Figure 12d). Thin walled epithelial cells in resin canals are also easily degraded during weathering, which creates small voids and, subsequently, microscopic checks in transverse surfaces (Figure 12b, c). Voids created by the erosion of ray parenchyma cells in softwoods enlarge beyond the original boundaries of the ray as a result of degradation and separation of the middle lamella above and below the ray (Figure 12g-j). Multiseriate rays in hardwoods such as beech, however, are more resistant to weathering than the surrounding tissue and remain intact during weathering and are readily detached from weathered hardwood surfaces (*78*). The extractives within softwood rays are also reported to be resistant to degradation, even though the rays themselves are not (*75*). Microchecks at growth ring boundaries and within rays and resin canals may enlarge or coalesce to form macroscopically visible checks (*59*).

Macroscopic

The most obvious macroscopic features of weathered wood are its gray color and rough surface texture caused by the presence of checks and the differential swelling and erosion of woody tissues (*6*).

Wood exposed to the weather changes color very rapidly. Light colored woods, including most coniferous species, darken in color initially and become yellow or brown due to the accumulation of photodegraded lignin constituents in the wood (79). Dark colored woods that are rich in phenolic extractives may fade initially before becoming yellow or brown (79). Irrespective of these initial color changes, wood exposed outdoors for 6 to 12 months (depending on climatic conditions) becomes gray as extractives and photodegraded lignin fragments are leached from the wood resulting in surface layers that are rich in cellulose (12). Weathered wood typically has an unattractive dark gray and blotchy appearance due to the presence of fungal hyphae, spores and pigments within surface wood layers (6). Wood exposed outdoors in coastal (exposed to salt) or very dry environments where microbial activity is restricted, often is an attractive silvery-gray color (6). In some circumstances the color of weathered wood is preferred

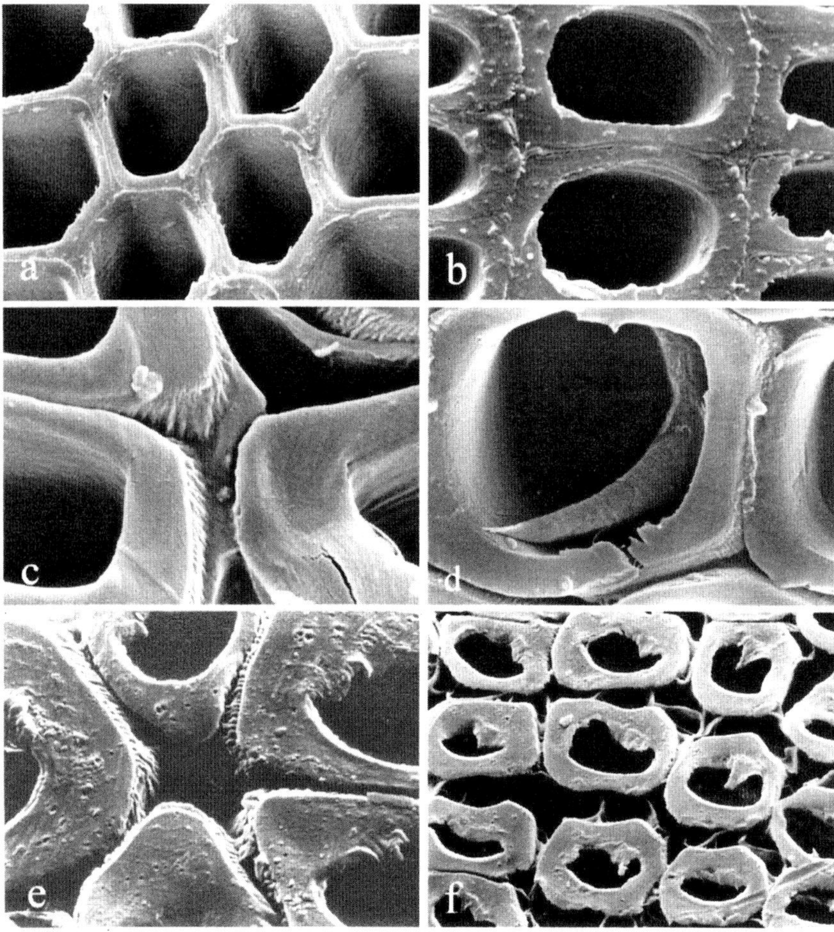

Figure 10. Microstructural changes at transverse surfaces of radiata pine following natural weathering in Canberra, Australia; (a), unweathered earlywood tracheids, x 935; (b), unweathered latewood tracheids, x 1343; (c), latewood weathered for 30 days showing erosion of the middle lamella and delamination of the S_1 layer of the secondary wall, x 3570; (d), latewood weathered for 30 days showing delamination of the S_3 layer of the secondary wall, x 1725.5; (e), latewood weathered for 40 days showing erosion of the middle lamella in a cell corner and between tracheids, x 2975; (f), latewood weathered for 40 days showing erosion of middle lamellae, delamination of S_1 layer of secondary wall and partial separation of tracheids, x 943.5; (g), earlywood weathered for 200 days showing erosion and thinning of secondary walls, x 1164.5; (h), latewood weathered for 200 days showing erosion and distortion of tracheids, x 853. (Reproduced with permission from reference 59. Copyright 1989 Institute of Wood Science.) Continued on next page.

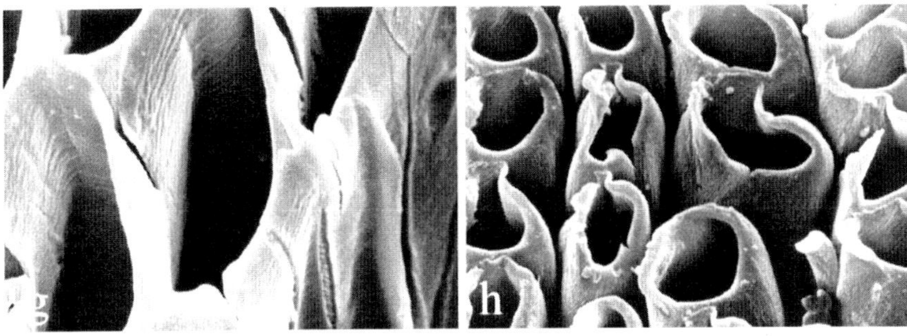

Figure 10. Continued.

to that of fresh or finished wood. A good example of this is the use of weatherboards for the construction of 'New England type' barns, where the wood may be artificially weathered prior to building construction to give the building an aged appearance in keeping with its rural surroundings (*80*).

The lustre of wood may decrease during weathering as the surface becomes rougher and the scattering of light becomes more diffuse (*79*). These increases in surface roughness are due to preferential erosion of low-density earlywood (*73*). The erosion of wood during exterior exposure is inversely proportional to wood density (*81*), and when the rate of erosion of different wood species during accelerated weathering is plotted against the logarithm of their density a linear relationship is obtained (*45*). This relationship is similar to that found for the penetration of light into wood and its density (*39*). Increased penetration of light and depth of photodegradation may thus explain why low density species, such as western red cedar (*Thuja plicata* D. Don), erode at a rate of 12 mm per century when exposed vertically facing south in the Northern Hemisphere, whereas comparable figures for higher density softwoods such as Douglas-fir (*Pseudotsuga menziesii* (Mirb.) Franco) and high density hardwoods are 6 mm and 3 mm, respectively (*6*). The erosion of wood during weathering is also influenced by the topology of the wood surface (*82*, *83*). Preferential erosion of earlywood may expose a greater surface area of denser latewood to light resulting in higher than expected erosion of such tissue (*83*).

Weathered wood often contains checks which can penetrate wood to a depth of several millimeters and extend longitudinally for hundreds of millimeters (*84*). A check is defined as a longitudinal separation that extends into wood to a depth that does not exceed 75% of the thickness of the board (*85*). In contrast, a crack can penetrate completely through a board and lead to its separation into two pieces. Both checks and cracks are usually much longer than they are wide (*84*).

The macroscopically visible checking that develops in wood exposed outdoors has its origins in the microchecking described above that causes

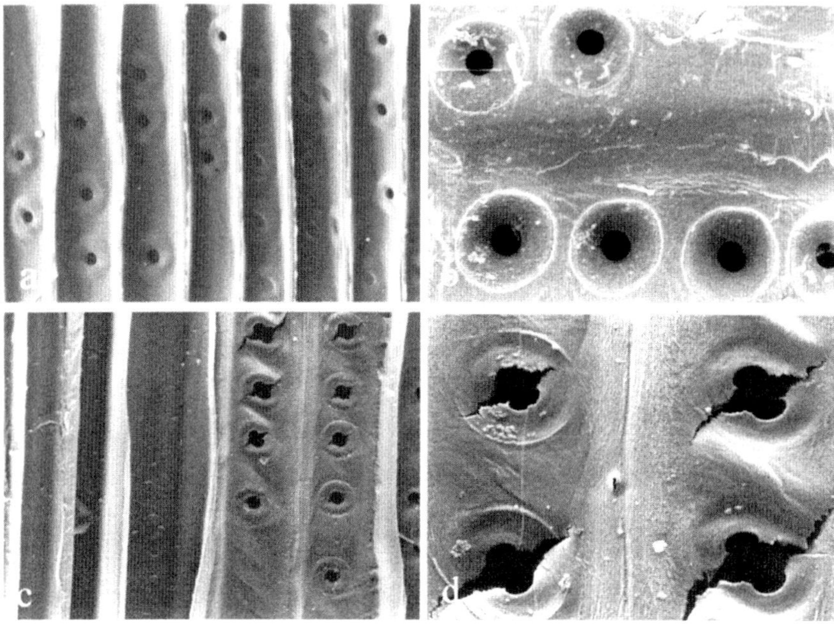

Figure 11. Microchecking of radiata pine tracheids during natural weathering of radial longitudinal surfaces; (a), bordered pits aligned in single files (uniseriate) in unexposed earlywood, x 357; (b), bordered pits with circular apertures in unexposed earlywood, x 1096.5; (c), progressive microchecking of bordered pits in earlywood tracheids following 30 days exposure to natural weathering, x 297.5; (d), microchecking of bordered pits in earlywood tracheids following 30 days exposure to natural weathering, note colonization of pit borders by bacteria (upper left), x 935; (e), loss of integrity of wood surface as a result of microchecking of bordered pits in tracheids and half bordered pits in rays (centre right) following 50 days exposure to natural weathering, x 106.25; (f), microchecking of the bridge of material between two bordered pit microchecks following 100 days exposure to natural weathering, x 867; (g), voids formed at a wood surface as a result of microchecking and loss of cell wall material following 200 days exposure to the weather, x 314.5; (h), microchecking of wood cell wall in the absence of bordered pits in wood exposed to the weather for 100 days, x 935. (Reproduced with permission from reference 59. Copyright 1989 Institute of Wood Science.) Continued on next page.

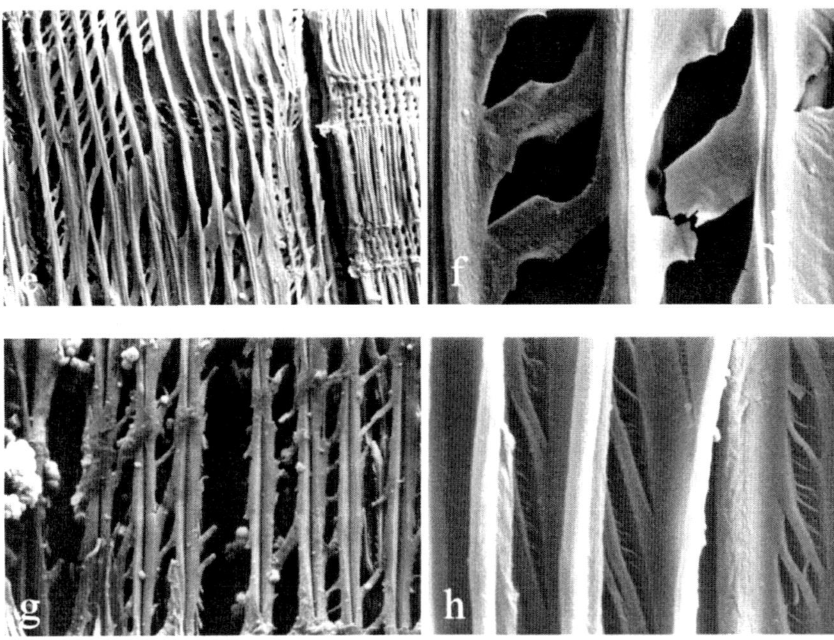

Figure 11. Continued.

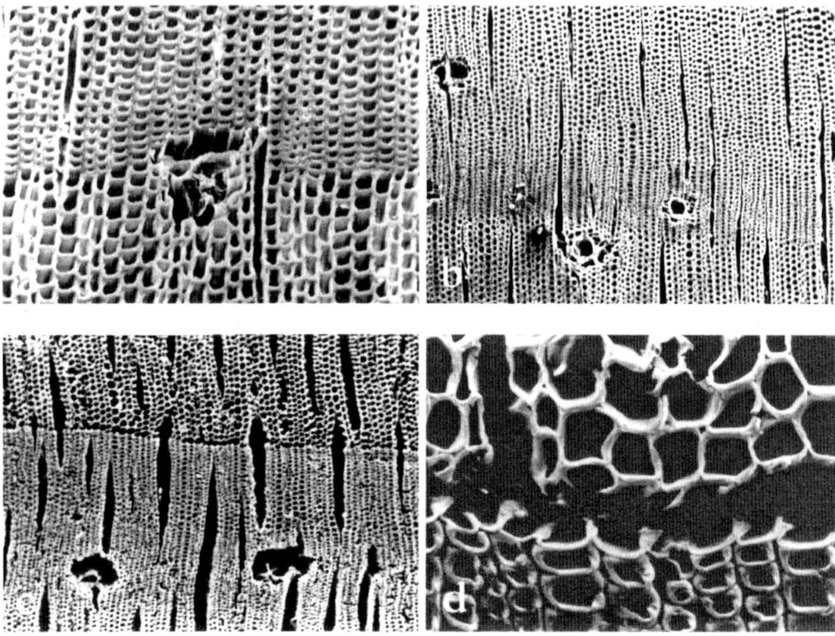

Figure 12. Microchecking of radiata pine originating in resin canals, rays and at growth ring boundaries during natural weathering of transverse surfaces (a-e) and tangential longitudinal surfaces (f-j); (a), growth ring boundary in unexposed wood showing earlywood (bottom) and latewood (top), note resin canal in center, x 100.3;(b), formation of small checks as a result of degradation and separation of rays following 20 days exposure to natural weathering, x 34.85; (c), enlargement of ray checks and formation of voids and checks within resin canals and at a growth ring boundary, respectively, following 100 days exposure to natural weathering, x 29.75; (d), failure of earlywood tracheids at the interface of a check formed at a growth ring boundary following 100 days exposure to natural weathering, x 306; (e), large checks formed within rays and at the growth ring boundary following 200 days exposure to natural weathering, x 130.05; (f), ray in unexposed wood, x 544; (g), degradation of thin walled parenchyma cells in a ray following 30 days exposure to natural weathering, x 654.5; (h), voids created by erosion of thin walled parenchyma cells in rays and separation of middle lamella above and below rays following 50 days exposure to natural weathering, x 103.7; (i), large void created by erosion of thin walled parenchyma cells in a ray following 50 days exposure to natural weathering, x 103.7; (j), enlargement of ray checks following 2 years exposure to natural weathering, x 95.2. (Reproduced with permission from reference 59. Copyright 1989 Institute of Wood Science.) *Continued on next page.*

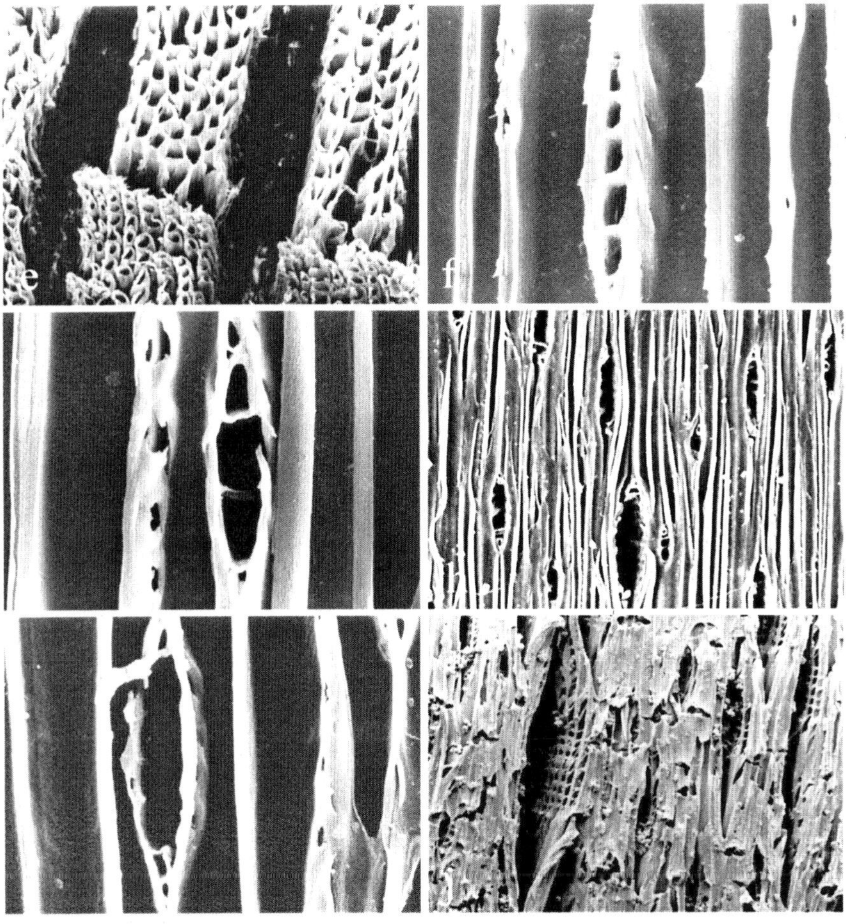

Figure 12. Continued.

separation of different woody tissues (*59*). Hence, the large checks that cause growth rings to separate on end grain surfaces of softwoods (Figure 1) result from microchecking at the interface between latewood and earlywood (*59*). Checking at tangential softwood surfaces exposed to the weather is more numerous, and the checks are larger than those found on similarly exposed radial surfaces (*57*, *86*). In the former, checks develop as a result of degradation and separation of rays particularly in the latewood whereas checks at radial surfaces often develop at the interface between latewood and earlywood. Checks often develop within the resin ducts found in *Pinus* species (*59*). In hardwoods, checks develop within rays and in vessels. There is considerable variation between species in the pattern of checking. For example, aspen (*Populus tremula* L) develops a high number of relatively short checks when exposed outdoors, whereas spruce (*Picea* spp.) develops fewer, longer, checks (*85*). Western red cedar boards develop fewer, smaller, checks than similarly exposed lodgepole pine (*Pinus contorta* Dougl. var. *latifolia* Wats.) and Douglas fir boards.

Protection of Wood

The ability of protection systems to restrict the weathering of wood surfaces depends on the properties of the wood substrate, particularly its density, permeability and shrinkage characteristics, and also on the effectiveness of treatments at preventing the photodegradation of wood and the development of unbalanced surface stresses that cause checking. Wood preservative systems are generally designed to prevent biological organisms from destroying wood's structural constituents. Some wood preservatives also restrict certain aspects of the weathering process, but such positive attributes are generally indirect by-products of one or more of the preservative's components rather than being specifically sought after when the preservative was formulated. The exception to this rule was the development of hydrophobic CCA preservatives containing wax or oil, which were designed to restrict the surface checking of wood, and also contained chromium (VI), that can photostabilize wood (*4*). Chromium containing wood preservatives have been banned in many parts of the World and hence this treatment option is generally no longer available. It would be desirable to develop preservatives systems that are just as effective as hydrophobic CCA formulations because of the size of the market for preservative treated wood in which appearance is important. Knowledge of how wood can be protected from degradation by light and water is useful in the context of developing such preservatives. The following sections examine the photoprotection of wood and the ability of current treatments to restrict the weathering of wood. Excellent reviews are available on treatments to protect wood from water and the reader is referred to these publications for further information on this subject (*87*, *88*).

Photoprotection

Photoprotection of wood can be achieved by reflecting or harmlessly absorbing the light responsible for photodegradation, terminating free radicals that degrade wood's constituents, or modifying lignin in wood to make it less susceptible to photodegradation. All of these approaches have been used to protect wood from photodegradation.

Reflectors and pigments

As early as 1871 Rayleigh demonstrated that the ability of small particles to scatter light was inversely proportional to the fourth power of the wavelength (*89*). In other words, the shorter the wavelength the greater the scattering by smaller particles. Thus small particles below a certain size are able to scatter UV light while having little effect on the visible component of the spectrum. This property of small particles underpins the use of minute particles of titanium dioxide and iron and zinc oxides as 'transparent' photoprotective agents for coatings applied to wood. Transparent synthetic iron oxides are often used in stains and are available as red, orange or yellow crystals (*90*). They differ from iron oxide pigments as they are smaller (0.01 to 0.15 μm) and transmit light in the visible spectrum while screening UV radiation. Such screening of UV radiation can be achieved at a concentration of $2g/m^2$ or 2% for a coating applied at $100g/m^2$ (*90*). Furthermore, transparent iron oxides are reported to retain their ability to protect wood from photodegradation over longer periods of exterior exposure compared to organic (benzotriazole) UV absorbers (*90*). One obvious disadvantage of iron oxides, in applications where it is essential to retain the natural appearance of wood, is their colour. Hence, there is also interest in using micronised titanium dioxide and zinc oxides as photoprotection agents for wood (*91*). Titanium dioxide is a highly effective white pigment that is commonly added to opaque paint. Micronised titanium dioxide, a few nanometers in diameter, does not scatter visible light, but has the capacity to protect wood from UV light when it is in the rutile form and treatments are used to reduce its photocatalytic effect (*91*). Micronised titanium dioxide has been shown to provide good protection to wood surfaces when used on its own at application rates of 0.5 to 1.0% or in combination with iron oxide (*89*). There is also interest in using micronised zinc and cerium oxides to protect wood from photodegradation (*50*). The potential advantages of zinc oxide compared to titanium dioxide for this application is that it is more protective against UV radiation and is less white at a given concentration (*92*).

UV absorbers

UV absorbers absorb incident UV radiation, undergo tautomeric conversion and dissipate the energy as non-radiative heat. A wide range of UV absorbers are available commercially, including benzophenone, benzotriazole, and triazine UV absorbers (*93*) (Figure 13). Within these basic types there is further variation in molecular weight and types of substituent groups. The absorption characteristics and light fastness of the UV absorber are determined by the chromophoric group, whereas the volatility, polarity and resistance of the UV absorber to leaching are controlled by its molecular weight and the nature of substituent groups (*94*).

UV absorbers are widely used to improve the resistance of plastics and coatings to photodegradation. They have also found use as additives to prevent photodegradation of wood finishes and the underlying wood. An early study found that treatment of Douglas fir veneer with a dibenzoylresorcinol UV absorber reduced the production of gaseous and volatile photodegradation products formed during irradiation of the veneer with artificial UV light (*41*). More recently it was found that pre-treatment of wood surfaces with the benzotriazole UV absorber Tinuvin 1130 was very effective at restricting the color change of Taiwania (*Taiwania cryptomeriodes* Hay.) heartwood exposed to fluorescent UV light (*95*).

The permanence and effectiveness of UV absorbers at preventing the photodegradation of polymers can be improved by chemically bonding the UV absorber to the polymer, and this approach has also been used protect wood from photodegradation. The majority of research in this area has employed either benzophenone or triazine UV absorbers containing an epoxy group. These UV absorbers can be permanently bonded to wood at high temperatures in the presence of an amine catalyst. Grafting of epoxy-functionalized UV absorbers reduced the erosion of unfinished wood during artificial accelerated weathering, and mass and tensile strength losses of wood veneers during natural weathering (*96-98*). Grafting of 2-hydroxy-4(2,3-epoxypropoxy)-benzophenone to wood was found to be as effective as chromium trioxide at restricting mass losses of thin wood veneers during natural weathering and was better than chromium trioxide at reducing tensile strength losses of veneers (*97*). XPS spectra indicated that the photoprotective effect of grafting on mass losses was due to the protection of lignin at exposed surfaces (*97*). Grafting of epoxy-functionalized UV absorbers to wood has also been found to improve the performance of clear coatings on modified veneer surfaces (*97*, *98*). An alternative route to grafting UV absorbers to wood is to employ UV absorbers

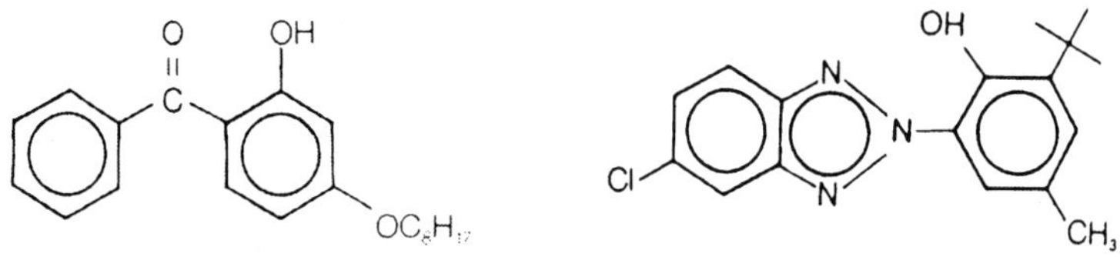

Figure 13. Chemical structures of typical (a) benzophenone and (b) benzotriazole UV absorbers used to photostabilise polymers and wood (Reproduced with permission from reference 59. Copyright 1989 Institute of Wood Science.)

containing an isocyanate group. It was reported that the photoyellowing of wood can be retarded by grafting an isocyanate modified benzotriazole UV absorber to wood, and pre-treating the surface with radical scavengers such as polyethylene glycol or hindered amine light stabilizers (*99*).

UV absorbers are effective at restricting the photodegradation of wood, but their effectiveness can be further enhanced by combining them with chemicals that terminate the free radicals which cause photo-oxidation of wood (*93*).

Radical scavengers

A range of compounds, including hindered amine light stabilizers, phenolic antioxidants, heavy hydrocarbons and glycols, are able to scavenge or terminate free radicals, and some of these have been employed in polymer and wood photoprotection systems.

Hindered amine light stabilizers are thermally and light stable derivatives of 2,2,6,6-tetramethyl piperidine that provide photo-protection to wood when they are used in combination with UV absorbers (*93*). They are also able to protect polymers from photodegradation at relatively low concentrations (*100*). The mechanisms by which hindered amine light stabilizers photostabilize wood have not been studied in any detail, however, much more is known about how they photostabilize polymers such as polyethylene and polypropylene. In general terms hindered amine light stabilizers photostabilize polymers by terminating the free radicals involved in photo-oxidation (*101*). Model studies of the interaction of hindered amine light stabilizers with simple hydroperoxides, peroxy and acylperoxy radicals have yielded important insights into their modes of action (*102*). These studies have shown that hindered amine light stabilizers trap acylperoxy radicals, converting them into carboxylic acids and in the process are converted to nitroxyl radicals (*102*). These nitroxyl radicals are extremely efficient at scavenging a range of free radicals including alkyl, alkoxy and alkylperoxy radicals (*103*). Nitroxyls are rapidly regenerated from

N-alkyloxy hindered amines by reacting with acylperoxy radicals (*102*). Quenching of excited polymer oxygen complexes (*101*) and chelation of transition metal impurities (*104*) have also been proposed as means by which hindered amine light stabilizers protect polymers from photo-oxidation.

Hindered amine light stabilizers are available with a range molecular weights and different substituent groups (R1 and R2) (Figure 14). The radical scavenging effectiveness and basicity of the molecule are determined by the substituent group at R1 whereas secondary properties, including solubility, thermal stability and migration resistance, are influenced by the group at R2 (*94*). The migration resistance of hindered amine light stabilizers is also influenced by their molecular weight and, accordingly, oligomeric hindered amine light stabilizers have been developed for demanding applications such as fibers and extracting environments (*105*). Polymer graftable hindered amine light stabilizers have also been developed that are highly resistant to leaching and volatilization (*105*).

Pre-treatment of wood surfaces with the hindered amine light stabilizer Tinuvin 292 restricted color changes of Taiwania heartwood exposed to fluorescent UV light, and a synergistic effect was observed between Tinuvin 292 and the UV absorber Tinuvin 1130 on heartwood color stability (*95*). Recently Rogez and coworkers described the synergistic effects of combinations of UV absorbers and hindered amine light stabilizers on the performance of clear coatings on wood, and the progress that has been made in developing more effective combinations of UV absorbers and hindered amine light stabilizers for the photoprotection of wood and clear finishes (*93*, *106*). Hindered amine light stabilizers are also useful inhibitors of iron oxidation and water degradation of wood (*107*).

Hindered phenolic antioxidants are traditionally used to terminate free radicals originating from heat-induced degradation of coatings (Figure 15). They are similar in function to hindered amine light stabilizers, so they can be used to retard the degradative effects of UV light on wood (*100*). Unlike hindered amine light stabilizers, however, they are non-regenerative and decrease in concentration during the photostabilization process.

Lignin modification

Most wood species contain 20 to 30% lignin, which is unevenly distributed across the wood cell wall and may reach concentrations as high as 90% in the middle lamella. Lignin strongly absorbs UV light and the formation of aromatic radicals is the key step that leads to depolymerisation of woods structural constituents on exposure to light. Modification of lignin to make it less susceptible to photodegradation is an obvious route to increasing the photo-

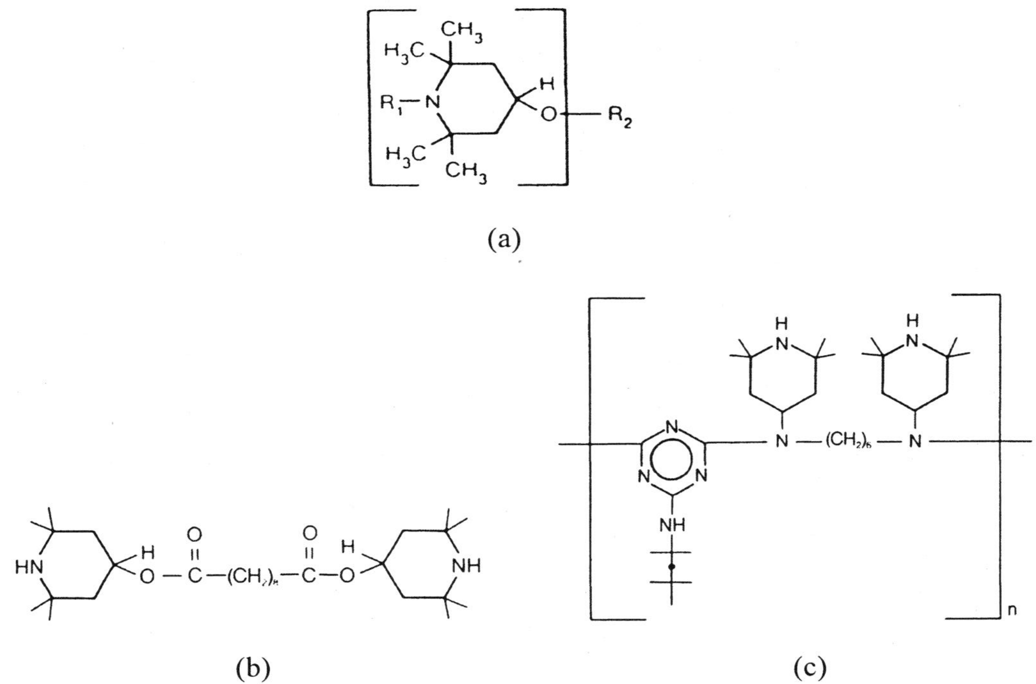

Figure 14. Generalized structures of hindered amine light stabilizers (HALS); (a) 2,2,6,6-tetramethyl piperidine; (b) low molecular weight HALS (Tinuvin 770); (c) oligomeric HALS (Chimasorb 944)

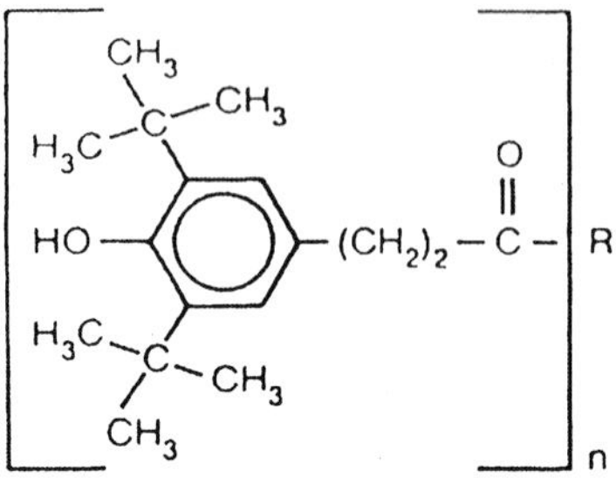

Figure 15. Structure of a phenolic antioxidant (Irganox series)

stability of wood. Treatments that fall into this category include inorganic metal compounds, most notably chromic acid but also copper and cobalt chromates, ferric chloride and nitrate, and various manganese, titanium and zirconium compounds (*76*, *108*, *109*).

Studies of the reaction of chromic acid and other inorganic compounds with simple lignin models have yielded insights into how chromic acid modifies lignin to make it less susceptible to photodegradation. Pizzi used a dilute solution of guaiacol as a lignin model to study the reaction of chromium (VI) with wood (*110*). He suggested that chromium (VI) formed an insoluble complex with lignin (*110*). A re-examination of the reaction of guaiacol with excess aqueous chromic acid found that guaiacol oligomers were bound or cross-linked by hydroxylated trivalent (presumably octahedral) chromium species to form complicated three dimensional high molecular weight polymers (*111*). Further studies of the reaction of guaiacol and 2,6-dimethoxyphenol with metal oxidants have confirmed that the reaction of chromic acid with guaiacol results in the formation of an amorphous chromium (III) complex (*112*). Oxidation of 2,6-dimethoxyphenol with chromic acid resulted in the formation of an amorphous chromium (III) coerulignone complex. It is therefore plausible that chromic acid oxidizes lignin phenol groups, resulting in the formation of chromium (III) quinone complexes at wood surfaces. Quinone and quinone derivatives are efficient antioxidants and UV protectants that have been used to photostabilize polymers (*113*, *114*). It is possible therefore that stable lignin-quinone complexes account for the photostability of chromic acid treated wood surfaces.

Pre-treatment of wood surfaces with chromic acid significantly improves the photostability of wood and, predictably, when used as a wood surface pre-treatment increases the service life of natural finishes used outdoors. Chromic acid treatment reduces the checking of simple and bordered pits during exposure of wood to UV light and natural weathering (*76*), and the surface erosion of wood during artificial accelerated weathering (*108*). Chromic acid treatment also imparts dimensional stability to wood, reduces its hygroscopicity and improves the finishing properties of wood (*115*). Pre-treatment of wood with hexavalent chromium compounds results in a green coloration, which is undesirable when the aim of finishing is to preserve the natural appearance of wood. Hexavalent chromium compounds are also carcinogenic, a fact that has prevented their commercial use as photostabilizing compounds for wood (except in Japan where pre-treatment of wooden doors with chromic acid was used in the 1980s to enhance the weathering resistance of acrylic-urethane finishes) (*116*). Trivalent chromium compounds are less toxic than chromic acid, but they are also less effective at protecting wood from photodegradation (*115*). Treatment of wood with iron (III) chloride reduced the microchecking of pits during the exposure of wood to artificial accelerated weathering (*76*), but it was ineffective at protecting wood from natural weathering (*117*, *118*). Treatment of wood surfaces with oxidative manganese compounds, such as manganese (III) acetate dihydrate or potassium permanganate, is able to protect wood from photodegradation. However, these compounds are less effective than chromic acid and also discolor the treated wood (*109*). Titanates and zirconates are colorless and there has been some interest in their use as photoprotective agents for wood (*109*). Recently it was shown that tetrabutyl, tetraisopropyl, and ethylhexyl titanate enhanced the tensile strength of wood both before and after weathering possibly due to the formation of complexes between the titanates and cellulose in wood. However, the titanates were unable to restrict veneer weight losses during natural weathering, indicating that they did not form stable complexes with lignin. Zirconates such as tetrapropyl and tetrabutyl zirconate were even less effective at protecting wood from weathering than titanates (*109*).

Protective Treatments

Surface coatings and penetrating finishes

The most common method of protecting wood from weathering and photodegradation is through the use of a wide range of coatings such as paints, varnishes, stains and water-repellents (*6*). Wood coatings are generally classed as either film-forming or penetrating, and the latter can be used as a pre-treatment or as a final finish (*6*). Film-forming finishes, such as paints, contain pigments that screen wood from solar radiation and, because they form a barrier over the wood surface, they also prevent surface wetting and erosion (*6*). Feist points out that a correctly applied and maintained paint system, including a primer and at least two top-coats, can greatly reduce the deleterious effects of weathering on wood (*6*). However, a problem with clear film-forming finishes on wood is their loss of adhesion during weathering due to failure of the underlying wood (*119*, *120*). The inability of opaque finishes to bond to weathered wood has been documented in several studies (*60*, *121-127*). Paints are often less effective than penetrating finishes at controlling decay and dimensional movement and, therefore, they often perform better on wood that has been pre-treated with a water-repellent preservative (*6*). The high maintenance requirements of paints when they fail and their tendency (when used on non-durable timber) to trap water and encourage decay has led to the increased use of penetrating water-repellent stains as a means of protecting wood used outdoors (*6*).

Penetrating finishes typically contain a hydrophobe such as wax, as well as an oil- or resin-based binder which penetrates the wood and cures. Unlike paints, which impose a physical barrier, water-repellents rely on the formation of a hydrophobic coating which raises the contact angle of the treated wood and applied water droplets to over 90°, preventing water from being readily taken up by the surface or sub-surface capillaries (*128*). Water-repellents reduce moisture absorption so they impart a certain degree of dimensional stability to wood. Fungicides and mildewcides are often added to water-repellents to create water repellent preservatives capable of retarding the growth of micro-organisms on finished wood surfaces (*129*). Effective penetration of end grain by water-repellents is essential to obtain good performance from finished joinery (*130*). During exterior exposure, however, the hydrophobic system eventually breaks down due to the effects of the weather and the presence of impurities and imperfections in the external and internal water-repellent coatings.

Penetrating stains are water repellent preservatives that contain a variety of additives to restrict the weathering of wood, including pigments and UV stabilizers. Penetrating stains tend to fail on wood during exterior exposure through cracking of the wood substrate and erosion of pigments from wood surfaces (*131*, *132*). This leads to discoloration of the finish through loss of pigmentation and accumulation of atmospheric pollutants (dust and dirt) (*133*). Semi-transparent stains contain pigments that partially obscure the wood surface, and hence they reduce the amount of light reaching the wood. At pigment concentrations of 8.4%, stains reduce erosion of wood during accelerated weathering by 65%. The use of water-repellents in the formulation provides added protection, but photodegradation cannot be prevented completely (*134*). Stains applied as successive coats may form high-build films similar to paints and are semi-transparent to opaque, thus obscuring the wood surface. Their behavior and mechanism of failure then becomes similar to that of paints (*135*). Stains provide protection against weathering for 3-6 years depending on wood species and surface texture, type and quantity of stain applied to the wood and degree of exposure to the weather.

Paints and to a lesser extent, stains, modify the appearance of wood. For end uses where it is important to retain woods natural color or texture the wood can be finished with a clear coating. Clear film-forming finishes, although they often contain UV stabilizers and a biocide, are limited in their ability to protect wood from weathering because they transmit visible light and some UV light that can degrade the underlying wood surface (*119*, *120*). Hence, they perform poorly on wood used outdoors, and invariably fail by peeling and cracking within one to three years of application. One means of increasing the performance of clear finishes on wood is to photostabilize the underlying wood surface prior to application of the clear finish using UV absorbers, hindered amine light stabilizers, or chromic acid and related inorganic compounds.

Wood Preservatives

Inorganic compounds used as components of some wood preservatives can also provide photoprotection to wood. Feist and Williams examined the ability of a CCA wood preservative to reduce the weathering degradation of unfinished wood and improve the durability of semitransparent and solid-color stains applied to wood (*136*). CCA provided long-term protection against weathering-induced erosion and greatly extended the lifetime and durability of a partially UV-transparent stain (*136*). The copper component of ammoniacal copper quat (ACQ) and copper azole wood preservatives also enhances the resistance of wood to weathering. Experimental studies have shown that both ACQ and copper azole restrict the rate of delignification of wood during exterior exposure (*137*, *138*). In contrast, metal-free organic wood preservatives provide little photoprotection to wood unless pigments or photostabilizers are added to the formulation. Alkylammonium compounds, which are found in some wood preservatives, may even accelerate photodegradation (*139*, *140*). This effect may be related to their ability to catalyze delignification of wood (*141*).

Chemical Modification

The weathering stability of wood can be improved by chemical modification, which involves covalently bonding chemicals to the polymeric constituents of wood and, in certain cases, bulking the cell wall. Chemical modification usually improves the dimensional stability of wood by reducing its hygroscopicity and keeping the wood cell wall in a swollen state. Chemical modification of wood by methylation, acetylation, or alkylation improves the color stability of wood during weathering because 'blocking' the phenolic hydroxyl groups retards the formation of quinones (*142*). Accordingly, acetylation of wood to weight gains of 10-20% reduces the photo-yellowing (*143*, *144*), checking (*145*) and erosion of wood exposed to natural or artificial weathering (*146*). The ability of chemical modification to reduce checking and erosion of wood during weathering may be explained by the increased dimensional stability and hydrophobicity of the modified wood. Acetylation of Scots pine (*Pinus sylvestris* L.) to low weight gains of 5 to 10%, however, has been shown to reduce the photostability of modified wood (*66*). At low weight gains, preferential substitution of lignin phenolic hydroxyl groups occurs. This may prevent the termination of free radicals via the formation of quinones, thus increasing photodegradation. When wood is acetylated to higher weight gains the photostability of acetylated wood is increased (*66*). At higher weight gains substitution of cellulosic hydroxyl groups occurs which may provide protection to wood because cellulose derivatives are generally less susceptible to photodegradation than unmodified cellulose. Thus acetylation can protect cellulose to some degree (*146*), but it is unable to photostabilize lignin (*142*, *147*, *148*) (Figure 16). Similar findings have been reported for wood esterified with dicarboxylic acid anhydrides (*149*).

Butyrylation of wood with butyric anhydride inhibits the photodiscoloration of wood (*150*). Treatment of wood surfaces with acetic or succinic anhydride reduced the formation of chromophores at the surface of UV-irradiated China fir wood, whereas treatment of wood with phthalic anhydride accelerated photodiscoloration (*151*). Modification of wood with butylene oxide, methyl isocyanate, or butyl isocyanate at higher weight gains (around 25% or more) did not photostabilize wood (*152*, *153*). In contrast, chemical modification of wood with a maleic acid-glycerol mixture enhanced the resistance of wood to weathering (*154*). Similarly, benzoylation of wood with benzoyl chloride (to high weight gains) is also effective at photostabilizing wood, including lignin (*155*, *156*). It was suggested that the ability of benzoylation to photostabilize wood was due to the absorption of UV light or termination of free radicals by benzoyl groups (*155*).

Thermal Modification of Wood

Thermal modification of wood at elevated temperatures under nitrogen, oil or steam increases the dimensional stability and decay resistance of wood and reduces certain strength properties (*157*). These changes result from the cleavage and degradation of hemicelluloses, depolymerization and condensation of lignin and the formation of a host of low molecular weight degradation products including organic acids, formaldehyde, furfural and aldehydes (*157*). Thermally modified wood is being produced commercially in Europe and is used for a range of exterior products including decking, garden furniture and fencing that are subjected to weathering (*157*). Most studies performed todate have concluded that thermally modified wood is more resistant to natural weathering than unmodified wood. Thermal modification of wood under nitrogen at 175 to 240°C for 2 to 4 hours reduced the erosion of wood during artificial accelerated weathering (*158*) and improved its color stability (*159*).

Lignin at the surface of heat treated wood is reported to be less susceptible to photodegradation than lignin in unmodified wood (*160*) and this may partly explain why thermally modified wood exposed to natural weathering is less susceptible to colonization by *A. pullulans* (*158*) than unmodified wood. The increased weathering resistance of thermally modified wood has been explained by its improved dimensional stability and water-repellency that restricts the leaching of photodegraded lignin and hemicelluloses from exposed surfaces (*158*). It has also been suggested that thermal condensation of lignin may also contribute to the increased photostability of thermally modified wood (*160*).

Monomer/Polymer and Resin Treatments

Treating wood with polymerizable monomers improves its weathering durability (*146*, *152*, *153*, *161*). Methyl methacrylate monomer impregnation and subsequent polymerization in wood cell lumens reduced the rate of moisture sorption and erosion of wood during accelerated weathering (*152*). The combination of acetylation followed by methyl methacrylate monomer treatment further increased the weathering durability of wood (*152*). Butylene oxide, methyl- or butyl-isocyanate modification of wood followed by methyl methacrylate monomer treatment also enhanced the resistance of wood to UV light and accelerated weathering (*152*, *153*), despite findings that modification of wood with butylene oxide, methyl- or butyl-isocyanate on their own was ineffective at enhancing the photostability of wood (*152*, *153*).

Recent research on the use of polymer treatments to restrict the photodegradation of wood have focused on the use of water-soluble thermosetting amino resins or textile cross-linking agents. Impregnation of Japanese cedar with low molecular weight phenol formaldehyde resin improved the photostability and resistance to cracking of the treated wood (*162*). Treatment of wood with melamine formaldehyde resin is also reported to protect wood from photochemical degradation (*163*). A range of textile cross-linking agents, including tetraoxane, gluteraldehyde, glyoxal and the *N*-methylol compound 1,3-dimethylol-4,5-dihydroxyethyleneurea (DMDHEU), have been tested to determine whether they can improve the weathering resistance of laminated veneer lumber manufactured from radiata pine (*164*). DMDHEU was the most effective compound at restricting photodiscoloration and checking of the treated composites. A more recent study has confirmed that DMDHEU reduced structural changes at wood surfaces exposed to weathering, but the treatment was ineffective at preventing delignification caused by UV light (*165*).

Impregnation of wood with polyethylene glycol decreased the generation of carbonyl groups and degradation of lignin in wood exposed to UV light. However, the polyethylene glycol was itself degraded by UV light and it was suggested that this would limit its ability in the long term to photostabilize wood

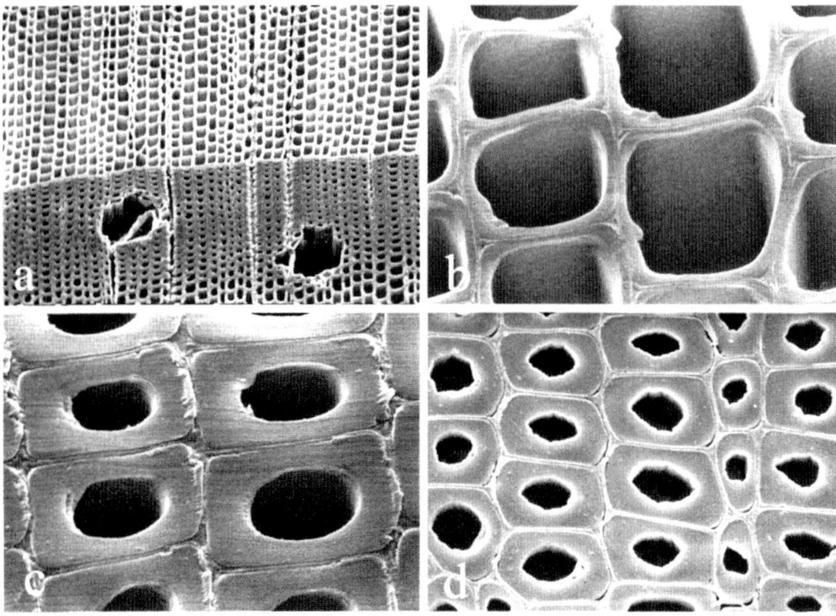

Figure 16. Structural changes at transverse surfaces of Scots pine following acetylation to 20% weight gain and natural weathering; (a), growth ring boundary in unexposed wood showing earlywood (top) and latewood (bottom), note two resin canals in latewood, x 111.35; (b), unweathered earlywood tracheids, x 935; (c), unweathered latewood tracheids, x 1343; (d), acetylated latewood tracheids showing swelling of the S_3 layer of the cell wall (next to cell lumen), x 722.5; (e), acetylated latewood tracheids showing swelling of cell wall layers, note swelling of middle lamella, x 1870; (f), ray checking in acetylated wood exposed to natural weathering for 35 days, note that cells have maintained their radial orientation, x 85.85; (g), unmodified latewood weathered for 35 days showing erosion of middle lamellae, x 1275; (h), acetylated latewood weathered for 35 days showing erosion of middle lamellae, x 722.5; (i), unmodified latewood weathered for 100 days showing erosion and distortion of cell walls, x 722.5; (j), acetylated latewood weathered for 100 days showing preservation of secondary cell wall and radial alignment of cells, x 722.5. (Reproduced with permission from reference 66. Copyright 2000 Institute of Wood Science.) *Continued on next page.*

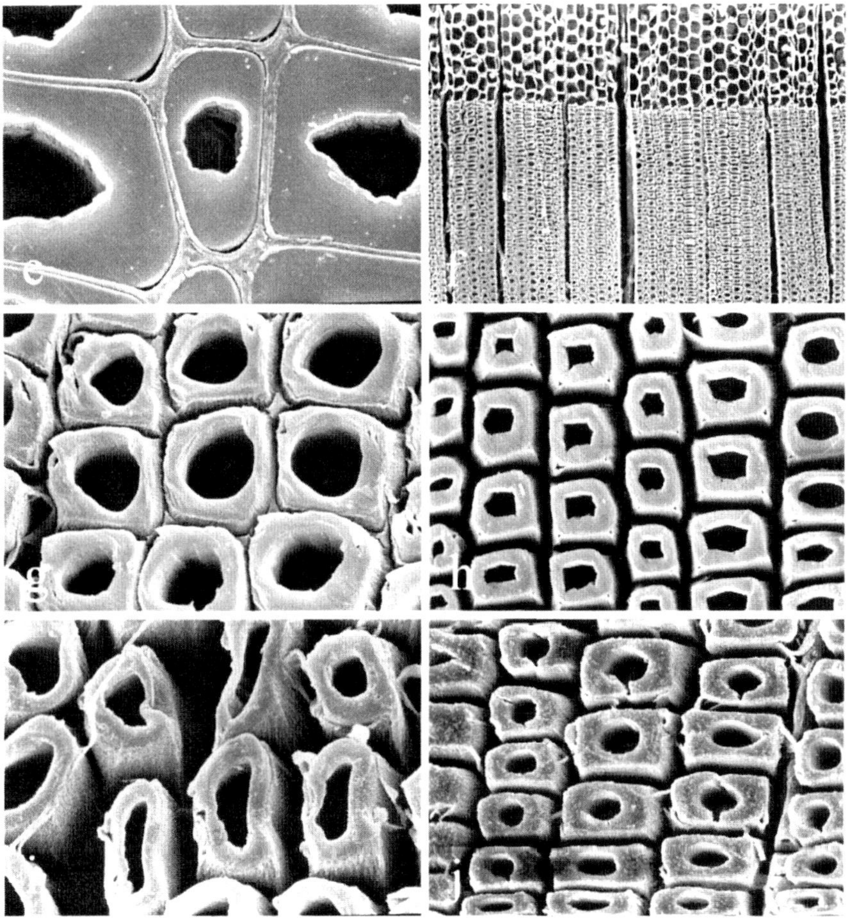

Figure 16. Continued.

(*166*). Nevertheless, pre-treatment of Douglas-fir plywood with a 10% solution of PEG greatly reduced film failure of clear finishes on panels exposed outdoors for two years (*167*).

Wood-Plastic Composites

Wood plastic composites containing wood flour or fibres and thermoplastic polymers are more resistant to weathering than solid wood, because the polymer matrix is hydrophobic and it is also less susceptible to photodegradation than wood. Hence, there is less change in the appearance of wood plastic composites when they are exposed outdoors and they require less maintenance. Nevertheless, such products are susceptible to surface weathering and there is interest in the development of photoprotective treatments for them. The addition of a photoactive pigment (rutile titanium dioxide) to a polyvinyl chloride-wood-fiber composite enhanced the UV stability of the composite (*168*). A zinc-ferrite pigment and a benzotriazole UV absorber (Tinuvin 328) also reduced undesirable changes in crystallininty, stiffness and cracking of a high density wood-flour-polyethylene composite during accelerated weathering (*169*, *170*). These studies concluded that a UV absorber and pigment were more effective at restricting the photodegradation of wood-polymer composites than hindered amine light stabilizers (*169*, *170*). Grafting of the epoxy functionalised benzophenone-type UV absorber, (2-hydroxy-4-(2,3-epoxypropoxy)-benzophenone), to wood fibers used in wood plastic composites also improved the color stability of the composite during accelerated weathering (*171*).

Future Prospects

The missing element in the new generation of organic-only wood preservatives and protection systems currently under development is the ability of the treatments to prevent the surface degradation of wood caused by the weather. It would be desirable for reasons outlined in the introduction to this review to develop an economical treatment that could prevent both the biological attack of wood and the physico-chemical degradation of wood surfaces caused by the weather so that treated wood can compete more effectively with lumber substitutes entering the market or under development. Weathering is a surface phenomenon and superficial treatments can be effective at restricting weathering. Many consumers, however, are no longer willing to apply surface treatments to treated wood and re-apply them periodically. Therefore, in the future, wood protection systems may need to consist of two phases; (a), modification of wood to dimensionally stabilize and protect it from biological attack; and secondly (b) a factory-applied surface treatment to restrict weathering and impart additional desirable characteristics to wood such as attractive color or surface texture. Greater attention to wood selection and the processing of lumber prior to these treatments could also be used to enhance the exterior performance of lumber by reducing the tendency of wood to check. Taken together these measures could create new wood-based materials that match market requirements for attractive, durable and low maintenance exterior residential building products.

References

1. Alderman, D., Smith, R., Araman, P. *For. Prod. J.* **2003**, *53*, 38-45.
2. McQueen, J., Stevens, J. *For. Prod. J.* **1998**, *48*, 86-90.
3. Belford, D. S., Nicholson, J. *Proc. Am. Wood Preserv. Assoc.* **1969**, *65*, 38-51.
4. Fowlie, D. A., Preston, A. F., Zahora, A. R. *Proc. Am. Wood Preserv. Assoc.* **1990**, *86*, 11-21.
5. Green, C. *Fine Homebuilding* **2005**, *172* (June/July), 44-49.
6. Feist, W. C. In *Archaeological Wood: Properties, Chemistry, and Preservation*; Rowell, R. M., Barbour, R.J., Eds.; ACS: Washington DC, 1990; pp 263-298.
7. Borgin, K. *Meded. Fak. Bosb, Univ Stellenbosch.* 1970, 4, pp. 6.
8. Derbyshire, H., Miller, E. R. *Holz Roh-Werkst.* **1981**, *39*, 341-350.
9. Evans, P. D. *Holz Roh-Werkst.* **1996**, *54*, 200.
10. Evans, P. D. *Polym. Deg. Stab.* **1989**, *24*, 81-87.
11. Qayyum, M. M., Davis, A. *Polym. Deg. Stab.* **1984**, *6*, 201-209.
12. Arndt, U., Willeitner, H. Holz Roh-Werkst. **1969**, *27*, 179-188.
13. Derbyshire, H., Miller, E. R., Turkulin, H. *Holz Roh-Werkst.* **1995**, *53*, 339-345.
14. Derbyshire, H., Miller, E. R., Turkulin, H. *Holz Roh-Werkst.* **1996**, *54*, 1-6.
15. Evans, P. D., Schmalzl, K. J., Michell, A. J. In *Cellulosics: pulp, fibre and environmental aspects*; Kennedy, J. F., Phillips, G. O., Williams, P. A. Eds.,. Ellis Horwood. Chichester, 1993; pp 335-340.
16. Kleinert, T. N. *Holzforschung Holzverwertung* **1970**, *22*, 21-24.
17. Evans, P. D., Banks, W. B. *Holz Roh-Werkst.* **1988**, *46*, 427-435.
18. Evans, P. D., Banks, W. B. *Holz Roh-Werkst.* **1990**, *48*, 159-163.
19. Wengert, E. M. *For. Prod.* J. **1965**, *15*, 263-68.
20. Derbyshire, H., Miller, E. R., Turkulin, H. *Holz Roh-Werkst.* **1997**, *55*, 287-291.
21. Mawson, D. *The Home of the Blizzard. Being the Story of the Australasian Antarctic Expedition, 1911-1914.* William Heinemann, London. 1915 p 86.
22. Spedding, D. J. *J. of Appl. Chem.* **1970**, *20*, 226-228.

23. Raczkowski, J. *Holz Roh-Werkst.* **1980**, *38*, 231-234.
24. Williams, R. S. *For. Prod. J.* **1987**, *37*, 37-38.
25. Park, B. S., Furuno, T.; Uehara, T. *Mokuzai Gakkaishi* **1996**, *42*, 1-9.
26. Schmidt, E. L., French, D. W. *For Prod. J.* **1976**, *26*, 34-37.
27. Schoeman, M. W., Dickinson, D. J. *Mycologist* **1997**, *11*, 168-172.
28. Schoeman, M. W., Dickinson, D. J. *Lett. Appl. Microbiol.* **1996**, *22*, 129-131.
29. Sell, J., Wälchli, O. *Mat. u. Org.* **1969**, *4*, 81-87.
30. Sudiyani, Y., Horisawa, S., Chen, K., Doi, S., Imamura, Y. J. *Wood Sci.* **2002**, *48*, 542-547.
31. Aalto-Korte, K., Lauerma, A., Aalanko, K. *Contact Dermatitus* **2005**, *52*, 36-38.
32. Schmolz, E., Brüders, N., Daum, R., Lamprecht, I. *Thermochimica Acta* **2000**, *361*, 121-129.
33. Urban, K. *MSc Thesis*, UBC, Vancouver, BC, Can. 2005.
34. Desai, R. L., Shields, J. K. Int. Biodet. Bull. **1971**, *7*, 11-13.
35. Hon, D. N. S., Feist, W. C. *Wood Fiber Sci.* **1993**, *25*, 136-141.
36. Turkulin, H., Derbyshire, H., Miller, E. R. *Holz Roh-Werkst.* **2004**, *62*, 307-312.
37. Horn, B. A., Qiu, J., Owen, N. L., Feist, W. C. *Appl. Spectrosc.* **1994**, *48*, 662-668.
38. Feist, W. C., Hon, D. N. S. In *Chemistry of Solid Wood*; Rowell, R. M., Ed.; ACS: Washington DC, 1984; pp 401-454.
39. Kataoka, Y., Kiguchi, M., Evans, P. D. *Surf. Coat. Int.* **2004**, *87*, 187-193.
40. Evans, P. D., Wingate-Hill, R., Barry, S. *For. Prod. J.* **2000**, *50*, 59-64.
41. Kalnins, M. A. *U.S. For. Prod. Lab. Res. Rept.* **1966**, *57*, 23-60.
42. Hon, D. N. S. Polym. *Plastics Technol. Eng.* **1979**, *12*, 159-179.
43. Hon, D. N. S. *J. Polym. Sci. Polym. Chem. Ed.* **1975**, *13*, 347-361.
44. Bos, A. *J. Polym. Sci.* **1972**, *16*, 2567-2576.
45. Kataoka, Y., Kiguchi, M., Fujiwara, T., Evans, P. D. *J. Wood Sci.* **2005**, 51 (in press).
46. Gutsche, C. D., Pasto, D. J. *Fundamentals of Organic Chemistry: Structures and Reactions of Organic Compounds.* Prentice-Hall: Englewood Cliffs, NJ, 1975; 942 pp.
47. Hon, D. N. S., Ifju, G. *Wood Sci.* **1978**, *11*, 118-127.
48. Hon, D. N. S., Feist, W. C. *Wood Sci.* **1981**, *14*, 41-47.
49. Schmid, S., Webster, R. D., Evans, P. D. *Int. Res. Group Wood Pres. Doc.* **2000**, 20186, 9 pp.
50. George, B., Suttie, E., Merlin, A., Deglise, X. *Polym. Deg. Stab.* **2005**, *88*, 268-274.
51. Kishino, M., Nakano, T. *Holzforschung* **2004**, *58*, 558-565.
52. Sandermann, W., Schlombom, F. *Holz Roh-Werkst.* **1962**, *20*, 245-252.

53. Schultz, T. P., Templeton, M. C. *Holzforschung* **1986**, *40*, 93-97.
54. Schoemaker, H. E., Harvey, P. J., Bowen, R. M., Palmer, J. M. *Febs Lett.* **1985**, *183*, 7-12.
55. Schniewind, A. P. *For. Prod. J.* **1963**, *13*, 475-480.
56. Skaar, C. *Water in Wood.* Syracuse University Press, Syracuse, New York, 1972, 218 pp.
57. Yata S. In *High-Performance Utilization of Wood for Outdoor Uses.* Imamura, Y. Ed. Wood Research Institute, Kyoto University, Kyoto, Japan, 2001, pp. 65-70.
58. Urban, K., Evans, P. D. *Int. Res. Group Wood Pres. Doc.* **2005**, 20313, 10 pp.
59. Evans, P. D. *J. Inst. Wood Sci.* **1989**, *11*, 172-181.
60. Evans, P. D., Thay, P. D., Schmalzl, K. J. *Wood Sci. Tech.* **1996**, *30*, 411-422.
61. Evans, P. D., Michell, A. J., Schmalzl, K. *J. Wood Sci. Tech.* **1992**, *26*, 151-163.
62. Hon, D. N. S., Chang, S. T. *J. Polym. Sci.* **1984**, *22*, 2227-2235.
63. Hon, D. N. S. *J. Appl. Polym. Sci.* **1984**, *29*, 2777-2784.
64. Leary, G. J. *Tappi* **1967**, *50*, 17-19.
65. Leary, G. J. *Tappi* **1968**, *51*, 257-260.
66. Evans, P. D., Wallis, A. F. A., Owen, N. L. *Wood Sci. Tech.* **2000**, *34*, 151-165.
67. Kitamura, Y., Setoyama, K., Kurosu, H. In *Wood Processing and Utilization* Kennedy, J. F. Phillips, G. O. Williams, P. A., Eds., Ellis Horwood, Chichester, Sussex, UK, 1989, Vol. 51, pp 387-392.
68. Browne, F. L., Simonson, H. C. *For. Prod. J.* **1957**, *7*, 308-314.
69. Bamber, R. K., Summerville, R. *J. Inst. Wood Sci.* **1981**, *9*, 84-88.
70. Yata, S., Tamaru, T. *Mokuzai Gakkaishi* **1995**, *41*, 1035-1042.
71. Futó, L. P. *Holz Roh-Werkst.* **1974**, *32*, 303-314.
72. Kuo, M., Hu, N. *Holzforschung* **1991**, *45*, 347-353
73. Borgin, K. *J. Inst. Wood Sci.* **1971**, *5*, 26-30.
74. Miniutti, V. P. *For. Prod. J.* **1964**, *14*, 571-576.
75. Miniutti, V. P. U.S. For. Prod. Lab. Res. Rept. **1967**, *74*, 1–32.
76. Chang, S. T., Hon, D. N. S., Feist, W. C. *Wood Fiber Sci.* **1982**, *14*, 104-117.
77. Coupe, C., Watson, R. W. Record Ann. Conv. British Wood Preservers Assoc. **1967**, *2*, 37-49.
78. Kŭcera, L. J., Sell, J. *Holz Roh-Werkst.* **1987**, *45*, 89-93.
79. Hon, D. N. S., Minemura, N. In *Wood and Cellulose Chemistry.* Hon, D. N. S., Shiraishi, N. Eds., Marcel Dekker, New York. 1991, pp. 395-454.
80. Anon. Wood & Wood Prod. **1976**, *81*, 33-35.
81. Sell, J. Feist, W. C. *For. Prod. J.* **1986**, *36*, 57-60.

82. Williams, R. S., Knaebe, M. T., Sotos, P. G., Feist, W. C. *Wood Fiber Sci.* **2001**, *33*, 31-42.
83. Williams, R. S., Knaebe, M. T., Feist, W. C. *Wood Fiber Sci.* **2001**, *33*, 43-49.
84. Evans, P. D., Donnelly, C. F., Cunningham, R. B. *For. Prod. J.* **2003**, *53*, 1-6.
85. Flæte, P. O., Høibø, O. A., Fjærtoft, F., Nilsen, T-N. *Holz Roh-Werkst.* **2000**, *58*, 135-139.
86. Sandberg, D. *Holzforschung* **1999**, *53*, 355-364.
87. Stamm, A. J. *Wood and Cellulose Science.* The Ronald Press Co. New York, pp. 312-342.
88. Rowell, R. M., Banks, W. B. U.S. For. Prod. Lab. Res. Rept. **1985**, *50*, 1–24.
89. Blackburn, S. R., Meldrum, B. J., Clayton, J. *Faerg och Lack Scandinavia* **1991**, *37*, 192-196.
90. Sharrock, R. F. *J. Coat. Tech.* **1990**, *62*, 125-130.
91. Schulte, K. *Asia Pacific Coat.* J. **2001**, *14*, 28-29.
92. Pinnell, S. R., Fairhurst, D., Gillies, R., Mitchnick, M. A., Kollias, N. *Dermatologic Surgery* **2000**, *26*, 309-314.
93. Hayoz, P., Peter, W., Rogez, D. *Progress Org. Coat.* **2003**, *48*, 297-309.
94. Rogez, D. CIBA-GEIGY Australia Rept. **1990**, 20 pp.
95. Chang, S. T., Wang, S. Y., Su, Y. C. *Holzforschung* **1998**, *52*, 13-17.
96. Williams, R. S. *J. Appl. Polymer Sci.* **1983**, *28*, 2093-2103.
97. Kiguchi, M., Evans, P. D. *Polym. Deg. Stab.* **1998**, *61*, 33-45.
98. Kiguchi, M., Evans, P. D., Ekstedt, J., Williams, R. S., Kataoka, Y. *Surf. Coat, Int. Part B: Coat Trans.* **2001**, *84*, 263-270.
99. Grelier, S., Castellan, A., Desrousseaux, S., Nourmamode, A., Podgorski, L. *Holzforchung* **1997**, *51*, 511-518.
100. Rabek, J. F. *Photostabilization of Polymers: Principles and Application.* Elsevier, London, 1990, 595 pp.
101. Gugumus, F. *Polym. Deg. Stab.* **1993**, *40* 167-215.
102. Klemchuk, P. P., Gande, M. *Polym. Deg. Stab.* **1988**, *22*, 241-274.
103. Brede, O., Goettinger, H. A. *Angewandte Makromol. Chem.* **1998**, *261/262*, 45-54.
104. Fairgrieve, S. P., MacCallum, J. R. *Polym. Deg. Stab.* **1984**, *8*, 107-121.
105. Malik, J., Ligner, G., Avar, L. *Polym. Deg. Stab.* **1998**, *60*, 205-213.
106. Rogez, D. *Polym. Paint Col. J.* **2001**, *191*, 33-35.
107. Hussey, B. E., Nicholas, D. D. *Proc. Am. Wood Preserv. Assoc.* **1985**, *81*, 169-173.
108. Black, J. M., Mraz, E. A. *U.S. For. Prod. Lab. Res. Rept.* **1974**, *232*, 40 pp.
109. Schmalzl, K. J., Evans, P. D. *Polym. Deg. Stab.* **2003**, *82*, 409-419.
110. Pizzi, A. *J. Appl. Polym. Sci.* **1980**, *25*, 2547-2553.
111. Schmalzl, K. J., Forsyth, C. M., Evans, P. D. *Wood Sci. Tech.* **1995**, *29*, 307-319.
112. Schmalzl, K. J., Evans, P. D. *Polym. Deg. Stab.* **2003**, *82*, 399-407.

113. Sabaa, M. W., Madkour, T. M., Yassin, A. A. *Polym. Deg. Stab.* **1988**, *22*, 205-222.
114. Yassin, A. A., Sabaa, M. W. *Polym. Deg. Stab.* **1982**, *4*, 313-318.
115. Williams, R. S., Feist, W. C. *For. Prod. J.* **1988**, *38*, 32-35.
116. Ohtani, K. *Chrom. Rev.* **1987**, *8*, 4-7.
117. Evans, P. D., Schmalzl, K. J. *Holzforschung* **1989**, *43*, 289-293.
118. Evans, P. D., Pirie, J. D. R., Cunningham, R. B., Donnelly, C. F., Schmalzl, K. J. *Holzforschung* **1994**, *48*, 331-336.
119. Singh, A. P., Dawson, B. S. W. *IAWA J.* **2003**, *24*, 1-11.
120. Macleod, I. T., Scully, A. D., Ghiggino, K. P., Ritchie, P. J. A., Paravagna, O. M., Leary, B. *Wood Sci. Tech.* **1995**, *29*, 183-189.
121. Ashton, H. E. *J. Paint Technol.* **1967**, *39*, 212-224.
122. Desai, R. L. *Can. Dept. Fish. & For. Bi-monthly Res. Notes* **1967**, *23*, 36-37.
123. Boxall, J. Buildg. Res. Est. (UK) Information Sheet. **1977**, *20/77*, 1-2.
124. Underhaug, A., Lund, T. J., Kleive, K. *J. Oil & Col. Chem. Assoc.* **1983**, *66*, 345-350.
125. Williams, R. S., Winandy, J. E., Feist, W. C. *J. Coat. Tech.* **1987**, *59*, 43-49.
126. Williams, R. S., Plantinga, P. L., Feist, W. C. *For. Prod. J.* **1990**, *40*, 45-49.
127. Williams, R. S., Feist, W. C. *For. Prod. J.* **1993**, *43*, 8-14.
128. Borgin, K. *Norsk. Skogindustri* **1965**, *15*, 507-521.
129. Feist, W. C., Mraz, E. A. *For. Prod. J.* **1980**, *30*, 43-46.
130. Voulgaridis, E. V., Banks, W. B. *Holzforschung* **1983**, *37*, 261-266.
131. Kiguchi, M., Kataoka, Y., Doi, S., Mori, M., Hasegawa, M., Morita, S., Kinjo, M., Kadekaru, Y., Imamura, Y. *Mokuzai Hozon* **1996**, *22*, 17-25.
132. Kiguchi, M., Suzuki, M., Kinoshita, T., Kawamura, J. *Materials Life* **1997**, *9*, 188-195.
133. Kiguchi, M., Suzuki, M., Kinoshita, T., Kawamura, J. *Mokuzai Kogyo* **1997**, *52*, 612-617.
134. Feist, W. C. *For. Prod. J.* **1988**, *38*, 41-44.
135. Hilditch, E. A., Crookes, J. V. Record Ann. Conv. British Wood Preservers Assoc. **1981**, 59-66.
136. Feist, W. C., Williams, R. S. *For. Prod. J.* **1991**, *41*, 8-14
137. Liu, R., Ruddick, J. N. R., Jin, L. *Int. Res. Group Wood Pres. Doc.* **1994**, 30040, 9 pp.
138. Cornfield, J. A., Hale, M., Fellis, G. *Int. Res. Group Wood Pres. Doc.* **1994**, 20023, 15 pp.
139. Jin, L., Archer, K., Preston, A. *Int. Res. Group Wood Pres. Doc.* **1991**, 2369, 10 pp.
140. Zhang, X. MSc Thesis, UBC, Vancouver, BC, Can. 2003.
141. Tanczos, I., Schmidt, H. *J. Wood Chem. Tech.* **2002**, *22*, 219-233.
142. Kalnins, M. A. *J. Appl. Polym. Sci.* **1984**, *29*, 105-115.

143. Tarkow, H., Stamm, A. J., Erickson, E. C. O. U.S. For. Prod. Lab. Res. Rept. **1946**, 1593, 15 pp.
144. Plackett, D. V., Dunningham, E. A., Singh, A. P. *Holz Roh-Werkst.* **1992**, *50*, 135-140.
145. Dunningham, E. A., Plackett, D. V., Singh, A. P. *Holz Roh-Werkst.* **1992**, *50*, 429-432.
146. Feist, W. C., Rowell, R. M., Ellis, W. D. *Wood Fiber Sci.* **1991**, *23*, 128-136.
147. Kiguchi, M. *J. Jap. Int. Res. Cent. Agric. Sci.* **1997**, *31*, 147-154.
148. Torr, K. M., Dawson, B. S. W., Ede, R. M., Singh, J. 1996. *Holzforschung* **1996**, *50*, 449-456.
149. Evans, P. D. *Holz Roh-Werkst.* **1998**, *56*, 294.
150. Chang, S. T., Chang, H. T. *Holzforschung* **2001**, *55*, 255-259.
151. Chang, S. T., Chang, H. T. *Polym. Deg. Stab.* **2001**, *71*, 261-266.
152. Rowell, R. M., Feist, W. C., Ellis, W. D. *Wood Sci.* **1981**, *13*, 202-208.
153. Feist, W. C. and R. M. Rowell. In *Graft Copolymerization of Lignocellulosic Fibres*, Hon, D. N.-S. Ed. ACS Symposium Series, Washington, D.C. 1982, 187, pp. 349-370.
154. Fujimoto, H. In *Chemical Modification of Lignocellulosics*. Plackett, D. V., Dunningham E. A. Eds. New Zealand Forest Research Institute Bull, Rotorua, New Zealand. **1992**, 176, pp 87-96.
155. Evans, P. D., Owen, N. L., Schmid, S., Webster, R. D. *Polym. Deg. Stab.* **2002**, *76*, 291-303.
156. Pandey, K. K. Proc. 3rd Int. Symp. on Surfacing and Finishing, 24-26 Nov, Kyoto, Japan, **2004**, pp. 75-86.
157. Evans, P. D. *For. Prod. J.* **2003**, *53*, 14-22.
158. Feist, W. C. and Sell, J. *Wood Fiber Sci.* **1987**, *19*, 183-195.
159. Ayadi, N., Lejeune, F., Charrier, F., Charrier, B., Merlin, A. *Holz Roh-Werkst.* **2003**, *61*, 221-226.
160. Nuopponen, M., Wikberg, H., Vuorinen, T., Maunu, S. L., Jaemsae, S., Viitaniemi, P. *J. Appl. Polym. Sci.* **2004**, *91*, 2128-2134.
161. Desai, R. L., Juneja, S. C. *For. Prod. J.* **1972**, *22*, 100-103.
162. Sudiyani, Y., Ryu, J. Y., Hattori, N., Imamura, Y. In *High-Performance Utilization of Wood for Outdoor Uses*. Imamura, Y. Ed. Wood Research Institute, Kyoto University, Kyoto, Japan, 2001, pp. 85-96.
163. Rapp, A. O., Peek, R. D. *Holz Roh-Werkst.* **1999**, *57*, 331-339.
164. Yusuf, S., Imamura, Y., Takahashi, M., Minato, K. *Mokuzai Gakkaishi* **1995**, *41*, 785-793.
165. Xie, Y., Krause, A., Mai, C., Militz, H., Richter, K., Urban, K., Evans, P. D. *Polym. Deg. Stab.* **2005**, *89*, 189-199.
166. Ohkoshi, M. *J. Wood Sci.* **2002**, *48*, 394-401.
167. Kiguchi, M., Kataoka, Y., Doi, S., Mori, M., Hasegawa, M., Morita, S., Kadekaru, Y., Imamura, Y. *Mokuzai Hozon* **1997**, *23*, 10-17.
168. Matuana, L. M., Kamdem, D. P., Zhang, J. *J. Appl. Polym. Sci.* **2001**, *80*, 1943-1950.

169. Stark, N. M., Matuana, L. M. *J. Appl. Polym. Sci.* **2004**, *94*, 2263-2273.
170. Stark, N. M., Matuana, L. M. *J. Appl. Polym. Sci.* **2003**, *90*, 2609-2617.
171. Kiguchi, M., Kataoka, Y., Kaneiwa, H., Akita, K., Evans, P. D. *Proc. 5th Pac. Rim Bio-based Comp. Sym.*, 10-13 Dec, Canberra, Australia. **2000**. pp 145-150.

Introduction and Wood Deterioration Overview

Section Breaker 1

Figure 5.3. A wasp (Vespula vulgaris L.) removing wood from the surface of weathered pine; note the lighter areas where insects have removed weathered wood

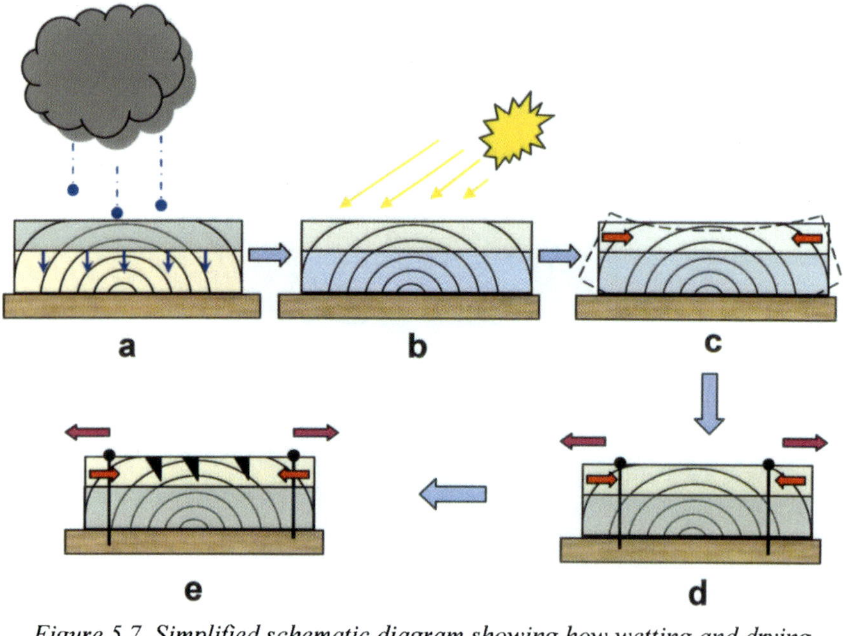

Figure 5.7. Simplified schematic diagram showing how wetting and drying generates surface stresses that cause restrained wooden decking boards to check during exterior exposure

Efficacy Testing of Treated Wood

Section Breaker 2

Molds and Mold Health Issues

Section Breaker 3

Chemical Biocides, Termite Control Methods, and Biocide Depletion

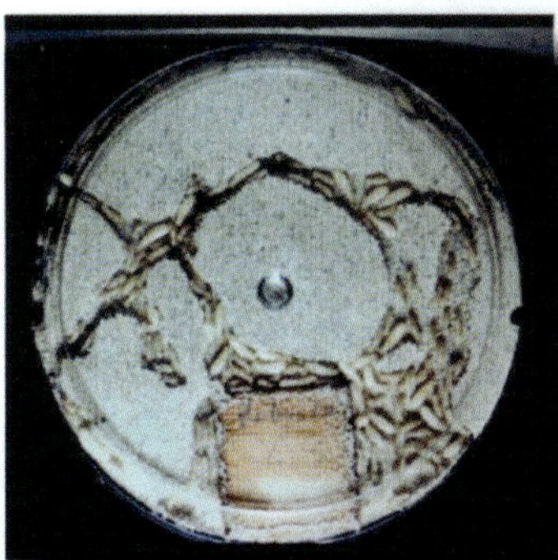

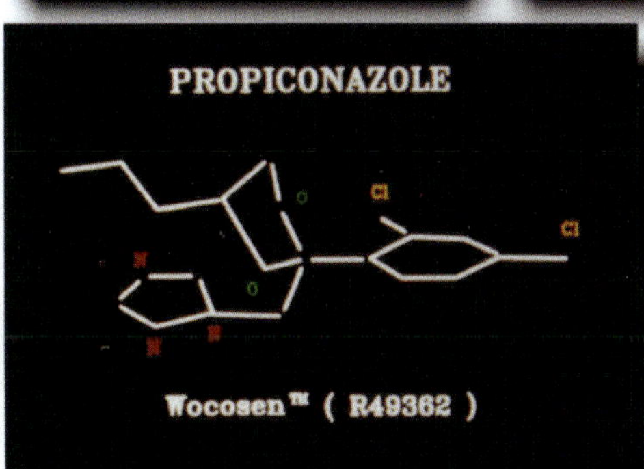

Section Breaker 4

Non-Biocidal Chemicals and Processes to Protect Wood

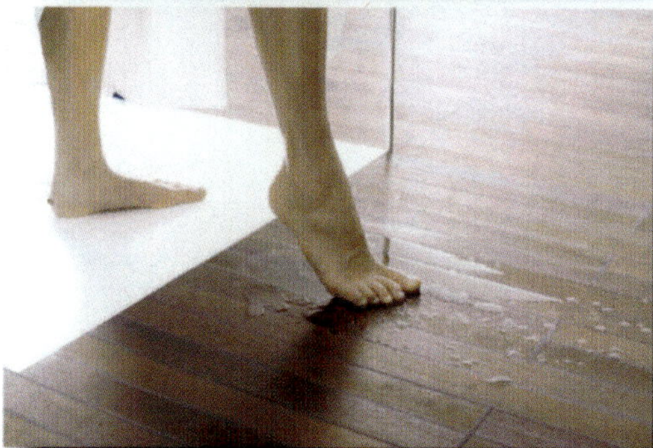

Section Breaker 5

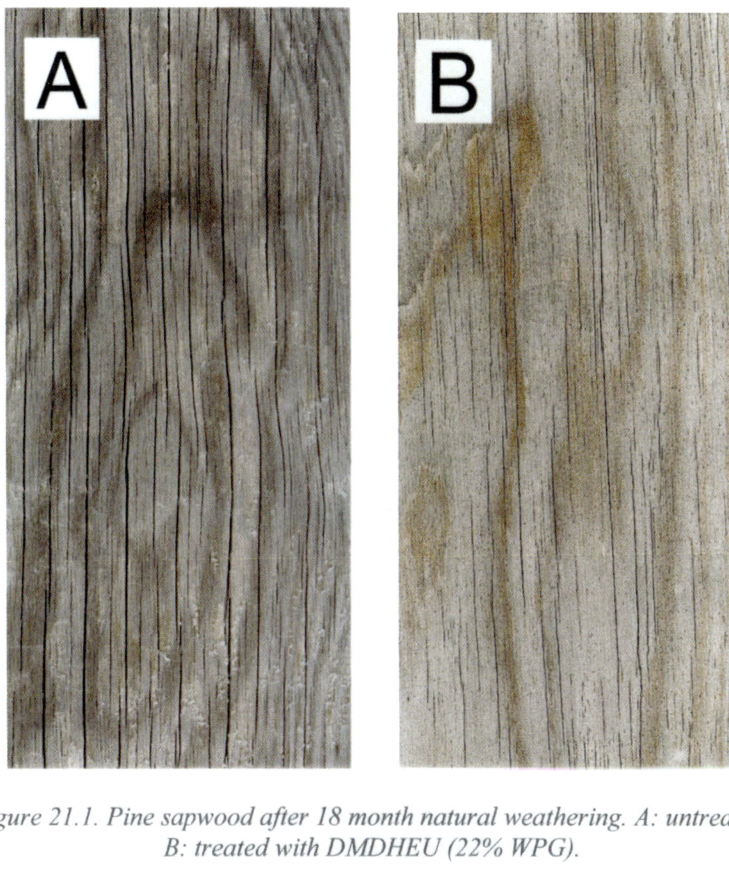

Figure 21.1. Pine sapwood after 18 month natural weathering. A: untreated; B: treated with DMDHEU (22% WPG).

Chemical Wood Preservative Systems

Section Breaker 6

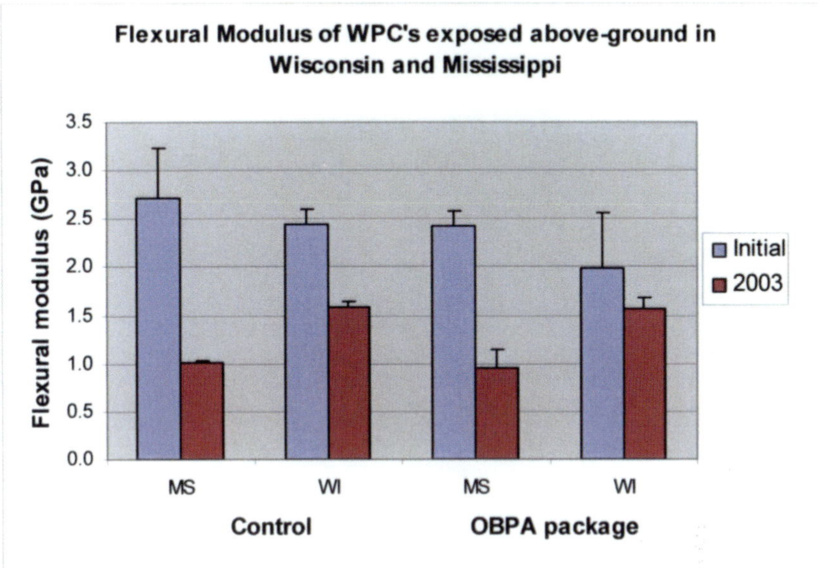

Figure 29.2. See complete caption in chapter 29. (Previously published in Conference Proceedings, Progress in Woodfiber-Plastic Composites, Toronto, Canada, 2004, 1-13)

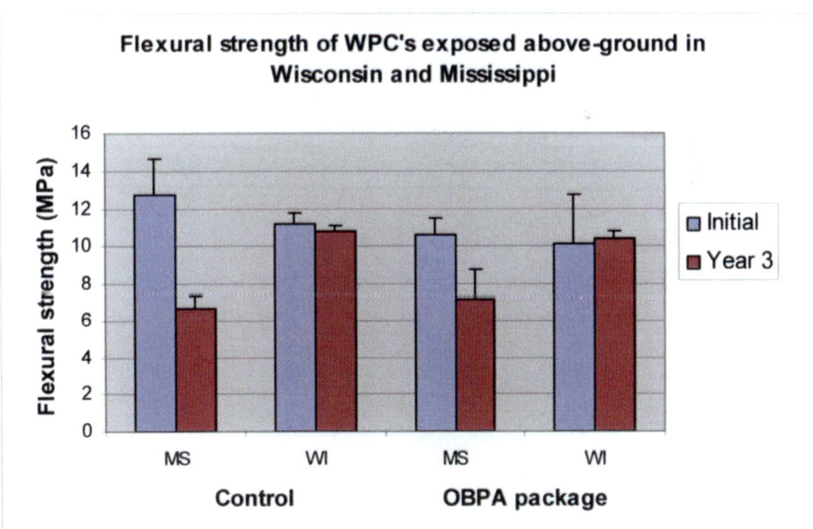

Figure 29.3. See complete caption in chapter 29. (Previously published in Conference Proceedings, Progress in Woodfiber-Plastic Composites, Toronto, Canada, 2004, 1-13)

Figure 29.4. Extruded WPC: a) unexposed; b) above-ground exposure for 3 years in Saucier, Mississippi; c) in-ground exposure for 3 years in Saucier, Mississippi.

Figure 29.5. Failure break of a WPC field stake after exposure in-ground for 3 years in Saucier, Mississippi.

Registration and Approval, and Environmental Disposal Issues of Wood Preservatives

Section Breaker 7

Worldwide Trends in Wood Preservation

Section Breaker 8

Efficacy Testing of Treated Wood

The top photograph is courtesy of Darrel D. Nicholas ad the bottom photograph is courtesy of Tor P. Schultz both of Mississippi State University. (See page 3 of color inserts.)

Chapter 6

Regional Biodeterioration Hazards in the United States

Stan T. Lebow and Terry Highley

Forest Service, Forest Products Laboratory, U.S. Department of Agriculture, Madison, WI 53726

It is well recognized that the rate of wood biodeterioration varies by geographic region across the United States. However, our understanding of the relationship between geographic location and biodeterioration hazard remains largely qualitative. This paper reviews the current state of the knowledge on biodeterioration hazard zones, including a discussion of the basis for, and experiences with, the classic Scheffer climate index for aboveground exposure. The discussion is expanded to cover biodeterioration hazard zones for wood placed in ground contact and the geographic implications for attack by termites and other insects. Finally, this paper discusses the relevance of biodeterioration hazard zones to the development and interpretation of durability tests for wood products.

It has been recognized for centuries that exposed wood deteriorates more rapidly in warm, wet climates than in cold and/or dry climates. Historically, the use of wood as a construction material mirrored this effect, with greatest use occurring in northern latitudes. In the past, the recognition of the geographical differences in biodeterioration hazard was largely based on practical experience, as the causal relationship between fungi and decay was not established until the late 1800s by the German researcher Robert Hartig (*1*). We now have a greater understanding of how temperature and moisture affect the survival and growth of wood-attacking organisms. This knowledge has led to the development of climate indexes and biodeterioration hazard zones for North America.

However, our understanding and application of the relationship of geographic location and climate to biodeterioration hazard remains largely qualitative. For more than 50 years, the widespread use of highly effective wood preservatives such as chromated copper arsenate (CCA), creosote, and pentachlorophenol has allowed builders to ignore regional differences in biodeterioration hazards. When developing new preservatives, the variation in regional biodeterioration hazards is acknowledged simply by testing these systems in the most severe hazard zones of their intended markets. This assures that the preservative treatments are more than sufficient to protect wood in other regions, but it can lead to use of unnecessarily high preservative concentrations. More recently, emphasis has shifted towards developing less toxic preservatives and/or employing lower biocide retentions. As part of this trend, there is increasing interest in matching the extent of preservative protection to the product's end-use and the regional biodeterioration hazard. Accomplishing this objective will require a better understanding and more quantitative application of regional biodeterioration rates.

This paper reviews the current state of the knowledge of biodeterioration hazard zones and includes a discussion of the basis for, and experiences with, the classic Scheffer climate index. The discussion considers biodeterioration hazard zones for wood placed above ground or in ground contact, and the geographic implications for attack by termites and other insects. Finally, this paper discusses the relevance of biodeterioration hazard zones to the development and interpretation of efficacy tests for durable wood products.

Biodeterioration Organisms

In this section, biodeterioration organisms are briefly described in the context of regional biodeterioration hazards.

In most applications for wooden construction materials, decay fungi are the most destructive organisms. Fungi are microscopic threadlike organisms whose growth depends on mild temperature, moisture, and oxygen (*2*). Wood decay fungi are ubiquitous. Given suitable conditions, wood is vulnerable to attack by some type of fungus. Many species of fungi attack wood, with a range of preferred environmental conditions.

Wood decay fungi are often separated into three major groups: brown-rot, white-rot, and soft-rot. Brown-rot and white-rot fungi are both Basidiomycetes but they prefer different wood species and differ in the way they degrade wood. Nonetheless, the optimal environmental conditions for these fungi are fairly similar. Soft-rot fungi are Ascomycetes or Deuteromycetes. They generally prefer wetter, and sometimes warmer, environmental conditions than do brown- and white-rot fungi. Damage by soft-rot fungi resembles that by brown-rot fungi but is typically slower, with only the outer portion of the wood affected initially.

Termites follow fungi in extent of damage to wood structures in the United States. Their damage can be much more rapid than that caused by decay, but their geographic distribution is less uniform. Like decay fungi, the type and severity of termite attack varies by species. In the United States, termites are categorized as ground-inhabiting (subterranean) or wood-inhabiting (non-subterranean) (*2*). Most damage is caused by subterranean termites. The threat from these termites has increased with the spread of the non-native Formosan subterranean termite in some areas of the southeastern United States. Nonsubterranean termites are less damaging than subterranean termites because they have a narrower geographic range and degrade wood more slowly.

Other insects, such as powderpost beetles and carpenter ants, can cause notable damage to wood in some situations, but their overall significance pales in comparison to that of decay fungi and termites. Bacteria and mold can also damage wood, and several types of marine organisms degrade wood in seawater. On an economic basis, however, decay fungi and termites are by the far the most destructive pests of wood used in terrestrial applications. Because of their relative importance in wood deterioration, this chapter focuses on decay and termite hazards.

Factors That Determine Regional Hazards

The two greatest influences on regional biodeterioration hazard are temperature and moisture. The growth of most decay fungi is negligible below 2°C and relatively slow from 2°C to 10°C. The growth rate then increases rapidly, with most fungi having an optimum growth rate between 24°C and 35°C (*1,2*). A few wood decay fungi prefer temperatures in the 34°C to 36°C range; these fungi are commonly found in wood exposed to sunlight or in chip piles (*3*). Soft-rot fungi typically tolerate warmer temperatures than do brown- and white-rot fungi, but care must be taken in making broad generalizations about temperature preferences as there is great variation between closely related taxa (*1*). The growth rate of decay fungi declines steeply at high temperatures, with little growth above 40°C and no growth above 46°C. In most locations and applications in the United States, the lower end of this temperature range has the greatest effect on fungal growth. In the north, temperatures may be too low for the growth of decay fungi during several months of the year, and conditions may be only intermittently favorable during other times. Practical experience has indicated that decay progresses more rapidly in warmer regions of the United States. Although temperatures on the surface exposed to sunlight can exceed those favored by decay fungi, the inner portions of wood products are usually cooler. Decay tends to develop more rapidly in wood in shaded locations, but this is usually associated with a slower rate of drying rather with protection from excessive heat.

The role of moisture in biodeterioration, especially by decay fungi, cannot be overemphasized. Decay fungi require wood moisture content of at least 20% to sustain any growth, and higher moisture content (over 29%) is required for initial spore germination (*1–3*). Decay fungi cannot colonize wood with moisture content below fiber saturation (average of 30% moisture content). Free water must be present. Most brown- and white-rot decay fungi prefer wood in the moisture content range of 40% to 80%. Growth at lower moisture content is much slower; typically, wood with less than 25% moisture content cannot be attacked unless the fungus has another source of moisture nearby. Previously established fungi are not necessarily eliminated at even lower moisture contents. Once established, decay fungi produce water as a metabolic product of wood decomposition. This metabolic water may extend the decay period in poorly ventilated areas. Decay fungi have been reported to survive (without further growth) for up to 9 years on wood with around 12% moisture content (*3*). As moisture content exceeds 80%, void spaces in the wood are increasingly filled with water. The subsequent lack of oxygen and build-up of carbon dioxide in free water limits fungal growth (*4,5*). Soft-rot fungi, however, tolerate higher moisture contents.

As with temperature, it is the lower end of the moisture content limitations that has the greatest impact on regional decay hazard. Humidity alone is not sufficient to raise wood moisture content to the level needed for decay, although an equilibrium moisture content (EMC) of over 20% can occur in cool, moist climates (*6*) (Figure 1). In some types of applications, such as a swimming pool enclosure, the combination of humidity and condensation can wet wood sufficiently for decay to occur. Air is able to hold more moisture at warm temperatures, lowering the relative humidity and EMC. (*6*). Humidity does play a key role in slowing the drying of wood once it is wetted. The drying rate also depends on the length of dry periods between wetting and on construction details that affect the uptake of free water and the loss of water vapor from the wood.

Temperature affects both the extent of activity and geographic distribution of termite species within the United States. The natural range of native

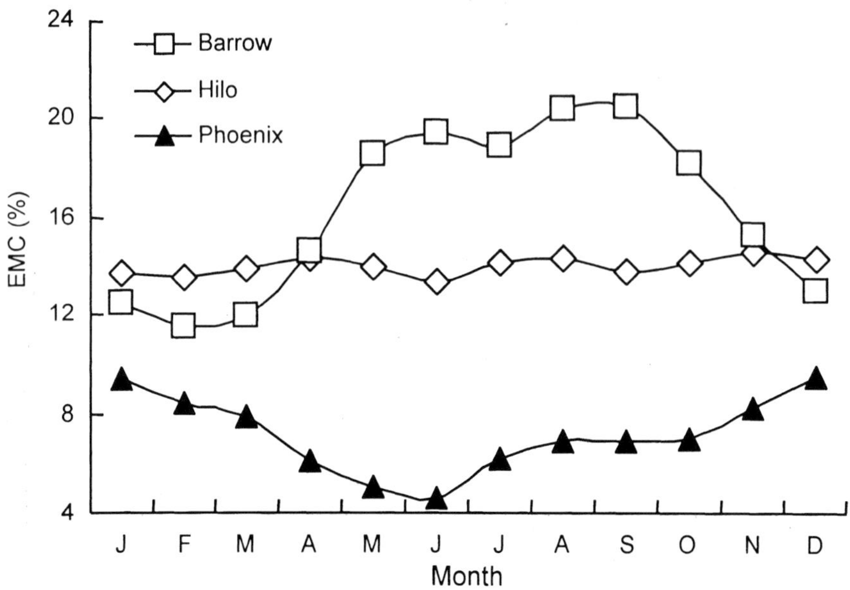

Figure 1. Examples of equilibrium moisture content (EMC) of wood exposed outdoors and protected from precipitation in Barrow, Alaska; Hilo, Hawaii; and Phoenix, Arizona.

subterranean termites is generally limited to areas where the average annual temperature exceeds 10°C, although termites have been found farther north where human activity creates pockets of warmer temperatures (*7*) (Figure 2). Within much of the range of termites in the midwestern and eastern United States, insect activity above ground gradually declines and little activity occurs in the winter (*8*). Termites become inactive when the temperature falls below freezing; in cold climates they may burrow more than 1 m into the ground to avoid prolonged freezing temperatures (*7–9*). A recent study found that native subterranean termites (*Reticulitermes favipes*) could not maintain normal physiological function at temperatures below 1.0°C to 4.9°C (*10*). Formosan subterranean termites are thought to be even less tolerant of cold temperatures, although the northern limits of their distribution in the United States have not yet been established. For example, Hu and Appel (*10*) reported that Formosan subterranean termites were unable to function at temperatures below 7.2°C to 9.0°C. Maximum temperatures for normal function were reported to be between 44.8°C and 45.9°C for Formosan termites and between 43.5°C and 44.9°C for native subterranean termites (*10*). There is less research on the optimum temperature for termite feeding activity. Fei and Henderson (*11*) reported that the rate of wood consumption by Formosan subterranean termites was

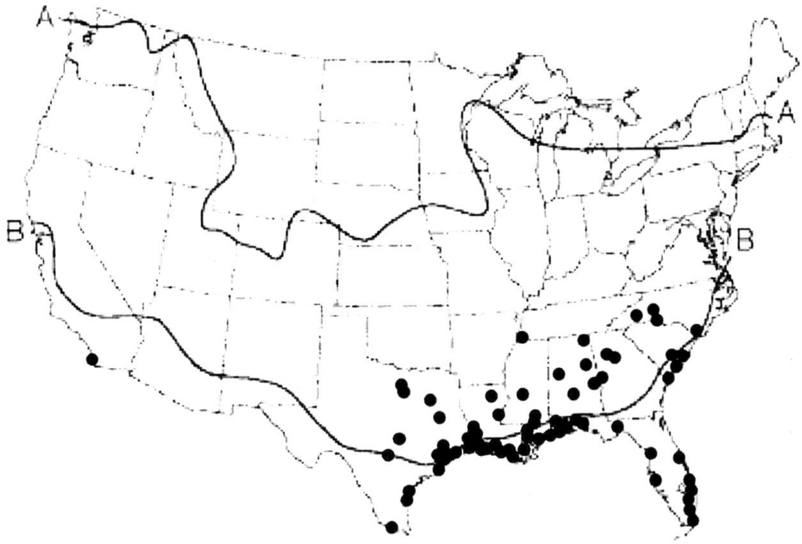

A Northern limit of recorded damage by subterranean termites
B Northern limit of damage by drywood termites
• Location of Formosan subterranean termite infestations

Figure 2. Approximate distribution of termite species causing the most damage in the United States. Localized populations of subterranean termites have been reported in more northerly regions. Adapted from Highley (2).

approximately doubled when the temperature was increased from 20°C to 30°C. Further increase in temperature to 33°C resulted in only a minor increase in the rate of wood consumption and, in some cases, a decrease.

The net effect of temperature on termite degradation of wood is similar to that of decay fungi: conditions are most favorable in regions with warmer climates. The temperature effect may be more extreme for termites than for fungi, however, as some regions of the northern United States have virtually no risk of termite attack.

The effect of moisture on termite attack varies with species. To some extent, the type of termite and its dependence on moisture does vary with climate, but it is a loose correlation. Dampwood termites require wood with high moisture content and typically only attack wood in direct contact with the ground. As a result, they have a relatively minor impact on wooden structures. The high moisture requirements of dampwood termites coincide with their preferred habitats in the northwestern United States and southern Florida, but these termites are found in the southwestern United States as well.

Native subterranean termites require moisture to prevent desiccation, but they can attack wood with moisture content well below the fiber saturation point by building shelter tubes. The shelter tubes are built across open areas to reach wood above ground, and the termites periodically return to the soil to replenish their water supply. Native subterranean termites are widely distributed in the southern two-thirds of the United States; their distribution is less uniform along the Pacific Coast (Figure 2). Formosan subterranean termites also require a source of moisture to attack wood above ground but are less reliant on proximity to soil for survival. They may establish colonies on upper floors of buildings if a consistent source of moisture is present.

Drywood termites are so-named because they are able to survive in wood above ground and can often derive sufficient moisture solely from the wood. They are commonly found in southern California, Arizona, and coastal areas from South Carolina to Texas (*7*).

Regional biodeterioration hazard is shaped by other factors besides moisture and temperature. Some of these, such as elevation, exert influence primarily through their effect on temperature and moisture. Others, such as soil properties, may be interrelated with moisture but also independently influence biodeterioration. Native subterranean termites, for example, generally prefer sandy soil over a clay base (*7*), and soil properties have been reported to strongly affect both the type and severity of fungal attack. Nilsson and Daniel (*12*) found that soil type can influence the relative abundance of brown-, white-, and soft-rot fungi. Nicholas and Crawford (*13*) reported that addition of composted wood to a forest soil increased both the fungal biomass in the soil and the decay rate of untreated pine sapwood stakes. Factors such as quantity of light and atmospheric pressure have also been theorized to affect rates of deterioration, but little is known about these effects (*1*).

Quantification of Regional Biodeterioration Hazard

Recognition of regional variation in deterioration hazard, and its possible importance in predicting durability of wood products, has led to several efforts to quantify hazard zones. Perhaps the most widely used and recognized of those efforts is the Scheffer index (*14*). The Scheffer index was developed in an effort to correlate climatic conditions to the decay rate of wood used above ground and fully exposed to the weather. In constructing his model, Scheffer assumed that temperature and moisture would be far more important than other climate factors. He considered various temperature and moisture parameters, and eventually chose mean monthly temperature and number of days each month with at least 0.25 mm of precipitation. Because fungal growth becomes negligible below 35°F (1.6°C), Scheffer subtracted 35°F from the mean monthly temperature. Negative values were converted to zero. He chose days with precipitation instead of precipitation volume on the premise that duration of wetting was the most significant factor. The "days of precipitation" value was somewhat arbitrarily reduced by 3 to keep the index for the driest regions of the United States near zero. The annual sum of the temperature and precipitation products for the entire year was divided by 30, so that most index values fall in the range of 0 to 100. Examples of the annual Scheffer index for various locations are shown in Table I.

Table I. Examples of Scheffer Index (Annual) for Various U.S. Locations

Location	*Index*	*Location*	*Index*	*Location*	*Index*
Atlanta, GA	67	Houston, TX	77	Philadelphia, PA	50
Chicago, IL	46	Long Beach, CA	4	Phoenix, AZ	7
Boston, MA	51	Miami, FL	131	Seattle, WA	50
Denver, CO	33	Mobile, AL	99	Yuma, AZ	0

The climate index is defined as:

$$\text{Climate index} = \frac{\sum_{Jan.}^{Dec.}\left[(T-35)(D-3)\right]}{30}$$

where T is mean monthly temperature (°F) and D is number of days in month with at least 0.01 inches (0.25 mm) of precipitation.

Using his model, Scheffer produced the widely published geographical contour map of decay potential (Figure 3) (*14*). This map allows the reader to quickly identify areas of high, low, and moderate decay hazard for wood used above ground. As expected, the areas of the continental United States with the highest decay potential are in the southeast, although a small pocket of moderately high decay potential can be found in the Pacific Northwest. Areas of the intermountain west and southwest have the lowest decay potential.

Scheffer verified, or evaluated, the model based on untreated "post-rail" and "flooring" specimens exposed above ground in three locations: Madison, Wisconsin; Corvallis, Oregon; and Saucier, Mississippi (southern) (*14,15*). Madison and Corvallis have very similar annual decay indexes (39 and 42, respectively), while Saucier has a decay index of 96. Scheffer calculated an estimated time to failure at each location, and an "average yearly increase in decay rating." He found that the rate of decay was essentially identical in Madison and Corvallis, but that it was 1.9 times (post-rail) or 2.5 times (flooring) faster in Saucier than in Madison. These relative decay rates are in good agreement with the relative decay indexes. Verification at more locations

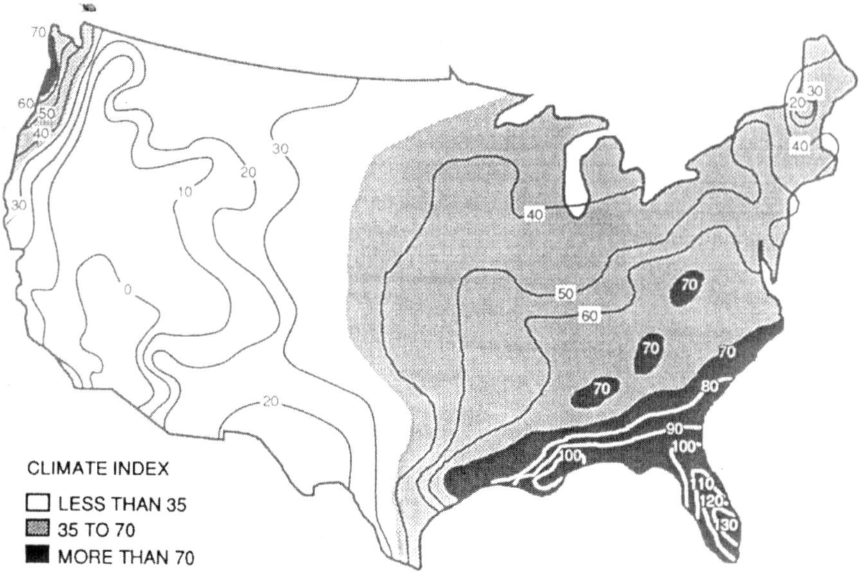

Figure 3. Scheffer climate index map of United States, as adapted from Scheffer (14).

would have been desirable, but little data are available for comparing above-ground decay rates with matched samples.

When calculated on a monthly basis, the Scheffer index produces some interesting comparisons. Although Madison and Corvallis have essentially equivalent annual indexes, the manner in which they reach those values is quite different. In Madison, the index is largely controlled by temperature and peaks during the summer. During several months each year the average temperature is below 1.6°C, which results in an index of zero (Figure 4). In contrast, the index in Corvallis is largely controlled by precipitation. The index reaches zero for 2 months in the summer because there are 3 or fewer days with precipitation. In Saucier, the rain pattern is similar to that in Madison but the index is higher because of the warmer temperatures.

As Figure 4 shows, the Scheffer index may underestimate decay potential in some situations because it is based on monthly averages. In Madison, for example, the temperature exceeds 1.6°C on many days in November and March, and fungal activity may be "non-zero." This is particularly true of wood exposed in south-facing locations where the sun might heat the wood to well above the ambient temperature. Similarly, it is doubtful that the progression of decay in wood in Corvallis stops for 2 months each summer. In larger lumber or poles, for example, water retained in the wood could allow decay to proceed, whereas Scheffer's model assumes rapid drying. Wood placed in contact with the ground

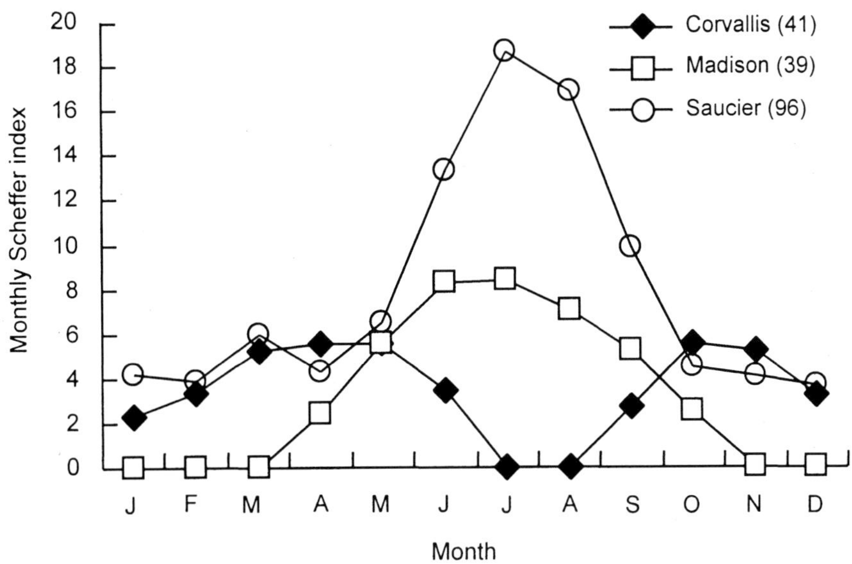

Figure 4. Monthly Scheffer index values for Corvallis, Oregon; Madison, Wisconsin; and Saucier, Mississippi. Legend shows annual index values.

also stays moist for much longer periods. Scheffer recognized this and discussed possible modifications of the formula for wood used in ground contact (*14*). For example, in the case of the dry summer months of Corvallis, Scheffer suggests that the model could be adapted to soil contact by averaging the indexes for the months immediately prior to and following the summer months (May and September, for example). This would eliminate the dip that occurs during the summer for Corvallis.

Another type of decay hazard map was developed by the Rural Electrification Administration (REA) of the U.S. Department of Agriculture (*16*). The REA compiled durability data on millions of utility poles installed in rural electric systems across the United States. Based on these data, it divided the 48 contiguous states into five deterioration zones (Figure 5). Scheffer's map of decay potential (Figure 3) and the REA deterioration zones (Figure 5) are generally in good agreement. However, the two approaches show several notable differences in identifying hazard zones. Unlike Scheffer's map, the REA hazard map is based on wood that was treated with preservative and placed in ground contact. These qualities may lend the REA study greater applicability for the relationship between hazard zone and the development of wood preservatives used in ground contact. Conversely, the mixture of wood species and preservatives represented in the REA data can make comparison of decay hazard zones more complex. The wood species used for poles varied by region;

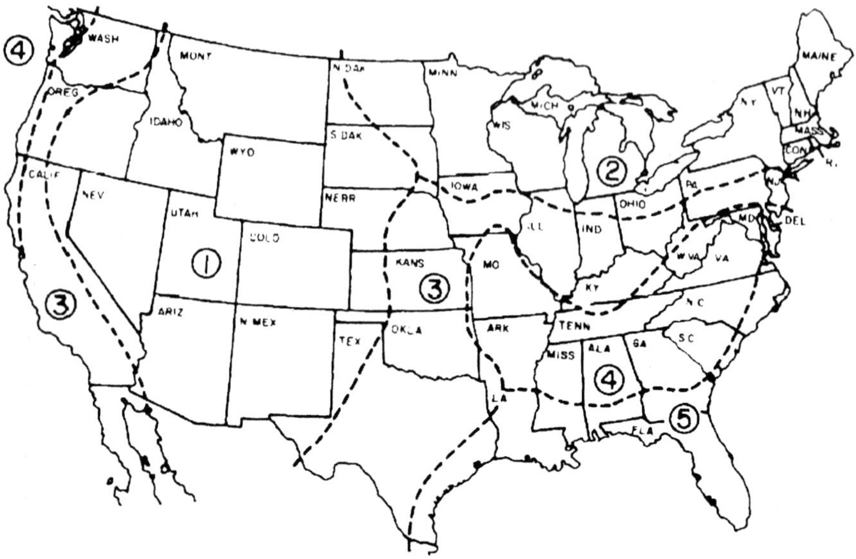

Figure 5. Deterioration zones for wood utility poles as defined by the REA (16). Hazard is least severe in zone 1 and most severe in zone 5. (Reproduced from reference 16.)

most poles used in the western United States were Douglas-fir, western redcedar, or lodgepole pine, whereas most poles in other areas were Southern Pine. In addition, a range of preservatives was used, although the majority of poles were treated with creosote. The value of the REA pole data was recognized by the American Wood Preserver's Association (AWPA), which added Alaska and Hawaii to the map and incorporated this information into Commodity Specification D: Poles (*17*). Section 1.4.1 of the standard refers the reader to the REA Deterioration Zone map for assistance in determining the retention level needed to protect utility poles. This is one of the few cases where AWPA standards recognize and refer to differences in regional biodeterioration hazards. It should be noted, however, that the REA divisions do not actually quantify the differences in rate of deterioration, as was attempted with Scheffer's index. For example, the REA divisions are not meant to imply that wood exposed in zone 4 will deteriorate twice as quickly as wood exposed in zone 2.

There have also been efforts to develop models of regional deterioration hazards in Australia. Foliente et al. (*18*) described efforts to model both aboveground and in-ground decay using methods somewhat similar to that of Scheffer (*14*). Their model for aboveground decay utilizes data on average annual temperature, average annual rainfall, and humidity. The inclusion of humidity is thought to help account for differences in the drying rate of wood after wetting. The in-ground model is a function of average annual temperature and rainfall. To account for soil moisture, an additional factor is added for the number of months in a year in which rainfall is less than 5 mm (*18*).

Review of Durability Data Across Hazard Zones

Aboveground Exposure

More than three decades after Scheffer's classic publication on aboveground decay hazards, there remains relatively little comparative data on aboveground decay rates in different regions of the United States. Most aboveground data are generated during development and evaluation of new preservatives, and, in those cases, researchers tend to test only in high hazard areas. One exception to this trend is data published by Highley (*19*) comparing the durability of wood exposed aboveground in Madison, Wisconsin, and Saucier, Mississippi. This large study included a range of hardwood and softwood species exposed unpainted as 1.9- by 7.6- by 15.2-cm specimens nailed together at their centers to form a cross. The deterioration rate in these specimens was relatively slow, even in southern Mississippi, because the design and small size of the specimens prompted rapid drying. Based on Scheffer's index, the rate of decay in Saucier (index 98) was expected to be approximately twice that of Madison (index 39). However, this was not the case for the softwoods evaluated (Figure 6), where the estimated service life was only slightly greater in Madison than in Saucier. Southern Pine sapwood, for example, had an estimated life of 13 years in Madison and 10 years in Saucier. For the hardwoods, however, the relative service life at the two locations was closely predicted by the Scheffer index. The difference in zone effect between softwoods and hardwoods is noteworthy and may reflect differences in the types of fungi predominant at each site. The results of an earlier study support this conclusion. When decay fungi were isolated from the specimens, only half as many fungal species were found in Wisconsin as in Saucier; little overlap occurred between the most commonly isolated fungi (*20*).

Another comparison of Southern Pine specimens exposed above ground reported a much greater regional difference (*21*). In that study, specimens were exposed in Hilo, Hawaii (Scheffer index an extremely high 312), and near Charlotte, North Carolina (index 64). L-joint specimens reportedly decayed at least three times faster in Hilo than in Charlotte. Specimens exposed in Hilo declined to 70% soundness in 18 months and had all failed within 3 years. The relative decay rates in this study correlate fairly well with that predicted by the Scheffer index.

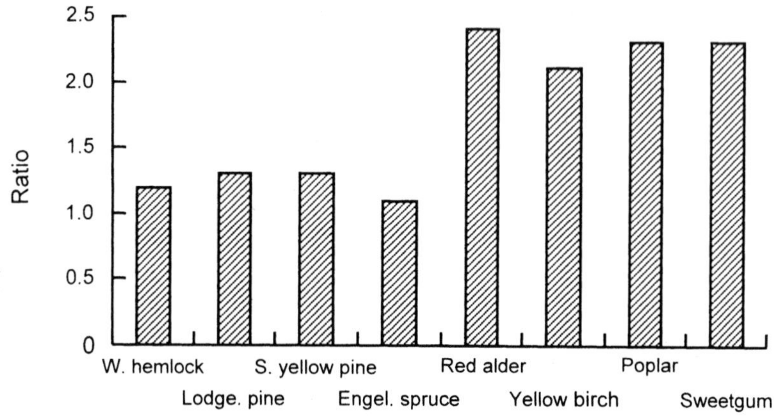

Figure 6. Ratio of estimated life in Madison to that in Saucier for various wood species exposed above ground.

Samples exposed well above ground are not usually attacked by termites, which can greatly alter the rate of biodeterioration. When conditions are favorable for termite attack, wood samples can fail much more rapidly than they do when exposed to decay alone. The rate of deterioration is also a function of the type of termite. Samples of softwood species exposed 60 to 80 mm above ground in Lake Charles, Louisiana, were severely degraded by Formosan subterranean termites within 6 months and by native *Reticulitermes* spp. within 12 months (*22*).

Ground Contact Exposure

As is the case for aboveground evaluations, most ground-contact data are generated during the development of new wood preservatives, and these tests tend to be conducted in areas with high deterioration hazard. The primary exception is the Forest Service exposure site in Madison, Wisconsin. Data on small (19 by 19 by 457 mm) Southern Pine sapwood stakes in Madison (REA deterioration zone 2), Starkville, Mississippi (zone 4), and Saucier, Mississippi (zone 5) showed good agreement with the expected relative rates of deterioration for those locations (Figure 7). Forest Service data (*23*) on untreated Southern Pine and lodgepole fence posts also indicate a trend of shorter life expectancy in higher deterioration hazard areas (Figure 8). However, the lodgepole pine exposure sites were concentrated in areas with relatively low decay hazard (mostly the intermountain west or Madison). The correlation between deterioration hazard and longevity within those low hazard regions was relatively poor.

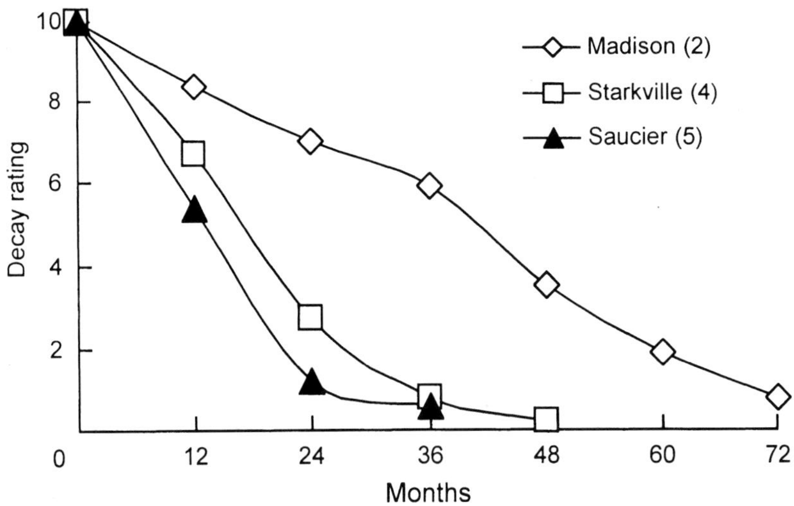

Figure 7. Average decay ratings for stakes exposed in Madison (REA deterioration zone 2), Starkville (zone 4), and Saucier(zone 5).

These data show how variability in service life increases as regional deterioration hazard decreases. As might be expected, in areas where deterioration is slower the deterioration rate becomes harder to predict. This problem is also demonstrated by comparing the longevity of sets of 19- by 19- by 457-mm Southern Pine sapwood stakes exposed in Madison and Saucier over three decades (Figure 9). The average life of the stakes was 5.4 years in Madison but only 2.3 years in Saucier. This difference in durability corresponds well to the Scheffer indexes (Figure 3) and REA deterioration zones (Figure 5) for those locations. However, the range of average stake longevity was also greater in Madison (3.6 to 7.4 years) than in Saucier (1.5 to 3.6 years). The cause of the difference in longevity of Southern Pine stakes in Madison is unclear. It could be related to variations in weather patterns over the period, or to differences in soil conditions within the Madison plot. Differences in wood properties are also a possibility, but this effect was not evident in the matched sets of stakes exposed in Saucier. There was essentially no correlation ($R^2 = 0.03$) between the longevity of matched sets of stakes at the two sites.

Challenges in Applying the Hazard Zone Concept to Preservative Use and Development

Little recognition is currently given to regional biodeterioration hazard in the development and use of treated wood. Regional differences are

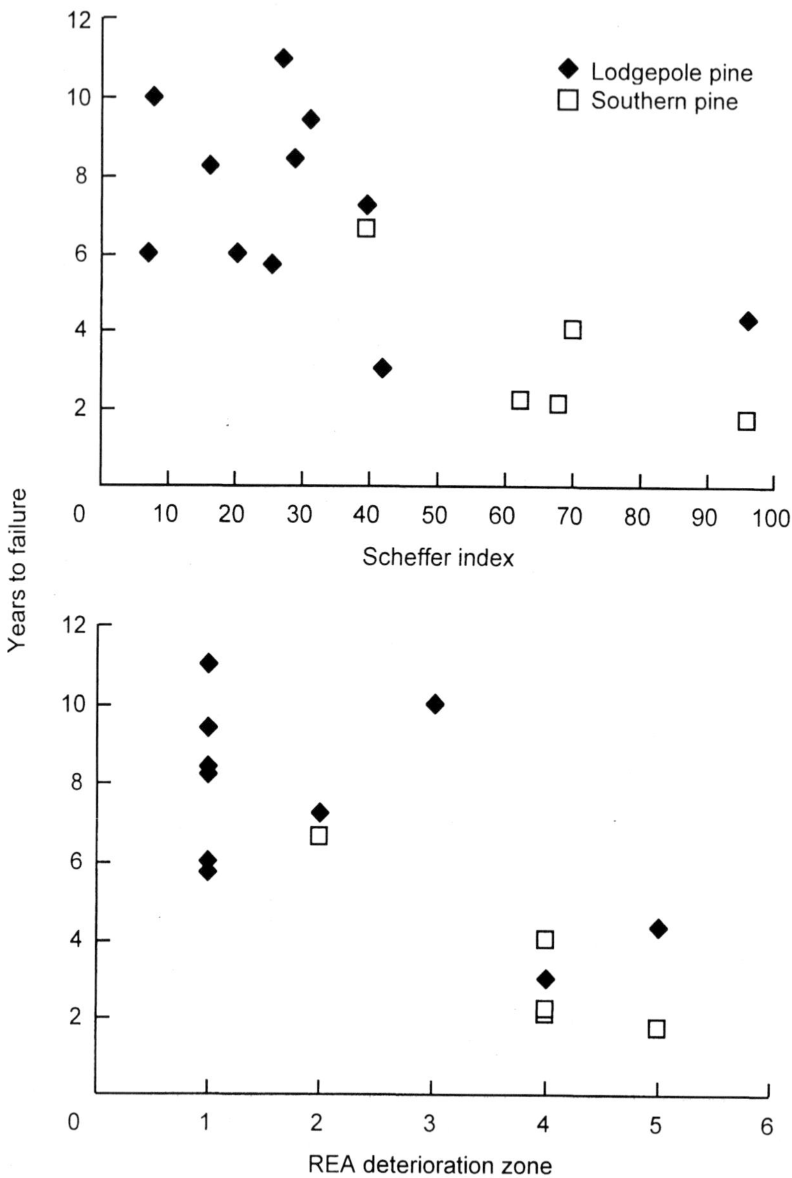

Figure 8. Durability of untreated pine fence posts relative to Scheffer index and REA deterioration zone.

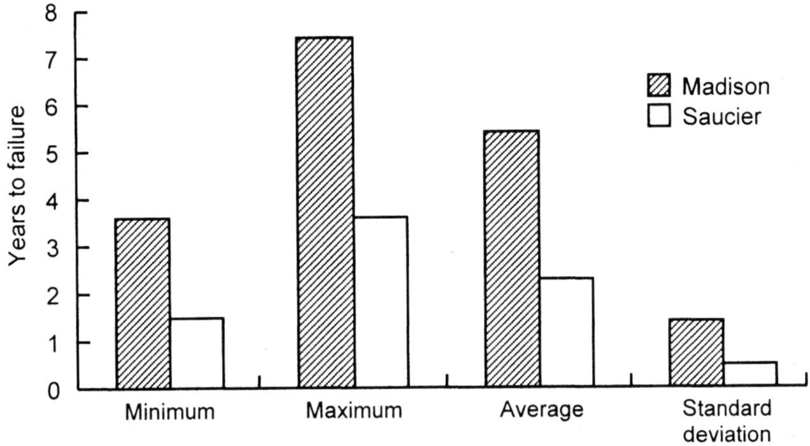

Figure 9. Durability of matched sets of Southern Pine stakes exposed in Madison or Saucier from 1939 to 1971.

acknowledged simply by evaluating new preservatives at test sites in the most severe deterioration zones. This approach is logical because in most cases it ensures that preservatives will perform well in regions with lower deterioration hazards. However, this "worst case" approach also means that the wood is over-protected in much of the United States. In northern regions, more chemical is used than is actually needed; in some areas, it is possible that less expensive or less toxic chemicals could be substituted for current formulations. As the cost of preservative formulations increases and as society's acceptance of chemicals decreases, it would seem advantageous to tailor preservative treatments more closely to the extent of biodeterioration hazard in a region.

However, there are serious challenges to matching preservative formulations to regional deterioration hazards. As discussed previously in this chapter, considerable variability can occur in the durability of untreated wood exposed in areas with similar deterioration hazard. This variability in deterioration rate increases in the northern regions of the United States, although differences in the distribution of termite populations can also cause variability in southern regions. The deterioration rate is also a function of wood species, although to some extent this problem can be ignored because the treated wood market is dominated by a single species group (Southern Pine). Thus, predicting the service life of untreated wood based on region is difficult enough. The addition of preservative to the wood appears to further increase variability.

In a comparison of matched sets of treated stakes exposed in multiple locations, the longevity of untreated controls corresponded well to the expected deterioration hazard of the site but the performance of the treated samples was less predictable (*24*) (Figure 10). For example, the durability of sodium pentachlorophenol was substantially greater than that of the other preservatives in the Panama Canal, but wood treated with this biocide did not perform as well as the other preservatives in Madison. All the treatments were less durable in Bogalusa, Louisiana (Scheffer index 89) than in Jacksonville, Florida (index 98) or Saucier, Mississippi (index 96). Moreover, wood treated with sodium pentachlorophenol and fluor chrome arsenate phenol deteriorated as or more rapidly in Madison than in Jacksonville. The reason for the durability difference between these treatments across these sites is not known. At least two factors influence the performance of preservative treatments: permanence and the presence of tolerant organisms.

Preservative Permanence

Climatic and exposure conditions can potentially contribute to regional differences in preservative permanence within the wood. In some cases, the effect is similar to that of, and may compound, the climate effect. For example, a preservative that is water soluble will tend to leach more quickly in the same types of climate that favor decay and termite attack. In other cases, such as a preservative that is susceptible to ultraviolet degradation, the effect of climate might be most severe in areas with relatively low deterioration hazard. Regional effects on preservative permanence in wood placed in ground contact are not well understood. Leaching in soil has been shown to be a function of soil chemistry and microorganisms, but the complexity of the soil system has made these relationships difficult to define (*25–28*). The relationship between exposure environment and preservative permanence is further complicated by the variability in depletion between similarly treated and exposed samples (*27*).

Preservative-Tolerant Organisms

Some regions have organisms that are tolerant to preservative components. Biocides with excellent fungicidal properties may have little or no insecticidal properties, rendering the wood vulnerable to attack by termites or other wood-boring insects. This risk tends to correspond to regional deterioration hazard because insect attack is most severe in the southern regions Even within those zones, however, preservative tolerance can vary. An example of this problem is the erratically distributed Formosan subterranean termite in the southern United States. These termites are more tolerant of preservatives than are native termite species (Table II), but their presence may not be known until after attack has occurred (*22*).

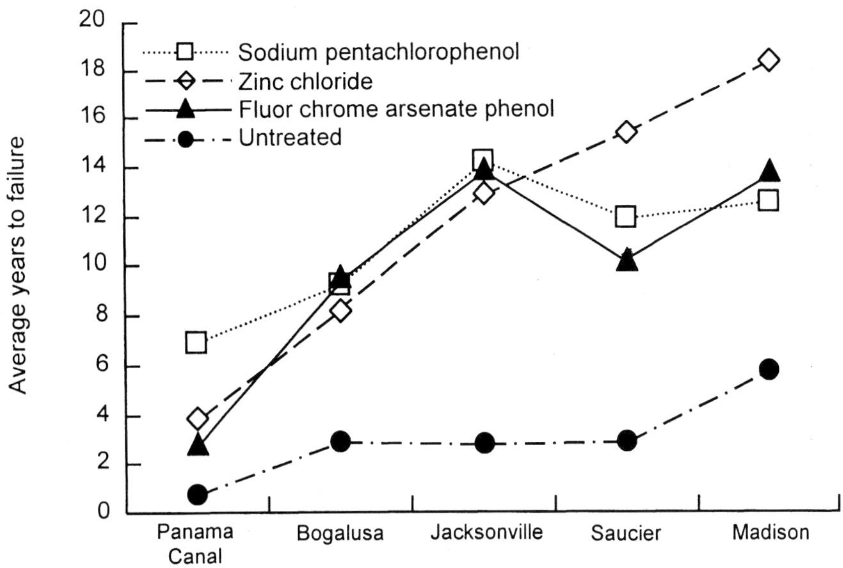

Figure 10 . Relative durability of matched sets of treated stakes at five exposure sites.

Copper tolerance by some types of fungi is a similar problem, although the presence of or colonization by these fungi is even less predictable than that of Formosan subterranean termites. Copper-tolerant fungi often colonize only one or two stakes within a exposure plot, or may rapidly destroy a stake that has not shown signs of colonization for more than a decade (*29*). The geographical distribution and concentration of these fungi is not well understood. Even less understood is the role that non-wood-degrading organisms, such as bacteria, molds, and other fungi, may play in altering preservatives so that they are less effective against wood-degrading organisms.

Local Variations in Deterioration Hazard

Perhaps the greatest difficulty in tailoring preservative treatment to regional deterioration hazard is local variability in conditions and in types of applications for treated wood. As discussed previously, localized populations of organisms such as the Formosan subterranean termite or copper-tolerant fungi can create problems for some types of preservatives. Localized deterioration zones also exist. Even in the driest climate substantial deterioration hazard may exist in low-lying areas, and irrigation can create artificial deterioration hazards.

Table II. Degradation Comparison of Treated Specimens by Formosan or Native *Reticulitermes* Termites at Lake Charles, LA

Preservative Formulation	*Retentio n(kg/m³)*	*Rating[a]*	
		Formosan termite	*Reticulitermes* spp.
Boarx–copper	0.60	5.7	8.5
Borax–copper	1.20	6.6	8.7
Borax–copper	2.40	7.4	8.8
Borax–copper	3.53	7.8	8.9
Disodium octaborate tetrahydrate	4.19	7.7	9.3
Chromated copper arsenate (CCA)	6.13	9.2	9.7

[a]Rating of 10 denotes sound and 0 denotes failure.

Similarly, conditions that slow the rate of drying (e.g., shade, construction design) and deposition of leaves or other organic matter can create areas where wood is exposed to a greater deterioration hazard than might be predicted for a particular climate. The nearly endless variety of the ways in which wood is cut and used also creates a broad range of potential moisture conditions. In a recent study in Minnesota, Schmidt and Jordon (*30*) noted that millwork components in some buildings had completely failed within 5 to 7 years, while untreated specimens exposed above ground suffered only slight to moderate decay during an equivalent period. In this case, condensation and moisture retention appear to have accelerated decay beyond that expected by climate alone. Many localized conditions cannot be controlled or predicted, even by the retailer of the treated product. As with the broader differences in regional deterioration hazards, the response to this problem has typically been to evaluate and develop preservatives for worst-case conditions.

Summary and Research Needs

It has long been recognized that wood exposed in some areas of the United States, such as the Southeast or Hawaii, deteriorates more rapidly than does wood exposed in other locations. Regional differences in deterioration rate are linked to factors such as temperature, precipitation, and insect population. Temperature and precipitation have been recognized as the primary factors and have been used in attempts to model and quantify rates of deterioration, especially in aboveground applications. A review of the data revealed that although models such as the Scheffer index often correlate well with observed rates of biodeterioration, there remains variability that is difficult to quantify or predict. In untreated wood, factors such as wood species, type of application, and localized insect population can influence the rate of deterioration. Variability is greatest in regions with relatively low deterioration hazard. Further variability is introduced when preservative-treated wood is evaluated because individual preservatives may be affected differently by regional conditions and wood degrading organisms. Finally, there remains the problem of localized deviation from regional hazard because of irrigation, shading, drainage, condensation, or other factors, particularly those that affect the rate at which wood dries.

These sources of variability within and between regions have made it difficult to tailor the development and use of treated wood for specific regional biodeterioration hazards. As in the past, preservatives continue to be evaluated under high-hazard conditions and used as if these conditions exist throughout the United States. As a result, preservative concentrations or the spectrum of preservative toxicity may be greater than necessary in many parts of the country. This situation will continue until research quantifies the deterioration rate under a broader range of conditions. Specifically, data are needed on the deterioration rate of wood exposed above ground in the northern two-thirds of the United States. Research is also needed on moisture content and deterioration rate for a broader range of exposure scenarios and wood dimensions. Even with such research, however, conservative assumptions will still be required to account for localized increases in biodeterioration hazard.

References

1. Zabel, R. A.; Morrell, J. J. *Wood Microbiology: Decay and its Prevention*; Academic Press, Inc.: San Diego, CA, 1992.
2. Highley, T. L. *Wood handbook—Wood as an Engineering Material*. Ch. 13. Biodeterioration of Wood. Gen. Tech. Rep. FPL–GTR–113. U.S. Department of Agriculture, Forest Service, Forest Products Laboratory: Madison, WI, 1999.
3. Morris, P. *Understanding Biodeterioration of Wood in Structures*. Forintek Canada Corp.: Vancouver, British Columbia, 1998. (www.durable-wood.com).
4. Highley, T. L.; Bar-Lev, S. S.; Kirk, T. K.; Larsen, M. J. *Phytopathol.* **1983**, *73*, 630–633.
5. Scheffer, T. C. *Can. J. Botany* **1986**, *64*, 1957–1963.
6. Simpson, W. T. *Equilibrium Moisture Content of Wood in Outdoor Locations in the United States and Worldwide*. Res. Note FPL-RN-0268. U.S. Department of Agriculture, Forest Service, Forest Products Laboratory: Madison, WI, 1998.

7. Moore, H. B. *Wood Inhabiting Insects in Houses: Their Identification, Biology and Control.* Interagency agreement IAA-25-75. USDA Forest Service and Department of Housing and Urban Development, 1979.
8. Cabrera, B. J.; Kamble, S. T. *Environ. Entomol.* **2001**, *30*, 166–171.
9. Esenther, G. R. *Ann. Entomol. Soc. Am.* **1969**, *62*, 1274–1284.
10. Hu, X. P.; Appel, A. G. *Environ. Entomol.* **2004**, *33*, 197–205.
11. Fei, H.; Henderson, G. *Environ. Entomol.* **2002**, *31*, 509–514.
12. Nilsson, T.; Daniel, G. Doc. No. IRG/WP/1433. *Int. Res. Group on Wood Preserv.*: Stockholm, Sweden, 1990.
13. Nicholas, D. D.; Crawford, D. *Wood Deterioration and Preservation.* Ch. 16. Goodell, B; Nicholas, D. D.; Schultz, T. P. Eds.; ACS Symposium Series 845. American Chemical Society: Washington, DC, 2003.
14. Scheffer, T. C. *Forest Prod. J.* **1971**, *21*, 25–31.
15. Scheffer, T. C.; Verrall, A. F.; Harvey, G. *Forest Prod. J.* **1963**, *13*, 7–13.
16. Rural Electrification Administration. 1973. Pole performance study. Staff Report. U.S. Department of Agriculture: Washington, DC, 1973.
17. *Book of Standards.* Use Category System U1-04. User Specification for Treated Wood. Commodity Specification D: Poles. American Wood Preservers Association: Selma, AL, 2004.
18. Foliente, G. C.; Leicester, R. H.; Wang, C-H.; Mackenzie, C.; Cole, I. S. *Forest Prod. J.* **2002**, *52*, 10–19.
19. Highley, T. L. *Int. Biodeter. Biodegrad.* **1995**, *35*, 409–419.
20. Eslyn, W. E.; Highley, T. L.; Lombard, F. F. Forest Prod. J. **1985**, *35*, 28–35.
21. Zahora, A. In: *Proceedings, Enhancing the Durability of Lumber and Engineered Wood Products.* Forest Products Society: Madison, Wisconsin, **2002**, 81–86.
22. Lebow, S. T; Shupe, T.;Woodward, B.; Crawford, D. M.; Via, B; Hatfield, C. A. *Wood Fiber Sci.* Submitted for publication.
23. Gjovik, L. R; Davidson, H. L. *Service Records on Treated and Untreated Fence Posts.* FPL-RN-068. U.S. Department of Agriculture, Forest Service, Forest Products Laboratory: Madison, WI, 1975.
24. Crawford, D. M.; Woodward, B. M.; Hatfield, C. A. *Comparison of Wood Preservatives in Stake Tests- 2000 Progress Report.* Res. Note FPL-RN-02. U.S. Department of Agriculture, Forest Service, Forest Products Laboratory: Madison, WI, 2002.
25. Crawford, D.; Fox, R.; Kamden, P.; Lebow, S. T.; Nicholas, D.; Pettry, D.; Schultz, T.; Sites, L.; Ziobro, R. IRG/WP 02-50186. *Proceedings, Int. Res. Group on Wood Preserv.*, 33rd Annual Meeting, Cardiff, United Kingdom, 2002.
26. Ruddick, J. N. R. *Mater. Org.* **1992**, *27*, 135–146.
27. Schultz, T. P.; Nicholas, D. D.; Pettry, D. D. *Holzforschung* **2002**, *56*, 125–129.

28. Wang, J.-H.; Nicholas, D. D.; Sites, L. S.; Pettry, D. E. Doc. No. IRG/WP/98-50111. Int. Res. Group on Wood Preserv., 1998.
29. Lebow, S. T.; Hatfield, C. A.; Crawford, D.; Woodward, B. *Proceedings, American Wood-Preservers Association,* **2003**, *99*, 142–149.
30. Schmidt, E. L.; Jordon, B.A. *Proceedings, American Wood-Preservers Association,* **2004**, *100*, 145–149.

Chapter 7

Concepts in the Development of New Accelerated Test Methods for Wood Decay

Darrel D. Nicholas[1] and Holger Militz[2]

[1]Department of Forest Products, Mississippi State University, Mississippi State, MS 39762–9820
[2]Institute of Wood Biology and Wood Technology, University of Gottingen, Busgenweg 4, D–37077 Gottingen, Germany

The rapid transition in the wood preserving industry toward the use of non-arsenical wood preservatives has emphasized the need for rapid test methods to evaluate new preservative systems. In this chapter progress toward the goal of developing accelerated test methods is presented. One of the key elements in accelerated test methodology is the use of accurate, quantitative methods for measuring the extent of decay. In this area bending elasticity and compression tests have been fully developed for small test specimens and dynamic MOE methods for larger samples show promise. Progress has also been made in developing accelerated soil contact and above ground decay test methodology.

Introduction

For the past two decades the wood preserving industry has been undergoing a major transition which has been driven mainly by environmental and health issues associated with the major wood preservatives. In the early 80's the Environmental Protection Agency (EPA) imposed regulations on the end use of products treated with the major wood preservatives—chromated copper arsenate (CCA), creosote and pentachlorophenol. These regulations shifted the emphasis from use of the oilborne preservatives for residential products to waterborne CCA, which proceeded to capture the major residential treated wood markets. This activity precipitated the need for research on development of new wood preservatives that were less toxic and more environmentally friendly. Additional restrictions on use of CCA for residential products around the turn of the century placed additional emphasis on the need for new wood preservative systems. Fortunately, at that point in time sufficient research had been conducted on two new waterborne preservative systems, anime copper quat (ACQ) and copper azole (CA) that could replace CCA for the residential treated wood product market. These preservatives are both copper rich systems with ACQ being composed of amine copper plus quarternary ammonium compounds and CA being composed of amine copper plus triazoles.

During this transition it became apparent that our test methods used to evaluate the efficacy of preservatives against wood decay microorganisms required excessively long exposure periods before sufficient data was available to confirm the viability of a given preservative. These methods consist of both laboratory and field tests, but at present none of the laboratory tests are capable of predicting long term performance of wood preservatives. Both above ground and ground contact exposure tests are used and these tests require a minimum of around five years exposure before definitive data can be obtained. Clearly, research is needed in this area to shorten the time frame needed for developing new as well as modified wood preservative systems. Some progress has been made in this area and a review of these developments along with suggestions for further improvement will be addressed in this chapter.

Methods of Detecting and Quantifying Wood Decay

One of the key elements that transcends all wood preservative evaluation schemes is the measurement of the extent of decay in test specimens. Presently, visual observation is normally used to detect and measure the extent of wood decay in field test specimens. This method suffers from being subjective, rather than quantitative and does not detect early decay in laboratory tests. Mass loss is another method often used to detect and measure the extent of wood decay. Shortcomings of this method are associated with: 1) difficulty in making adjustments for variation in wood moisture content, 2) loss of wood preservatives, 3) inability to make adjustments for the biomass weight gain as a result of fungal colonization, 4) does not differentiate between localized decay from decay in the entire sample. These factors negate the possibility of detecting the early stages of decay. Furthermore, in larger samples decay generally occurs in limited areas so mass loss of the entire sample is not representative of the damage. Because of these limitations and other factors the use of mass loss is essentially limited to laboratory soil block and agar block tests.

Another approach to detecting and measuring the extent of wood decay is based on changes in mechanical properties of wood as it is attacked by fungi. The bending, compression and torsional properties of wood are very sensitive to biodeterioration and can be used to develop quantitative data in decay studies (*1*). In this regard, static bending has been shown to be a viable method of measuring the extent of decay in laboratory decay tests (*2-5*). However, use of static bending to determine the MOE has limitations because the test must be conducted in the laboratory. Another approach in determining MOE of test samples is the use of dynamic methodology. This method was originally developed on the basis of ultrasonic pulse excitation methodology (*6*). More recently a MOE_{dyn} method based on resonant vibration excitation technology has been developed (*7,8*). This methodology is of interest because static and dynamic MOE values are highly correlated (*7-10*) it has fewer sample size and configuration restrictions and the equipment can be used for on-site measurements.

Immunodiagnosis is another method that can be used to detect the presence of wood decay fungi (*11*). This method is very sensitive and has the potential to detect incipient stages of decay. However, this method does not provide information on the progressive development of decay, so has limited potential in wood decay test methodology. Nevertheless, coupling innunodiagnosis with other methods such as bending stiffness could possibly provide an enhanced picture of the wood decay process.

In recent years near infrared (NIR) spectroscopy has been used for many applications that involve rapid analysis of various materials. Preliminary studies with decayed wood indicate that NIR may be useful in detecting and quantifying biodeterioration of wood . By coupling NIR analysis with multivariate regression or principal components analysis, it has been demonstrated that there is a good correlation between NIR analysis and both mass loss and compression strength loss of brown-rotted wood. However, additional studies are needed to determine whether or not this method is capable of detecting the early stages of decay as well as other types of fungi before the potential of this method is clarified.

Above Ground Decay Tests

Presently there are no laboratory tests available that can be used to provide a reliable evaluation of candidate wood preservative systems. Consequently, we must rely on field tests which require an inordinate amount of time to produce definitive results. The major problems with current test methods are associated with:

- Lack of fully developed, reliable, accurate quantitative methods for measuring wood decay
- The extreme variability in decay rates of replicate samples used in a given test
- The long delay between initial exposure of test specimens and the onset of detectible decay

As mentioned above, of all the methods considered for measuring the extent of wood decay in test samples, changes in mechanical properties has been shown to offer the most promise. Both compression strength and bending strength (elasticity) have been shown to be parameters that can be used in assessing the extent of wood decay. However, applying these techniques to above ground test samples is a challenge. One approach recently proposed was to incorporate thin sticks as a component of the lap-joint test unit (*1*). Another possible approach is to use bundles of thin sticks as the test specimens (*12*). Still another possibility is the use of dynamic MOE rather than static bending as a way to measure decay. This method shows good potential and can be more readily adapted to larger test specimens than is possible with static bending. As a result of this progress it now appears that reasonably good tools are available to quantitatively measure decay and all that is lacking is the development of test specimens that enhance the decay process and are compatible with these instruments that accurately measure the extent of decay.

Variability in Decay Rates

The extreme variability in decay rates among replicate untreated test samples is vividly illustrated in Figure 1. This represents typical data obtained for three individual L-joints in an above ground test and shows that while some samples decay rapidly others remain sound for a number of years. Needless to say, this variation in decay rates among replicate samples complicates the assessment of candidate wood preservatives because it makes it difficult to accurately compare the results with the performance of known wood preservatives. Furthermore, if one were to repeat the test with a new set of samples different results during the early stages of the field trial would more likely than not be obtained because of the extreme variability.

The question is, what is the cause of this extreme among sample variation? In a previous paper (*1*) it was proposed that variation in wood moisture content and wood extractives were likely associated with the variation in decay rates. When wood is exposed above ground, rain is the major source of moisture that is essential for decay. The amount of water trapped and retained by the wood is largely influenced by wood permeability and test unit design. In a recent study (*12*) it was shown that decay rate was highly correlated with wood permeability. This finding suggests that at least some of the variability associated with above ground tests could be avoided if permeability measurements were used to screen the selection of test material.

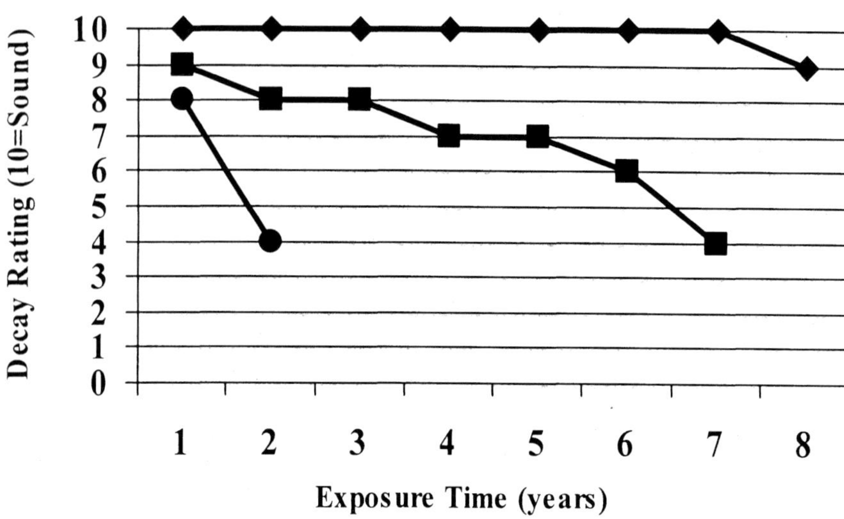

Figure 1. Comparative Decay Rates for 3 untreated L-Joints Exposed in Hilo, HI for 8-Years

Design of the test units is also very important and can have a major impact on trapping and retaining rain water. In this respect, the present lap-joint design specified in AWPA Standard E-16 is not satisfactory. The incorporation of the lap component in the 2 x 4 sample does permit water movement into the center of the unit, but this only traps a minimal amount of water when it rains. Furthermore, the openings along the sides and bottom of the unit promote fairly rapid drying. One way to overcome this design fault is to use split PVC pipe as clamps along both sides of the sample. This modification, along with foam rubber strips on the sides of the racks which partially seal the ends increases water retention in the samples. However, the bottom opening in the lap-joint is still exposed and permits drainage and drying. Hence, further modification of the lap-joint is needed. One possible modification is shown in Figure 2. This is a three piece unit that not only improves water retention, but also provides components that can be readily evaluated for decay by the dynamic MOE instrument.

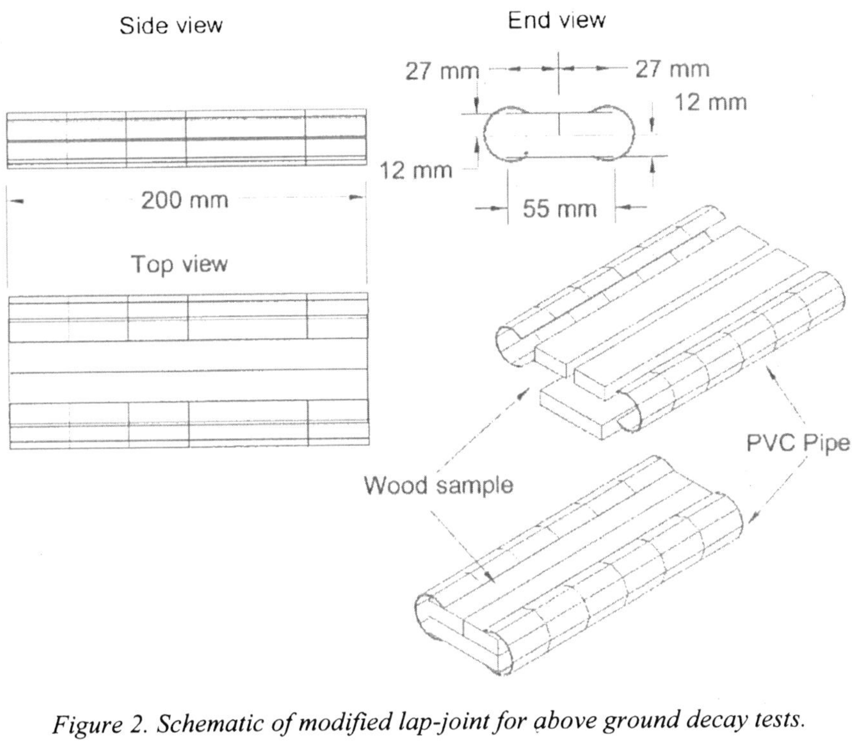

Figure 2. Schematic of modified lap-joint for above ground decay tests.

Another concept in above ground testing is the Ground Proximity Test (*13*) . In this test the samples are placed on top of hollow concrete blocks that are in contact with the ground. The concrete blocks retain water for fairly long periods after rainfall and this keeps the wood moist for extended periods of time. By covering the unit with shade cloth, water evaporation is depressed somewhat and helps maintain adequate wood moisture levels.

The wood sample configuration—small elongated blocks—in this test is such that decay measurement could be made with the dynamic MOE instrument. Another conceivable improvement to this method would be to fill the hollow concrete blocks with a material that absorbs and retains water for extended periods of time. Such a modification should increase the time that the wood has sufficient moisture for decay to proceed.

Another possible modification to this method is to use larger test samples (20 x 30 x 300mm) that are arranged in 10 stacks of 40 pieces. Both treated and untreated samples are intermingled in the stacks to increase the severity of the test. This arrangement also offers the possibility of comparing the performance of samples exposed directly on the concrete to samples in the middle and top of the stacks. The samples placed directly on the concrete attain higher moisture content levels which promotes soft-rot, whereas basidiomycetes predominate in the other samples.

Effect of Wood Extractives

Even though sapwood is normally used in above ground tests, the wood is not entirely free of extractives. Both hydrophilic and hydrophobic extractives are present in pine sapwood. The hydrophilic extractives are composed mainly of simple sugars, pectins and possibly minor amounts of nitrogen containing compounds. In the initial stages of decay these compounds are assimilated by bacteria and mold/stain fungi. Variation in the amounts and composition of these extractives may influence the overall wood decay process. Indeed, a recent study (*12*) indicates that removal of the hydrophyllic extractives reduces the rate of decay to a minor extent.

The hydrophobic extractives in pine sapwood are principally composed of terpenoid monoterpenes, resin acids, fats and fatty acids along with low levels of flavonoids and stilbenoids (*14*). The composition and relative amounts of these extractives are undoubtedly variable and probably influence the overall decay process. Indeed, the influence of sapwood extractives on above ground decay rates of pine was recently reported (*12*). In this study bundles of thin sticks were used as the test samples and the extent of decay after above ground exposure was measured by bending stiffness. Prior to exposure, one set of sticks was extracted with hot water, a second set was extracted sequentially with methanol/cyclohexane and the third set was un-extracted. The results of this study after various exposure periods showed that hot water extraction reduced the decay rate slightly, whereas, the methanol/cyclohexane extraction significantly increased the decay rate.

The above study demonstrates the utility of using bending stiffness as a quantitative measure of wood decay. It also clearly demonstrates that sapwood extractives have a major impact on above ground wood decay rates and as such are probably a major causal factor involved in the extreme variation in decay rates of untreated pine sapwood. These results suggest that extraction of test samples to remove the hydrophobic extractives before further processing would result in an accelerated and more uniform decay rate of sets of test samples. Furthermore, by coupling extraction with pre-screening for wood permeability it should be possible to greatly reduce the variability in decay rates of above ground test units.

Soil Contact Decay Tests

Many applications for treated wood require that wood be placed in contact with soil, which represents a more severe exposure compared to above ground exposures. This is mainly because soil contains large microbial populations that inoculate wood with decay microorganisms, serves as a water reservoir and also serves as a nutrient source. The soil environment also optimizes conditions that promote depletion of biocides by leaching and biodegradation mechanisms. Despite these favorable conditions, the time required to evaluate new and modified preservatives systems is lengthy and methods of accelerating this process are of interest.

One approach to accelerating soil contact decay tests is to use soil beds in the laboratory that are maintained at near ideal conditions for wood decay (*1*, *15*). By using small test specimens and maintaining the wood moisture content in the range of 40 to 80%, reasonable acceleration of decay can be achieved. Despite these attributes, accelerated soil contact decay tests have seen only limited use in the past because of concerns about the relevance of the test results in relation to actual field exposure. Nevertheless, some studies have shown that a good correlation between laboratory and field trials exists when the same soils and appropriate lab conditions are used (*15,16*). In these studies, the acceleration factor in lab tests appears to be in the range of 2-3 fold. In further development of the soil bed concept, it has been shown that the wood decay rate can be further enhanced by the addition of composted wood to the soil (*17*). The addition of composted wood adds wood degrading microorganisms to the soil along with some essential nutrients and accelerates the initiation of the decay process. In this study decay was measured by compression strength of the test samples. Although this method provides a good measure of the extent of decay, it is destructive and does not permit evaluation of comparative decay rates.

The concept of accelerating wood decay by amending soil with wood compost is appealing and, consequently, has been evaluated further in a recent study. In this study thin sticks (3 x 14 x 150mm long) were used so that decay could be measured by changes in elasticity (MOE), which permitted repeated measurements in order to establish decay rates. The test was carried out in small plastic cups with only the center portion of the sticks exposed to the soil. The comparative decay rates for the untreated samples in the two soil substrates is shown in Figure 3. From this data it is apparent that amending the soil with composted wood accelerates the rate of wood decay in untreated wood. This study also showed that the decay rate in treated wood was also accelerated. A soil microbial community analysis was also carried out in this test and showed that adding the compost did not alter it appreciably. Consequently, it appears that this may be a viable method for achieving further acceleration of a soil contact decay test.

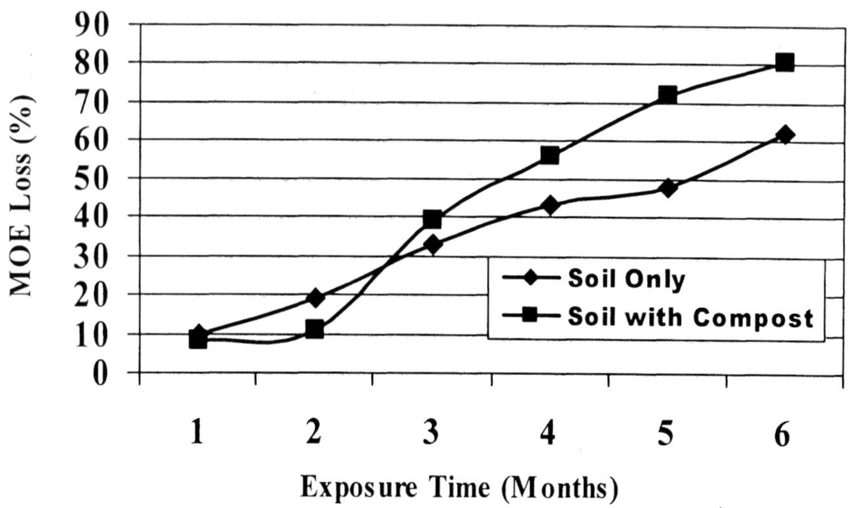

Figure 3. Decay rate, as measured by MOE loss, of untreated southern yellow pine sapwood sticks exposed in soil with and without compost

References

1. Nicholas, D.D.; Crawford, D. In: ACS Symposium Series 845. *Wood Deterioration and Preservation: Advances in Our Changing World*; Goodell, B.; Nicholas, D.D.; Schultz, T.P. eds.; American Chemical Society, Washington, D.C. 2003, pp. 288-312.
2. Gray, S.M. *Internat. Res. Group on Wood Preservation* **1986**, IRG/WP 2269.
3. Stephan, I.; Grinda M.; Rudolph D. *Internat. Res. Group on Wood Preservation* **1998**, 98-20149
4. Crawford, D.M. M.S. Thesis, Department of Forest Products, Mississippi State University, Mississippi State, MS, 1994
5. Grinda, M.; Goller, S. *Internat. Res. Group on Wood Preservation* **2005**, 05-20319.

6. Ross, R.J.; DeGroot, R.C.; Nelson, W.C.; Lebow, P.K.; Pellerin, R.F.; In: *Proceedings of the Internat. Wood Engineering Conference*; Gopu, V.K.A., Ed; LA State University., **1996**, 3:213-217.
7. Machek, L.; Militz, H.; Sierra-Alvarez, R. *Internat. Res. Group on Wood Preservation* **1998**, 98-20139.
8. Machek, L.; Militz, H.; Sierra-Alvarex, R. *Wood Sci. and Technol.* **2001**, *37*, 411-417.
9. Pellerin, R. *Forest Products J.* **1965**, *15*, 93-101.
10. Grinda, M.; Goller, S. *Interna.l Res. Group on Wood Preservation* **2005**, 05-20320.
11. Clausen, C.A. In: ACS Symposium Series 845. *Wood Deterioration and Preservation: Advances in our Changing World*; Goodell, B.; Nicholas, D.D.; Schultz, T.P. eds.; American Chemical Society: Washington D.C., 2003, pp. 326-336.
12. Nicholas, D.; Schultz, T.; Sites, L.; Buckner, D. *Internat. Res. Group on Wood Preservation* **2005**, 05-20310
13. American Wood Preservers' Association Standards *2004*; American Wood Preservers' Association, Selma AL
14. Koch, P. 1972. *Utilization of Southern Pines*. U.S. Department of Agriculture, Forest Service, Agriculture Handbook No. 420, 1972.
15. Nicholas, D.; Archer, K.J. Forest Products Society Proceedings No. 7308. *"Wood Preservatives in the 90's and Beyond"*. Sept. 26-28, 1994 pp. 110-115.
16. Polman, J.E.; Michon, S.G.L.; Militz, H. *Internat. Res. Group on Wood Preservation* **1992**, IRG/WP/2384.
17. Nicholas, D.; Borazjani, H.; Schultz, T. *Internat. Res. Group on Wood Preservation* **2004**, 04-20284

Chapter 8

Evaluating the Durability of Wood-Based Composites

Glenn M. Larkin and Peter E. Laks

School of Forest Resources and Environmental Science, Michigan Technological University, Houghton, MI 49931

Wood-based composites are increasingly used as substitutes for solid lumber in construction. Many evaluation methods for solid wood durability evaluation have been or can be adapted for this purpose, provided that the need for dimensional stability of the composites is accommodated. Improved methods to evaluate their durability need to be developed, discussed, and standardized in order to keep up with the pace of wood-based composite adoption into the market place. There is also a need for methods which are unique to wood-based composites.

Wood-based building products can be divided into groups based on solid wood (lumber, trusses, shakes, glulam, etc.), veneer-based composites [plywood, laminated veneer lumber (LVL), and parallel strand lumber (PSL)], particle-based composites [oriented strandboard (OSB), particleboard, and laminated strand lumber (LSL)], and Wood-Plastic Composites [a specialized type of particle-based composite in which wood particles are coupled to thermoplastics instead of bonded to adhesives]. Over recent decades, production of solid wood products has remained relatively steady, while particle-based wood composite production has dramatically increased. (*1*). For example, the production of OSB increased from approximately 235 million cubic feet to 780 million cubic feet between 1990 and 2003 (*2*). During the same period, the production of lumber and plywood oscillated around 7.5 billion and 1.5 billion cubic feet, respectively. The production growth for other wood-based composite materials has been equally dramatic. Wood-based I-Joist production increased nine-fold between 1990 and 2003 (*3*), and wood plastic composite production increased thirteen-fold in the seven years between 1995 and 2002 (*4*). These production increases have been accompanied by the substitution of wood-based composites as construction materials of choice, supplanting lumber, treated lumber, and plywood.

The first substitutions of wood-based composites for lumber and plywood occurred in applications with minimal hazards from fungal or insect attack: interior sheathing and flooring in dry parts of buildings. More recent uses include exterior sheathing, siding, and decking which carry significant hazards from fungal and insect attack, and from dimensional instability induced by exposure to moisture. The response has been to treat the composites with wood preservative systems that can effectively protect them against biodeterioration. Other additives are sometimes added to protect against degradation by abiotic means (e.g. dimensional changes and UV radiation). The ever increasing popularity of wood-based composites as construction materials necessitates a regimen for evaluating their durability in a broad spectrum of protocols that address both the testing of wood preservative efficacy and the performance of actual products like I-Joists, trusses, siding, and decking.

Durability, for the purposes of this discussion, is defined as the ability to last for a long time without deterioration, or longevity of a material/object in its intended use. Here, durability has a holistic meaning that includes resistance to:

- Decay and insect attack (biological),
- Dimensional change and water absorption (physical),
- Strength losses (mechanical), and
- Loss or degradation of active ingredients (chemical)

It is necessary to consider the different types of durability and to recognize they act in synergy to affect material longevity. To fully evaluate the durability of wood-based composites, a holistic approach is needed. For example, a wood-based composite that has swelled and has high moisture content is more likely to become decayed, lose strength, and leach wood preservatives than one which has not swelled. It is well known that increases in thickness from swelling induced by moisture cycling lead to increased exposed surface area and void volume for fungal colonization. The absorbed water facilitates fungal growth, which in turn raises the likelihood of biological degradation. Dimensional change and water absorption affect wood-based composite durability far more than solid wood durability. Table 1 outlines the differences in physical properties that generally make wood-based composites less durable than solid wood.

The scope of this chapter is limited to North American industry and evaluation practices.

Table 1. Properties of Wood-Based Composites Compared to Solid Wood

Property	*Wood-based Composites*	*Solid Wood*
Dimensional Stability	Poorer	Better
Surface Area	Increased	Decreased
Water Absorption	Increased	Decreased
Durability	Poorer	Better

Durability Tests for Wood-based Composites

Most durability tests for wood-based composites are adapted from protocols long established in the evaluation of solid wood durability in the laboratory and the field. They are generally taken from standardized methods published by AWPA (*5*), ASTM (*6*), or other national trade associations and authoritative bodies (*7*). Although a good starting point, these methods do not always reflect how wood-based composites will be used or the differences between these composites and solid wood. When choosing a test protocol the question, "what is the intent of the durability evaluation" should arise. If data on preservative performance is the aim, then model wood-based composite systems may be used (e.g. aspen waferboard). If data on product performance is needed, then the "real" products must be used. Preservative performance is often evaluated for above ground applications using AWPA Method E16-98 (*8*) and a model waferboard substrate, but wood plastic composites (WPC) are usually exposed as test decks to better reflect product end use. In either case, the intended application should be used to develop an appropriate evaluation protocol. Unfortunately, one size does not fit all. The AWPA Use Category System (see chapter by Laks) has been developed as a guide to help develop evaluation protocols.

Laboratory Durability Tests

Table 2 lists laboratory tests that may be useful in evaluating the durability of wood-based composites. The true value of these tests is that they can quickly screen candidate wood preservatives for suitability in field exposures. They are not a substitute for field evaluations of durability.

Table 2. Laboratory Durability Tests for Wood-Based Composites

Durability Type	*Evaluation Method*	*Standardized Method* [a]
Biological	Soil Block Decay Test	AWPA E10-01 (*8*)
	Mold Test	AWPA Proposed [c]
	Termite Test	AWPA E1-97 (*8*)
	Soil Bed (Fungus Cellar)	AWPA E9-97 (*8*)
Physical	Dimensional Stability	ASTM D1037 (*9*)
		ASTM D790M (*10*)
	Water absorption	ASTM D1037 (*9*)
		ASTM D570 (*10*)
Mechanical	Internal Bond [b]	ASTM D1037 (*9*)
	M.O.E. & M.O.R.	ASTM D1037 (*9*)
		ASTM D790M (*10*)
	Compression	ASTM D1037 (*9*)
Chemical	Aquoeus Leachability	AWPA E11-97 (*8*)
	Soil Leachability	AWPA E20-04 (8)
	Corrosivity	AWPA E12-94 (*8*)

[a] Numbers in parentheses are references, [b] Internal Bond, a relative measure of adhesive performance, is the one method listed which is unique to wood-based composites. [c] Standard has been proposed, but not yet accepted.

Biological Tests

Soil Block Decay Test

The soil block decay test is designed to access the durability of preservative treated and untreated wood and wood-based composites using monocultures of basidiomycete wood destroying fungi. The test is run in a similar manner for solid wood and wood-based composites. Specimen sizes may differ slightly. The individual specimens are incubated for a prescribed time in the presence of the test organism, at optimal temperature and humidity for fungal growth. Performance is generally measured in terms of test specimen weight loss on an oven dry basis.

Mold Test

Laboratory mold tests of wood and wood-based composites are designed to evaluate the resistance to a mixed-species inoculum of ascomycete and fungi imperfecti mold and stain fungi. There are a number of permutations of the test (*8, 11*), but the most common method exposes test specimens in a "humidity cabinet" of sorts for a prescribed time at elevated temperature and 100% RH. Mold and stain growth is generally measured as percent surface coverage. The test is run in a similar manner for wood-based composites and solid wood.

Termite Test

Laboratory termite tests are designed to evaluate either the toxicity and/or repellency of preservative treated wood and wood-based composites against termites (*8*). Generally, *Reticulitermes flavipes* (Eastern subterranean termite) and/or *Coptotermes formosanus* (Formosan subterranean termite) are tested in North America. Termite mortality and test specimen weight loss are the typically measured variables. The test is run similarly for composites and solid wood.

Soil Bed (Fungus Cellar) Test

The fungus cellar test is designed to simulate field ground contact decay tests (see below), but in an accelerated manner. Decay stakes are exposed to soil beds in a controlled environment with frequent evaluations. The traditional soil bed test is conducted in a bin with non-sterile soil (AWPA E9-97), but in recent years, variations have been developed to allow better control of soil conditions and alternatives to the traditional visual ratings of durability performance. For example, Nicholas has devised a soil bed procedure based on PVC tubes in which only the middle section of the stake is in soil contact (*12*) and Ilman devised a soil bed that allowed *in situ* nondestructive evaluations during soil bed exposure (*13*). The test is run similarly for composites and solid wood.

Physical Tests

Laboratory dimensional stability and water absorption tests are particularly important to the evaluation of wood-based composites durability. They, along with the internal bond test (see below) are predictive of whether a wood-based composite substrate can withstand field exposure unprotected from direct precipitation.

Dimensional Stability

A major concern in the field testing of wood-based composites is thickness swell. If a test substrate experiences too much thickness swell, it may also experience excessive preservative loss from leaching, delamination from exceeding the internal bond (failure of the adhesive), increased fungal colonization due to increased void volumes, and difficulty with visual inspections due to "softness" from both decay and void spaces. Figure 1 shows a comparison of thickness swelling between plywood and OSB in which the OSB

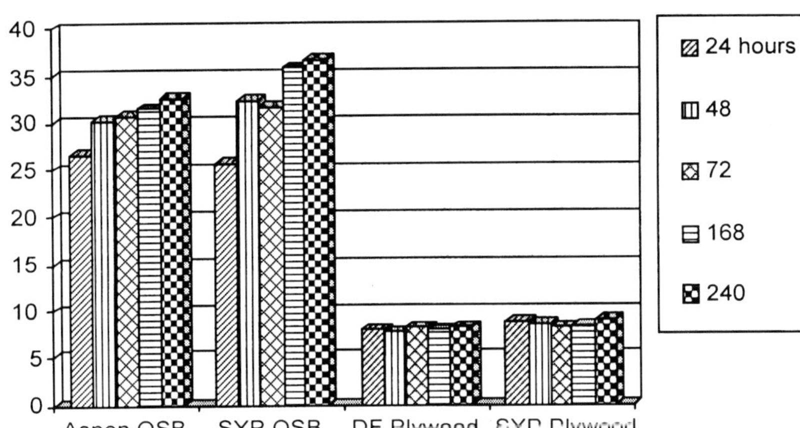

Figure 1. Thickness swell (in percent) measured at the corners of 152 x 152 mm specimens of plywood and OSB over 240 hours. OSB adhesive content is 3%. Taken from reference 14.

thickness continued to increase over the entire study while the plywood thickness quickly stabilized (*14*). The majority of the dimensional increase occurs in the first twenty-four hours, approximately 78% and 69%, respectively. High adhesive content has been shown to lead to improved dimensional stability. A field study of untreated aspen waferboard showed that it is possible to prepare "model" composite substrates that will remain dimensionally stable for an extended period (see Figure 2) exposed to direct precipitation (*15*). Ground contact decay tests have been conducted for seven years with particleboard (*16*) and eleven years with waferboard (*15*) at high adhesive contents, as well as with WPC (*17*).

Figure 2 shows that thickness swell of aspen waferboard remained statistically unchanged over a 48-month exposure to direct precipitation. The magnitude of the thickness swell after 12 months results from heavy precipitation

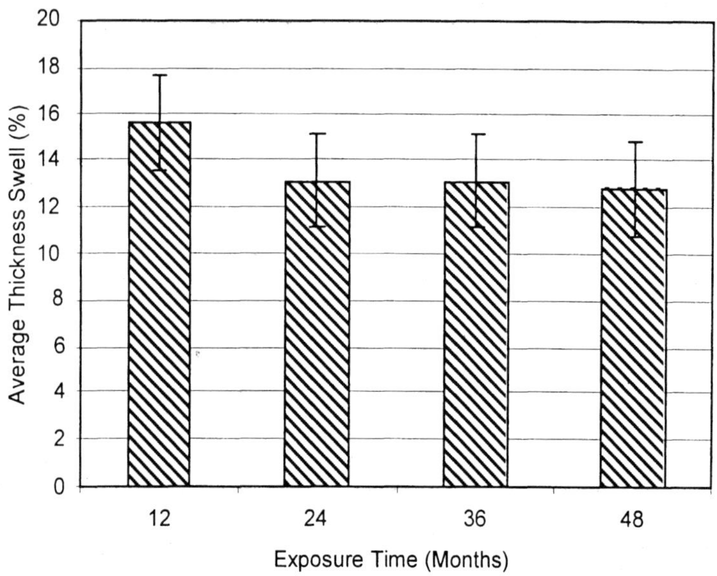

Figure 2. Aspen waferboard lap joint thickness swell over 48 months of field exposure at the MTU Mauna Kea Field Test Site near Hilo, HI. Adhesive content is 7% polymeric diphenyl-methyl-di-isocyanate (pMDI).

at the time of the measurement. High adhesive content allows for long term field evaluations of preservative systems in wood-based composites.

Mechanical Strength Tests

Internal Bond Test

Internal bond (IB) is a fundamental measure of adhesive performance that serves as a practical predictor of how well a wood-based composite will resist delamination when subjected to the forces induced by moisture cycling in field tests. Figure 3 shows a lap joint that experienced delamination failures after only six months of test exposure near Gainesville, FL. Composites with low IB's are not suitable for direct exposure to rainfall and moisture cycling in above-ground and ground contact field tests. Based on many years of field tests and correlations of delamination failures in the field with pre-exposure internal bonds, a minimum IB of 690 kPa, 100 psi is recommended for exposure of wood-based composites in field tests (*15*).

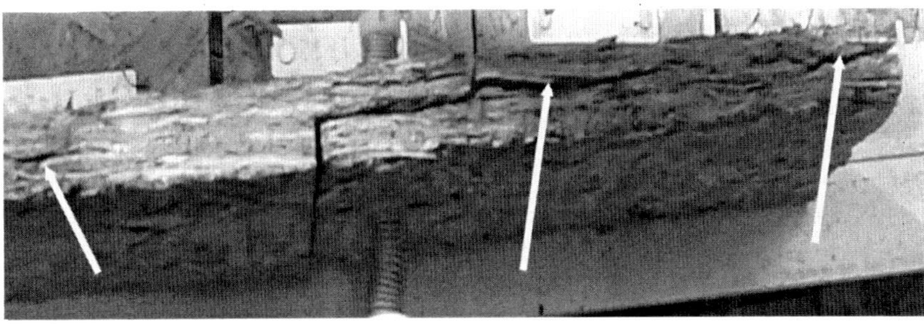

Figure 3. Aspen waferboard lap joint exposed for six months near Gainesville, FL. Delamination failures are due to excessive thickness swell and a low internal bond (~ 520 KPa, 75 psi). Arrows point to areas of delamination. Photo by G. M. Larkin

Other Mechanical Strength Tests

Researchers have periodically looked for alternative measures to weight loss to evaluate the biological durability of wood and wood-based composites. For example, Schmidt *et al.* studied strength loss as a measure of preservative performance in aspen waferboard, and found a good correlation between compression strength and weight loss after specimen exposure in a soil block test (*18*). Nicholas *et al.* have also been active in using mechanical strength tests to develop accelerated evaluation methods (*12*). More recently Illman *et al.* used nondestructive methods to evaluate OSB exposed in a modified decay test that allowed *in situ* stress wave measurements with minimal disturbance of the samples (*13*). They concluded that nondestructive evaluation successfully detected decay in OSB.

Chemical Tests

Leachability Test

Water leachability tests are designed to be a "worse case" aqueous extraction of wood preservatives from wood and wood-based composites under laboratory conditions. Their value lies in preservative performance evaluations and in selecting preservative systems for field exposure. Composites with highly leachable wood preservatives are more likely to fall beneath their activity threshold and become more susceptible to decay and insect attack than low leachable preservatives.

Recently, a soil leachability test has been added to the AWPA Standards that is analogous to the water based test, but aimed at predicting preservative loss in ground contact (*8*). These tests are run similarly for wood-based composites and solid wood.

Field Durability Tests

As stated above, the intent of a durability evaluation should drive the choice of protocol selected. Table 3 lists field tests that may be useful in the evaluation of wood-based composites. They fall into two categories: general-purpose tests for preservative performance and application-specific tests for product performance.

Table 3. Field Durability Test for Wood-Based Composites

Test Type	*Exposure*	*Evaluation Method*	*Standardized Method* [a]
General	GC[b]	Decay Stake Test	AWPA E7-01 (*8*)
		Biocide Depletion Test	AWPA E7-01
	AG[b]	Lap Joint Test	AWPA E16-98 (*8*)
		Biocide Depletion Test	-------[c]
	GP[b]	GP Decay Test	AWPA E18-04 (*8*)
		GP Termite Test	-------
Specific	AG	Siding Exposure Test	-------
		L-Joint Test	AWPA E9-97 (*8*)
		Deck Exposure Test	-------

[a] number in parenthesis is reference. [b] GC = ground contact, AG = above ground, GP = ground proximity. [c] no standard.

General Tests

All of these tests are run similarly as for solid wood, sometimes with modifications to accommodate the differences in properties between the substrates. The important thing is to assure that exposed wood-based composites have sufficient physical and mechanical durability to withstand the rigors of field testing so that meaningful biological and chemical durability information may be obtained on preservative performance.

Decay and Depletion Stake Tests

Ground contact evaluations are run on composites to obtain "accelerated" preservative performance data and because of possible uses of wood-based composites in AWPA Use Category 4 applications (see chapter by Laks) like railway ties and utility poles (*19*). These tests are possible with high enough adhesive content to prevent premature delamination. Modification of the stake size helps reduce edge and thickness swell effects. The same stakes could be used for decay and depletion studies. Decay tests are conducted following AWPA E7-01 guidelines and visual evaluations for both fungal and insect attack are adequate in the case of wood-based composites. Optional nondestructive and destructive mechanical properties tests and weight loss may also be used. Figure 4 shows that long term decay tests to evaluate preservative performance in wood-based composites are achievable with the above considerations. The data shown is for aspen waferboard with 7% pMDI adhesive content exposed near Gainesville, FL for eleven years. Note the dose response to zinc borate preservative loading.

WPC stakes have also been evaluated recently in field decay tests (*17*). Thickness swell and water absorption were significant factors in these studies. These physical properties coupled with their high plastic content which can mask decay make visual evaluations an inadequate measure of WPC performance in field stake tests. Therefore, other measures of durability (e.g.: post exposure strength and density losses) are needed. It is also critical that all surfaces are either machined or extruded to avoid dimensional instability.

Decay stake tests are usually run in conjunction with a ground contact depletion test in which treated wood-based composite stakes are exposed and evaluated for biocide loss via chemical analysis.

Lap Joint Decay and Biocide Depletion Tests

The lap joint test is a general above ground protocol that can be used with wood-based composites to evaluate preservative performance following the guidelines of AWPA E16-98. Evaluations are visual and use the AWPA E7 scale. It is necessary to control lap joint thickness swell and to carefully consider all types of durability during visual evaluations.

Thickness swell may be controlled, to an extent, by adhesive type and content. By using a spring and nylon bolt to hold the lap joint together, residual

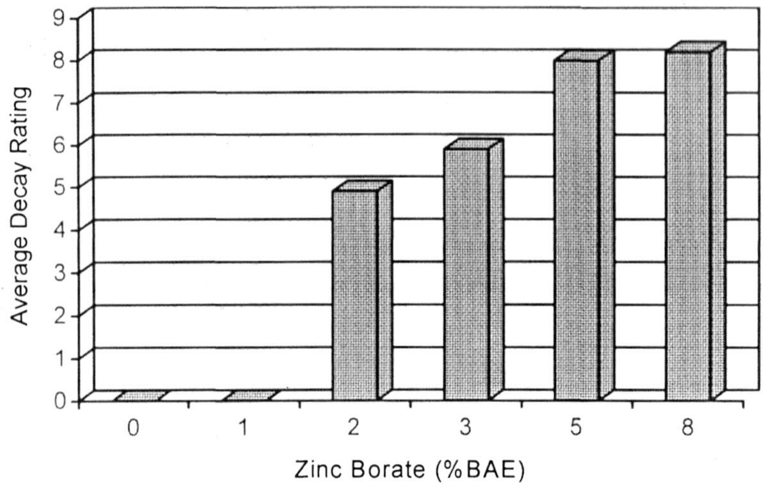

Figure 4. Aspen/pMDI Waferboard Decay Stakes Exposed for 11 Years near Gainesville, FL. Adhesive content = 7%. Unpublished Data (Laks & Larkin 2005). BAE is Boric Acid Equivalents

thickness swell is accommodated. The optimal total spring load is approximately 9 Kg (20 lbs.) which is sufficient to hold the lap joint together and still be able to easily open it for inspection. Figure 5 shows a decayed OSB lap joint with the nylon bolt and spring assembly.

Evaluations of wood-based composite lap joints are complicated by the wide range of moisture contents that are encountered during inspections. The specimens may be "bone dry" with considerable voids, the result of significant irreversible thickness swell, or "mushy" with high moisture content where it is difficult to distinguish softness due to decay from that caused by wetness. One way to possibly avoid some of these pitfalls is to evaluate only within the lap region, except when there is a complete failure of the specimen. Lap joint tests are often run in conjunction with an above ground depletion test in which treated wood-based composite blocks are exposed and evaluated for biocide loss via chemical analysis.

Ground Proximity Decay and Termite Tests

A ground proximity decay test was recently added to the AWPA Standards (*8*). It uses a platform of paver bricks to support test specimens and is topped with a porous, fabric covered frame that allows direct precipitation to wet the specimens, as well as reduces evaporation (drying). We have modified this method by replacing the cinder block platform with an acrylic grid that rests 25

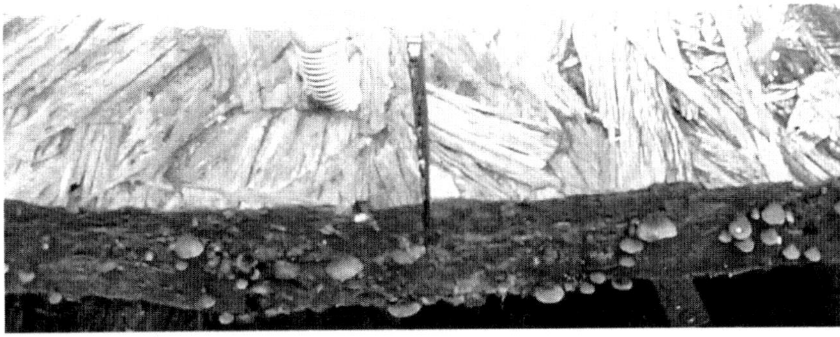

Figure 5. An untreated control OSB lap joint exposed for two years near Hilo, HI showing the combined effects of moisture cycling and decay (lap joint creep). The structures on the side of the lap joint are basidiomycete fruiting bodies that indicate the extent of decay. Photo by G. Larkin

mm above the ground. The mesh frame cover is replaced with an opaque polypropylene cover that completely protects specimens from precipitation. All moisture comes from soil respiration and humidity condensation. In some climates, there is sufficient moisture to support ongoing decay. Studies are ongoing to determine the utility of this modified method.

This procedure is also a modification of the MTU ground proximity termite test method, which is described and discussed elsewhere (*20*). In addition to the ground proximity termite test, there are "lunch box" (*21*), in-ground trench (*22*), and "farm" methods (*23*) that might be suitable for wood-based composites.

Application-Specific Tests

Product performance for wood-based composites must be evaluated in application-specific protocols. Specimens are exposed in field tests that reflect their use. The number of possible variations is limited only by the number of different products, the practicality of meaningfully testing them, and the imagination of the researchers. Three commonly used test protocols are siding, L-Joint, and deck exposures.

Siding Exposure Test

In a siding exposure, test specimens of wood and wood-based composite siding are tested in a configuration that simulates their installation on the exterior of buildings. This usually entails using an exposure test rack with a plywood backer board to simulate sheathing, and some type of fastener system to keep specimens in place. Depending on the objectives of the test, samples may be of various sizes and may be exposed at any angle, facing in any direction. North facing specimens tend to dry more slowly than those facing in other directions and tend to deteriorate more quickly. Fieldsite geography and microclimate may effect this trend. It is common to run tests at 90° and 45° to the ground. The former simulates actual use and the latter accelerates deterioration. Visual rating for decay and insect attack, and measurements of water absorption and thickness swell are the most common evaluation methods. Wood-based composites and soild wood are evaluated similarly. Figures 5 shows a typical exposure rack with specimens of waferboard siding manufactured at MTU. These specimens were generated from aspen with 3% pMDI adhesive. The front face and edges were coated with

Figure 5. Aspen siding specimens exposed at 45° (foreground) and 90° (background) near Hilo, HI. The 45° exposure angle provides accelerated durability data compared to the 90° exposure (see Figure 6). The specimens shown have similar preservative treatments and have been exposed for the same time interval. Photo by G.M. Larkin

a white tinted, waterborne exterior acrylic latex paint. Figure 6 shows the utility of multiple exposure angles to accelerate decay. The 45° specimens decay more quickly because more precipitation falls directly on the face of the specimens and is captured by the backerboard. This in turn leads to greater water absorption and thickness swell while remaining wet for longer periods than the 90° exposure. The average visual decay ratings for untreated specimens along the drip-edge of some of the specimens are shown in Figure 5. If periodic depletion analysis, along with water absorption and thickness swell measurements, are conducted in tandem with visual evaluations of preservative treated specimens, it may be possible to estimate the service-life of siding specimens using a combination of accelerated and "real use" data.

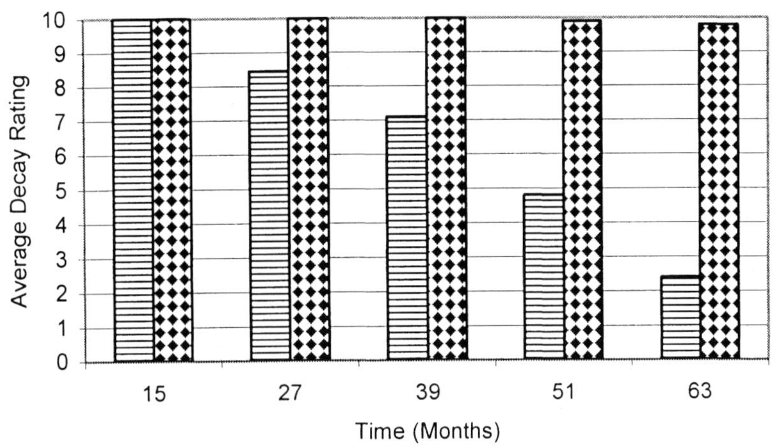

Figure 6. Average decay rating of untreated, coated aspen waferboard siding specimens exposed, facing East for 63 months, at the MTU Mauna Kea Field Test Site near Hilo, HI. The 45° (stripes) specimens have decayed steadily over the exposure time, whereas the 90° (checkered) specimens show minimal biodeterioration. Unpublished data (Laks & Larkin 2005).

L-Joint Test

The L-Joint test was designed to evaluate the durability of millwork components like window and door frames. The details of the test are presented in AWPA Standard E9-97 (*8*). Dimensional stability, water absorption, decay and mold resistance, and insect attack are the most commonly measured parameters. This test is run similarly for wood-based composites and wood.

Deck Exposure Test

The deck exposure test entails construction and exposure of small decks, and would be appropriate for WPC and other exterior decking materials. Photo stability, dimensional stability, water absorption, decay and mold resistance, and insect attack are the most commonly measured parameters. Visual decay ratings for WPCs are not useful, as discussed above, so post exposure measurement of strength properties is recommended. Tests are generally conducted similarly for WPC and wood.

Conclusions

Wood-based composites are increasingly important as construction materials. In expanding their use to higher AWPA Use Categories, there is need for durability data. Evaluating the durability of wood-based composites is not as straight forward as it is for solid wood, but can be accomplished when the properties of these composites are accommodated. Fortunately, most test protocols for solid wood can be adapted for use with wood composites with these caveats:

- Evaluating the durability of a wood-based composite should be driven by the intent to test either preservative performance in a model composite or an actual "finished" product. Select general tests for evaluating preservative performance. Select application-specific tests for finished products.
- Consideration of dimensional stability is critical to the successful evaluation of wood-based composites in tests modified from solid wood protocols. Adhesive type and content greatly affect the dimensional stability of wood-based composites and their suitability for field exposures. Generally, dimensional stability is dependent on adhesive content.
- Multiple evaluation methods will become more necessary in the future. Field exposures evaluated with visual rating for decay and insect attack, and physical property measurements could be combined with nondestructive evaluation in the field and/or post exposure mechanical properties tests in the lab. In using mechanical tests to evaluate wood-based composite biodeterioration, results must be carefully interpreted in the light of weight loss data in order to assure accurate conclusions. Factors like specimen variability (*24*) and moisture history need to be considered in data analysis.

References

1. Laks, P.E. Protection of Wood-based Composites. Proceedings: *American Wood Preservers Association* **2004**, *100*, 78-82.
2. *2003 Yearbook: Forest Product Market Prices and Statistics* Random Lengths: Eugene, OR, **2003**.
3. *E171 Yearbook* APA: The Engineered Wood Association Tacoma, WA **2005**; 33.
4. Morton, J.; Quarmley, J.; Rossi, L. In Proceeding: *The Seventh International Conference on Woodfiber-Plastic Composites* Forest Products Society: Madison, WI, **2004**, 3-6.
5. American Wood Preserver's Association, Selma, AL
6. American Society for Testing and Materials, West Conshohocken, PA

7. Foliente, G.C.; Leicester, R.H.; Wang, C.; MacKenzie, C.; Cole, I. *Forest Products Journal* **2002**, *52* [1], 10-19.
8. *American Wood Preserver's Association Standards 2005*: Selma, AL
9. *Annual Book of Standards* American Society for Testing and Materials: West Conshohocken, PA **2002**, Vol 04.10, pp 142-171.
10. *Annual Book of Standards* American Society for Testing and Materials: West Conshohocken, PA **1999**, Vol 08.01.
11. Downey, A. *Resin Review* **1991**, *41*, 10-11
12. Nicholas, D.D. In *Wood Deterioration and Preservation: Advances in Our Changing World* Goodell, B.; Nicholas, D.D.; Schultz, T. , Eds; ACS Symposium Series 845; American Chemical Society: Washington, DC, **2003**; 288-312.
13. Illman, B.L.; Yang, V.W.; Ross, R.J.; Nelson, W.J. Nondestructive Evaluation of Oriented Strand Board Exposed to Decay Fungi *International Research Group on Wood Preservation* **2002** Doc. IRG/WP/02-20423
14. Laks, P.; Richter, D.; Larkin, G.; Keranen, E. *Mold and Decay Resistance of Common Sheathing Materials* Durability and Disaster Mitigation in Wood Frame Housing. PATH. Madison, WI. November 1-2, 1999. (Poster)
15. Unpublished data (Laks & Larkin 1999)
16. Hall, H.J.; Gjovik, L.R.; Schmidt, E.L.; Gertjejansen, R.O.: Laundrie, J.F. *Forest Products Journal* **1987**, *37* [4], 49-53
17. Verhey, S.A. ; Laks, P.E. ; Richter, D.L. ; Keranen, E.D. ; Larkin, G.M. *Forest Products Journal* **2003**, *53* [5], 67-74.
18. Schmidt, E.L. ; Hall, H.J. ; Gertjejansen, R.O. ; DeGroot, R.C. Biodeterioration and strength reductions in preservative treated aspen waferboard. *International Research Group on Wood Preservation* **1983** Doc IRG/WP/2195
19. Gaby, L.I. *Forest Products Journal* **1986**, *36* [11/12], 61-62.
20. Laks, P.E. Field Testing with the Formosan Subterranean Termite. Proceedings: *American Wood Preservers Association* **2005**, 101, (in Press)
21. Preston, A.F. *Above Ground Field Tests: U.S. Experiences* (Oral Presesentaion) COST Meeting, Antibes, France February 7, 2001.
22. Lenz, M.; Creffield, J.W.; Evans, T.A.; Kard, B.M.; Vongkaluang, C.; Sornnuwat, Y.; Preston, A.F. International Research Group on Wood Preservation **2003** Doc IRG/WP/03-20261
23. Smith, R.W.; Amburgy, T.L.; Henderson, G.; Ring, D.R. *Forest Products Journal* **2004**, *5* [5], 26-28
24. Curling, S.F.; Winandy, J.E.; Carll, C.; Micales, J.A.; Tenwolde, A. *Holzforschung* **2003**, *57*, 8-12.

Molds and Mold Health Issues

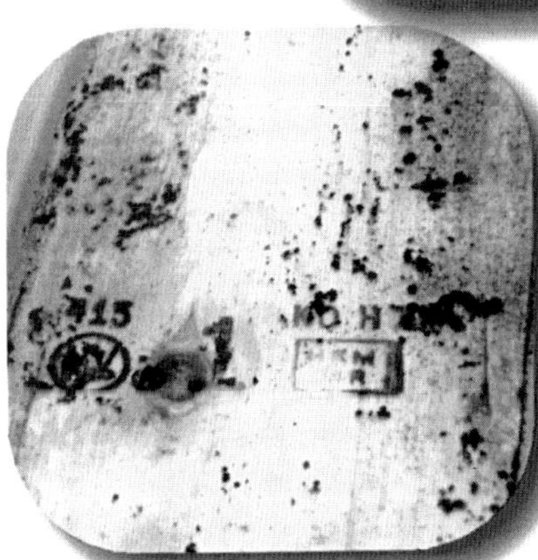

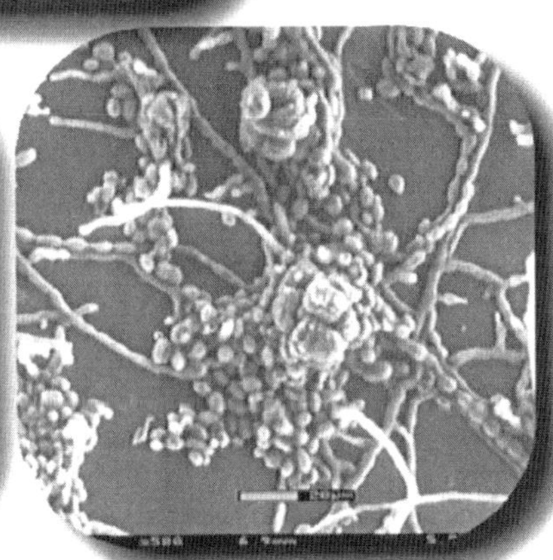

The first and third photographs are courtesy of Barry Goodell of University of Maine and the second photograph is courtesy of Steve Quarles of the University of California Cooperative Extension. *(See page 4 of color inserts.)*

Chapter 9

Mold Growth in Structures: An Overview

Stephen L. Quarles

Division of Agriculture and Natural Resources, University of California Cooperative Extension, Richmond, CA 94804

The advent of litigation related to indoor air quality in structures has made non-decay fungal growth on wood and wood-based materials more than an aesthetic issue for the consumer. Although mold growth on non-wood based materials occurs, growth on wood-based materials has been a critical component of indoor air quality litigation in commercial and residential structures. The predominance of wood-based materials in construction, and the use of both initially wet and dry wood-based products in structures, has resulted in some specific problems regarding fungal growth on these products, and fungal growth that can be "expected" in wood-framed structures. The common use of wood-based products has also lead to some misconceptions regarding the growth requirements of the non-decay fungi, particularly with regard to minimum moisture requirements and wood constituents that can be utilized as a food source. The objectives of this chapter are to present field-based observations regarding fungi that are commonly found on wood-based materials, and review research-based information on growth requirements.

Durability issues for wood and wood-based construction materials used in buildings have traditionally focused on structural factors. Regarding fungal infestations, damage from decay fungi was of primary importance because of the impact these organisms have on reductions in the strength of wood-based construction materials. In a literature survey that reviewed research on the strength loss associated with decayed solid wood, Wilcox (*1*) reported reductions of 50% or more, in brown rotted wood and depending on the particular mechanical property, with mass losses of 5% or less. In evaluating the impact of decay fungi on structural performance the extent of decay must also be considered. Since decay progresses from one location to another within a wood sample, the process isn't uniform and therefore advanced decay can be observed locally without structural failure occurring in the member. In these situations the loading distribution on the member is also critical.

Mold and sapstain fungi were previously considered an aesthetic issue, almost always with a negative connotation. However, in the case of certain sapstain (blue stain) fungi in selected products, the stain could be considered a desireable feature. With the recent inclusion of personal injury components associated with mold growth in building construction and materials litigation though, the perception and importance of mold has changed dramatically.

The objectives of this paper are to review the growth requirements of mold and decay fungi and discuss related implications that have been observed in field work and interaction with experts in mold-related litigation. Fungal growth on 'green' (non-kiln dried) and rewetted wood products will also be discussed.

Growth Requirements

The general growth requirements for wood-inhabiting fungi are well known and include a favorable temperature, sufficient oxygen, a food source (materials with accessible carbon containing compounds) and sufficient moisture (*2*). Two other requirements, favorable pH and the availability of certain essential elements, can also be included in this list. Control of fungal growth in service is usually accomplished by controlling moisture, or where that isn't feasible, by adding biocides to the product.

Although there is general agreement that decay fungi require liquid water for growth to initiate and propagate, there has been some disagreement regarding the minimum amount of moisture required for mold growth. Similarly, whereas there is no question regarding the strength reducing capacity of decay fungi, there has been some disagreement as to whether mold fungi can structurally degrade wood. This is particularly true where a mold fungus has been reported to produce cellulase, such as some *Trichoderma* species. Some cellulases are actually a multienzyme complex that can decompose accessible cellulose (*2*); however, some organisms produce only one type of cellulase (endo or exo) and thus they may have limited capacity to completely degrade cellulose even if it is accessible. These two factors will be discussed in more detail in the following sections.

Moisture

Zabel and Morrell (*2*) state that 'free water on the surfaces of the cell lumina' is required for fungal growth. Free water on the surface impies that at least locally, the moisture is above the fiber saturation point, typically taken to be approximately 28% (oven dry basis). This condition also indicates that the relative humidity, again at least locally, is 100% (i.e. at saturation), with an associated water activity, a_w, of 1. Water activity is measured specifically by the water vapor pressure of a material comparted to the water vapor pressure of pure water at the same conditions (*3*), which leads to confusion when it is used in a manner which equates it to a relative humidity index. Other documents clearly state that mold growth on the surface of wood can occur at humidity conditions far less than saturation. For example, Flannigan and Miller reported minimum water activity levels (a_w) of 0.8 and lower for some mold fungi (*4*). These results are in agreement with those of Viitanen, and other research cited by Viitanen (*5*). TenWolde (6) reported the ability of mold fungi to grow at moisture contents less than the fiber saturation point of wood, implying again that free water is not required. Based on these contradictory results, it is perhaps understandable that there is confusion in the the non-refereed technical journals and trade publications regarding minimum moisture content. As an example, Wemhoff (*7*) reported that 'mold begins to form at about 17%, and rotting begins at about 27%'. At 70°F (21°C), the 17% moisture content would occur at a relative humidity between 80 and 85%, in line with the published 0.8 minimum a_w requirement (*4 - 6*). The 27% value is the approximate fiber saturation point.

In an informal survey of forest products pathologists, conducted by this author in 2001, there was general agreement among the respondents that liquid water was required at the location where the fungus was active (*8*). Most of the respondents indicated that one potential reason for the confusion regarding minimum moisture content and relative humidity requirements lies in the difficulty in accurately measuring local moisture content and relative humidity at the micro scale, particularly in wood cell lumens. The moisture content usually reported in studies is a bulk or average value (*5, 6*), measured at some distance from the specimen. Depending on the dimensions of the test specimens, there can be moisture gradients present that would mask the actual moisture content at the surface. Another factor commonly discussed by the respondents was the difficulty in accurately controlling ambient relative humidity at elevated levels, and the sensitivity of the moisture content of wood at these conditions. For example, at 70°F (21°C), the nominal equilibrium moisture content is 20.6% at a relative humidity (RH) of 89%, but it is at the fiber saturation point at 95% (RH) (*9*). Although the fiber saturation point is technically reached at a relative humidity of 100%, this information is indicative of the difficulty in accurately measuring moisture content in wood at elevated humidities. Finally, some of the respondents thought that the presence of the gelatinous hyphal sheath could explain some of the apparent contradictions in the literature regarding moisture content requirements. In this case, the hyphae would be in intimate contact with the food source.

The argument for the liquid water requirement is based on the external digestion process that occurs with fungi. Water serves as a diffusion medium for enzymes and other metabolites produced by the fungus to reach the carbon-based food source, and for break down products to return to the hyphae. The argument for elevated moisture contents, but at levels lower than the fiber saturation point, lies in the concept of 'available water' (*10*). In this case, the more loosely held bound water in the secondary sorption layers at given sorption sites in wood would have thermodynamic properties closer to that of liquid water, thereby allowing the necessary diffusion process to occur. Some of the sorption theories (e.g., BET) and models that have been developed assume thermodynamic properties of some secondary sortion layers equal to or similar to liquid water, particularly at the moisture contents achieved at elevated relative humidity levels (*11*).

Another factor that could have led to confusion regarding minimum relative humidity levels is related to the substrates used in research where these levels are determined. For example, Clarke, et al. (*12*) reported that hightly xeroplilic mold fungi can survive at humidity levels as low as 75%. In this study, samples were collected in a mold infested house by pressing contact agar plates on the affected material, then incubating the plates in the laboratory. The agar plates were evaluated over the course of the study, and where growth occurred, the mold fungus was identified. Given the use of agar plates as the growth substrate, with its accompanying high water content and the associated accessibility of the agar media, the determined minimum relative humidity absolutely has to be lower that that determined with mold growth on wood and wood-based products where the food wouldn't be as accessible. Experiments conducted on agar could be evaluating, for example, the sensitivity to dehydration of surface hyphae and the fruiting portions of the fungus. Other studies have also used growth on agar media to evaluate minimum relative humidity conditions required for growth (*13*), but as indicated, studies have also been conducted on wood where minimum humidity levels of approximately 80% have been reported (*5*). However, even given the results reported by Clarke, et al. regarding minimum relative humidity (*13*), they still report that the most important factor affecting fungal growth on building materials is the local relative humidity and temperature because of their combined impact on the availability of free water for the fungus.

Accessibility of the Food Source

As previously stated, multienzyme complexes are known to be associated with the degradation process of cellulose and hemicellulose in wood (*2*) but non-enzymatic systems are known to be required in association with these enzymes for decay to occur (see Goodell et al. chapter, in this book). Some confusion has occurred, particularly among industrial hygienists, in interpreting potential structural damage to wood-based products from mold fungi that produce cellulases. The presumption is that when cellulase is produced, structural degradation of wood will occur. However, Cowling and Brown (*14*) reported that the enzymes associated with the decomposition of wood were too large to penetrate its gross cellular structure, implying that cellulase enzymes alone were not sufficient to decompose wood. In addition, mold fungi do not control the pH of their micro environments, nor do they produce appropriate non-enzymatic agents which are required for cellulose decomposition. A summary of the findings of Cowling and Brown (*14*, *15*) is shown in Table 1. This work clearly shows that non-enzymatic reactions must accompany, or precede the enzymatic reactions that lead to the depolymerization of cellulose and hemicellulose. As stated by Goodell (*16*), 'The absence or presence of a key component ... may be all that separates degradative ability or mechanism for lignocellulose degradation.'

The data provided in Table 1, combined with the increased understanding of the underlying mechanisms associated with decomposition of wood by decay fungi, help explain the possible sources of food when mold fungi are found on wood-based construction materials. These results clearly show that when mold growth is observed on solid wood products, that this growth is occurring either: 1) on sapwood, where sugars and other low molecular weight carbohydrates are found in parenchyma (storage) cells, or 2) on other organic debris that has settled or been applied to the surface (in the case of some coatings) and which is being used as the food source. With the increased use of faster grown trees, and with the associated relative increase in sapwood content, mold growth can more readily be observed on plywood and oriented strandboard (OSB) products when these are wetted. For aspen-based OSB, this would be particularly true because of the aspen's large sapwood zone. Industrial hygienists involved in indoor air quality litigation have speculated that the adhesive used in OSB has resulted in the observed mold growth, but this, and other more directed research, has indicated that the adhesives commonly used OSB don't contribute to mold growth (*17*).

These results also help explain the occurrence of mold growth on gypsum wall board (Figure 1). Discussions in the field have centered on mold growth on the paper faces commonly used on gypsum products. A factor that some don't consider is the contribution of the starch-based adhesive used to bond the paper to the gypsum substrate. Starch can be readily metabolized by most mold fungi and does not require even the production of cellulase enzymes. The larger pore

Table 1. Size of gross cellular structure of cell wall capillaries in three materials.

Material	Size of opening (Å)	
	Median	Maximum
Water-swollen Wood	10	35
Cotton	5	75
Wood Pulp	25	150

NOTE: Units are Angstrom, Å. The smallest cellulases are approximately 125°A in diameter by 140°A.
SOURCE: Data obtained from Reference 13.

size of the pulp fibers may allow more water to be held in capillaries and the combination of greater access to the fiber and the inclusion of the starch make this an better growth medium for mold fungi.

Common Mold Fungi In Buildings

California and other regions in the western United States have traditionally used non-kiln dried ('green') lumber in construction. Wet lumber is solid piled at the lumber mill and shipped either to the lumber yard or the construction site. In either of these cases, only the lumber on the edge of the unit can even partially dry until the unit is unbundled and the lumber installed in the structure. The moisture content of the sapwood of 'green' lumber is usually in excess of 100% (oven dry basis), and therefore the available moisture is more than adequate to support fungal growth (*18*). With the transition in recent years to second growth trees for most of our construction lumber, a greater proportion of studs and joists contain sapwood and therefore they have more starch and sugars in the sapwood parenchyma cells. These components are a ready food source for the mold fungi. The time lumber spends in a solid unit can be up to four to six months, which is sufficient time for fungal growth to occur. As a result, it is common for new construction to have fungal growth 'built in' (Figure 2).

This author conducted a survey of fungal growth on 'green' lumber at selected construction sites and lumber yards in northern and southern Califorinia in 2001 – 2003. The survey consisted of taking either bulk (short end cuts from the lumber) or tape lift samples from the surface of the lumber. For tape lift samples, transparent tape was pressed against the fungal growth on the lumber surface, and then attached to the inside of a plastic zip-lock bag. These samples were sent to an American Industrial Hygiene Association (AIHA) Accredited Laboratory for identification of mold growth. The most commonly identified fungal species on the 'green' lumber were sapstain fungi in the *Ceratocystis/ Ophiostoma* group. Occasionally *Gonatobotrium* sp. was also identified.

Figure 1. Stachybotrys atra growing on gypsum wall board. The source of water was a leak at the deck to wall interface located above this ceiling space.

Figure 2. The use of non-kiln dried ('green') lumber in construction in California and other portions of the western United States has has resulted fungal growth being 'built in' in many buildings. Note the abrupt transition in fungal growth on this stud, coinciding with the transition from sapwood (fungal growth) to heartwood (no fungal growth).

Gonatobotrium is a mold fungus that lives parasitically on *Ceratocystis/ Ophiostoma*.

Many sapstain fungi have evolved to be dissiminated via bark beetles, and therefore they have developed sticky spores, probably as an adaptation to insect dispersal (*18*). For this reason, evidence of sapstain fungi (pigmented hyphae throughout the sapwood portion of the cross section) is often observed in the log yard, and in fallen trees in the forest. Pigmented hyphae can also be seen in the sapwood portion of processed lumber, as well as dark colored spores on the lumber surface, also limited to the sapwood. Even after the unit is unbundled and the lumber surface dries, these spores are not considered to be as readily airborne as the spores of other mold fungi whose spores are transmitted predominantely via air currents. However, because the spores of fungi in the *Ceratocystis/Ophiostoma* are indistinct, and therefore not readily identifiable alone (i.e., without the fruiting body), they could often be missed or combined in a generic category. Because the spores of these sapstain fungi are held in a sticky mass once released from the asci, many industrial hygienists consider the health implications negligible regarding the common occurrence of these fungi in residential construction.

In order to avoid the occurrence of sapstain fungal growth on 'green' lumber, some lumber mills have started applying anti-sapstain chemical treatments, either in a spray system or by dipping lumber and timbers in a vat. The antisapstain chemicals are intented to limit sapstain fungal activity until the bundled lumber can get to the lumber yard or construction site. Kiln drying is also an option, however, in the west the drying capacity of available kilns doesn't meet the production needs. 'Green' Douglas-fir is still desired by some contractors, reportedly because it is easier to work with (e.g., easier to drive a nail through it). 'Green' lumber is still readily available.

Dry kilns are sometimes used to perform a heat treatment procedure on lumber intended for use in packaging (pallets, etc). Kiln temperatures reached during Heat Treatment (HT), a sterilzation process applied to lumber products used in packaging applications for international trade, are adequate to kill any fungal spore on, or in, the lumber during the drying process. An 'HT' grade stamp indicates that the lumber core reached a temperature of 56°C (about 135°F) for at least 30 minutes, but does not imply that the moisture content of the lumber has been reduced to any specified level. Surface temperatures reached on material with either HT or Kiln Dried (KD) stamp assure that sapstain fungi was killed while in the kiln. If this material is rewetted, newly deposited mold fungi whose spores are dissiminated by air currents will grow. In order to demonstrate this occurrence, this author has conducted fungal growth surveys, again using tape-lift samples, on lumber exhibiting a KD or HT stamp, and also showing evidence of fungal growth. Whereas fungi in the *Ceratocystis/Ophiostoma* group were consistently found on 'green' lumber, they were not observed on KD/HT stamped lumber. A number of mold fungi were identified on these samples For example, a few different kinds of mold were identified on the the stud shown in Figure 3, including *Trichoderma* sp.,

Penicillium/Aspergillus sp., and *Cladosporium* sp. This finding points out the importance of storing HT or KD under cover in the lumber yard and on construction sites. Whereas it is generally thought that the health risk presented by *Ceratocystis/Ophiostoma* sp. is very minor, the same cannot be said regarding exposure to other mold fungi (see Chapter by Jellison in this book).

Summary

Information and data from scientific and non-peer reviewed technical journals was reviewed. Of particular interest was information regarding the minimum moisture requirements necessary to support growth of mold and decay fungi, and enzymatic and non-enzymatic processes that would affect the ability of mold and decay fungi to utilize two of the basic chemical components found in wood (cellulose and hemicellulose) as a food source. Some of the experimental factors that could have led to the differences in published minimum moisture requirements were discussed. These included the difficulty in measuring moisture content and relative humity at a very local level, and the variability of data obtained using agar media to simulate conditions for mold growth on wood and wood-based products.

Results of fungal growth surveys on lumber in lumber yards and on construction sites was also discussed. Sapstain fungi in the *Ceratocystis/ Ophiostoma* group were found almost exclusively on 'green' lumber. Occasionally the mold fungus *Gonatobotrium* sp. was also found. Fungal growth on kiln-dried (KD) or heat-treated (HT) lumber that was rewetted was always from a mold species whose spores were disseminated by air currents.

Acknowledgements

This author gratefully acknowledges the useful suggenstions and contributions of Professors Barry Goodell (University of Maine) and Susan E. Anagnost (Syracuse University of New York) resulting from their review of this manuscript.

References

1. Wilcox, W. W. 1978. Review of Literature on the Effects of Early Stages of Decay on Wood Strength. Wood and Fiber 9(4):252-257.
2. Zabel, R.A. and J.J. Morrell. 1992. Wood Microbiology: Decay and Its Prevention. Academic Press, Inc. New York. 476 pp.
3. Shaughnessy, R.J., P.R. Morey, and E.C. Cole. 1999. Prevention and Control of Microbial Contamination. In: Bioaerosols: Assessment and Control. Ed: J. Macher. American Conference of Governmental Industrial Hygienists, Cincinnati, OH. Chapter 10.

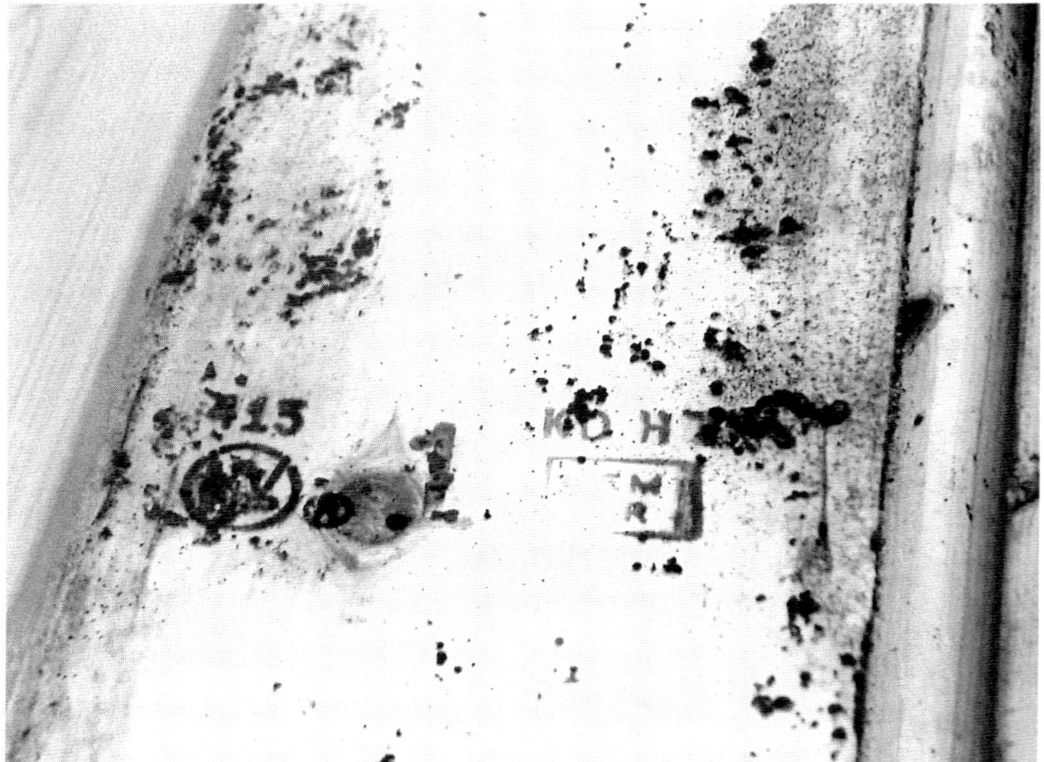

Figure 3. The mold fungus Trichoderma sp. was identified on the surface of this 2x4 that was in an outside storage area in a lumberyard and rewetted by rain. Note the "KD HT" stamp on the lumber.

4. Flannigan, B. and J.D. Miller. 1993. Indoor Humidity and the Building Envelope, In: Bugs, Mold & Rot II. W.B. Rose and A. TenWolde, Eds. National Institute of Building Sciences, Washington, DC. pp. 43-50.
5. Viitanen, H.A. 1997. Modelling the Time Factor in the Development of Mould Fungi – the Effect of Critical Humidity and Temperature Conditionson Pine and Spruce Sapwood. Holzforchung. 51(1) 6-14.
6. TenWolde, A. 1993. Indoor Humidity and the Building Envelope, In: Bugs, Mold & Rot II. W.B. Rose and A. TenWolde, Eds. National Institute of Building Sciences, Washington, DC. pp. 37-41.
7. Wemhoff, P. 2001. Mold, A Poltergeist. Home Energy. 18(1):19-23.
8. Personal Communication. 2001. Professor Barry Goodell, University of Maine, Orono, ME. Dr. Peter Laks, Michigan Technological University, Houghton, MI. Dr. Paul Morris, Forintek Canada Corporation, Vancouver, Bristish Columbia, Canada, Professor Elmer Schmidt, University of Minnesota, St. Paul, MN. Professor Wayne Wilcox, University of California Berkeley, Richmond, CA.
9. Dry Kiln Operator's Manual. 1991. W.T. Simpson, Ed. United States Department of Agriculture Forest Service Forest Products Laboratory. Madison, WI. 274 pp.
10. Personal Communication. Professor John Straube, University of Waterloo, Waterloo, Ontario, Canada.
11. Skaar, C. 1988. Water in Wood. Springer-Verlag, New York. 263 pp.
12. Clarke, J.A., C.M. Johnstone, N.J. Kelly, R.C. McLean, J.A. Andeson, N.J. Rowan, and J.E. Smith. 1999. Building and Environment. 34:515-521.
13. Nielsen, K.F. 2002. Mould Growth on Building Materials. Secondary Metabolites, Mycotoxins, and Biomarkers. Ph.D. Thesis. Danish Building and Urban Research. Hørsholm, Denmark.
14. Cowling, E. and W. Brown. 1969. Structural Features of Cellulosic Materials in Relation to Enzymatic Hydrolysis. Advances in Chemistry Series 95. American Chemical Society, Washington, DC. pp. 152-187.
15. Green III, F. and T.L. Highley. 1997. Mechanism of Brown-Rot Decay: Paradigm or Paradox. International Biodeterioration & Biodegratation. 39(2-3):113-124.
16. Goodell, B. 2003. Brown-Rot Fungal Degradation of Wood: Our Evolving View. In: Wood Deterioration and Preservatation: Advances in Our Changing World, B. Goodell, D.D. Nicholas, and T.P Schultz, Eds. ACS Series 845. American Chemical Society, Washington D.C. pp 97-118.
17. APA – The Engineered Wood Association. 2005. Laboratory Test of Mold Growth on Wood Adhesives. Form No. TT-103. Tacoma, WA. 3 pp.
18. Wood Handbook: Wood as an Engineering Material. 1999. Reprinted from Forest Products Laboratory General Technical Report FPL-GTR-113. Forest Products Society, Madison, WI. FPS Catalogue No. 7269.
19. Webster, J. 1970. Introduction to Fungi. Cambridge University Press. Cambridge, Great Britain. 424 pp.

Chapter 10

The Biology and Microscopy of Building Molds: Medical and Molecular Aspects

Jody Jellison[1], Barry Goodell[2], and Geoffrey Daniel[3]

[1]Department of Biological Sciences and [2]Wood Science and Technology, The University of Maine, Orono, ME 04469
[3]Department of Wood Sciences, Swedish University of Agricultural Sciences, Box 7008, Uppsala, Sweden

Fungi are ubiquitous and essential to the process of organic matter recycling. Molds are microfungi, often associated with damp substrates, and characterized by thread like growth and microscopic spores produced on or in microscopic fruiting structures (*1*). Molds can grow on building materials, but unlike the growth of some macrofungi (notably the basidiomycetes), mold growth does not generally result in wood weakening or decay (*2*). Mold growth does result in a decrease in the aesthetic appeal of wood products and often indicates significant moisture and water issues within a structure. More importantly, mold growth within buildings may, under certain circumstances, pose potential risks to human health. It is this last possibility that has resulted in numerous lawsuits and legislation concerning indoor air quality in recent years. In 2003, 27 state legislatures considered more than 60 items of indoor air quality legislation with 18 passed into law (*3*). In 2004 there were approximately 10,000 mold related lawsuits pending. Policy and litigation however, are often not science-based, so it is appropriate to consider what we know about molds in the indoor environment.

The purpose of this brief review is to introduce information on the biology and physiology of mold fungi, examine the production of fungal allergens and toxins, and consider factors that may impact mold colonization of buildings. A brief overview of possible health implications and of methods used to detect and identify mold fungi is also included.

What are the Fungi?

Fungi belong to their own special kingdom as they differ from both plants and animals. Fungi are eukaryotic, lack chlorophyll, have absorptive nutrition, typically reproduce by spores, and have walls containing chitin. There are more than 100,000 fungal species in the world, with at least 1,000 commonly found in the United States (*1*). Molds and other fungi make up about 25% of the earth's biomass (*4*). The vegetative fungal body is structurally simple and well suited for penetration and proliferation. Molds reproduce by spores. Spores can be less than 5μm in diameter, small enough to be inhaled into the lungs. Asexual spores are called conidia and are important for reproduction, dispersal and survival of the mold. Fungi may exist as single cells such as the yeasts, but typically are multicellular. The living body of a fungus is called the mycelium that is made up of a branching network of filaments known as hyphae. The mycelia are usually hidden in a food source such as wood and we often only know they are present when they develop mushrooms or other fruiting structures. Some fungi only produce microscopic fruiting bodies and we rarely notice them. Fungi are classified based upon phylogeny, reproduction strategies, and morphology. The higher fungi are called the Amastigomycota. Within the Amastigomycota are the Zygomycetes, Ascomycetes and Basidiomycetes (groups which are based upon sexual spore-producing structures) and the Deuteromycetes, sometimes called the fungi imperfecti, which is an artificial grouping often allied to the ascomycetes but with no known sexual state.

The fungi are integral components of recycling processes essential for ecosystem stability. The fungi can also be both diverse and beneficial in their diversity. Interesting fungi include one of the largest living organisms in the world, a specimen of the fungus *Armillaria ostoyae* that covers 1,500 acres in Washington State and may be 400-1000 years old. This fungus grows primarily underground and forms root like structures known as rhizomorphs that help it grow through the soil. Noticeable evidence of this fungus includes the clumps of mushrooms that emerge after fall rains (*5*). Economically important fungi include edible mushrooms and many other foods prepared with the aid of fungi. The yeast *Saccharomyces cerevisiae* is used to ferment sugar to alcohol and carbon dioxide, the process used to make beer and wine and also to make bread rise. The fungi *Aspergillus oryzae* and *A. sojae* are used in the production of foods such as soy sauce and miso. We also use fungi to produce flavorings, vitamins and enzymes, to mature many cheeses, and in the production of antibiotics (*6*).

Although many fungi are useful to people, some fungi are harmful. A few fungi are known to induce human diseases called mycoses as reviewed below. Some fungi produce toxins called mycotoxins, and some fungi are allergens (*7,8*). These aspects of fungal interactions with humans are discussed in this chapter.

Basic Biology

Fungi feed by absorbing nutrients from the organic material that they live in or on. They [partially?] digest their food before they absorb it by secreting extracellular acids, non-enzymatic metabolites and hydrolytic enzymes. Growth occurs via elongation of the hyphal tips. Spores germinate forming first germtubes and then mycelia. The hyphae elongate and branch and colonize the food substrate. Asexual reproduction occurs via specialized spore-bearing hyphae. Fungi are classified by their sexual state, when possible. Molds are primarily imperfect fungi although many are allied to sexual states primarily in the Ascomycetes. Asexual mold spores are referred to as conidia and are important for reproduction, dispersal and survival. Conidiogenesis is the transformation of part of a cell into a separable spore (Fig. 1). The ways in which spores are formed, the arrangement of spores, and spore shape, size, color and septation are all useful in classifying molds and identifying mold spores (Fig. 1). The molds are considered to be multi-celled with genomes of $30x10^6$. The fungal genome is relatively small and has limited repeats.

Because fungi are eukaryotic cells and share many structural and physiological similarities with human cells, mycoses are difficult to treat and pharmaceutical specificity for control is both important and difficult to achieve. Fungal characteristics, which can be used as targets of antifungal therapy, include the unique cell wall, cell membrane, and characteristics of fungal DNA synthesis. Ergosterol rather than cholesterol is the principal sterol in fungal cell membranes. This allows estimation of fungal biomass based upon ergosterol detection and quantification and also provides a potential target for pharmaceuticals. Unlike mammalian cells, fungi have a cell wall whose synthesis can be selectively disrupted. The fungal cell wall is composed of surface mannoproteins, a layer of ß1,3 and ß1,4 glucans. (ß1,3 glucans are inflammatory and 10-30% of asthmatics are sensitive to them.) The layer next to the membrane is chitin. The cell membrane includes ergosterol and ß1,3 glucan synthase. Glucan synthase is localized in the apical tips and growth is apical in the fungi (*9*). Echinocandins can act at the apical tips of hyphae to inhibit wall formation. Echinocandins are active against *Aspergillus* species. Activity against other yeast and molds is less well described or variable. Some compounds may be selectively activated by fungi, arresting DNA synthesis and can also act as anti-fungal agents (*9*).

Secondary Products and Toxicity Reactions

Microbial Volatile Organic Compounds

Fungi produce secondary metabolites, which may include microbial volatile organic compounds (MVOC's) and mycotoxins. MVOC's are an odor-associated blend of alcohols, ketones and other chemical substances. Common

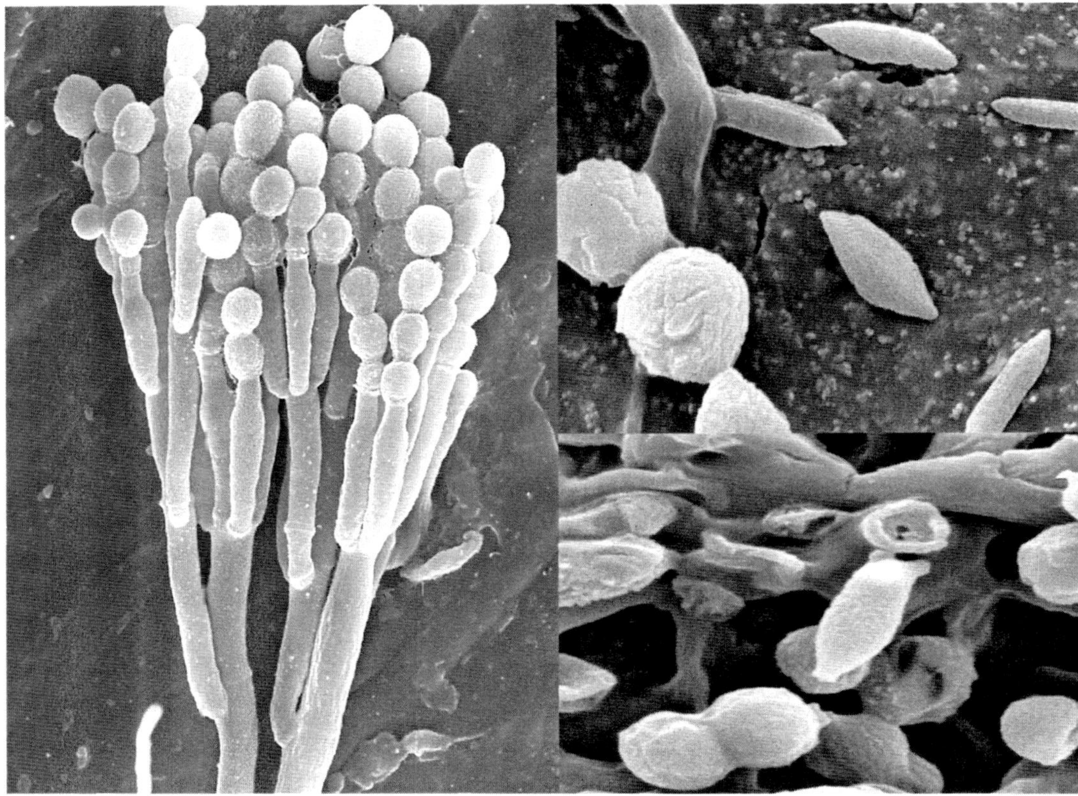

Figure 1. Penicillium *fruiting bodies (left).* Aureobasidium *spores (right).*

MVOC's include 3-methyl-1-butanol, 1-hexanol, 1-octen-3-ol, 2-heptanone, and 3-octanone (*10*). Ethanol is also an MVOC and can enhance the effectiveness of fungal toxins (*11*). MVOC's accompany microbial activity but no single MVOC has been found to be a reliable indicator of fungal colonization of building materials, and characteristic MVOC's are not always produced on wood constituents (*12*). MVOC production depends upon nutrient availability and humidity. Korpi et al.,(*10*) analyzed volatile compounds from mixed microbial cultures (134 fungi and 1 actinomycete) on building materials to see if MVOC's were differentially present in contaminated materials. MVOC's accompanied microbial activity but no single compound was a reliable or specific indicator of colonization by a specific species (*10*).

Mycotoxins

Mycotoxins are chemical products of fungi that have the capability to damage animal health and contaminate crops. Mycotoxins are small peptides, between 200-800 mass units, that can be produced by fungal hyphae and spores growing on damp buildings and other substrates (*13*). Production depends on substrate, water activity level, temperature and other environmental factors (*14,15*). Mycotoxins can interfere with cellular respiration, with carbon and lipid metabolism, and in some cases directly bind to DNA and RNA. Generally, mycotoxins are complex molecules and, unlike MVOC's, are not volatile. A single mold species may produce several mycotoxins and a given mycotoxin may be produced by several different mold species. Not all strains of a given species of fungus necessarily produce mycotoxins and many mycotoxins are not produced until exponential growth of the colony occurs. Mycotoxins entering the body orally can be teratogneic or carcinogenic. Inhaled mycotoxins can be immunosuppressive (*16*). Isolation of a mold species that is capable of producing mycotoxins does not mean mycotoxins are present, or present at an active level. Many species never produce toxins in quantities which effect human health. Others produce toxins only under specialized environmental conditions.

Stachybotrys chatarum is one of the few known building molds to produce mycotoxins that can effect humans. Other mold genera which contain species shown to produce mycotoxins under selected environmental conditions include *Aspergillus, Cladosporium, Penicillium and Trichoderma* (*14*). *Aspergillus versicolor,* for example, has been shown to produce the mycotoxin sterigmatocystin on naturally infested building materials (*17,18*). Repeated human ingestion of aflatoxin (a type of mycotoxin produced by *Aspergillus spp.*) is associated with consumption of contaminated cereal grains and has been linked to liver cancer. Ergot alkaloids, such as the cyclic peptides from Amanitin, are also toxic upon ingestion. *Fusarium spp.* produce a mycotoxin with estrogenic activity. One group of significant mycotoxins is the trichothecenes.

Inhalation of trichothecenes may cause protein synthesis inhibition and immunosuppression (*11*). Trichothecenes are produced by multiple mold genera including *Aspergillus, Cladosporium, Penicillium, Stachybotrys,* and *Trichoderma.* Production is influenced by fungal isolate and environmental factors. Trichothecenes inhibit protein synthesis by interacting with the 60S ribosomes and have a direct deleterious effect on rapidly dividing leukocytes. They can cause skin irritation, vomiting, hemorrhage and convulsions. This group includes T-2 toxin, investigated as a potential biological weapon, which inhibits antibody response and hypersensitivity (*16*). Other known mycotoxins include satratoxin which destroys bone marrow cells, and a carcinogenic toxin produced by *Aspergillus versicolor.* Most toxicity studies have focused on ingestion or inhalation of spore concentrations many orders of magnitude greater than those encountered under normal environmental conditions.

Molds on Building Materials

Mold is ubiquitous in the environment and extensive mold growth is found in areas of high moisture and relative water availability. Water must be available for hydrolysis and diffusion, and it also functions as a solvent for metabolism. Wood, coatings and other construction materials can provide a carbon source and appropriate substrate for mold growth. Mold genera often found on wood and other building materials include the genera *Trichoderma, Aspergillus, Penicillium, Philocephala, Cladosporium, Alternaria, Chaetomium, Stachybotys, Aureobasidium*, and yeasts (Fig. 2).

Molds utilize low molecular weight carbohydrates and other accessible materials. Growth rates on building materials such as pine, plywood and gypsum have been monitored using fungal species such as *Alternaria alternata, Aspergillus niger, Auerobasidium pullulans* and *Cladosporium cladosporioides.* The differential growth of these fungi on the substrates has been partially explained by substrate nutritional content (*19*). The production of mycotoxins by building molds has also been specifically examined. Nielson (*14,15*) studied gypsum chipboard inoculated with selected fungi and observed the production of toxic chaetoglobins by *C. globosum*. Mycotoxin production was not observed for *Paecilomyces* spp. or *Aspergillus ustus*, although both are known to be capable of toxin production. Trichothecene production was observed for four of five *Stachybotrys* isolates but not for any of the *Trichoderma* isolates tested. Shirakawa et al. (*20*) also found mycotoxin production to be environmentally dependent.

Control of Mold in Buildings

As noted above, mold growth is highly moisture dependent. Primary colonizers such as *A. versicolor* will grow at water activities of 0.65-0.70.

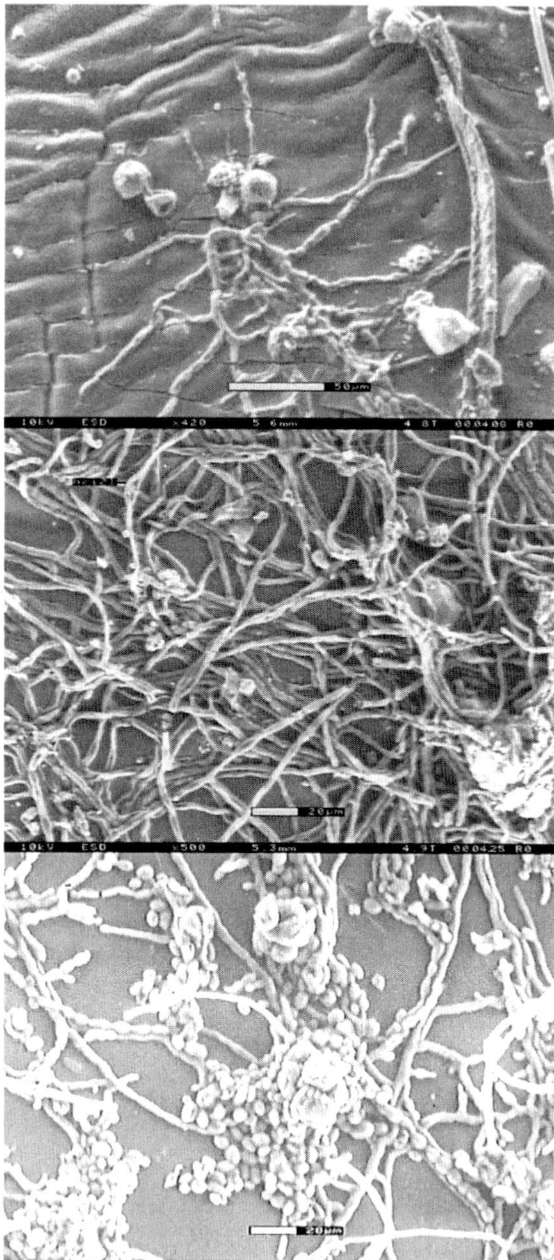

Figure 2. Substrate colonization of construction wood and wood finish coatings by molds.

Secondary colonists such as *Cladosporium* spp. grow at water activity levels of 0.8-0.9 and tertiary colonists such as *Stachybotrys* and *Trichoderma* grow at water activities greater than 0.9-0.95 and are often associated with roof leaks, water pipe leaks or other significant water issues (*14,15*). Water activity levels, however, represent a somewhat artificial measure of moisture availability and thus may inadequately represent the potential for colonization of building material substrates.

Mold mycelia tends to grow on the surface of the wood and do not penetrate deeply down into the wood cells. The terms mold and stain are often used to refer to the same fungi. If these fungi are growing only on the surface they can be brushed, washed, or simply planed off the wood for control. Applications of surface bleach will temporarily control their action. Hyphae are usually not colored, while spores may be colored and are what causes mold-associated discolorations that are typically black, green, or brown. Permeability problems, like those occurring with stain fungi and bacterial attack, can occur. Other problems associated with mold colonization include the detoxification of wood preservatives (*21-25*). For example, species of *Penicillium* are reported to detoxify mercury compounds and *Trichoderma* can detoxify sodium fluoride (*21,22,26*).

Mold growth on materials can be inhibited by borates (*27*). However, biocides can also influence the toxicity of mold spores in cases where protection is incomplete. Murtoniemi et al. (*17*) found microbes that were grown on wetted plasterboard produced bioactive compounds. They examined *S. chartarum* grown on 13 modified plasterboards, with and without biocidal additions and monitored spore cytotoxicity and pro-inflammatory activity. Fungal growth and bioactivity of spores were influenced by minor changes to the core or liners of the plasterboard materials. Interestingly, incomplete biocide protection increased spore toxicity of the remaining molds (*18*).

Significant concerns about indoor air quality have often focused on the presence of molds and mold toxins. Exposure standards for molds and mycotoxins do not currently exist. Much of the work on mold toxicity has focused on the environments of animal- or agricultural workers, where mycotoxin exposure is magnitudes greater than that found in mold-infestested buildings (*16*). Current literature has yet to provide compelling evidence that mold exposure at levels found in mold-contaminated indoor environments results in adverse health effects in otherwise healthy individuals, although allergenic effects have been documented. Mold infestation can, however, be an aesthetic issue and an indicator of significant moisture and water issues within buildings. Molds can also be a significant health issue for immunocompromized individuals and those with certain pre-existing medical conditions.

Health Implications

Molds are ubiquitous and mold presence is not an apriori threat to human health. High concentrations of mold spores can, however, sensitize humans, ultimately producing allergic reactions, and serious mold induced pathology may be seen in immunocompromised individuals. Molds can effect human health in three major ways: via allergic sensitization and immune response; infectious growth in or on the body; and toxic disruption of cellular function and/or interaction with DNA (*16*). A complete medical discussion of fungal allergic and pathogenic effects is beyond the scope of this review, but examples of allergic reactions and selected human mycoses are provided below.

Fungi are often mentioned in the context of "sick building" syndrome, the cause of significant public concerns. Sick building syndrome is associated with headaches and allergic-like reactions to unspecified stimuli, with symptoms apparently arising from allergens or chemicals within the interior of buildings. Molds are always present but the association of specific organisms with reported symptoms is often times difficult to confirm. In 1999 Jacobs et al. (*28*) surveyed homes in 24 cities in the US and found nearly 50% had moisture problems. They found that molds belonging to the genera *Penicillium, Alternaria,* and *Cladosporium* were frequently present. In a 1991 Canadian study (*29*), 38% of houses examined were found to be damp or have mold growth. From these 38%, 50% of the inhabitants had a higher occurrence of lower respiratory symptoms and 25% a higher occurrence of upper respiratory symptoms than inhabitants of other buildings. Epidemiological studies, however, are confounded by multiple uncontrolled factors with potential health impacts. Nielson (*14,15*) found that proteins gave an allergic reaction within minutes of exposure and also stimulated non-IgE mediated histamine release. 1-3 glucans also trigger inflammatory reactions (*30*). Antibodies against mold and mycotoxins were shown to be elevated in individuals after exposure to toxigenic fungi in water damaged buildings. Levels of IgG, M A and E antibodies vs 7 molds and 2 mycotoxin antigens were determined by ELISA for 40 control individuals and 40 mold-exposed patients. IgE levels differed only for the mold genus *Aspergillus* and the mycotoxin satratoxin. Problems were identified in individuals who were predisposed genetically or immunologically, and become sensitized and allergic. Of the 48 subjects exposed to attack, four had elevated antibody levels which indicated possible allergic responses. Researchers, however, also found an elevation of mold-specific antibodies in selected healthy controls not previously exposed to experimentally-elevated mold levels. Young et al. (*31*) noted that glucans had been implicated in organic toxic dust syndrome.

Huttunen et al. (*32*) studied the production of proinflammatory mediators elicitated by indoor air bacteria and fungal spores in mouse and human cell lines. *Aspergillus versicolor, Penicillium spinulosum,* and *Stachybotrys chartarum* were examined for the ability to stimulate production of inflammatory mediators in macrophage lines. Exposure to fungal strains did not markedly increase nitrous oxide production indicating a lack of cytotoxicity. Cytokine production for IL-6 was slightly increased by exposure to *Stachybotrys chartarum.* Spores were less cytotoxic to human cell lines than to mouse cell lines. Their data suggested that bacteria in water-damaged buildings should also be considered as causative agents of adverse inflammatory effects in humans.

Beyond the allergic or inflammatory reactions that may be stimulated by molds, additional interactions between humans and molds include direct infection. Normally molds do not directly infect or cause significant harm to the human body. Although most fungi grow well at body temperature, humans are typically not infected. AIDS, cancer, organ transplants or complications from taking corticosteroids all predispose humans to fungal infection and, in addition, patients often become more susceptible when taking antibiotics or immunosuppressive drugs, particularly subjects with leukemia or AIDS who are taking drugs to suppress their immune system because of organ transplantation. Most patients with a deadly fungal disease do not have a fully functional immune system. Although there are drugs to treat fungal infections, these drugs can have debilitating side effects because they are often toxic to people as well as fungi. Serious health issues associated with mycoses include infections which are superficial (affecting the skin, hair and nails), subcutaneous (affecting the muscle and connective tissue immediately below the skin), and systemic (invasive involving the internal organs). In addition, allergic reactions affecting lungs or sinuses in patients with conditions such as chronic asthma, cystic fibrosis or sinusitis may also occur. Numerous fungi have been implicated as causal agents in serious health disorders of a chronic nature. These include disorders ranging from dermatomycoses such as athlete's foot to serious respiratory diseases such as those associated with *Cryptococcus neoformans* and systemic mycoses such as Valley Fever caused by *Coccidiomyces immitis* (*6*). These organisms are commonly present in the environment and are not specific to mold infestation of buildings.

Molds in the News

Stachybotrys chartarum

Stachybotrys is one of the least common airborne fungi and is found in building materials only 1-2% of the time. It has, however, developed a reputation as a well-known producer of mycotoxins and been the focus of much media attention. In 1994 the Center for Disease Control (CDC) published an article linking the deaths of infants with *Stachybotrys chartarum* (also known as *S. atra*). This organism is known to cause hemorrhaging in horses exposed to moldy straw, but no antibodies to *Stachybotrys* were found in human victims. The situation is further confounded by the fact that not all *Stachybotrys* isolates (only approx. 30%) are toxic. The clinical symptoms in exposed infants were, however, similar to veterinary symptoms such as fatal hemorrhaging, dermatitis and severe respiratory tract irritation seen in animals exposed to moldy straw infected with *Stachybotrys*. Similar clusters of hemorrhaging in humans, however, were not always associated with exposure and some heavily exposed infants showed no symptoms. The CDC's initial article caused significant concern among health officials and the general public. Most molds found in buildings are, however not *Stachybotrys*, and *S. chartarum* can produce mycotoxins under some, but not all environmental conditions. Studies have confirmed the ability of some isolates to cause symptoms at high dosages in laboratory animals (*32,33*). It is less clear how significant the organism is as a factor for affecting human health under normal conditions. The CDC in 2000 revised its position, saying "at present there is no test that proves an association between *Stachybotrys chartarum (Stachybotrys atra)* and particular health symptoms." (*34*).

Aspergillus spp.

Species of the genus *Aspergillus* can cause problems to human health as pathogens, mycotoxin producers and allergens. The mycotoxin ochratoxin produced by *Aspergillus* is teratogenic and causes kidney damage. *Aspergillus* is a genus of around 200 molds and is found worldwide. We normally breathe in 100-200 *Aspergillus* spores daily. Spores are in soil, air, bedding, fans, food, walls, compost etc. There are 185 species known, 20 of which can cause infection. *A. flavus, A. fumigatus, A. glaucus, A. nidulans, A. niger*, and *A. terreus* are the most commonly cited as being problematic. *Aspergillus* spp. grow well at 37° C and produce proteolytic enzymes such as elastase and proteases. *Aspergillus* causes more infections worldwide than any other mold and the genus causes pulmonary infections that can become invasive and spread throughout the body. Aspergillosis is an infection caused by the inhalation of the *Aspergillus* spp. conidiospores. Spores can cause allergic reactions that may cause permanent lung scarring. The symptoms, which include fever, coughing, chest pain and/or breathlessness, can occur in many other illnesses, so diagnosis can be difficult. Usually only patients with already weakened immune systems or who suffer other pre-existing lung conditions are susceptible. Those most at risk of acute invasive aspergillosis include some cancer and leukemia patients, those on chemotherapy and transplant patients. Also at risk are individuals showing immune hyper-reactivity that may suffer from allergic aspergillosis or allergic sinusitis (*35,36*).

Zygomycetes

Zygomycetes are a fungal group characterized by non-spetate or minimally septate hyphae, asexual spore production in sporangia and sexual reproduction resulting in zygospore formation. Zygomycosis is an angiotropic (blood vessel invading) infection produced by various members of the Zygomycete Order Mucorales. These fungi grow well at the average temperature of the human body, 37°C, and they are characterized by rapid growth and a cottony texture. The infections typically involve the rhino-facial-cranial area, lungs, GI tract, skin or other organ systems. The most common genera that cause zygomycoses are

Rhizopus and *Absidia. Rhizopus* produces non-septate or sparsely septate hyphae,that are quite broad (6-15 um in diameter). *Rhizopus* spp. produces sporangia arising opposite rhizoids. The primary *Rhizopus* species that cause health problems are *R. arrhizus, R. oryzae, R. microsporusva,* and. *R. rhizopodiformis*. *Absidia* produces broad and non-septate fungal hyphae that grow at temperatures up to 52°C. It is widely distributed in nature but only one species of the genus (*A. corymbifera*) is a pathogen of humans. When present, sporangiophores arise from points between the rhizoids. Acute zygomycosis is associated with acidic diabetes, and it is also found in severely burned patients and those with immunosuppression diseases (*37,38*).

Patient Diagnosis and Treatment

Diagnosis of fungal diseases often starts by detecting galactomannan antigens in the patient's blood serum. Culturing and identification of molds from the sputum or microscopic examination for hyphae infected tissues mounted in 10% KOH is also commonly performed. Treatments include amphotericin B and oral itraconazole for *Aspergillus* infections, and for zygomycosis, the surgical removal of dead tissue, and intravenous amphotericin B therapy for 10-12 weeks. Amphotericin, a fermentation product of *Streptomyces nodusus*, is an antibiotic particularly effective for systemic mycoses. Amphotericin binds sterols in the fungal cell membrane creating transmembrane channels causing subsequent electrolyte leakage. It is active against most fungi except *A. terreus*. Azoles have also been used for a variety of fungal infections and they function by disrupting fungal ergosterol synthesis. In fungi, the cytochrome P450-enzyme lanosterol 1,4 α demethylase is responsible for the conversion of lanosterol to ergosterol. Azoles bind to lanosterol 1,4α-demethylase, thus inhibiting the production of ergosterol. There is some cross reactivity with mammalian cytochrome p450 enzymes, and issues exist associated with drug interactions and the impairment of steroid production. Other drug types which interrupt ergosterol synthesis include allylamine (Terbinafine) and morpholines (Fenpropimorph, Amorolfine) (*39*).

Mold Detection in the Environment

Currently, a universal protocol to test for the presence of all molds is not available (*40*). ASTM assays have been used to test specific building materials, coatings or other substrates for their ability to support mold growth. For example, ASTM G-21-96 (*41*) can be used to determine the resistance of specific materials to fungal surface growth. G-21 uses a small environmental chamber, where a sample is sprayed with a spore mixture and incubated at greater than 85% relative humidity. The samples are visually rated 0-4 for colonization. This is generally used to test mold resistance against mixed cultures but can also be modified to determine inhibition of specific organisms. ASTM D3273 (*42*) is used to evaluate mold on the surface of interior coatings and also uses a small environmental chamber. ASTM D5590 (*43*) examines the mold resistance of paint films and is a 4-week plate assay. ASTM 5588 (*44*) examines the microbial conditions of paint and the sterility of raw materials using swab samples and culturing onto plates. Other ASTM tests are also available. In addition, the American Wood Preserver's Association (AWPA) is currently developing a laboratory test based upon previous ASTM tests. This new test would entail placing wood or wood composites in a controlled temperature and humidity circulating air test chamber where they are subject to fungal spore colonization for the duration of the test. Acceptable substrate treatments would allow no more than 10 % surface mold growth after eight weeks. No single assay however, is entirely predictive of a material's mold resistance.

Mold detection and prevalence on building materials can be estimated using plate assays, spore counting under microscopic examination and immunological assays. Biomass can be quantified via ergosterol assays (*45*) or microbial cleavage of a fluorescein (*46*). Reeslev *(46),* in quantifying mold biomass on gypsum board, compared mold biomass, ergosterol and enzyme activity (based on beta-N-acytylhexosamidase) and showed a good correlation between ergosterol levels and enzyme activity. Specific metabolites can be quantified using GC-MS or HPLC, and MVOC sampling on carbon-based filters can be used to indicate fungal colonization. Mycotoxins can be quantified using bioassays, chemical assays and immunological assays. When necessary, identification to species can often be achieved by microscopic examination of asexual or sexual spores and fruiting structures.

The enzyme-linked immunosorbent assay (ELISA), developed for use with mold fungi by production of antisera made to fungal cell wall components, fungal metabolites and allergens, can be used to monitor mold levels (*47*). Alternately, ELISA can be used to monitor mold antibody levels in a patient's blood. PCR can be used for the detection and identification of specific species. Fungal species identity can be determined using species-specific primers and PCR amplification or more general primers followed by restriction enzyme analysis and sequencing. Detection of wood colonizing molds for example can be done using PCR amplification using primer pair ITS 1/ITS4 (internal transcribed spacer I, 5.86 ribosomal DNA and ITS II). High copy number ITS regions are highly variable but the primers can be fungal specific. Amplification from colonized wood has been achieved for organisms such as *Aureobasidium pullulans, Ceratocysistis pilifera, Phialocephala fusca, Philophora mutabilis, Trichoderma reesei, T. viride,* and *Chaetomium globosum* (*48*). Specific sequence amplification yields taxonomically, but not necessarily physiologically, relevant information.

Fungal toxicity and immunoreactivity can be examined using multiple assays targeting: Interleukin-6 (an inflammatory mediator), cytotoxicity of mouse macrophage cells' tumor necrosis factor (TNF) and nitric oxide production (via the Griess diazotization reaction). TNF is a proinflamitory factor associated with immune system activation. Macrophages are important in innate immunity and clearance of fungi. Nitrous oxide is a mammalian secretory protein that initiates host defense and is associated with activation of the immune system. The Griess Diazotization Reaction is based upon a multistep colorimetric reaction to produce an azo dye chromophore detected at 540 when a stable product of NO oxidation is present. NO is an indication of cytotoxicity because macrophages can be activated to produce nitric oxide synthetase in the presence of fungi. An enzymatic "kit" for the nitrous oxide system (NOS) can then oxidize L-arginine to yield L-citrulline and NO gas, an important inflammatory mediator. Cytotoxicity is determined using vital dyes and microscopic analysis. Interleukin-6, and TNF can be detected using ELISA based assays (*18*). An overview of relevant methodology for fungal exposure assessment is given by Pasanen (*47*).

Conclusions

Fungi are biologically unique and dependent upon external nutrition and moisture for their metabolic needs. Reproduction and distribution is via sporulation and the spores can at times reach high concentrations in both indoor and outdoor environments. Molds are ubiquitous throughout the environment, and wherever moisture is present they can grow readily on wood and other building materials. Mold growth decreases the aesthetic appeal of wood products and typically indicates significant moisture and water issues within a structure. Mold growth within buildings may also, under certain circumstances, pose potential risks to human health, particularly to individuals with pre-existing medical conditions. Toxicity can differ among fungal isolates, however, and will vary under differing environmental conditions. Substrate/mold species interactions can determine fungal metabolite production and toxicity, and toxicity cannot be inferred merely from mold presence. In evaluating the significance and options for control when confronted with mold contamination it is necessary to take into consideration the specific fungal isolates involved and their physiology and biology in determining any potential course for remediation.

Acknowledgements

The authors gratefully acknowledge the technical and editorial contributions of B. Kitchin and A. Ostrofsky and the financial support of the Maine Agricultural and Forestry Experiment Station. This is publication XXX of the Maine Agricultural and Forestry Experiment Station.

References

1. Hawksworth, D.L.; Kirk, P.M.; Sutton, B.C.; Pegler, D.N. *Ainsworth and Bisby's Dictionary of the Fungi* 8th Edition, International Mycological Institute CAB International , Oxon U.K., **1995**.

2. Daniel, G. In *Wood Deterioration and Preservation: Advances in our ChangingWworld. ACS Symposium Series 8452003* (Eds. Goodell, B., D. D. Nicholas and T. P. Schultz.) ASC Washington DC **2003**, pp 34-72.
3. Ross, A. S. . *Engineered Wood J.* **2004**. http://www.apawood.org/level_befm?content=pub_ewj_arch_f04_mold.
4. Robbins, C.; J. Morrell. *WWPA Mold, Housing and Wood.* Sept. 27, **2002**, http://www.wwpa.org/lumberandmold. Htm
5. Smith, M.I. Bruhn, J.N., Anderson, J.A. *Nature* **1992**, *356*, 428-431
6 Hudler, G. W. *Magical Mushrooms, Mischievous Molds.* Princeton University Press. **1998**.
7. Chapman, J.A.; Terr, A.I.; Jacobs, R.L.; Charlesworth, E.N.; Bardana, E.J. *Ann. Allergy, Asthma and Immunology* **2003**, *91*(3), 222-232.
8. Khalili, B., Bardana, E.J. *Ann. Allergy, Asthma and Immunology* **2005**, *95*(3), 239-246.
9. Griffin, D.H. *Fungal Physiology*, 2nd Edition Wiley-Liss, New York **1994**, 458pp.
10. Korpi, A.; Pasanen, A-L.; Pasanen, P. *Appl. Environ. Microbiology.* **1998**, *64*, 2914-2919.
11. McNeel, S. and Kreutzer, R. *Health and Environmant Digest* **1996**, *10*(2), 2-12. 12.
12. Wilkins, K. Larsen, K., Simkusm *Environmantal Science and Pollution.* **2003**, *10* (4), 206-208.
13. Sorenson, W. *Environmental Health Perspective* **1999**, *107*(3), 469-472.
14. Nielsen, K.F.; Gravesen, S.; Nielsen, P.A.; Andersen, B.; Thrane, U.; Frisvad, J.C. *Mycopathologia.* **1999**, *145*, 43-56.
15. Nielsen, K.F.; Thrane, U.; Larsen, T.O.; Nielsen, P.A.; Gravesen, S. *Internat. Biodeterioration Biodegrad.* **1998**, *42*, 9-16.
16. Robbins, C.A.; Swenson, L.J.; Nealley, M.L.; Gots, R.E; Kelman, B. *J. Applied Occupational and Environ. Hygiene* **2000**, *15*, 773-784.
17. Murtoniemi, T.; Nevalainen, A.; Suutari, M.; Hirvonen, M.-R. *Inhalation Toxicology* **2002**, *14*, 1087-1101.
18. Murtoniemi, T.; Nevalainen, A.; Hirvonen, M.-R. *Applied and Environ. Microbiology.* **2003**, *69*, 3751-3757.
19. Laks, P.E.; Richter, D.L.; Larkin, G.M. *Forest Products J.* **2002**, *52*, 41-44.
20. Shirakawa, M.A.; Selmo, S.M.; Cincotto, M.A.; Gaylarde, C.C.; Brazolin, S.; Gambale, W. *Intern. Biodeterioration Biodegrad.* **2002**, *49*, 293-298.
21. Merrill, W., French, D.W. *Phytopathology* **1966**, *56*, 301-303.
22. King, B., Eaton, R.A., Baecker, A.A.W. *Internat. Research Group on Wood Preservation.* **1978**, Document No. IRG/WP/177.
23. Brown, F. L. *J. Forest Products Research Society* **1953**, *3*, 67-69.
24. Stranks, D. W. and M.A. Hulme. *Material und Organismen Symposium* Berlin-Dahlem **1975**, *3*, 346-353.
25. Verrall, A.F. *J. Agricultural Res.* **1949**, *78*, 695-703.
26. Zabel R. A., and J.J. Morrell. *Wood Microbiology: Decay and its Prevention.* Academic Press., NY., **1992**.
27. Fogel, J.L.; Lloyd, J.D. *Forest Products J.* **2002**, *52*, 38-43.

28. Jacobs, D. E., Friedman, W., Ashley, P., annd McNairy, M. *The healthy homes initiative: A preliminary plan (Full Report) U.S. Department of Housing and Urban Development, Office of Lead Hazard Control,* Washington, D.C. **1999**, 1-82.
29. Dales, R., Burnett, R., Zwanenburg.. *Annual Review Respiratory Disease.* **1991**, *43*, 505-509.
30. Vojdani, A.; Campbell, A.W.; Kashanian, A.; Vojdani, E. *Archives of Environ. Health* **2003**, *58*, 324-336.
31. Young, S.-H.; Robinson, V.A.; Barger, M.; Porter, D.W.; Frazer, D.G.; Castranova, V. *J. Toxicology and Environ. Health, Part A.* **2001**, *64*, 311-325.
32. Huttunen, K.; Hyvärinen, A.; Nevalinen, A.; Komulainen, H.; Hirvonen, M.-R. *Environ. Health Perspectives* **2003**, *111*, 85-91.
32. Mason, C.D.; Rand, T.G.; Oulton, M.; MacDonald, J.; Anthes, M. *Toxicology App. Pharmacology* **2001**, *172*(1), 21-28.
33. Hudson, B.; Flemming, J.; Sun, G.; Rand, T.G. *J. Toxicology Environ. Health Part A.* **2005**, *68*(15),1321-1335.
34. CDC. 2000. *Update: pulmonary hemorrhage/hemosiderosis among infants---Cleveland, Ohio, 1993-1996.* MMWR March 10, **2000**, 49(09), 180-4.
35. The Sanger Institute. *Aspergillus fumigatus Genome Project.* 2004. accessed January13,2005. http://www.sanger.ac.uk/Projects/A_fumigatus/
36. Engelhart, S.; Loock, A.; Skutlarek, D.; Sagunski, H.; Lommel, A.; Färber, H.; Exner, M. *App. Environ. Microbiology.* **2002**, *68*, 3886-3890.
37. Larone, D. H. *Medically important fungi- A guide to Identification, 3rd Ed.* ASM Press. Washington, D.C. **1995**.
38. Ribes, J. A., C. L. Vanover-Sams, D. J. Baker. *Clinical Microbiologyl. Review.* **2000**, *13*, 236-301
39. Borgers, M.; Degreef, H.; Cauwenbergh, G. *Current Drug Targets* **2005**, *6*(8), 849-862.
40. McLeister, D. *Door and Window Magazine* **2004**, August.
41. ASTM. Standard practice for determining resistance of synthetic polymeric materials to fungi. *American Society for Testing and Materials. Standards on Materials and Environmental Microbiology.* **1996**, G21-96.
42. ASTM. Standard Test Method for Resistance to Growth of Mold on the Surface of Interior Coatings in an Environmental Chamber. *American Society for Testing and Materials.* **2000**. D3273-00.
43. ASTM. Standard Test Method for Determining the Resistance of Paint Films and Related Coatings to Fungal Defacement by Accelerated Four-Week Agar Plate Assay. *American Society for Testing and Materials.* **2001**, D 5590-00, 639-642.
44. ASTM. Standard Test Method for Determination of the Microbial Condition of Paint, Paint Raw Materials, and Plant Areas. *American Society for Testing and Materials.* **1997**, D 5588-97, pp630-633..
45. Bjurman, J. *Internat. Biodet. Biodegradation* **1999**, *43*, 1-5.
46. Reeslev, M.; Miller, M.; Nielsen, K.F. *App. Environ.l Microbiology.* **2003**, *69*, 3996-3998.
47. Pasanen, A.-L. *Indoor Air.* **2001**, *11*, 87-98.
48. Jasalavich, C., Ostrofsky, A., Jellison, J. *App. Environ. Microbiology,* **2000**, *66*, 4725-4734.

Chapter 11

Molds and Moldicide Formulations for Exterior Paints and Coatings

K. Hansen

Wood Business Unit, Troy Corporation, 8 Vreeland Road, P.O. Box 955, Florham Park, NJ 07932–0955

Various fungicides are used to control mold in wood products. However, bioactive compounds are facing increasing restrictions worldwide, and this trend will continue. Further, many differences are seen in molds that colonize woody materials around the world. The impact of these differences on moldicide or coating formulation strategies is discussed, along with the water borne and solvent borne biocide combinations currently used to control molds. Other themes discussed include: Molds that have been identified at different sites around the USA and world, and the microclimate effect on mold growth; construction factors to minimize mold growth; data from treated wood panels that were monitored monthly throughout the year for mold growth. In addition, the issue of mold growth correlated to the amount of active ingredient still present in the treatment and the inherent variation in the test results typically observed which together make it difficult to predict treatment service life is discussed. Future moldicide development is also discussed.

Molds, Staining Fungi and Black Yeasts in the Environment on Wood and Coated Exterior Surfaces.

Molds are a term used to describe a large group of different fungi having a well-marked mycelium or spore mass (*1*). Some molds belong to the Class Zygomycetes, such as Mucor and Rhizopus, and cause spoilage of bread and other types of food. Other molds belong to the Ascomycetes, and are mostly seen in their imperfect stages. Molds are able to grow on most carbon containing materials such as wood, leather, plastics, food, paints and many other substrates. On wood and wood treated with paint or wood protection products, staining fungi and black yeasts may appear. Though they taxonomically belong to the Ascomycetes, as many of the molds do, they are often regarded as belonging to separate groups because of distinct differences in their growth pattern. The staining fungi have a high concentration of dark to black hypha growing among the wooden cells, and the wood will appear dark blue to black stained. The black yeasts can grow and look like mold fungi but will often have stages where conidia multiply with buds in yeast like way.

Figure 1. Black mold on a gable in New Jersey. The portion to the right has been cleaned.

Wood is only rarely destroyed by mold growth. For aesthetic reasons, however, many people invest a lot of effort trying to remove and renovate attacked surfaces. One simple strategy could be to apply a black stain or paint on the substrate's surface, and by this way camouflage growth of fungi with dark or black mycelium. However, when lighter colors are desired, more is needed than just trying to camouflage molds growing on the surface. Any paint or penetrating coating applied to the surface needs to possess an inborn resistance against mold growth, which can be achieved by adding a suitable fungicide, or mixture of fungicides, to the formulations. The particular biocide(s) employed are dependent on the location, and the molds that are typically observed in that location. Other factors include the microclimate, the particular application and the additives used in the formulation. For example, hydrophilic products such as cellulose thickeners, surfactants and some pigments increase the amount of fungicide needed. The physical structure of a film itself can result in more water attachment to the surface, which increases the risk for mold attacks. This is especially true for films with a rough surface or many pinholes.

Molds in the USA and Around the World on Exterior Coated Surfaces

If a hydrocarbon-containing product is not sterilized, sealed, frozen or dried, mold fungi by chance will find their way to it sooner or later. This is understandable, due to the huge number of spores and conidia a fungus can produce. A Petri dish containing solidified malt agar and inoculated with the mold fungus *Aspergillus niger* can easily produce 100 million conidia in less than a week. Many other species of fungi are producing similar vast numbers of conidia that easily are carried long distances through the air, even over oceans (*2*). Conidia from *Cladosporium herbarum* e.g. are found in air samples throughout the world (*3*, *4*). Therefore if the right conditions are available a mold will invade a space that is not already taken. *Aspergillus fumigatus* is another cosmopolitic mold with hygroscopic conidia that distributes well in the air masses. This mold had even found its way to a penguin rookery in Antarctica (*5*). Many different mold species are therefore available in the form of conidia in the air at any time. Which species that actually will have success colonizing a newly exposed surface is dependant partly on the substrate and the nutrients provided and partly on the microclimate. The microclimates can vary considerably over short distances or among racks within a test site, and even from panel to panel. For example, the amount of solar irradiation absorbed by a surface depends primarily on its color, which results in a temperature variation between different colors. At one location in Canada the average temperature difference over a year was found to be about 10°F between horizontal placed white and black panels (*6*). This means that water will evaporate slower from a white panel than a black panel and fungi will have water available for a longer time period on the white panel. Therefore, other issues being equal, molds may grow better on a white panel than on an adjacent black panel.

As part of this work, the frequency of fungi, algae, cyanobacteria and lichens on painted/coated surfaces were compared at three test sites in the USA, the east coast (Sparta, NJ), the west coast (Corvallis, OR) and in the south (Miami, FL). Samples of growth were taken from 234 exposed panels. Samples were taken with wetted cotton swabs, or, when possible, pieces of the mold containing film were removed. The swabs were brought to the lab in a transportation medium and the films were transported in sealed plastic bags. Samples were collected from areas with significant growth (more than one cm^2). A species was only counted once on each panel. Identifications were carried out on material taken directly from the collected sample, using a compound light microscope and with reference to relevant literature (*7-18*). The frequency of appearance was listed in Table I.

The most prevalent molds that grew on the coated panels belonged to the genera *Aureobasidium* and *Cladosporium*, followed by *Nigrospora*, *Alternaria*, *Curvularia*, and *Penicillium*. However, major differences were observed in species between test sites. At the Sparta, NJ site *Aureobasidium pullulans* had absolute dominance, occurring in about 90 of the isolates, while the Miami, FL and Corvallis, OR sites had more differentiated growth with two *Cladosporium* species as the most frequently identified molds. Curvularia lunatus and *Nigrospora sp.* were only isolated in Oregon. Together with climatic variations between the sites, the structure and the type of ground located under the test racks also had an impact on the microclimate of the panels. The Miami test racks were placed on a huge lawn where mowed grass was not removed. The Sparta site was on gravel and the Corvallis site in shade over bare soil where some soiling was seen on the test panels.

Based on the molds identified, biocide producers or paint formulators should choose among the following fungi for their lab evaluations: *Aureobasidium pullulans*, *Cladosporium herbarum*, *C. cladosporioides*, *Nigrospora sphaerica* and *Alternaria alternata*. At least *Aureobasidium*, *Cladosporium* and *Alternaria* are commonly used in laboratory testing all over the world. *Trichoderma* is used in some testing especially for applications or locations with high humidity. *Aspergillus niger* is usually not seen on outdoor test panels but appears frequently as an indoor fungus. The fungus is one of the favorites in lab testing not because it is especially relevant, but probably because it is very easy to work with.

Laboratory research to develop a biocide formulation to control all possible worldwide molds will result in extensive work and often lead to confusion as well; e.g., shall a promising biocide be dropped if it fails in laboratory tests against a relatively uncommon organism? One strategy could be to concentrate work around five to eight species that appear with high frequency throughout the world, but keeping a door open for local troublemakers as well.

Table I. Identified organisms and their frequency at three sites in the USA

Mold, algae, lichens, bacteria	*Frequency in pct. of identified*		
	Corvallis Oregon	*Miami Florida*	*Sparta New Jersey*
Molds			
Alternaria alternata	4.4	7.5	2.2
Aureobasidium pullulans	84.5	15.1	88.8
Chaetomium globosum	0.9	1.9	0.0
Cladosporium cladosporioides	15.7	37.7	0.0
Cladosporium herbarum	10.5	0.0	6.7
Curvularia lunata	2.6	7.5	0.0
Drechslera biseptata	0.9	0.0	0.0
Drechslera sp.	0.9	1.9	0.0
Nigrospora sp.	0.9	1.9	0.0
Nigrospora sphaerica	6.1	15.1	0.0
Penicillium sp.	0.9	3.8	0.0
Phoma sp.	0.9	0.0	0.0
Phyrenochaeta sp.	0.9	1.9	0.0
Scytalidium lignicola	0.9	1.9	0.0
Sporidesmium vagum	0.9	0.0	0.0
Stemphyllium sp.	0.9	0.0	0.0
Sterile white mycelium	0.9	1.9	0.0
Trichoderma viride	0.9	0.0	0.0
Basidiomycetes			
Schizophyllum commune	0.0	1.9	0.0
Green Algae			
Actinochloris terrestris	0.0	0.0	1.1
Neospongiococcum sp.	1.7	0.0	0.0
Stichococcus sp.	0.9	0.0	0.0
Cyanobacteria			
Gloeocapsa sp.	2.6	0.0	0.0
Scytonema sp.	0.9	0.0	0.0
Lichens			
Evernia prunastri	0.9	0.0	0.0
Hypogymnia physodes	0.9	0.0	1.1
Usnea subfloridana	0.9	0.0	0.0

To cover as many markets as possible, it would be beneficial to identify common denominators among molds growing around the world, and then use these as test organisms. Towards this end, molds were collected and identified from 5 countries representing the temperate, the sub-tropical and the tropical climate zones on 4 continents. The most common molds identified from various areas are shown in Table 2.

Based on the above, if a global set of molds was to be chosen as test fungi, then *Alternaria*, *Cladosporium*, *Phoma* and *Aureobasidium* would be a reasonable set. For formulations intended for the tropics, *Phoma* should not be omitted as a test fungus. This mold is frequently seen in temperate climates; however, it probably does not have the same importance as in the tropics. It should be recognized that developing biocides that are effective against specific fungi might well prove an effective strategy to eliminate the growth of those fungi on target substrates. However, if biocidal specificity is too narrow, eliminating one fungus may simply create an environment where a different less common mold species, that is more resistant to the biocide, is created. Caution should therefore be exercised in developing selective biocidal treatments and it is not the intent of this chapter to suggest this strategy.

How to Avoid Mold Growth

If free water is kept away from the wood, the moisture content will usually remain below 20% and, consequently, there will be no mold growth. There are various ways to keep water away from construction: Avoid flat roofs, because there is a risk that over the years the roof membrane will become damaged and water will easily find a way into the construction. It is important to be careful with the site grading when gutters and downspouts are installed and later to keep the gutters clean. Flashing around windows, doors and chimneys must also be installed in a way that water traps are avoided.

Keeping water from penetrating into the wood through paint or other wood coatings and water repellent film can be difficult, since water vapor and even free water are able to penetrate directly through most films. Wood is hydrophilic and thus will absorb moisture through the film during periods of high humidity while water will evaporate during dry periods. This will result in swelling of the wood in times of high rainfall and humidity, and shrinking of the wood when the humidity is low. Wood will shrink from a few percent to more than 10 percent when the wood goes from saturated to completely dry, depending on the wood species and the direction of measurement relative to the grain orientation.

In reviewing water penetration properties of common wood coatings, acrylic films tend to allow more water to penetrate through the film than alkyd-based films. Polyurethane coatings can be totally impermeable; however, if the film

Table 2. Most Important Molds Identified Throughout the World

Molds	*Seelze (Germany)*	*Miami, Oregon, Sparta, (USA)*	*Kuala Lumpur (Malaysia)*	*Kabinburi, Trang, (Thailand)*	*Mongagua, Gravatai Rio G. Belem (Brazil)*
Alternaria	7	3	15	4	7
Aureobasidium		44		7	6
Cladosporium	24	25	8	11	6
Curvularia					7
Fusarium					6
Nigrospora		5			
Penicillium	2				
Phoma			18	59	13
Stemphyllium	2				
Other molds and algae	65	23	52	19	55

cracks due to normal swelling and shrinking, water will have easy access to the wood, where it will accumulate underneath the film, leading to more wood swelling and fungal growth.

The film itself can contain readily available carbon food sources to sustain fungal growth, and it is therefore often highly vulnerable to mold growth. A typical film contains compounds such as softeners, emulsifiers, defoamers, thickeners and detergents that provide a nutrient source for mold growth. This is the reason the first attacks on an unprotected film are often observed after only a few weeks of exposure. Then there often follows a period with less vigorous growth, since many of the formulation additives have been leached out and the substrate is depleted of nutrients. When the film starts to crack and fail after two to three years, nutrients from the wood underneath become accessible and a new wave of mold attacks occurs.

Biocides

When all constructional precautions have presumably been taken, it is still important to incorporate a coating or finish to exposed wood to inhibit mold growth. Adding fungicide, water repellent and UV protectant to a coating formulation best accomplish the task of protecting exposed wood. The list of available permitted fungicides is, for environmental reasons, being reduced yearly. Table 3 shows the most common current fungicides used to protect wood and paint films against mold fungi:

Table 3. Biocides commonly used for wood preservation or film protection

Wood	*Film*		*Name of Compound*
Fungi	*Fungi*	*Algae*	
X			Boric Acid
X			2-bromo-2Nitropropane-1-3diol (Bronopol)
X			α-[2-(4-Chlorophenyl)ethyl]-α-(1,1-dimethylethyl)-1H-1,2,4-triazole-1-ethanol (Tebuconazole)
X			Copper naphthenate
X		X	4-5-dichloro-2-n-octyl-4-isothiazoline-3-one (DCOIT)
X	X		1-[2-(2.4-dichlorophenyl)4-propyl-1.3-dioxolan-2-ylmethyl]-1H-1,2,4-triazole (Propiconazole)
X			Di-sodium octaborate tetrahydrate (DOT)
X	X		Diiodomethyl-p-tolylsulphone
X	X		3-iodo-2-propynyl butylcarbamate (IPBC)
X	X		Methyl benzimidazole-2-ylcarbamate (carbendazim)
		X	N-(3,4-dichlorophenyl)-N,N-dimethylurea (Diuron)
		X	N-cyclopropyl-N-(1,1-dimethylethyl)-6-(methyltio)1,3,5-triazine-2,4-diamine (Irgarol)
X	X		N-dichloroflurometh yl thio-N,N-dimethyl-N-phenylsulfamide (Dichlofluanid)
X	X		N-dichlorofluoromethyl thio-N,N-dimethyl-N-tolylsulfamide (Tolylfluanid)
X	X		N-trichloromethyl thiophthalimide (Folpet)
	X		2-octyl-2H-isothiazol-3-one (OIT)
X			Quaternary ammonium compounds (BKC, DDAC, TMAC)
		X	N^2-tert-butyl-N^4-ethyl-6methylthio-1,3,5-triazine-2,4-diamine (Terbutryn)
	X		Tetrachloroisophthalonitrile (Chlorothalonil)
X	X		2-(thiazol-4-yl)benzimidazole (Thiabendazole)
X			2-(thiocyanomethylthio)benzthiazole (TCMBT)
	X	X	Zinc bis (dimethyldithiocarbamate) (Ziram)
X			Zinc borate
X			Zinc naphthenate
	X		Zinc oxide

Solvent and Water Based Biocide Systems

A perfect biocide would be a compound that would be easy to dissolve in a liquid and when deposited at its target substrate, would withstand the sun and leaching by water. Most organic biocides can be easily dissolved completely in a solvent-based formulation. Compromises are sometimes seen, for example, when only 0.5% of a biocide can be dissolved in a formulation that requires 1.1% a.i. to work. If 1.1% is used, the product will contain both dissolved and undissolved biocide. This is often seen when Folpet is the biocide of choice. Some biocides can only be used in solvent based formulations, such as Dichlofluanide and Tolylfluanide that hydrolyze upon contact with water. Practical knowledge about how to utilize a biocide optimally increases with the length of time the compound is commercially employed. For example, certain fungi may build up resistance against a biocide, or it is found that the biocide may be weak against a particular type of molds. As a consequence, more than one biocide is often used in a formulation. However, not all biocides can be mixed with each other. In Tables 4 and 5 the biocides most often are used in mixed formulations are listed.

Table 4. Fungicide-combinations for solvent-based systems

Copper naphthenate + Dichlofluanide
Coper naphthenate + Folpet
Copper naphthenate + Tolylfluanide
Dichlofluanide + Propiconazole
Dichlofluanide + Tebuconazole
Folpet + IPBC
IPBC + Propiconazole
IPBC + Tebuconazole
IPBC + Propiconazole + Tebuconazole
Propiconazole + Tebuconazole
Propiconazole + Tolylfluanide
Tebuconazole + Tolylfluanide

Other Organisms That Grow on Wood Surfaces: Cyanobacteria, Green Algae and Lichens

In reviewing the growth of mold fungi on garden and other exterior building elements exposed to high moisture, cyanobacteria, green algae and lichens must

Table 5. Fungicide-combinations for water-based systems

Fungicide combination
Carbendazim + Chlorothalonil
Carbendazim + DCOIT
Carbendazim + IPBC
Carbendazim + OIT
Carbendazim + Quaternary ammonium compounds
Carbendazim + Zinc oxide
Carbendazim + Zinc pyrithione
Carbendazim + Ziram
Chlorothalonil + DCOIT
Chlorothalonil + IPBC
Chlorothalonil + OIT
Chlorothalonil + Propiconazole
Chlorothalonil + Zinc oxide
DCOIT + OIT
DCOIT + OIT + Zinc oxide
DCOIT + IPBC
IPBC + OIT
IPBC + Quaternary amm. Comp.
IPBC + Quaternary amm. Comp. + Propiconazole
IPBC + Thiabendazole
IPBC + Propiconazole
IPBC + Propiconazole + Tebuconazole
IPBC + Zinc oxide
IPBC + Zinc pyrithione
IPBC + Ziram
OIT + Zinc oxide
Propiconazole + Quaternary amm. Comp.
Thiabendazole + Ziram
Zinc oxide + zinc pyrithione

also be discussed. Cyanobacteria and green algae often play a major role in the initial colonization of surfaces that are moist for long periods. Later in the succession process as when surfaces have been exposed for years, lichens also appear. Without the aid of a microscope, it is often impossible to determine if a growth consists of Cyanobacteria or mold or if both groups are represented at the same time, since both organisms often appear dark green to black when dry (*19*). Figure 2 shows part of a bench with growth that looks like it was attacked by mold fungi. However, examination by microscope showed that the growth was a mixture of both mold and blue green algae.

Figure 2. Garden bench with mixed growth of both mold and cyanobacteria

Algaecides are normally added to a formulation together with a fungicide and, for waterborne systems, a bactericide as well. The five algaecides listed in Table 3 can be regarded as herbicides. The formulations, therefore, must be of a quality that will minimize leaching as much as possible to avoid damage to the surrounding vegetation. They are, in general, easy to formulate, although an algaecide such as Terbutryn must be formulated with care to avoid odor problems due to its methyl thio group.

Formulations

Because very few actives are available today, only a few combinations are possible. To further improve performance, strategies that involve adding non-biocidals need to be considered:

- Avoiding hydrophilic ingredients in the paint formulation.
- Reducing/eliminating pinholes by using efficient bubble breakers and reducing the viscosity.
- Ensuring that enough of the product is applied on the substrate for an adequate film thickness; e.g. 80 – 100 μm.

- Applying on a clean surface. Wood unprotected for only a few days will be loaded with spores and conidia that may grow later when conditions are right.
- Keeping the film as dry as possible by incorporating wax, silicones or other water repellents into the formulation.
- Protecting the biocide by incorporating UV-absorbers in the formulation.
- Using pigmented systems, and remembering that the color choice may affect the surface temperature, and thus have an influence on which microorganisms may attack the film.
- Most acrylic films are more difficult to protect than alkyd-based films, due to differences in water permeability.

Testing of Stains and Paints Against Mold Fungi in Outdoor Exposure Tests

A common way to compare coatings is to apply paints and wood protection products on wood panels. These are then exposed to outdoor conditions on racks at a test site. The panels are inspected periodically for the paint's physical performance and for the degree of mold growth. For example, a three-foot panel can be divided into four parts. The first portion is coated with paint without biocides, and remaining portions are coated with paint plus 0.1%, 0.3%, and 0.5% of the chosen fungicide. Normally two replicates are used and the panels are inspected periodically over the next two years and rated for fungal growth. In the US, most field test ratings for field performance follow the ASTM standard D3274 Revision 95 with the scale going from 0 to 10, where 0 is completely overgrown and 10 is free of growth. In Europe a reversed scale is used that goes from 0 to 5, where 0 is free of growth and 5 is completely overgrown.

There is no "pass" or "fail" criterion with these scales, and most evaluators are uncomfortable with that. It is therefore typical for evaluators to use a "pass-fail" point, which in the US is typically "7" and in Europe "2". The definition of the "pass-fail" point will vary from company to company, but is often defined as the point where a layman may think it is time to clean or repaint the panel. Using this technique, the "pass-fail" point will soon be the most important point on the scale. The problem, however, has always been to agree on the exact borderline between "pass" and "fail", as the visual rating is subjective and each person has a different standard for when a surface is no longer acceptable.

Complicating the issue, inherent test variations are normally large in outdoor exposure tests (*20*). This is often seen when blank controls are randomly placed on a rack and ratings may vary from "2", almost completely overgrown, to "9", almost clean. As a consequence of such a variation among controls, the whole test may not be valid. To overcome the problem with very heterogeneous data sets, it is necessary to randomize samples using procedures such as a Completely Randomized Design (*21*, *22*). Using this method the number of replicates is determined by the variation in results between samples of same treatment. This, of course, cannot be known before the test has been completed and it is therefore necessary to rely on earlier results to estimate the number of replicates needed, which will often be around 5 and never less than 3.

To obtain useful data, outdoor exposure can take from as little as one month to more than three years, depending on how severe the conditions are at the test site. Over the test period the degree of differentiation in mold growth between treatments will increase until a maximum is reached (Point of Maximum Differentiation). Later, when the average rating (ASTM) for the set falls to around four, the difference between treatments will decrease again and eventually disappear when the samples become totally overgrown.

Seasonal Variations in Ratings

In climates with distinct seasons, panel ratings often appear to improve over the winter months. This is because the fungi stop growing in the cold season and get rinsed off by snow and rain. To see this effect demonstrated, panels can be inspected at short intervals, such as monthly. Following the winter season, the molds will again begin to grow and ratings will start to decrease again shortly after springtime (Fig. 3). Typically in the second season less improvement is seen during the winter months, and the drop in ratings is more dramatic during the summer and the following autumn.

Comparison Between Rating Methods

It is often desirable to compare the visual subjective evaluation to a more objective evaluation by relying on a quantitative measurement of the biocide level in the film. High Performance Liquid Chromatography (HPLC) is a common analytical method for organic biocides. Neutron Activation Analysis (NAA) requires only a small sample and is precise, but one drawback is that it only measures the elements present rather than the active compound. In some situations, the biocidal compound itself may degrade but the specific compositional element may still be present on the substrate surface. An alternative method is the use of bioassays, where a fungal response to a sample containing a biocide is measured. The sample, which could be a disc cut out of a board with a plug cutter, is placed on nutrient agar with the painted surface toward the agar and sprayed with fungal conidia. The clear zone of inhibition between the growing mold fungus and the sample disc is measured. The method is well suited for biocides that are not strongly chemically bound to the substrate. Some leaching of the biocide into the agar is needed to obtain a measurable zone of inhibition (*23*). It is important to use a fungus that is very sensitive to the biocide, in order to differentiate between biocide concentrations. Typically the best results are obtained with fungi having minimum inhibitory concentration (MIC) values close to 1 ppm.

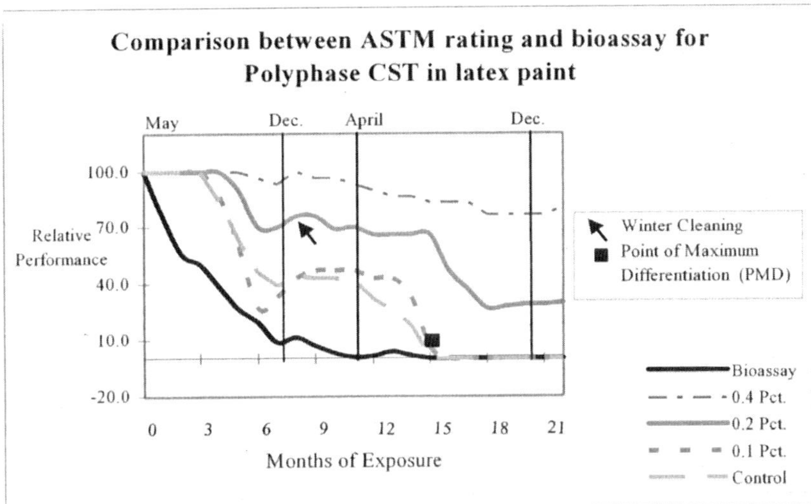

Figure 3. Polyphast CST is an UV-stabilized product containing 20% IPBC. This bioassay was carried out on a paint film initially containing 0.4% Polycast CST. Relative performance: The ASTM 1 to 10 rating scale was converted to relative performance by multiplying by 10, and the bioiassay was converted by regarding the maximum zone of inhibitaion as 100 and no zone as 0.

In Figure 3, the ASTM rating was compared to a bioassay test carried out with *Aspergillus niger* on 66.5 mm^2 discs cut from the exposed panels. *A. niger* was chosen as test fungus because it was easy to work with, and because it was very sensitive to even small concentrations of IPBC with a MIC value between one and two ppm. A concentration dependent response for Polyphase CST was observed; however, at a level of 0.1% the fungicide did not perform better than the control. Specifically, the bioassay showed no zone of inhibition after 16 months, indicating a low level of biocide in the film. However, 0.2 and 0.4% Polyphase CST still performed satisfactorily after 16 months. The reason was probably that even very low levels of the biocide were enough to prevent growth in a film where nutrients, such as emulsifiers, water-soluble carbohydrates and detergents had been leached out.

The bioassay method proved to be an easy and relatively precise method to monitor the biocide level in a film over time. The paint performed well even when the bioassay indicated a low level of biocide in the film. It was easy to find the "Point of Maximum Differentiation". It was easy to determine which months the fungi thrived as well and when molds were inactive during the winter months.

The Future

When the number of biocides started to shrink about 25 years ago, it was relatively easy to change to alternative biocides. Since approximately 10 years ago, however, very few new biocides emerged and, at the same time, fungal resistance developed against some of the most common biocidal compounds. This initiated a period of time where formulators began to combine the biocides available to ensure broad fungal efficacy, combat resistance or just to increase the general efficiency of a product. Regulatory trends suggest that there will be even fewer biocides permitted in the future; however, the need for biocides will remain. The consequences of this are that most paints and stains will rely on the same few actives for mold protection. However, this is not the end of development work. Formulators are already trying to enhance the quality of formulations to develop products that will provide long-lasting hygroscopic films with dependable and desirable properties while minimizing use of the remaining approved organic biocides.

References

1. Ainsworth, G.C. *Ainsworth and Bisby's Dictionary of the Fungi*; Commonwealth Mycological Institute, Kew, 1960.
2. Pady, S.M.; Kapica, L. *Mycologia* **1955**, *47*, 34–50.
3. Hudson, H. J. *Trans. Br. Mycol.Soc.* **1969**, *52*, 153-159.
4. Hyde, H. A., Williams, D. A. *Trans. Br. Mycol. Soc.* **1949**, *36*, 260–266.
5. Wicklow, D. T. *Can. J. Microbiol.* **1968**, *14*, 717–719.
6. Stephenson D. G. *Can. Building Digest* **1963**, *47*, 1–6.
7. Domsch, K. H. , Gams W. and Anderson T. H. *Compendium of Soil Fungi.* 1980, Academic Press, London.
8. Wang, C. J. K. and Zabel, R. A. *Fungi from Utility Poles*. 1990, American Type Culture Collection, Allen Press, Inc., Kansas.
9. larone, D. H. *Medically Important Fungi* 1995, ASM Press, Washington, D.C.
10. Ellis, M. B. *Dematiaceous Hyphomycetes*. 1993, CAB Int., Wallingford.

11. Ellis, M. B. *More Dematiaceous Hyphomycetes.* 1993, CAB Int., Wallingford.
12. Barron, L. B. *The Genera of Hyphomycetes from Soil.* 1977, Robert E. Krieger Publishing Co. Inc., New York.
13. Gams, W. *Cephalosporium – artige Schimmelpilze.* 1971, Gustav Fischer Verlag, Budapest.
14. Hanlin, R. T. *Illustrated Genera of Ascomycetes*, Vol. I and II. 1992, APS Press, St. Paul, Minnesota.
15. Arx Von, J. A. *The Genera of Fungi Sporulating in Pure Culture.* 1974, J. Cramer, Vaduz.
16. Holt G. J. et al. *Bergey's Manual of Determinative Bacteriology.* 1994, Williams and Wilkins, Baltimore.
17. Hanus, E.; Gärtner *Süsswasser Flora von Mitteleuropa.* 1988, Gustav Fischer Verlag, Stuttgart.
18. Prescott G. W. *Freshwater Algae*, 1978, McGrawhill, Boston.
19. Hansen, K. *PPCJ*, **1997**, *187*, 21–23.
20. Colon, I., Kuusisto, E.-L.; Hansen, K.; Location affects performance of biocide – containing paints; 2004, *PCI*, 11.
21. Steel, R. G. D. and Torrie, J. H. *Principles and Procedures of Statistics* 1960, McGraw-Hill Book Company, Inc, New York.
22. Snedecor, G. W.; Cochran, W. G. *Statistical Methods.* 1968, The Iowa State University Press, Iowa.
23. Morrell, J. J. *Wood and Fiber Sci.* **1987**, *19*(4), 388–391.

Chapter 12

The Changing Landscapes of Mold Litigation

Michael J. Pietrykowski, Constantine M. Panagotacos, and Marisa A. McKeown

Gordon & Rees LLP, 275 Battery Street, Suite 2000, San Francisco, CA 94111

In the late 1990's, concern about "toxic mold" transformed traditional construction law, where the mold claim was a minor and ancillary aspect of the suit, into high-stakes personal injury litigation. This new mold litigation involved expanding parties and claims and increased dollars. This chapter explores these past developments with a preview of future trends in mold litigation.

Mold Litigation: How It All Began

Mold claims are not a novelty as they have been a component of larger construction defect or property damage cases for many decades. These cases, falling within the purview of construction law, focused on the property damages resulting from water intrusion. The scope of mold abatement and the resulting cost were marginal because there was little to no concern about health and safety.

In the late 1990's, mold litigation took a new turn, focusing on mold's effect on humans. It began with a study in 1994 by the Center for Disease Control and Prevention (CDC) which investigated a series of pulmonary hemorrhages in infants in Cleveland, Ohio (*1*). The CDC issued a preliminary report that linked acute pulmonary hemorrhage among infants to exposure to *Stachybotrys* in homes. After the CDC implicated *Stachybotrys* exposure as the cause of the infants' illness, however, in 2000, the CDC took the very unusual step of retracting its endorsement of the earlier reports, citing faulty methodology (*2*). However, the "bell was rung" and this purported causal link between toxic mold and toxic responses in humans prompted significant response by the media and the public.

Public Response: More Cases, More Money

Following the release of the 1994 CDC study, a perfect storm unfolded that set the stage for mold litigation. A mold media frenzy was created by sensational news articles, popular movies, and reports of multi-million dollar lawsuits (*3*). The news media began reporting the problems associated with mold which created an inevitable "mushroom" effect. One objective report in a mid-page section of a newspaper quickly turned into front page news on the cover of national magazines like Newsweek and Time and featured on primetime television networks like CBS's 60 Minutes (*4*). The introduction of the internet made the instant transfer of information, often inaccurate, possible. Quickly, public perception had evolved, causing the public to think of mold as "toxic" and deadly.

In 2001, in the wake of the mold media frenzy, Melinda Ballard, a Texas plaintiff, obtained a record-setting mold jury verdict (*5*). Ballard filed suit claiming that her home had been infested by mold, particularly stachybotrys, after a series of plumbing leaks. A Texas jury awarded Ballard and her husband $32.1 million for her mold claim (*6*). When news of the verdict broke, a media frenzy continued, prompting mold claims to "flood" American courts in unprecedented numbers. Although mold litigation surged in warmer-climate regions like Texas and California, the litigation frenzy became national – as evidenced by the largest mold lawsuit of 2002, which involved a 400-resident apartment complex in New York City where the plaintiffs sought $8 billion in damages as part of a class action lawsuit (*7*).

Heavy media attention regarding the possible health effects of mold drew awareness to the legal issues raised by mold (*8*). As the news of "toxic mold" hit the public, the public responded by going to the courthouses with scores of lawsuits based on alleged mold-related injuries. For example, the number of mold claims to a large insurance carrier in Texas rose from 12 claims in 1999 to approximately 12,000 in 2001. From the beginning of 2000 to the end of 2001, mold filings in Texas increased sharply – up by 1,300% according to the Insurance Information Institute (*9*). Also according to the Insurance Information Institute, insurers paid $3 billion in mold claims in 2002, more than double the $1.4 billion paid in 2001 (*10*). Not only did the big verdicts and media attention spark more litigation, but the magnitude of the damage claims and verdicts also increased tremendously (*11*).

Expanding Parties and Claims

With the advent of the personal injury claims, the scope and type of parties involved in mold litigation expanded. Mold cases were being prosecuted by property owners, renters, students, teachers, office workers, judges, and a seemingly endless stream of new plaintiffs. Defendants are no longer just developers or subcontractors involved with building projects, but rather, defendants now include architects, landlords, tenants, property managers, industrial hygienists, homeowner's associations, abatement contractors, insurance companies and lawyers.

Simultaneously, the mold personal injury claims have greatly expanded the different types of claims and the potential for damages. Common causes of action in mold lawsuits include: negligence, concealment, breach of contract, fraud, emotional distress, construction defects, implied warranties of habitability and professional liability. Very few, if any, of these claims were present prior to the 1990's.

The Pendulum Swings Back

Human thought is like a monstrous pendulum, it keeps swinging from one extreme to the other (*12*). Not surprisingly, after nearly ten years of progressive growth in mold litigation, the pendulum began shifting back. Medical literature was published admitting the lack of studies to support the far reaching claims of toxic exposure. Given the lack of scientific evidence proving that mold causes toxic, rather than allergic, health effects, mold cases became more difficult to prove. Plaintiff's were having difficulty proving the necessary element of causation because some courts were refusing to accept expert testimony linking mold exposure to the symptoms of plaintiffs. The exclusion of expert testimony proved fatal to many plaintiffs' cases.

Evidentiary hurdles have caused major setbacks for even the most high-profile mold plaintiffs. Melinda Ballard, the Texas plaintiff who received the landmark $32.1 million mold verdict, ran into trouble when defendants challenged the verdict. On appeal, defendants argued that the large punitive damage award against the insurance provider was unconscionable. The appellate court held that expert testimony regarding mold-related personal injuries was inadmissible and the damages should be capped at $4 million in actual property damage. The appeals court left intact the $4 million award for actual damages but threw out $17 million for mental anguish and punitive damages (*13*).

Anatomy of a Mold Lawsuit

Plaintiff Strategy

The new mold litigation bred new and creative theories by plaintiffs' counsel. In recognition of potential problems associated with causation, talented plaintiffs' attorneys build up the damage claim and play on the fears and sympathies of the public and the jury. In this context, the natural ally of plaintiffs and their counsel is the media. The media (national, local, internet and even Hollywood) is constantly reporting, often exaggerating, about the "killer mold" in our homes, schools, work places, etc. (*14*).

Plaintiffs also focus on fraud, or like, claims in order to bolster damages. For cases of fraud, plaintiffs' counsel assert that a building owner or property manager knew, or should have known, that there was mold in the building and failed to advise the residents of the existence of mold and the corresponding health-related issues. Instead of a simple case with temporary exposure, general symptoms and relatively minimal damages, the case becomes a classic tale of the big defendant jeopardizing the health of the young and old for the purpose of economic gain.

Defense Strategy

For the defendant, the need to develop simple and appealing themes is as critical as it is to the plaintiff. A threshold decision for defendants is whether to accept the alleged damages such as toxic mold induced injuries as "real." In many cases, plaintiffs will make a showing of a temporal relationship between exposure and illness. However, this showing of exposure and subjective symptoms may be insufficient evidence to establish a prima facie case of causation. Courts require expert testimony to establish causation between the alleged exposure and injury (*15*). Plaintiffs, defendant's argue, are not entitled to proceed without competent expert testimony to establish the causal link (*16*). One major defense strategy, as explained below, is to challenge the reliability of plaintiffs' scientific evidence in order to cast doubt on plaintiff's alleged injuries.

Proving The Case: Evidentiary Hurdles

Scientific methods must meet certain standards, or the data derived from those methods cannot be admitted at trial (*17*). The critical question for mold related personal injury claims is whether there is sufficient scientific support to make the link between the alleged exposures and mold-related injury (*18*). The presence of elevated mold levels, alone, is not enough to clear this evidentiary hurdle. Likewise, the presence of an injury, even though it may be logically plausible, is not sufficient to allow a claim to proceed.

Rather, courts require that plaintiffs' claims be supported by acceptable medical and scientific evidence. The Supreme Court in *Daubert v. Merrell Dow Pharmaceuticals* established that expert witness testimony is only admissible if it is relevant and considered reliable in the scientific community (*19*). The Court recognized that trial judges serve as the "gatekeepers" of claims regarding the reliability of methods used by such experts, and judges shall preclude conclusions based on uncertain scientific methodology. In 1999, the Supreme Court further expanded this standard in *Kumho Tire v. Carmichael* (*20*). In Kumho Tire the Supreme Court determined that a trial judge's gate-keeping obligations to ensure that expert testimony rests on a reliable foundation applies not only to scientific testimony, but rather to all expert testimony based on technical or specialized knowledge (*21*).

State courts have followed suit in adopting standards requiring reliability prior to admissibility of expert testimony. In California, for example, the standard for admitting expert testimony was initially established in *People v. Kelly* (*22*). In Kelly, the California Supreme Court adopted the evidentiary standard set in a federal case called *Frye v. U.S.* (*23*). Pursuant to this decision, often referred to as the Kelly/Frye standard, scientific evidence is only admissible after a showing of reliability demonstrated by substantial agreement in the scientific community. Courts are required to ascertain an expert's methodology and determine whether the methodology has been generally accepted within the expert's scientific discipline. In short, the connection between plaintiff's exposure and the alleged injury must be supported by acceptable medical and scientific evidence. Courts may also require plaintiffs to establish their causal link in an evidentiary hearing before trial (*24*).

Several mold cases have turned on the issue of expert testimony. In many cases, courts have flatly excluded the testimony of experts that was based on speculation or otherwise unreliable scientific methods (*25*). In *Roberti v. Andy's Termite and Pest Control*, a Los Angeles court excluded expert testimony that Dursban, a chemical, could cause autism. The court explained that since the expert's opinion was not generally accepted in the scientific community it could not be admitted at trial under the Kelly/Frye test (*26*).

Construction Defect Actions

The biggest cause of litigation in residential construction is due to water intrusion: through roofs, windows, showers, ground/slab, HVAC, etc. In California, courts have held that the developer or builder, and sometimes the general contractor, of mass-produced homes with construction defects that cause water intrusion and fungal growth may be strictly liable to the homeowner and to any subsequent purchaser (*27*). Property damages can increase because they often include loss of use, loss of profit, repair, remediation, and personal property.

Other factors have lead to an increase in mold claims in the construction defect arena aside from the intense media coverage and ingenuity of the plaintiff's bar. One example is the changes in building materials and techniques in recent years as an emphasis has been placed on building energy efficient, air-tight homes. Another important factor is that while residential developers have concerns about litigation and avoiding legal costs, the primary focus is placed on closing home sales and continue rapid development in order to capitalize on the housing boom. In January of 2005, new home construction rose 4.7%, which was the highest level in two decades (*28*). The growth has not hit its peak either, residential construction over the next 10 years is projected to exceed the rate of residential construction during the last 10 years (*29*).

Landlord-Tenant

When a residential unit has sustained water intrusion and resulting fungal growth, the tenant may sue the landlord for a breach of the statutory warranty of habitability or the implied warranty of habitability. No written lease is necessary to raise such a claim (*30*). A residential tenant may assert other theories for recovery against landlords, such as: negligence (failure to maintain, inadequate repairs, failure to inspect), nuisance (loss of enjoyment of property and health, negligence per se (violation of building and housing codes or state health and safety laws), constructive eviction (when acts or omissions of landlord force tenant to move), retaliation (if landlord evicts after tenant complains of conditions, reports landlord to governmental agency or retains attorney) and fraud/misrepresentation (if landlord knowingly or negligently misrepresents or conceals the condition of the property). Moreover, there is an implied covenant of quiet enjoyment that may allow a commercial or residential tenant to sue the landlord for mold exposure (*31*).

Insurance Issues

In most circumstances, defendants will look to insurance as the source of payment for remediation, defense costs, settlements and verdicts. The threshold question, of course, is whether these claims are covered. Most of the "coverage" issues relating to mold claims involve the handling of the claim. That is, the issues involve: timely response to the claim, identifying and retaining the appropriate consultants, contractors and experts to handle the problem, and implementing the appropriate remedy to rid the exposure site of mold and the water source which feeds the mold.

Most significantly, many current insurance policies have a specific exclusion for mold and/or fungus. If such mold-related exclusions in insurance policies are uniformly enforced, then individual defendants will face independent responsibility for paying for and defending claims of mold damage. Without the help of insurance policies, the prohibitive costs of defending lawsuits may drive defendants into bankruptcy. However, plaintiffs will continue to bring mold claims as long as any defendants remain solvent. The mere existence of bankrupt defendants, as seen in the context of asbestos litigation, is not a deterrent to the filing of lawsuits.

Personal Injury Actions

Personal Injury mold plaintiffs can choose from a cornucopia of legal theories when filing their mold lawsuits: Strict liability, Negligence, Products Liability, Breach of Implied Warranty of Habitability, Breach of Expressed warranty, Fraud, Proposition 65 Violations, Conversion, etc. To date, mold lawsuits have taken many forms given the wide array of legal options available to plaintiffs.

Unlike with construction lawsuits requiring direct expense reimbursements, personal injury suits have the unfortunate task of quantifying the dollar amount of human illness. Damages in personal injury suits can quickly escalate when they incorporate pain and suffering, loss of income, loss of consortium, loss of household services, and other indefinite sums. Moreover, punitive damage claims can also significantly increase the personal injury lawsuit damages. Defendants can be pursued for allegations of wanton or malicious conduct, leading to astronomical damage claims that were virtually non-existent in prior construction lawsuits.

What Direction Will The Pendulum Swing?

So in which direction is the pendulum moving? Recent litigation has produced varied results as the science behind mold-related illness struggles to catch up to the litigation.

Despite the difficulty in overcoming the obstacle of insufficient evidence, there have been several large verdicts that could be an indicator that the pendulum has shifted again. For example, in 2003 a Florida mold case, *Henriques v. Archstone-Smith Operating Trust, et al.*, settled for $20 million (*32*). In that case, the plaintiff and 16 other current and former tenants of a 452-unit apartment building filed a proposed class action complaint seeking damages and injunctive relief for exposure to mold growing inside the apartments and in the common areas. The court certified the class and opened it up to any tenant or invited guest who has been exposed to mold growth. The plaintiffs alleged they suffered personal injuries, including eye irritation, immune system suppression, migraine headaches, dizziness, memory loss, sinus infections, pulmonary illness, kidney disorders, cognitive impairment and brain disorder (*33*).

Also in 2003, Contractors involved in the building of Economedes High School in Edinburg, Texas settled a mold lawsuit by agreeing to pay $20.9 million. Construction of the school was completed four years prior and administrators blamed mold contamination on construction defects. Defendant Landmark Organization agreed to pay $11 million of the settlement, while 16 other subcontractors agreed to pay the additional $10.7 million (*34*). In 2005, a Colorado case involving a cause of action against a homeowners association for mold and water damage resulted in a $2.6 million arbitration award (*35*).

Significantly, in 2002 a class of plaintiffs received a record-breaking settlement of $107.5 million after alleging that Behr wood-sealant caused mildew damage. Plaintiffs contended that after applying Behr wood coating products, wood surfaces suffered excessive mildewing, resulting in extensive property damage. Most significantly, the astonishing settlement did not involve claims of mold-related personal injury damages. Rather, the class of plaintiffs claimed that the Behr wood products failed to perform as warranted, causing property damages ranging from $14,500 to $38,000 per plaintiff (*36*). Plaintiffs claimed that the products caused damage to various wood surfaces, including decks, homes and fences. The court reached this large group settlement amount based on a damage calculation formula involving the type of wood used, the extent of molding on the wood products, and the amount of wood affected by the defective wood sealant. The case involved multiple plaintiffs in over a dozen states, resulting in this incredible class settlement to compensate the multiple plaintiffs (*37*). In November of 2005, a $22.6 million settlement was reached in a mold litigation case filed by the Gorman family against the Crenshaw Lumber Company and 16 other suppliers and contractors who had constructed their Manhattan Beach, California residence (*38*). At trial, plaintiffs alleged that the Crenshaw Lumber Yard failed to cover or otherwise protect stored lumber used in the construction of their $1.9 million beachfront residence, allowing it to get wet and grow mold. Further, the lumberyard failed to store the wood at least 12 inches off the ground. The lumber was said to have developed visible mold but was still used to build the Gorman home. The Gormans contended that their whole family suffered injuries because of exposure to toxic molds growing on framing studs that had been improperly stored at Crenshaw Lumber Co. Most significantly, the Gorman's young son, Kellen Gorman suffered from serious illnesses as a result of the mold exposure. Kellen was born in the Gorman's home in 1999 and lived in the subject property. From the time of his birth, until the time that plaintiffs moved from the mold-contaminated home three weeks after his birth, Kellen Gorman developed multiple injuries, including: autism, pervasive developmental disorder, metabolic problems and neurological deficiencies (*39*). Evidence was presented that Kellen underwent an MRI revealing brain lesions and cysts, allegedly caused by exposure to toxic mold in the lumber (*40*). The child, who was five years old at the time of trial, was said to function as an infant due to his injuries, and was said to require lifelong medical care. Kellen's father also complained of cognitive and neurological problems such as memory loss, which the plaintiffs' medical experts testified is associated with exposure to certain fungi. At trial, the court excluded 10 of the lumberyard's experts, including a toxicologist and microbiologist, on procedural grounds. As a result, the lumber company settled for $13 million during trial. Meanwhile, the other defendants settled prior to the commencement of trial for a total of $10 million in 2005. This case was one of the first successful mold lawsuits against a lumber yard - and the largest to involve a single-family home (41).

These cases exemplify the types of mold-related cases that can warrant astounding damages and settlements. Not only are these types of large settlements and verdicts still in play, but the number of claims continue to rise: the Insurance Information Institute, New York City, estimated that as of June 2005, over 10,000 lawsuits were currently pending across the country alleging mold-related injuries.

What's Next?

In 2001, the American Bar Association predicted that mold could surpass asbestos in case volume and value of awards (*42*). However, there are many reasons why mold litigation should not be considered the next asbestos or lead litigation. Asbestos and lead litigation were both supported by a relatively well-developed body of scientific knowledge regarding the toxicological effects these substances had on the human body. Accordingly, while exposure standards may not have initially been set at the time those who ultimately recovered from asbestos and lead suffered exposure, there was never any real doubt that standards could ultimately be set because of the availability of epidemiological data.

As the scientific and medical issues associated with mold evolve, so will the strategies in the mold litigation. For today, those battlegrounds focus on appropriate remediation and causation issues. Although it is uncertain how jurors and courts will respond to the many issues presented in the mold cases, it appears certain that mold litigation is here to stay and will continue to expand as to the number of claims, number of parties involved and, most significantly, the money involved.

The key battlegrounds for mold litigation revolve around scientific research. As more conclusive links tie mold exposure to human illness, mold plaintiffs will be able to solidify their personal injury claims with substantive evidence. Environmental protection agencies continue their efforts to educate the public and control mold exposure. Legislation regulating mold exposure levels continues to develop. Across multiple industries, including law, medicine and politics, mold is under scrutiny and development. This fascinating point in time creates a great opportunity for scientists, experts and attorneys involved in these cases to shape an evolving area of law. For scientists and experts involved in these cases, mold provides a chance to be at the center of, and responsible for, the development of entirely new areas of law (*43*).

References

1. CDC. Acute pulmonary hemorrhage/hemosiderosis among infants—Cleveland, January 1993-November 1994. *MMWR* 1994;43:881-3.
2. Centers for Didsease Control Update: Pulmonary Hemorrhage emosiderosis Among Infansts – Cleveland, Ohio, 1993-1996 (Mar. 10, 2000), available at http://www.cdc.gov/epo/mmwr/preview/mmwrhtml/mm4909a3.htm.
3. A few examples include the December 3, 1999 cover of USA Weekend titled, "The MOLD in your HOME may be DEADLY;" the December 4, 2000 article in Newsweek titled, A Hidden Health Hazard: Sneezing and sniffling? Maybe the problem isn't a cold but mold. It's more dangerous

than you think:" and an April 18, 2001 ABC World News Tonight article titled, "Moldy Schools: Are your kids getting sick at school?."

4. 60 Minutes: Sick Building Syndrome, February 7, 2000. (CBS Television)
5. *Ballard v. Fire Insurance Exchange*, No. 99-05252 (Dist. Ct., Travis Co., Texas, verdict rendered June 1, 2001) ($32 million award).
6. *Id.* (The $32 million in damages included $12 million in punitive damages, $5 million for mental anguish, and nearly $9 million for attorneys fees.
7. *Davis v. Henry Phipps Plaza South*, No. 116331/98 (New York Co., N.Y., Sup. Ct. Oct. 11, 2001).
8. High profile celebrities contributed to the legitimacy of mold claims by empowering victims and educating the jury pool. For example, Erin Brokovich filed a lawsuit alleging mold in her Southern California home and Ed McMahon filed suit alleging that mold was responsible for the death of his dog and his family's health problems.
9. Insurance Information Institute, TX: Estimated Total Number of Mold Claims, in Presentation: How Did We Get Here? Texas: Mold's Ground Zero (citing Texas Department of Insurance and Insurance Information estimates) available at http://www.iii.org/media/met/mold.
10. The New York Times on 12/11/2003.
11. One example of this is *Anderson v. Allstate Ins. Co.*, No. CIV-S-00-907-PAN (E.D. Cal. verdict Sept. 29, 2000), a Northern California jury awarded a 96-year-old man $500,000 in economic and emotional distress damages and $18 million in punitive damages after he sued Allstate, his insurance carrier, for refusing to pay for mold remediation and related repairs in his home. The verdict was eventually reduced on appeal, but highlights the response jurors had to the mold scare.
12. Field, Eugene. Love Affairs of a Bibliomaniac (Kessinger Publishing: 2004).
13. *Fire Insurance and Exchange v. Ballard*, No. 03-01-00717-CV, Texas Court of Appeals, (2003 Third District, Austin).
14. *Supra* notes 4 ,5 and 9.
15. *Cottle v. Superior Court* (1992) 3 Cal.App.4th 1367.
16. *Id.*
17. Federal Rules of Evidence § 701.
18. Mold Spore Counting: Is it Admissible Science?, [For the Defense?], J. Wylie Donald, August 2005 p.33.
19. (1993) 509 U.S. 579, 592-595.
20. (1999) 526 U.S. 137.
21. *Id.*
22. (1976) 17 Cal.3d 24.
23. (1923) 293 F.1013 (D.C. Cir.).
24. *Cottle v. Superior Court* (1992) 3 Cal.App.4th 1367.

25. Balakrishnan (Federal Court, LA, 1998) Plaintiff's experts excluded., Ballard (Texas, 2001) Plaintiff's defense experts excluded. Stroot (Delaware, 1999) Defendant's motion to exclude medical experts denied., Mazza (Sacramento, 2001) Defendant's motion to exclude plaintiff's medical expert opinions denied, Murray (San Diego, 2001) Defendant's motion to exclude testimony Re: neurological damage was successful.
26. *Roberti v. Andy's Termite and Pest Control* (2003) 113 Cal.App.4th 893.
27. *Kriegler v. Eichler Homes, Inc.* (1969) 269 CA2d 224).
28. New Residential Construction in January 2005. U.S. Census Bureau News Joint Release from U.S. Department of Commerce and the U.S. Department.
29. The State of the Nation's Housing 2004, Joint Center for Housing Studies of Harvard University.
30. See Cal. Civil Code §§ 1941, et seq.; *Green v. Superior Court* (1974) 10 Cal.App.3d 616.
31. Cal. Civil Code § 1927.
32. *Henriquez v. Archstone-Smith Operating Trust, et al.*, No. 02-21697 CA 02 (Fla. Cir. Ct. 11th Jud. Cir., Miami-Dade Cty.)
33. *Id.*
34. Knight Ridder Tribune Business News, Valley Morning Star on 12/18/2003.
35. Triton Development (Colorado, 2005).
36. Securities and Exchange Commission, Quarterly Report Regarding Masco Corporation, Commission File 1-5794, March 31, 2003.
37. *Id.*
38. *Kellen Gorman v. Crenshaw Lumber*, et al.
39. John Parker Sweeney, The Gorman Case: Anatomy of a Settlement, Mealey's Litigation Report, Vol. 6:3 (March 2006).
40. Susan Lillard-Roberts, $13 Million From Lumberyard Brings California Mold Settlement To $22.6 Million, Nov. 7, 2005, available at http://www.mold-help.org/content/view/691/.
41. Keeran, Collen, Mold Property Damage: Brain Injury Claims Settle for $23 Million, Mealey's Litigation Report (Oct. 31, 2005).
42. Stephanie Francisc Cahill, For Some Lawyers, Mold is Gold, ABA J., Nov. 15, 2001 (quoting plaintiff's lawyer as stating: "The use of asbestos isn't occurring anymore, and most of the asbestos products were done away with... With mold, it's naturally occurring, and the supply is endless"), available at http://www.rvclaw.com/aba-12-01.asp.
43. Expert Witness and Scientific Testimony Issues Concerning Mold Litigation in State and Federal Courts, Jeremy R. Larson, Foster Pepper & Shefelman PLLC, http://phys4.harvard.edu/~wilson/soundscience/mold/larson.html.

Chemical Biocides, Termite Control Methods, and Biocide Depletion

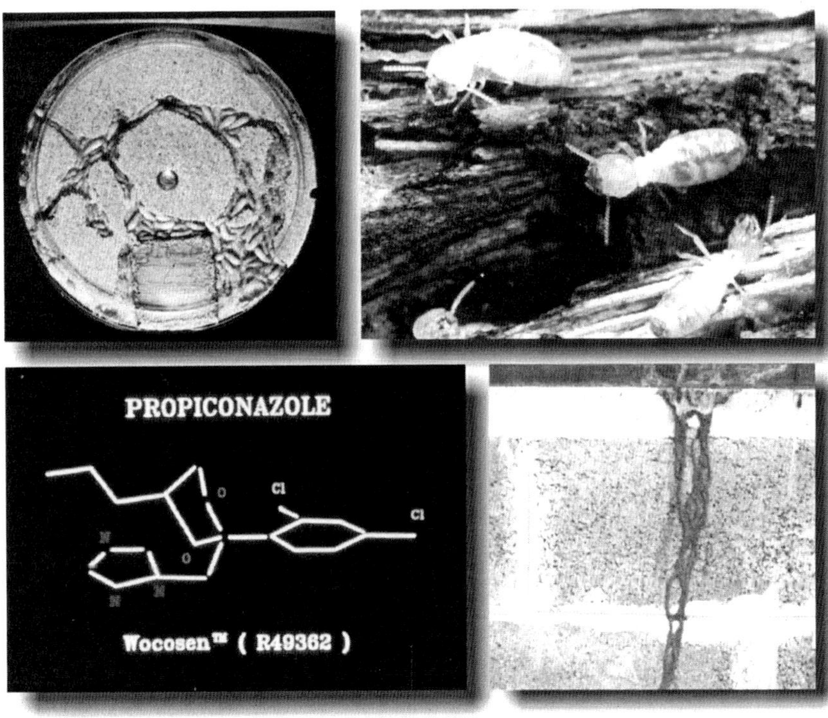

The first and third photographs are courtesy of Darrel D. Nicholas and the fourth photograph is courtesy of Terry Amburgey both of Mississippi State University. The second photograph is courtesy of Clemson University, USDA Cooperative Extension Slide Series, Bugwood.org. *(See page 5 of color inserts.)*

Chapter 13

Wood Preservative Fungicides and the American Wood Preservers' Association Use Category System

Peter E. Laks

School of Forest Resources and Environmental Science, Michigan Technological University, Houghton, MI 49931

Fungi are the primary biological hazard for wood products. Thus, the fungicide component of a wood preservative system is typically the most important constituent. The type of fungal hazard and the nature of the fungicide most appropriate to control that hazard will depend on the application for the treated wood product. The biological hazard for a given application has been systemized by the American Wood-Preservers' Association Use Category System. This is a useful tool in understanding how wood preservative fungicides are currently used. The chemical, physical, and biological characteristics of a fungicide determines its suitability for a given wood application. Important characteristics are efficacy/cost ratio, breadth of efficacy spectrum, activity against non-target organisms, stability, and leach resistance. Wood preservative fungicides are a very diverse group and are in the process of undergoing further change due to government regulations, environmental issues, and the demands of the marketplace.

Fungi that Attack Wood-in-Use

Fungi are a kingdom of organisms with a broad range of eco-niches. They are commonly defined as non-motile, filamentous or single-celled microbes, lacking chlorophyll, with chitinous cell walls, reproducing by spores, some of which produce large fleshy fruiting bodies (mushrooms) (*1*). With respect to wood-in-use, the most important general types of fungi are the basidiomycetes, soft-rot fungi, and the mold and stain species. There is some overlap between these groups, as species that are commonly classified as molds can also cause soft-rot decay under the appropriate conditions. A thorough discussion of the fungi that attack wood is beyond the scope of this paper, but some understanding of fungi is useful when discussing fungicides.

Fungi have a number of classic growth requirements, but three are especially important in determining whether a wood-based product will be prone to fungal attack. The first is wood moisture content. Under certain conditions, some fungi can grow on wood when the moisture content is less than 20% (*2*), however, most authorities state that a moisture content at or above the fiber saturation point (about 30% for temperate-climate commercial wood species) is necessary for significant fungal growth (*3*). The second important requirement for fungal growth is available nutrients. All wood contains the polymeric structural carbohydrates that basidiomycetes and soft-rot fungi utilize as a fixed-carbon source. Mold and stain fungi utilize low molecular weight carbohydrates, starches, fats, and fatty acids that have a more variable distribution in wood products. For example, the heartwood of many wood species contains little or none of these nutrients, making it quite resistant to mold and stain fungi. The third factor is the presence/absence of fungitoxic chemicals that inhibit fungal growth. These chemicals can either be naturally occurring extractives present in the heartwood of some wood species, or synthetic wood preservatives introduced on or into the wood using a treating process. An appropriate loading of an effective preservative can completely inhibit fungal activity.

Basidiomycetes are fungi that feed on the structural carbohydrate polymers in the wood (cellulose and the hemicelluloses). They include the mushroom-forming fungi. Some of the basidiomycetes, particularly the so-called white-rots, also produce enzymes that can degrade the third major wood structural polymer, lignin. Through their ability to attack these structural polymers, the basidiomycetes can rapidly reduce wood strength properties. Basidiomycete decay is characterized by softening and mass loss in a relatively large volume of the wood article. In untreated susceptible wood, the volume limitation is often the moisture content of the wood. The optimal moisture content for basidiomycete growth in wood depends on the fungal species and the investigator, but is typically in the 30-80% range (*4*). Some species have acquired the ability to transport water to drier portions of a wood article.

Basidiomycetes are the classic wood decay fungi and cause the most damage to wood-in-use.

Soft-rot fungi are lignolytic ascomycetes and deuteromycetes. Generally, they require a longer time to degrade wood compared to the basidiomycetes. Soft rot decay tends to dominate in treated wood under high wood moisture content conditions where the growth of basidiomycetes may be inhibited.

Mold fungi are imperfect fungi with colored mycelium and/or spores that grow on the surface of wood, utilizing low MW sugars and starch. Mold fungi do not cause significant strength loss in the wood. Their damage is aesthetic. Molds are commonly a problem on unweathered wood surfaces. For example, freshly-cut sapwood lumber is very prone to mold growth. To maintain a clean surface under warm conditions, either the lumber has to be dried quite quickly or a moldicide applied to the wood surfaces. Sapwood that has been pressure treated with common waterborne wood preservatives is also often initially susceptible to mold. This susceptibility usually is reduced as the exposed wood surfaces weather (e.g. the treated lumber is used as exterior decking), although older wood surfaces can support mold growth if nutrients such as pollen or other plant residues accumulate on the surfaces.

An understanding of the fungi that attack wood is important when discussing the agents used to control them – fungicides. It has been said many times that the most common wood preservative method is drying the wood and keeping it dry. This comes from an understanding of the basic moisture requirements of fungi. Similarly, there is no point in changing the availability of free sugars and starches if basidiomycetes are being targeted for control since basidiomycetes typically utilize the cellulose and hemicellulose components of the wood. An understanding of the basic biology of fungi is important in their efficient control.

The Use Category System

Wood products are used in a broad array of applications, ranging from furniture in a heated home where there is little or no biological deterioration hazard to marine exposure in a tropical climate. In the United States, this broad range of possible applications has been classified with the Use Category System (UCS) (*5*). This system was adopted by the American Wood-Preservers' Association in 1999. An understanding of the UCS is important in any discussion of wood preservative fungicides because it provides a means to group applications together that have similar fungicide performance requirements. By understanding the fungicide needs for a given Use Category, biocides can be specifically selected for that UC. Following is a summary of the UCS classifications:

UC1 – Interior Construction, Above Ground, Dry – These are applications protected from external and internal water sources. Typical examples are interior furniture and millwork, as well as studs, joists, subflooring, etc. Since, ideally, there should be no sources of liquid water contacting the wood, the moisture content of these materials should remain below the fiber saturation point (FSP) and, therefore, there should not be a fungal hazard in UC1. Termites, beetle larvae, and carpenter ants can attack relatively dry wood in the appropriate climates and settings, so UC1 applications should only experience an insect hazard. One of the most common insecticides used for pressure-treating UC1 commodities is disodium octaborate tetrahydrate (DOT, also abbreviated SBX). DOT is an effective fungicide as well as being an insecticide. Its use is an example of fungicide protection being provided when it is not really needed.

UC2 – Interior Construction, Above Ground, Damp – UC2 applications have limited protection from interior and exterior water sources. The typical example is a sill plate. This is the horizontal lumber that sits atop the foundation and then contacts the studs, joists, or rimboards in conventional construction. Water can be wicked up through the foundation and come in contact with the sill plates resulting in a decay hazard. Usually there is some kind of moisture protection between the foundation and the sill plate, but it may not be 100% effective. Another UC2 example is a window sash. During the winter in a cold climate, condensation is possible on the interior of the window leading to liquid water contact with the sash framing. Leachable preservative systems can be used in UC2 applications because there is no continual contact with liquid water that would provide the depletion route. Fungicides used in UC2 applications include DOT, organics such as 3-iodo-2-propynyl butyl carbamate (IPBC), and the common waterbornes such as the alkaline copper quat (ACQ) and copper azole (CA) systems.

UC3A – Exterior Construction, Above Ground, Coated and Rapid Water Runoff – The use category for above-ground exterior construction is split into two subgroups depending on whether the application has some protection from liquid water. The UC3A category is for commodities that have some limited protection from precipitation and are in applications where there is rapid draining of water from the surfaces. Painted siding is a good example of a UC3B commodity. The paint provides some protection from water absorption and the vertical orientation of the siding results in rapid runoff of any liquid water on the siding surfaces. Other UC3B products are exterior millwork and trim. A common preservative used in these applications is zinc borate, used as a preservative in a variety of exterior wood composite products (*6*).

UC3B – Exterior Construction, Above Ground, Uncoated and Poor Water Runoff - This is the other use category for above-ground exterior construction. In UC3B, the wood product has no coating and/or the commodity is oriented in a fashion that does not allow rapid water runoff. Examples are decking, railings, fence pickets, utility pole crossarms, and any uncoated wood used on the exterior of a building. This use category includes some of the most important applications for pressure-treated wood – most treated wood is used in exterior deck construction. A variety of wood preservatives are used in UC3B commodities, including the dominant ACQ and CA systems. Fungal decay and insect damage control is important for both UC3 subdivisions.

UC4A, B, and C – Ground/Fresh Water Contact - UC4 covers all ground-contact and fresh water applications. The three subcategories (UC4A, UC4B, and UC4C) increase in severity of exposure and/or criticality of the application in this sequence. UC4A is for non-critical treated wood products in contact with the ground or freshwater. Examples are fence/deck posts, landscaping timbers, utility poles in low decay hazard regions, and backyard dock pilings. UC4B covers ground contact applications in critical applications and/or when replacement of the treated wood is difficult. Examples are utility poles in moist temperate climates, wood foundations for houses, building poles, and sea walls subject to salt water splash. UC4C is for products in ground contact in severe climates and applications where replacement would be very difficult. Examples are utility poles in tropical and semitropical climates, and foundation pilings for commercial buildings. A broader range of wood preservative treatments are used in UC4 applications. Preservatives that are not allowed for residential use such as chromated copper arsenate (CCA), creosote, and pentachlorophenol are commonly used in non-residential UC4 applications. Railway crossties are another important commodity that fall within this general group. Similar to utility poles, the exact classification for the crosstie depends on the severity of the installation climate. Fungal decay control is critical for UC4 applications, although insect performance is important as well.

UC5A, B, and C – Salt/Brackish Water Contact – UC5 covers applications where the treated wood product is exposed to marine borer attack. The three subdivisions reflect the degree of exposure severity. UC5C is the most severe exposure where there is exposure to warm water organisms that are difficult to control with wood preservatives. The common preservatives used in UC5 applications are creosote, CCA, or a dual treatment of the two. Fungicide performance in UC5 is relatively unimportant because of the difficulty in controlling the mariner borers. Retentions needed for marine borers are many times higher that those needed for fungal or insect control.

Figure 1 graphically illustrates how the various types of biological degradation vectors correlate with AWPA use category. Different fungicides have different efficacy spectrums. It is important that the fungicide cover the

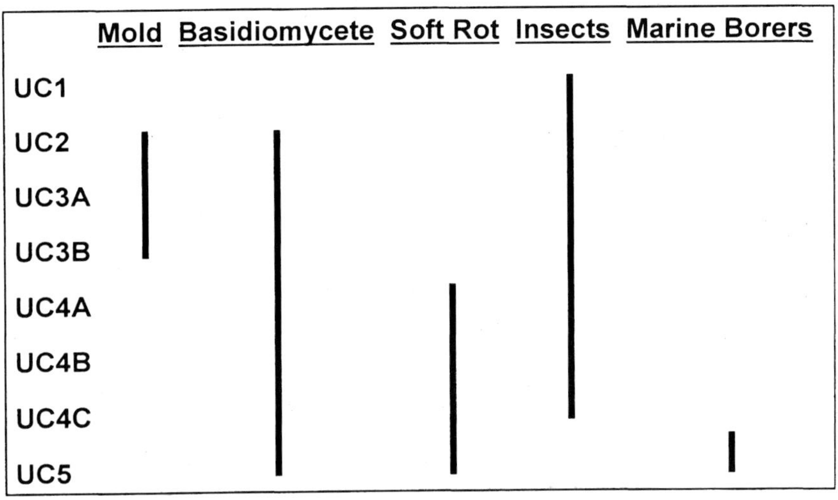

Figure 1. Biological hazard for wood products correlated with Use Category.

appropriate range of problem fungi when selecting a biocide for use in a product meant for a given use category. For example, in ground contact applications (UC4) efficacy against soft rots is necessary. If the fungicide used in a wood preservative system meant for ground contact does not have efficacy against soft rots, an additional fungicide with this activity should be included in the formulation. Conversely, if a preservative system has been designed only for above ground use, efficacy against soft rots is not required. This may lower the total cost of the system.

Characteristics of Wood Preservative Fungicides

Preservative-treated wood products typically have a long service life. As shown above in the discussion of the UCS, applications for treated wood are diverse, leading to a broad range of performance requirements. Following is a list and discussion of ideal wood preservative (WP) fungicide characteristics.

High Efficacy/Cost Ratio - A fundamental aspect of most building products is that they are relatively inexpensive. This comes from the very large volumes produced and competition from different materials. Depending on the specific application, pressure-treated wood has competition from materials based on steel, concrete, thermoplastics and others. Since the treated wood product has to be inexpensive, the efficacy/cost ratio for the preservative treatment has to be high. Copper II has dominated as the wood preservative fungicide of choice because of its reasonable intrinsic antifungal activity and very low cost compared to organic fungicides.

Broad UC Application Spectrum – As described above and depending on the application, WP fungicides have to control a broad range of target fungi. A single fungicide with an efficacy spectrum broad enough to cover all of the target fungi would obviously be most desirable as long as the other performance requirements are met. This is often difficult to achieve, however, leading to the common practice of mixing complementary fungicides in a formulation to achieve control of the entire range of problem fungi. The most obvious example of this practice is with the current generation of "copper plus cobiocide" formulations. Copper II has many desirable WP characteristics, but there are "copper-tolerant" basidiomycetes and soft rots that can attack wood treated with only copper (*4*). A quaternary ammonium salt or triazole is added to the ACQ or CA formulations, respectively, to control the copper-tolerants.

Benign to Non-Target Organisms – By definition, fungicides are biocides. It is most desirable, of course, that a wood preservative fungicide not affect non-target organisms, especially the human user of the treated wood. As a non-human example, birds perching or nesting on treated wood structures should not be negatively affected by contact with the treated wood.

Chemical/Physical Stability – Pressure-treated wood products typically have a long service life. Sill plates must last for the lifetime of the house in which they are installed, and utility poles are commonly described as having lifetimes of 30-40 years. Warranties for pressure-treated lumber are typically "Limited Lifetime", but since the average residential deck in the United States is replaced or modified 11 years after installation (*7*), the actual lifetime of use in this very important segment of the market is relatively short. Nevertheless, the fungicide has to be stable in the treated wood product over a long period of time. This stability takes a number of forms. The stability issue is more of a consideration for organic fungicides than inorganics. These must be stable to oxidation, heat, and interaction with other formulation components. Inorganic fungicides, such as copper (II) and borates, typically have fewer issues with stability, but there can still be loss of the active ingredient (a.i.) with time through water leaching.

Leach Resistance – In UC3 through UC5, resistance to loss from water leaching is a very important characteristic. In general, the leach resistance requirement increases in importance with increasing use category. UC1 and 2 are interior applications where the treated wood should never be exposed to significant amounts of liquid water, so leachable fungicides such as DOT can be used. There are two fundamental impacts of a.i. leaching from treated wood. First, the amount of a.i. in the wood is reduced over time. Eventually this will lead to reduction below a critical level and fungi will be able to attack the wood. Secondly, the a.i.(s) migrate into the surrounding environment and may cause significant contamination of surrounding soil or bodies of water. It is generally accepted, however, that some water mobility is necessary for the fungicide(s) in a conventional wood preservative system to perform.

Minimal Negative Effects on Wood Properties – Incorporating a chemical formulation into wood can have effects on many of its nondurability-related properties. The most important are strength, corrosivity, and fire properties. A good discussion of the effects on strength properties can be found in the Wood Handbook (*8*). Obviously, it is best that the fungicide(s) and other components of a WP formulation have a minimal negative effect on these properties.

Readily Available - A WP fungicide has to be registered with the U.S. EPA before it can be used commercially. It is also desirable from the points of view of the formulator, treater, and user that the fungicide be available from more than one chemical manufacturer. Not only does this typically reduce the cost of the a.i., it also means there are alternative suppliers of the chemistry in case one manufacturer goes out of business or simply decides not to manufacture that fungicide anymore.

Easy to Formulate – Pressure treatment of wood involves penetration to a significant depth within the wood structure. The actual required depth depends on the wood species being treated and the nature of the commodity being manufactured. These parameters are detailed in the AWPA Book of Standards (*5*). The a.i.(s) in the formulation have to be in a form that allows this penetration to occur. A true solution is, by far, the most common formulation type, but emulsions and small particle suspensions may become more important in the future. A WP fungicide has to have chemical properties that allow it to be formulated in an appropriate carrier that results in adequate penetration, but does not contribute negative characteristics as described above in this list. An obvious example is that the emulsifiers in an emulsion formulation should not promote leachability of the a.i.(s).

In many cases, the relative importance of these fungicide characteristics depends on the use category (ies) in which the treated product will be used. As described above, resistance to leaching is only important in the higher UCs. As another example, a reduction in strength properties may not be important if the treated wood is being used in a non-structural application.

Common Wood Preservative Systems and Fungicides

An excellent and detailed discussion of WP systems and biocides is given in Schultz and Nicholas (*9*). For more detail on individual fungicides and wood preservative systems, please see their paper. The purpose of this section is to provide more general information. Following is a list of WP systems in common use in the United States and Canada and their acronyms:

- Chromated Copper Arsenate (CCA)
- Creosote (CR)
- Pentachlorophenol in P9 Type A Oil (PCP-A)
- Alkaline Copper Quat (ACQ)
- Copper Azole (CA)
- Disodium Octaborate Tetrahydrate (SBX)
- Copper Naphthenate (CuN)
- Ammoniacal Copper Zinc Arsenate (ACZA)
- IPBC plus cobiocide(s) (IPBC+)
- Zinc Borate (ZB)

The major wood preservatives are commonly divided into first generation systems (CCA, creosote, PCP, ACZA) and second generation systems (ACQ, CA). Some of the others are not so easy to classify. For example, SBX has only developed as an important preservative in the United States during the last 15 years, but has been used in other parts of the world since the 1950's (*10*). Others of these systems have been developed for specific application niches. As mentioned above, zinc borate is commonly used as an in-process preservative for wood-based composites, while the IPBC+ systems are common as non-pressure applied treatments as used for manufacture of window parts or moldicides for OSB (*6*). The active ingredients in these basic WP types are listed in Table 1.

Copper (II) is a common fungicide found in many of the major wood preservative systems. It possesses many of the desirable fungicide characteristics listed above. In particular, its efficacy/cost ratio, UC application spectrum, stability, leach resistance, effects on wood properties, formulation ease, and availability are all excellent. Two notable issues are copper's weakness against copper-tolerant fungi, which requires the addition of a co-fungicide to copper-based formulations, and its possible effect on non-target marine organisms if it is used in UC5 applications (*11*). The second generation systems require a particularly high loading of copper into the treated wood (Table 2). This has lead to speculation that the next generation of wood preservatives may be similar to the current systems in that they are based on copper plus a cobiocide, but may require a lower level of copper in the wood.

Table 1. Fungicide and insecticide active ingredients found in commonly used wood preservative systems. Fungicides in parentheses may be present, depending on the exact formulation used.

	Preservative Active Ingredient(s)
CCA	Cu(II), As(V) oxides
CR	Phenols, pyridines, etc
PCP-A	Pentachlorophenol, oil components
ACQ	Cu(II), quaternary ammonium salts
CA	Cu(II), tebuconazole (BO_3^{-3})
SBX	BO_3^{-3}
CuN	Cu(II), naphthenates
ACZA	Cu(II), Zn(II), As(V) oxides
IPBC+	IPBC (propiconazole, tebuconazole, chlorpyrifos, permethrin, imidichloprid)
ZB	BO_3^{-3}

Table 2. Copper content (as Cu)) of wood preservatives in southern pine lumber when treated according to AWPA Standard U1-05 (in pounds per cubic feet, pcf) (5). Letter designations after the preservative acronym refer to the Type as defined in the AWPA Standards, e.g. CCA-C = CCA Type C.

	Retention (pcf)		
WP	Total	CuO	Cobiocide
CCA-C	0.4	0.074	0.14
ACQ-D	0.4	0.27	0.13
CA-B	0.26	0.25	0.0082

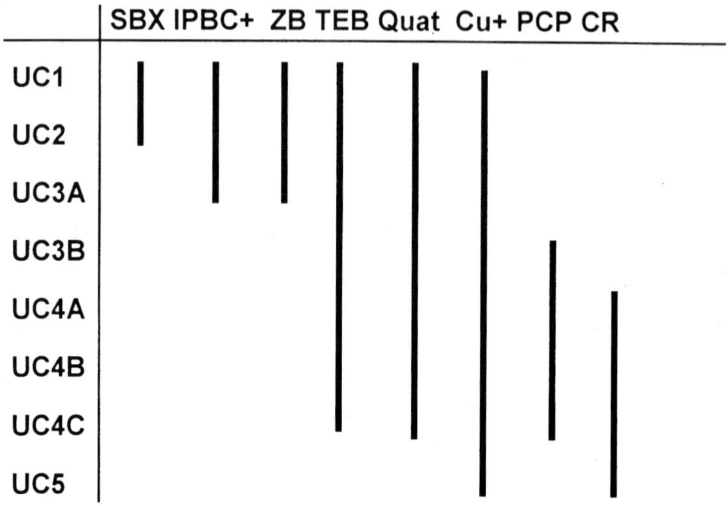

Figure 2. Major wood preservative fungicides correlated to AWPA use category. TEB = tebuconazole, Quat = quaternary ammonium salt, see text for additional abbreviations used.

Such systems could be referred to as generation 2B wood preservatives. These 2B systems would need to use either a higher loading of the cobiocide used in existing systems, or use a mixture of cobiocides along with the lower loading of copper.

The wood preservative fungicides can be classified in terms of the use categories in which they are used. Figure 2 illustrates this point for sodium borate, IPBC, zinc borate, tebuconazole, quaternary ammonium salts, copper II, pentachlorophenol, and creosote. Fungicide characteristics determine which use categories it is suitable for. SBX (disodium octaborate tetrahydrate) is readily leachable by liquid water. Hence, it is only suitable for use in UC1 and 2, where there is no significant leaching hazard. Creosote and pentachlorophenol are not suitable for use in or around the house because of odor and human toxicity issues, so they are restricted to heavy-duty applications found in UC3B through UC5. Examples of these applications are utility poles, railway ties, pilings, and commercial dock structures.

Conclusions

Wood preservative fungicides comprise a diverse group of chemistries. Technologies to preserve wood from fungal attack are literally hundreds of years old (*12*). At the present, many of these old technologies are used alongside newer WP systems that have been developed with modern environmental impact and human toxicity concerns in mind. The market for wood preservatives has also grown considerably over the last 25 years (13). All this has led to a very broad array of biocides that are available in the United States and Canada for wood preservative use. The AWPA Use Category System is a useful tool in classifying these preservatives and their component fungicide(s), and understanding how they are currently used. Suitability of a wood preservative fungicide for a given application depends on a large number of its characteristics. Some of the most important ones are efficacy/cost, breadth of efficacy spectrum, activity against non-target organisms, stability, and leach resistance. The marketplace for treated wood products continues to change due to competition from competing materials like steel and wood/thermoplastic composites, new environmental and human toxicity concerns, participation by chemical-producing companies that are new to this market, and the demands of the consumer. The fungicides used in wood preservative applications will also continue to change.

References

1. Hawksworth, D.L.; Kirk, P.M.; Sutton, B.C.; Pegler, D.N. *Ainsworth & Bisby's Dictionary of the Fungi*; CAB Interational, University Press, Cambridge, UK, Eighth Edition, 1995.
2. Morton, L.H.G.; Eggins, H.O.W. *Material und Organismen* **1976**, *11*, 279-294.
3. Highley, T.L. *Wood Handbook, Wood as an Engineering Material*; Forest Products Laboratory, Forest Service, U.S. Dept. of Agriculture: Madison, WI, **2002**, 283-297.
4. Eaton, R.A.; Hale, M.D.C. *Wood: Decay, Pests and Protection*; Chapman and Hall, New York, NY, **1993**; 76-110.
5. *Book of Standards*; Standard U1 05; American Wood-Preservers' Association: Selma, AL, **2005**, 5-7.
6. Laks, P.E. *Proceedings of the American Wood-Preservers' Association*; AWPA: Selma, AL, **2004**, V *100*, 78-82.
7. Truini, J. *Home Mechanix* **1996**, *92*(805), 12.
8. Green, D.W.; Winandy, J.E.; Kretschmann, D.E. *Wood Handbook, Wood as an Engineering Material*; Forest Products Laboratory, Forest Service, U.S. Dept. of Agriculture: Madison, WI, **2002**, 103-105.
9. Schultz, T.P.; Nicholas, D.D. In *Wood Deterioration and Preservation, Advances in Our Changing World*; Goodell, B; Nicholas, D.D.; Schultz, T.P., Eds.; ACS Symposium Series 845; American Chemical Society: Washington DC, **2003**; Chapter 26, pp 420-432.

10. Murphy, R.J. *Wood Preservation in the '90s and Beyond*; Proceedings No. 7308; Forest Products Society: Madison, WI, **1995**, 162-168.
11. Lebow, S.T.; Foster, D.O.; Lebow, P.K. *Forest Products Journal* **1999**, *49*(7/8), 80-89.
12. Freeman, M.H.; Shupe, T.F.; Vlosky, R.P. *Forest Products Journal* **2003**, *53*(10), 8-15.
13. Preston, A.F. *Forest Products Journal* **2000**, *50*(9), 12-19.

Chapter 14

Bioactive Compounds to Prevent Insect Degradation of Wood

Gregg Henderson

Department of Entomology, Louisiana State University AgCenter, Baton Rouge, LA 70803

The deterioration of wood by xylophagous insects has a large economic impact in countries where wood is used as a major structural material. Among the insects, subterranean termites are by far the most destructive. Formosan subterranean termites consume even creosoted telephone poles and salt water dams. Increasing the visibility of possible termite entry points into a home through proper construction and landscaping modifications can reduce the damage caused by this insect. Using various wood insecticidal treatments, non-repellent termiticides and baits can be effective, but information on termite biology is necessary to make these treatments even more efficient. If wood is to remain as a choice building material, the management of subterranean termites must be readily achievable and cost effective. Knowledge of termite behavior on wood treated with various active compounds will aid in our understanding of these interactions.

Research on the development and testing of new insecticidal wood preservatives has advanced rapidly in the wake of environmental and human health concerns over the heavy metal additives present in many weather-resistant treated wood products. A recent EPA allowance for use of a borate wood preservative as a "stand-alone" preventative treatment for non-exposed applications against termites (now approved in several states) puts a greater burden, possibly an unachievable one, on wood preservative performance. Incorporation of currently registered termiticides (several pyrethroids, imidacloprid and fipronil) into wood-based composites appears to be a promising new direction; one that is particularly inviting given that registrations for other insecticidal applications are already granted. Another anti-termite chemistry thrust is toward natural products, especially secondary plant compounds that are long-lived and have a reduced hazard to the environment. Some of the longer-lived sesquiterpene constituents (as opposed to more volatile monoterpenes) from plant extracts can effectively stop termites for over 1 year (*1, 2*). Likewise, an extract of corn bud oil, 2-acetonaphthone, has low water solubility and volatility and shows promise against termites (*3*). However, the industry had realized a 30 yr persistence in resistivity against termites with some of the heavy metal-laced wood treatments. Attempting to achieve this feat with natural chemistries will most likely result in failure. If these natural products are to make it to market, testing standards employed in the laboratory and field need a reevaluation, since, in some countries (eg., European standards), even etching of the treated wood by termites is considered a product failure. Possibly a more realistic assessment of a these wood products may b suggested in the Australian standards, where a failure is recorded when 5% or more of the treated wood is consumed (*4*). Another wood treatment research area needing attention are field tests. Current American Wood Preservers' Association (AWPA) field tests allow wooden stakes that serve as controls to remain in place even after being attacked by termites over many years. There is a concern that such a trial is using a biased control given that termites have found the food source (or location) and have been recruited to it, possibly for some colonies, for over 20 years! Additionally, understanding termite behavioral effects from wood treatments that include repellents, toxicants or feeding deterrents can help the industry's efforts to develop appropriate products for specific needs. Finally, accelerated field testing may be useful in complying with EPAs newly instituted 5-year testing requirements (still under consideration; M. Suarez, personal communication).

A Brief Introduction to the Major Insect Consumers of Structural Wood

The impact that wood destroying insects have on serviceable wood can be extreme and is influenced by multiple physiological factors (*5*). Many of the structural-infesting xylophagous beetles rely on a minimum level of starch and moisture typically available in sapwood and woods being degraded by fungi. Most structural infesting lytids and bostrichids gain nutrition from the starch in the wood and can not digest cellulose. Anobiids generally attack and consume the cellulose of softwoods (and some hardwoods) used as structural and non-structural timbers if moisture levels are adequate. The old house borer *Hylotrupes bajulus* (Coleoptera: Cerambycidae) is one of the few long-horned beetles to attack seasoned conifers. Of the Hymenoptera, two of the major wood destroyers (but not consumers) can be found among the carpenter ants (Formicidae: *Camponotus*) and carpenter bees (Apidae: *Xylocopa*). Although ants and bees inhabit but can not digest wood, wood treatments still can impact their health since the transfer of toxicants readily occurs during grooming of contaminated mouthparts and legs.

The Order Isoptera is the most damaging insect of structural and non-structural wood, mainly because they live in well organized societies that can number into the millions. In addition, they feed directly on cellulose (except for some species in the family Termitidae). Dampwood termites (Hodotermitidae) are able to infiltrate very wet wood and are found along the Pacific Coast and inland into New Mexico and Arizona. Drywood termites (Kalotermitidae) need a minimum wood moisture content of about 14%, and are found in warmer, more humid climates in the US (but are sometimes found even in Arizona and New Mexico). The subterranean termites (Rhinotermitidae) are the most damaging family of termites and the most widely distributed and catholic in their taste for wood. These termites (all three families listed above) rely on protozoans harbored in their intestinal tract for cellulose breakdown and digestion. Subterranean termites attack hardwoods and softwoods, fungus infested wood and sound wood, and although moisture and nutrients are essential in their environment as social insects they can manipulate their habitat to near optimal conditions. Nitrogen content, which is critically low in most cellulose sources, can be increased by attack from various fungi and as such will be preferred by many termite species. Some of the brown rot fungi produce the same chemical that termites synthesize as a trail pheromone, Z,Z,E, (3,6,8) dodecatrien-1-ol, and, as with the natural pheromone, cause termites to follow lines drawn with this extract. Termites also can move moisture into the galleries as they forage by carrying water in their crops from nearby sources, usually below ground.

Termites cause economic losses in the US of about 3 to 5 billion per year. Evidence of termites can instill unreasonable fear in a homeowner, and a belief that severe termite damage only takes several weeks or a couple months at most is pervasive. However, serious damage to wood generally requires several years of infestation. Nevertheless, homeowners are often willing to pay premium prices for the sense of security that preventative termite control treatments offer. Current control tactics require an integrated pest management strategy which can

Table 1. Comparison of Termite Treatments

	Repellant Chemical Barrier	*Wood Treatment Products*	*Non-Repellent Chemical Barrier*	*Termite Baiting System*	*Botanical Termiticides*
Residual Life of Treatment	After 5 years < 10 % remains	Mostly lifetime, but variable with type of application	After 5 years, < 10% remains* *Fipronil >5 years	Continuous process; applied as necessary	Relatively short; < 2 years
Relative Cost	Least Expensive	⟶			Most expensive
Advantages	Immediate protection, less expensive	Low toxicity, no need for soil treatments	Requires little active ingredient, immediate protection	Environmentally friendly, less distruptive and intrusive (no drilling required)	Natural, perceived as "safe"
Disadvantages	Must be applied carefully to ensure proper barrier, barriers may fail, termiticides break down in soil	More suitable for preconstruction treatment or as supplement to other treatments	Must be applied carefully to ensure proper barrier, barriers may fail, termiticides break down in soil	More expensive, continual monitoring, longer to take effect, passive control	Must be reapplied, not commercial-ized, volatiles may be of concern

include chemical barrier treatments, non-repellent soil treatments, bait systems, wood treatments and follow-up inspections (Table 1).

The most damaging of the Rhinotermitidae belong to the subterranean termite genus *Coptotermes*. In the US, *C. formosanus* will build carton nests both below and above ground composed of glandular and anal fluids, and pre-digested wood. The nest helps maintain proper conditions the egg laying queen (s) and her brood. First discovered in the continental US in the mid-1960s along coastal ports of Louisiana and Texas. *C. formosanus* was soon found in trees and even creosoted railroad ties, utility poles and salt water dams (Figs. 1-3). Because a colony can be 10X larger than a subterranean termites in the genus *Reticulitermes* and *Heterotermes* damage to a structure can occur almost 10X faster. From its ports of entry new infestations of *C. formosanus* have spread to 8 additional states, mainly through human transport of infested wood (out-of-service railroad ties) used in parking lots and yard landscaping. *C. formosanus* readily attacks living trees in addition to structural wood, builds large carton nests above, as well as below ground, and has colony populations that exceed by 10X most native subterranean colony populations. Homes can be severely damaged in 6 months and nearly destroyed within 2 years if an infestation is left unchecked. Foraging galleries of a single colony may connect to multiple food sources and can extend over an area of 100 meters (*6*).

Worthy of note here is the current situation in Louisana with the Fomosan subterranean termite. In New Orleans and in Lake Charles, LA, large populations of Formosan termites took-up residence over 50 years ago. The recent disaster of the weeks-long flooding in many parts of New Orleans and some parts of Lake Charles resulted from hurricanes Katrina and Rita. It is believed that the flooded conditions may have reduced termite populations in these areas. It was previously documented that *Reticultermes* spp. experienced a 77% population decline after a long-term flooding event in Georgia (*7*). However, the above ground nesting typical of *C. formosanus* may more than make-up for this drowning short fall, and it is believed that populations in the affected parishes will persist despite the long-term ground flooding. Meanwhile the massive amount of cellulose debris directly related to this disaster and as homes are subsequently demolished pose a threat for spread of this termite outside of the infested areas. To this end, on October 3rd, 2005, the LA Commissioner of Agriculture and Forestry imposed a quarantine to stop any movement of structural wood or tree debris out of the infested areas. This result has challenged the industry to develop novel wood treatment methods that will allow for offsite transport for burning, burying or slavaging wood for resale. Meanwhile, Texas concurrently imposed a quarantine on movement into the state of any wood coming from Louisiana and Mississippi was the first state to impose a strict quarantine on movement of wood by homeowners afflicted with Formosan termites.

Figure 1. Salt water dam at the Calcasieu River in Lake Charles that had to be replaced due to Formosan termite attack.

Figure 2. Creosoted telephone pole hollowed out by Formosan termites.

Figure 3. Used railroad ties in New Orleans being readied readied for use as landscaping timbers.

A Brief Account of Wood Treatments With Emphasis on Newly Developed Directions

Until relatively recently, most wood treatments relied on creosote, pentachlorphenol, arsenic and heavy metals like copper and chromium for protection against wood destroying insects and fungi. The first recorded use of a wood preservative can be found in the Old Testament where Noah was instructed to build his ark using pitch (*4*). Large restrictions placed our most popular wood treatment over the last decade, CCA (chromated copper arsenate), was fairly swift in the USA when arsenic was found in the sand around playgrounds that used arsenic-treated wood for the play equipment. ACQ (alkaline copper quat) and CA (copper azole largely replaced CCA for residential use. However, ACQ and CA contains large amounts of copper and already has been banned from use in some countries. Borate wood treatments have been used since the early 20th century but use is restricted to interior conditions due to leaching properties of the compound (*4*). The "Holy Grail" of wood treatments, a fixed, non-leachable boron wood treatment, has not yet been developed.

With the continuing development of wood composites, already registered insecticides that have been in use for years are now finding their way into the wood treatment market. These include several pyrethroids, imidacloprid (Preventol TM, Bayer Product Information) and fipronil (*8*) (Table 2 and 3). Natural wood constituents were realized early on to also possess insect antifeedent, repellent and toxicant properties (*5,9*). While naturally resistant woods are constantly being evaluated (*10*) considerably more effort is being put into identifying the extractives of naturally resistant plants potential as a treatment additive to susceptible but readily available and cheap materials already in use (Table 4).

Recently, several states have approved the use of BoraCare (Nisus Corporation) spray on treatments up to 3 ft from the foundation as the single (stand-alone) treatment to prevent termites from entering a structure (see for example, www.pctonline.come/News/news.asp?Id=3070). Published data on the value of borate wood treatments as a stand-alone termite treatment are mixed. Some data indicate that termites may be repelled from wood treated with borates but the majority of reports show that termites can bypass these treatments and attack untreated wood in other parts of the structure. It has also been noted that termites can penetrate spray-on borate treatments and may not ingest enough of the toxic wood to cause serious harm to the colony. Using a single preventative tactic using wood treatment that have anti-feedant properties (as the borates have) posses a serious challenge to the wood industry.

Table 2. Repellent chemical barrier treatments (synthetic pyrethroids).

Active Ingredient	Brand Name / Manufacturer	Structure	Action on Termites
Fenvalerate	Tribute®/AgrEvo	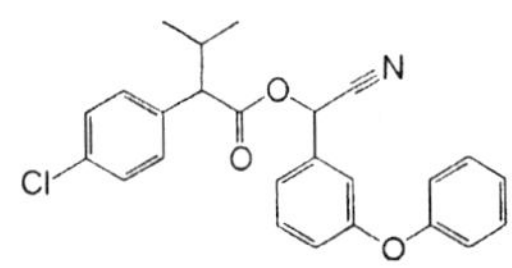	Axonic Poison Sodium Channel Blocker
Permethrin	Prelude®/Syngenta Permethrin Pro/ MicroFlo, Co. Permethrin TC/ Speckoz, Inc.	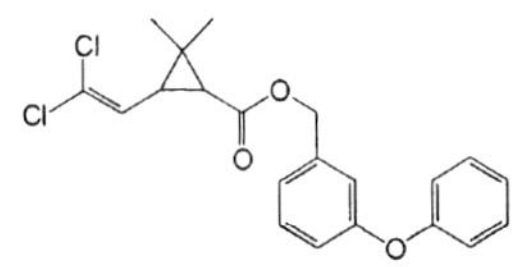	Axonic Poison Sodium Channel Blocker
Cypermethrin	Demon® TC/ Syngenta Prevail®/ FMC Corporation	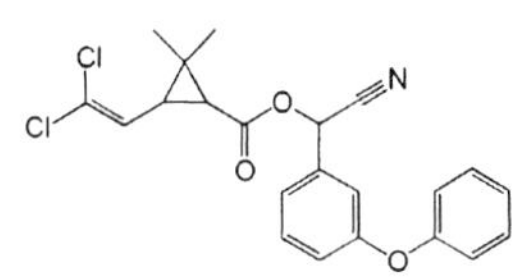	Axonic Poison Sodium Channel Blocker
Bifenthrin	Talstar One ®/ FMC Corporation	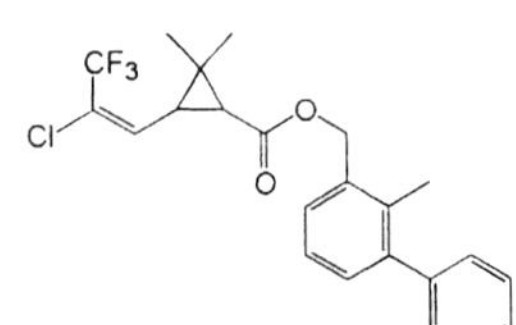	Axonic Poison Sodium Channel Blocker

Table 3. Examples of non-repellent liquid treatments. Note, chlorpyrifos has in the past been labeled as non repellent but most research indicates it is a repellent to termites. Also, New compounds are being continuously developed (this is only a partial list).

Active Ingredient	Brand Name / Manufacturer	Structure	Action on Termites/Class
Imidacloprid	Premise® 75/ Bayer	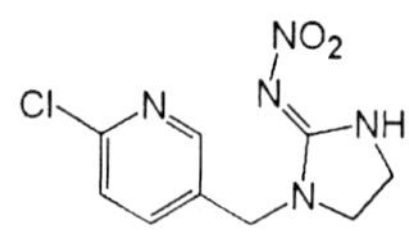	Chloride Channel Blocker/ Chloronictinyl
Fipronil	Termidor®/ BASF	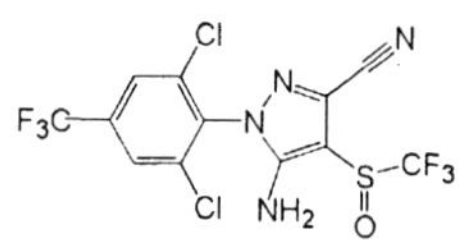	Chloride Channel Blocker/ Phenyl Pyrazole
Chlorfenapyr	Phantom®/BASF	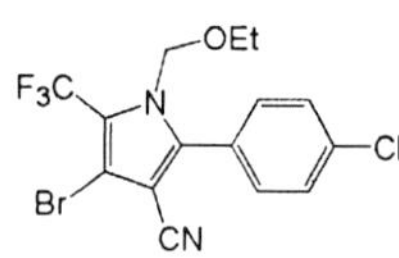	Stomach Insecticide Uncoupling Phosphorylation/ Pyrole
Chlorpyrifos	Cyren® TC/Cheminova Dursban™ TC/ Dow Agrosciences	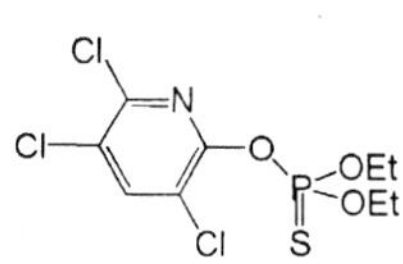	Cholinesterase Inhibitor/ Organophosphate

Table 4. Examples of some termiticides derived from plants.

Natural Source	Active Component	Structure
Tobacco Leaves (*Nicotiana tabacum*)	Nicotine	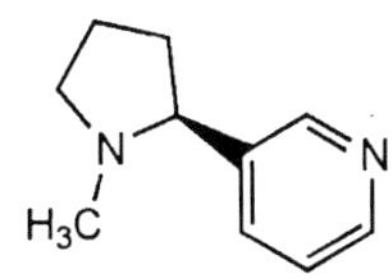
Catnip/ "Catmint" (*Nepeta cataria*)	Nepetalactone	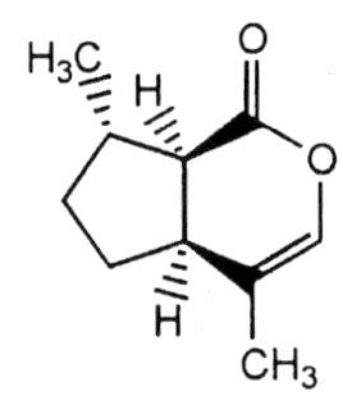
Red Peppers (*Capsicum annuum*)	Capsaicin	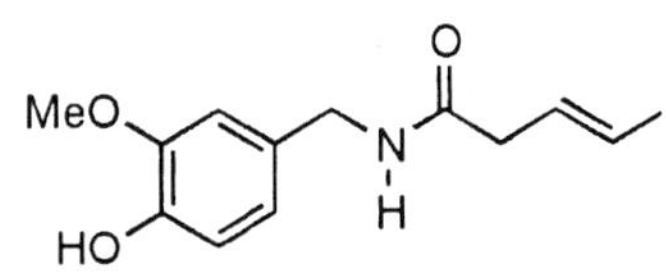
Sesame Oil (*Sesamum indicum*)	Sesamol	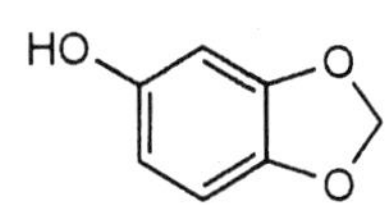
Corn Bud Oil (*Zea mays*)	2'-acctonaphthone	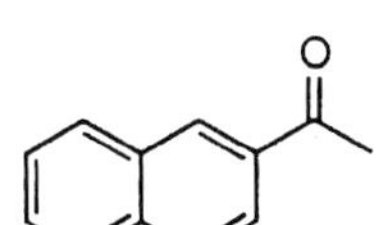
Alaskan Yellow Cedar (*Chamaecyparis nootkatensis*)	Nootkatone	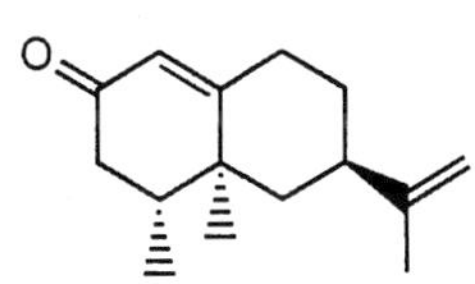

Behavioral Categories for Wwood Treatments

With the advent of newly developed wood treatments and the increased demand on the wood industry that some treatments be used as stand-alone termite preventatives, it is important to consider how termites react to different treatments. In Table 5 I attempted to provide behavioral categories for wood treatments presently available or in development. Tradtional treatments, like borates, CCA, CA and ACQ are feeding deterrents. These treatments protect the wood and require consumption of the treatment to affect termite survivorship.

Table 5. A behavioral categorization of some wood preservatives.

Behavioral Category	Definition	Routes to Toxicity	Examples
Repellent	Oriented movement away from source	Contact	Secondary plant compounds, pyrethroids
Attractant	Oriented movement toward source	Contact or Ingestion	Some organochlorines and chitin synthesis inhibitors
Deterrent	Will not feed on treatment	Ingestion	Borates, CCA, ACQ, CA, etc
Arrestant	Stops movement; causes aggregation	Ingestion	None presently used
Non-Repellent	No immediate change in behavior	Contact or Ingestion	Imidacloprid, fipronil, chlorphenapyr

As a preventative measure, this type of treatment will generally fail to keep termites from bypassing the toxicant and attacking wood elsewhere in the structure. Treatments that act as repellents or contact toxicants may be preferred over feeding deterents for wood treatments used as preventative treatments, since repellents will stop the movement of termites and contact toxicants kill them before they can get to untreated areas. Non-repellent termiticides and natural products that repel termites may offer the best protection.

Traditionally, wood treatments were examined foremost for their protection against fungi. However, while the pyrethroids, and non-repellent termiticides like fipronil and imidacloprid are efficacious against insects they do little against fungi and a fungicide must be co-added to the wood treatments for this protection. Treatment cost can double with the addition of a second chemical treatment.

Many of the essential oils or semiochemicals with low volatility have a fairly long termite repelling period and may also be efficacious against fungi. The volatile and semivolatile compounds of essential oils can be classified into four major groups; terpenes, benzene derivatives, hydrocarbons and miscellaneous compounds (*11*). Terpenes, organic compounds consisting of multiples of isoprene units are the dominant constituents of essential oils. Among them, the mono- and sesquiterpene derivatives predominate (*12-14*). Other examples include extracts of catnip, sesame seeds and corn bud oil show potential against termites (3, *15-16*; see also Table 4). However, studies have shown that naturally derived chemicals are generally less active then conventional (synthetic) insecticides (*17*). The bottom line is that if naturally derived chemicals are going to make it to the marketplace, then standards and/or applications that were approved for heavy-metal laced wood treatments must change. For example, eco-friendly European standards require that no etching of the wood take place, otherwise the treatment is considered a failure. Most newer treatments would fail this test, especially if some nibbling by termites is required for toxicity.

Currently the EPA has mandated testing requirements for treated wood to be at least 2 years but less than 5 (*18*). Most recently, however, the EPA has leaned toward a 5 year minimum though written documentation for this is not yet available. Testing procedures commonly follow a standard "ground stake test" as regularly performed by USDA-Forest Service and explained in the AWPA (American Wood Preservers' Association) standards for field testing of treated wood stakes. One of the important variables that is difficult to control in field tests of this sort is the presence of a relatively high termite population. Moreover, if the control stakes (untreated) are not attacked over several years there are no data to speak of. As a result, attack on untreated stakes is essential and, once attained, are maintained as diligently as possible to be able to have a comparison with treated stakes. Herein lies a major problem: once the control stake is attacked, even if it is replaced the control location remains forever biased due to the galleries of the subterranean termites under investigation (Figure 4). Under the current standards the only way to correct for this bias is to rerandomize treatments yearly. This is currently not done in the majority of cases. Field testing methodologies need further consideration. Drs. John French and Berhan Ahmed (Australia) have made a case for adopting a new standardized protocol for field and laboratory testing of termites (*19*). Although

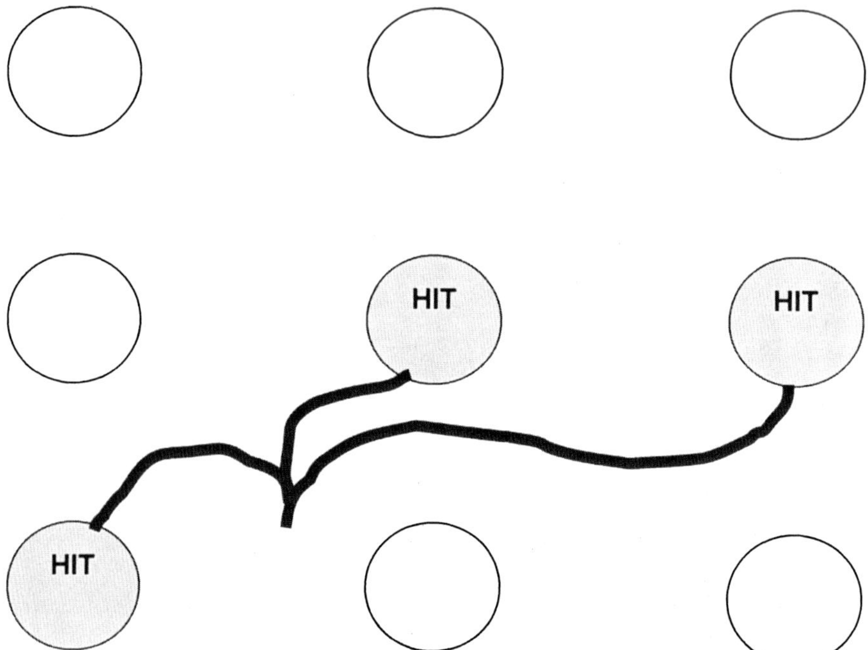

Figure 4. A schematic of wood placements in a field experiment. Initially all placements have an equal likelihood of attack. However once controls are attacked they remain biased for the duration of testing. Control (gray), treatment (white), line (gallery system made by termites.

they are particularly concerned with field situations of "uniform high termite hazard", for soil termiticide testing, their methodologies could be applied to testing wood treatments. In addition, accelerated field testing may be useful as has successfully been demonstrated using waterbound termite infested trees or field sites where termite populations are seeded in place (*20, 21*).

References

1. Maistrello, L.; Henderson, G.; Laine, R. A. *Pest Management Science* **2002**, *59*, 58-68.
2. Zhu, B.C.R., Henderson, G., Chen, F., Fei, H., Laine, R.A. *J. Chemical Ecology* **2001**, *27*, 1617-1625.

3. Ibrahim, S. A., Henderson, G., Fei, H., Laine, R. A. *Pest Management Science* **2004**, *60*, 746-754.
4. Ahmed, B. M., **2000**, Ph.D. Dissertation, University of Melbourne, School of Frestry, Melbourne, Victoria, Australia, 335 pp.
5. Becker, G., *Wood Science and Technology* **1971**, 5, 236-246.
6. King, E. G., *Biology of the Formosan subterranean termite, Coptotermes formosanus Shiraki, with primary emphasis on young colony development,* **1971**, Ph.D. Dissertation, Department of Entomol. LSU, 193pp.
7. Forschler, B.T., Henderson, G., *Environmental Entomology* **1995**, *24*, 1592-1597.
8. Kamden, D. P., Hope, J., Oudenweyer, J, McKenzie, M., *Forest Products J.* **2002**, *52*, 40-43.
9. Isman, M.B., *Pesticide outlook* **1999**, *10*, 68-72.
10. Grace, J. K., Yamamoto, R. T. *Forest Products J.* **1994**, *44*, 41-45
11. Ngoh, S. P., Choo, L. E., Pang, F. Y., Huang, Y., Kini, M. R., Ho, S. H., *Pesticide Science* **1998**, *54*, 261-268.
12. Rice, P. J., Coats, J. R., *Pesticide Science* **1994**, *41*, 195-202.
13. Zhu, B. C. R., Henderson, G., Yu, Y., Laine, R. A., *J. Agricutural and Food Chemistry* **2003**, *51*, 4585-4588.
14. Zhu, B. C. R., Henderson, G., Chen, F., Fei, H., Laine, R. A., *J Chemical Ecology* **2001**, *27*, 1617-1625.
15. Ibrahim, S., Kambham, S., Henderson, G., Jayasimha, J., *Proceedings National Conference on Urban Entomology* **2004**, *61-63.*
16. Peterson, C. J., Ems-Wilson, J., *J Economic Entomology* **2005**, *95*, 1275-1282.
17. Osbrink, W. L. A., Lax, A. R., Cantrell, C. L., *5th International Conference on Urban Pests* **2005**, 213-223.
18. USEPA Product Performance Guidelines, **1998**, OPPTS 810.3600 Structural Treatments, EPA 712-C-98-424, 4pp.
19. French, J R. J., Ahmed, B. M., *Sociobiology* **2005**, *46*, 551-562.
20. Henderson, G., Forschler, B. T., *Louisiana Agriculture* **1997**, *40*, 9-11.
21. Smith, W. R., Amburgey, T. L., Henderson, G., Ring, D. R., *Forest Products J* **2004**, *54*, 26-28.

Chapter 15

Termite Control from the Perspective of the Termite: A 21st Century Approach

R. Joseph Woodrow and J. Kenneth Grace

Department of Plant and Environmental Protection Sciences, University of Hawaii at Manoa, Honolulu, HI 96822

Termites, although essential recyclers of carbon in tropical ecosystems, are serious structural pests in urban environments. Traditional termite management was based largely on the use of biocides, which did not require an intimate knowledge of the organism. However, recent legal and socioeconomic pressures have forced many of the toxicants out of the marketplace. We present a review of the current and cutting-edge developments in termite management that are based upon a greater understanding of termite behavior and ecology.

Introduction

The termites (Order Isoptera) are an important insect group comprising approximately 2300 species (*1*). One could easily argue that termites are principally a beneficial group of insects owing to the fact that only 10% of these species are pests (*2*). Termites are the chief decomposers of cellulosic debris (trees and other plants) in the tropics, contributing to the development of humus, as well as bringing lower-level nutrients to the soil's surface (*3*). Their food can be a standing, dead tree, an old stump, buried wood, or your home. Although consuming dead wood is generally a beneficial function in nature, it puts termites in direct conflict with humans in our urban environment.

Individually, termites are blind, helpless insects. However, collectively they are unaffected by many environmental and manmade hazards. Some researchers have suggested that social insect colonies, such as those of termites (as well as ants, wasps, and bees) act not as collections of individuals, but as "superorganisms" (*3*). Like the many specialized cells within our own bodies, the different castes (workers, soldiers, reproductives), and even specialized groups within each caste, conduct specialized activities: foraging, temperature regulation, feeding, defense, reproduction, and brood care. Thus, when relieved of the necessity to perform other tasks a single termite queen can produce thousands of eggs per day. Because of this specialization, colonies of some subterranean termite species can exceed 1,000,000 workers (*4, 5*). Members of the colony are linked by a complex communication network that involves chemical (pheromone) and tactile (vibratory) stimuli. Communication facilitates the rapid recruitment of the colony to food sources, repellence of invaders and defense from pathogens and toxins.

Subterranean termites (Family Rhinotermitidae), as their name suggests, live beneath the soil surface. A colony can literally exist anywhere in the landscape totally hidden from view, while maintaining numerous foraging galleries, each leading to a separate food source. Foragers of some species can cover distances in excess of 100 meters (*4*) and can occupy an area as large as 3000m^2 (*6*). The cryptic and diffuse nature of termite colonies makes control a challenging proposition.

Although drywood termites are more common overall in structures than subterranean termites (*7*), the damage they cause is relatively minor when compared to the rapid and severe damage caused by large subterranean termite colonies. Su (*8*) estimated that 80% of the economic damage is due to subterranean termites. The most important subterranean termite species in the United States are *Coptotermes formosanus* in Hawaii and along the Gulf coast (*9*), *Heterotermes aureus* in the southwest, *Reticulitermes hesperus* in California, and *Reticulitermes flavipes* from the southeastern states north into southeastern Canada (*10*).

Economic estimates of damage from subterranean termites are problematic as they involve many different segments of the economy, including not only the pest control and construction industries, but real estate, insurance and legal as well. It would generally be agreed that damage in the United States easily exceeds $1.5 billion (*8*). In the state of Hawaii, a paradise for termites as well as humans, the Formosan subterranean termite costs the state at least $100 million annually (*11*); a 15-year-old estimate that can easily be doubled given the recent real estate boom.

Due to its sheer size, a subterranean termite colony needs to consume large quantities of biomass to maintain itself. While the total worker mass of a colony of *R. flavipes* can range from 7 to 10 kg (*5*), the foraging population of a *Coptotermes* colony can be as large as 34 kg (*4*). Grace (*6*) suggested that a large subterranean termite colony can be thought of as a moderately sized herbivore (an invisible, giant rodent gnawing on your home), consuming on the order of 1 kg (2.2 lb) of cellulose per day.

The potential damage wrought by these insects dictates requisite efforts to protect property. Control of subterranean termites in buildings has traditionally involved the use of persistent, broad-spectrum insecticides applied to the soil beneath structures. This persistence was once necessary to provide a barrier, which either killed termites attempting to cross it or repelled them from the structure. The organochlorine soil insecticides chlordane and dieldrin, for example, could keep termites out of structures for many decades (*12*). However the same features that made these treatments so effective, their persistence and broad-spectrum activity, also created environmental problems and a sociopolitical backlash that can still be felt throughout the pest control industry. Due to these environmental concerns these persistent insecticides have been removed from the market and other more environmentally-benign methods and materials have taken their place.

Chemical Barriers

The earlier practice of applying environmentally-persistent termiticides (organochlorines) was very effective and provided multiple decades of protection. Ideally, this traditional preventative strategy was implemented prior to construction, when termiticide could be thoroughly applied directly to the soil. Less persistent termiticides have now taken the place of the persistent materials, however. Thus, the overall approach to termite prevention has had to be modified because modern soil termiticides do not remain effective as long as the incredibly persistent chlordane.

Insecticides used as "barriers" around structures can be repellent (e.g., pyrethroids such as permethrin, cypermethrin, and bifenthrin), non-repellent and causing rapid mortality (e.g., chlordane and chlorpyrifos), or non-repellent with a delayed action (e.g., imidacloprid and fipronil). Su et al. (*13*) referred to these categories as Type I, II, or III toxicants, respectively. Chlordane and its immediate successor in the marketplace, chlorpyrifos, were not repellent; however, the accumulation of dead termites at the treatment boundary appears to cause a secondary repellency in the form of necrophobia (*13*). Repellent pyrethroid insecticides protect the structure by preventing penetration of the treated soil, while leaving the termite population largely intact. The non-repellent, delayed toxicants, such as imidacloprid and fipronil, are the latest tools for soil treatment. These newer insecticides appear to protect the structure primarily by killing a large portion of the termite population, rather than by any sort of direct or indirect repellent action.

Repellent toxicants are generally pyrethroids. Since they are overtly repellent to termites and probably do not actually kill many foragers, the barrier effect can be rather profound. The first generation pyrethroid permethrin is particularly persistent and has remained effective for over 10 years at some locations in field tests in Hawaii (*14*). Later generation pyrethroids, such as bifenthrin, are more biologically active than permethrin but are also less persistent, most likely due to their application at much lower concentrations.

Delayed toxicants, such as imidacloprid and fipronil, are toxic to termites at low concentrations. Unlike the traditional rapid toxicants these materials may require days or even weeks to kill. Thus, the paradox of using these delayed action termiticides is the fact that termites can readily penetrate the treated soil (*15*). Notwithstanding, several studies have indicated that termite penetration is actually more limited than is generally thought (*16*, *17*). Yet, the ultimate role of these materials may not be in the traditional sense as barriers, but rather their effect in reducing the overall termite population.

A population-level effect could reduce termite pressure on structures and possibly even resemble the effects of a termite bait application, as discussed later. Transfer of these materials from exposed donors to unexposed recipients in laboratory studies (*18*, *19*) suggests that they may reduce termite populations beyond the treated area, although Shelton and Grace (*18*) suggested that the transfer effect is likely limited in scope. The mechanism of transfer is unknown, but it is probably the result of grooming and/or trophyllaxis. Su (*20*) investigated the horizontal transfer of the delayed toxicants fipronil and thiamethoxam as compared with a typical bait toxicant, noviflumuron, in a linear, 50-m foraging arena. He found that while the bait toxicant had a universal impact throughout the arenas, termite mortality with the delayed toxicants was limited to a range of less than 5m from the site of application.

A recent development in termite control is the use of polymer sheets impregnated with repellent insecticides. In building construction, the polymer sheeting can act both as a moisture barrier and as a barrier to termites. The polymer may be either "leaky" or "non-leaky." A leaky barrier functions much as a slow-release insecticide formulation and slowly emits a small quantity of termiticide into the surrounding soil. This effect can extend the residual efficacy of insecticides, such as deltametrin, that would otherwise degrade in the moist soil environment (*21*). In the United States, a similar type of material containing lambda-cyhalothrin proved effective (*22*). However, concerns over handling and pesticide residues led to further development as a non-leaky barrier, in which the insecticide-impregnated plastic layer is sandwiched between two impermeable plastic layers. Tests with this three-layer barrier have shown that termite penetration of the plastic barrier ceases immediately when the middle layer is reached. Theoretically, these materials may be able to provide the level of long-term protection once conferred by the persistent organochlorine termiticides.

Physical Barriers

The loss of persistent, broad-spectrum termiticides also facilitated the development of various physical alternatives, each of which act as an obstacle between the termites' subterranean environment and manmade wooden structures. While physical barriers are certainly longer-lasting than insecticides applied to the soil, because they are not toxic or repellent termites will readily probe the surface of these materials in search of entry points. This means that these materials must be flawlessly installed to the extent that termites must venture into an area visible to inspection. Thus, barriers also may aid in termite detection, as the termites' cryptic nature is their greatest asset in avoiding control strategies. Conventional monolithic slab construction alone does not provide a significant barrier to termites, as they can come through plumbing and utility penetrations, or through cold joints and cracks that inevitably form in concrete slabs. Both wire mesh and particle barriers have been developed and commercialized to protect these common entry points.

Sand or gravel particle barriers take advantage of the relative size and strength of termite workers. This concept was first presented using silica sand (*23*), and later developed using crushed, basaltic rock in Hawaii (*24*). The commercial product now widely available in Hawaii is called Basaltic Termite Barrier, or BTB (Ameron HC&D, Honolulu). Tamashiro et al. (*24*) found that Formosan subterranean termites would not penetrate a layer of gravel particles with a size range of 2-2.5mm when the thickness was at least 10 cm thick. From the perspective of the termite, a larger particle size leaves large gaps that they can penetrate, while a smaller particle size can be manipulated with the mouthparts and thus excavated. Within the critical size range, the gaps between particles are too small to penetrate and the particles too large to be grasped and moved. BTB can be applied in the same way as normal gravel fill before a concrete slab is poured, and has been accepted into the building codes in Hawaii as a preconstruction termite treatment (*25*). Elsewhere in North America where the basalt matrix is not readily available, similar gravel barriers have been investigated but have not been commercialized. In Australia, however, a similar barrier of crushed granite was developed and is used commercially against *Coptotermes* species (*26*).

Perhaps the simplest barrier is a durable metal screen with a mesh size small enough to exclude termites. TermiMesh (TermiMesh Hawaii, Honolulu) is a marine-grade stainless steel mesh that can be used instead of chemical barriers. Field studies with this material have demonstrated that it can prevent termite foraging or penetration for extended periods (*25*, *27*). However, this is a relatively high-cost material, and cannot be used in large-scale installations under concrete slabs in the same manner as the particle barriers. Instead, the mesh is generally inserted only in critical areas, such as around plumbing penetrations, in cold joints and along retaining walls, all areas that are difficult to treat with insecticides.

Biological Control

Biological control is a process whereby natural parasites or predators are introduced or augmented to reduce or eliminate insect populations. Many successful biological control agents have been developed for insects other than termites, thus there has been a great deal of interest in their use against termites. A number of recent reviews of termite biological control have been conducted (28-32). Given the cryptic nature of termites, control is inherently difficult. Thus, it has been hypothesized that a biological agent introduced into a termite colony might be able to penetrate the inner sanctum and kill colonies that would otherwise be impossible to control. Unfortunately, a number of studies have demonstrated that termites, which live in an environment rich with all manner of pathogens, are adept at detecting these agents and responding defensively.

Lack of success with biological control has not been due to a paucity of potential agents. In a recent review, Myles (*32*) listed 2 viruses, 5 bacteria, 17 fungi, 5 nematodes and 4 mites that are potentially detrimental to termites. Specifically, Zoberi and Grace (*33*) identified a unique pathogenic fungi associated with *R. flavipes*, while 15 bacteria and one fungus have been associated with *C. formosanus* in Louisiana (*34*). The most plausible and potentially effective biological control agents to date have been either fungi or nematodes.

The most promising fungal pathogens are those that are endoparasitic, including the two common insect parasitic species *Metarhizium anisopliae* and *Beauvaria bassiana*. Numerous studies have shown that these fungal species readily kill termites under laboratory conditions, but field studies have not been as favorable (*29*). Laboratory studies in Hawaii (*35, 36*) indicated that even small doses of conidia applied for durations of less than one hour were lethal to termites. The goal of using these fungi was to treat a small portion of the foraging population which would then become infected and spread the infection to the remainder of the colony. The principle difficulty with these fungal species, however, was that they did not sporulate readily within dark termite galleries, and infection was limited to only those individuals immediately treated. While it seems evident that epizootics could be generated in the field, the concentrations of conidia required and the numbers of termites that would have to be exposed simply may not be practical under typical commercial field conditions.

Nematodes in the genera *Heterorhabditis* and *Steinernema* have shown promise as pathogens against termites. Wang et al. (*37*) investigated a number of nematode species against *R. flavipes* and *C. formosanus* in laboratory studies and found high levels of virulence for some nematode species at concentrations exceeding 200 nematodes per termite. *Heterorhabditis* spp. nematodes have been used to control residual populations of *Coptotermes* spp., however high nest temperatures (30°C) preclude their use against entire colonies (*31*). Additionally, Mankowski et al. (*38*) recently reported that termite workers of

Coptotermes species readily removed the nematodes *Heterorhabditis indica* and *Steinernema carpocapsae* by mutual grooming. These authors concluded that the dosages required for control in the field were not practical. As with some fungal applications, these high nematode concentrations may also be hampered by the fact that they are repellent to termites (*37*).

Limited field efficacy of biological control agents against subterranean termites is more the rule than the exception. Termites have a number of social, sanitary behaviors, and also physiological mechanisms, that allow them to avoid epizootic infections. Termites may avoid contact with fungal spores and toxin-contaminated surfaces by vacating contaminated gallery systems (*39*). Dampwood termites, *Zootermopsis* spp., have been reported to exhibit an immune response to pathogens (*40*) in a manner analogous to the immune response of a human to a bacterial or viral infection. Once a pathogen is detected, workers rapidly communicate alarm via head-banging (an acoustic signal) or other behaviors. Head-banging has been observed by *Z. angusticollis* in the presence of spores of *M. anisopliae* (*41*), and this vibratory activity causes unexposed individuals to rapidly flee the infected area. For those individuals already exposed, allogrooming (*42*, *38*); i.e., termites grooming each other, may also reduce the pathogenicity of agents that need to germinate on, or penetrate, the cuticular surface. Termite workers not only can detect pathogens but will also isolate or bury sick individuals away from the rest of the colony (*43*). There is also evidence that termites may be able to produce antibiotics and fungistatic compounds, such as naphthalene (*44*).

Although these biological agents have had limited efficacy in the field, success with biological control may not be with the agents alone but in combination with other strategies. As concluded by Culliney and Grace (*29*), biological control agents may come to "supplement" but not "supplant" established control strategies. Lenz (*31*) concluded that pathogens are more effective when the colony is weakened by age or insecticide application. Sublethal doses of imidacloprid have been linked to high rates of fungal infection, synergistically causing faster colony collapse than either agent alone (*45*). These combined strategies have yet to be developed into commercially viable products, however. A final strategy for successful microbial control may be to engineer more effective pathogens, as suggested by Grace (*30*) and discussed below.

Physical Control

All insects are poikilothermic; i.e., they are at the mercy of environmental extremes, as their body temperature closely follows that of their surroundings. A number of free-living insect species are adapted for living in extreme environments, such as desert-inhabiting tenebrionid beetles, or arctic chironomid midges. Unlike ants, which often forage in the open, termites must carry their cool, moist microclimate with them wherever they go. Termites, as individuals, are extremely sensitive to minor variations in the physical environment. It is only as members of a colony that can they modify their environment to maintain homeostasis.

Temperature extremes have been investigated in a number of studies as a means of control. Woodrow and Grace (*46*) found that termites are generally susceptible to high temperatures: 90-minute exposures of 42°C for subterranean termites and 45°C for drywood termites are fatal. Additionally, termites are not able to acclimate to rapidly-rising temperatures, as has been observed with many free-living insect species (*47*). Heat has been used successfully to control drywood termites in a commercial setting, as they are captives of the structure which they infest (*48*). However, thermal structural treatments are not effective for the control of subterranean termites because they have no effect on those individuals that are not present in the structure; i.e. the remainder of the colony, at the time of treatment.

Extreme low temperatures (achieved with liquid nitrogen) have also been investigated for control of drywood termites under simulated field conditions (*49*). Lewis and Haverty (*49*) determined that excessively high rates of liquid nitrogen (>200kg/m^3) were required to achieve 99% mortality. Additionally, there are numerous safety concerns that go along with the use of large quantities of a frozen, compressed gas. In commercial applications of liquid nitrogen, depletion of the oxygen in confined spaces, such as attics, is a recognized danger for applicators. As with high temperatures, it is difficult to conceive of a way to use this technology to control soil-inhabiting termite species.

Termite Bait Technologies

Modern baiting technologies are the ultimate example of use of the biology of termites against them. As discussed previously, one of the most confounding aspects of subterranean termite biology is the fact that colonies are cryptically located, making direct control all but impossible. Knowledge of foraging behavior has given rise to baiting strategies, because it has allowed us to utilize the foraging worker population as a means of introducing a toxicant to the colony at large. The goal of this technology is to reduce or eliminate termite populations around structures, and thus reduce the possibility of infestation. To understand the technology, it is important to understand the historical context.

Early research indicated that termites transfer food and other materials to their nestmates via a process called trophyllaxis. For the sake of this discussion, this transfer is particularly profound between foragers and nestmates, as only a portion of a social insect colony actually leaves the nest or visits any particular foraging site. While typical materials transferred include food, symbionts and hormones, this social sharing behavior is an inherent weakness that baiting can exploit. Termite foraging studies were advanced by the use of mark, release and recapture (MRR) methods (*50*). MRR is a process where a large number of foraging termites are captured in the field and brought back to the laboratory, where they are counted and fed a persistent dye marker, and then returned to the field site where they were initially captured. After releasing the marked termites, sampling is then again undertaken at subsequent intervals. The termite samples will contain a proportion of marked and unmarked individuals, and when compared to the original released quantity and location can be used to estimate the foraging range and size of the foraging population. Various studies established that the foraging territories of subterranean termites can exceed 3000m^2 meters, with foraging populations often in excess if 1,000,000 individuals (*4*, *5*,). This technique was also essential in the development of baits in another way: it established the efficacy of toxicants in the field; termite activity was monitored within the pre-established foraging range in non-baited monitors after applying a bait toxicant to ascertain a decline in the termite population or activity (*51*, *52*).

In addition to basic discoveries about termite foraging, an essential requirement for a baiting protocol was the discovery of an effective toxicant. It was reported in the 1970s that the organochlorine insecticide mirex could reduce foraging populations of subterranean termites (*53*). However, this insecticide was subsequently deregistered in the United States. The search for another viable toxicant took over a decade, leading one researcher to draw an analogy to the search for the Holy Grail (*54*) given the various obstacles to bait development.

A number of criteria must be met for a toxicant to be effective in a baiting scenario. Toxicants must first be slow-acting; i.e., toxicity must be delayed long enough for the material to be transferred to the remainder of the colony from foragers. Also, delayed toxicity will assure that there is no association between the toxicant and any subsequent mortality (*13*). Secondly, the material must not be repellent to foragers or it must be undetectable at toxic concentrations. Thirdly, the material must be stable enough in the termite's body as to be transferred between individuals and persist sufficiently long for toxicity to be expressed. Lastly, the toxicant must be effective at low concentrations, as one cannot reasonably expect foragers to consume large quantities of a bait in the field, given competing natural food sources.

Various materials have been investigated for use in baits, including fungi (*55*) and various chemical toxicants: sulfluramid (*56*), borates (*57*), hydromethylnon (*13*), abamectin (*58*), imidacloprid (*59*), and fenoxycarb (*60*); all with somewhat limited success, although several of these have in fact emerged as viable commercial products. The one class of insecticides that has emerged as most effective in termite baiting is the chitin synthesis inhibitors. Specifically, the benzoylphenyl ureas hexaflumeron, novaflumeron, diflubenzeron and lufenuron have been shown to be potentially effective, and are currently in use as commercial bait toxicants (*61*). These materials are true delayed toxicants, since mortality largely occurs only during molting, or other cuticle forming periods of the termite life cycle. As such, these insecticides, once consumed, are like time bombs that kill arbitrarily at a later time, with no association to the original source of exposure.

Subterranean termite baiting often involves two distinct processes: monitoring (first to detect termites, and second to establish efficacy of the treatment) and toxicant baiting. Commonly, the first step in a baiting protocol is to establish termite foraging activity in a monitoring station. This involves installing monitoring stations around structures, with some portion of the stations below ground, into which some form of cellulose food is placed. Periodic checks of the monitors are made to determine whether termites are present and actively feeding. Once termites are observed in the monitor, a toxic bait matrix is placed into the station in place of the edible cellulose. A number of variations of the process can be found among commercial systems. Some systems do not differentiate between monitoring and baiting, placing baits into inactive stations either with or without wooden attractants around the perimeter of the station.

The apparent efficacy of subterranean termite baiting systems has been accompanied by a certain degree of controversy due to the difficulty of directly observing colony mortality as a result of bait toxicant consumption. However, this does not mean that baiting is not effective, just that it is extremely difficult to establish the absence of an exceedingly cryptic organism. In a recent review, Su and Scheffrahn (*62*) reported that among 16 published field studies of the Sentricon Colony Elimination System (Dow AgroSciences, Indianapolis), total cessation of activity was reported in 89% of 53 colonies representing eight termite species. However, of necessity these claims about colony elimination/mortality are based largely on termite foraging activity in monitoring/bait stations. Some longer-term studies have observed reinvasion of monitors (*63*), which would be expected over time, as new colonies move into a given area. This underscores the importance of continued monitoring following bait treatments.

The controversy over how to best define the efficacy of baiting systems has led to a great deal of recent, productive research into termite tunneling and foraging behavior. For example, Campora and Grace (*64*) demonstrated that subterranean termites would continue to forage outward for new food sites from sites that they had recently discovered. This work reinforced the likelihood of foraging termites finding bait stations placed in the field, even though the colony was also feeding on natural food sites, and possibly manmade structures as well, at that location.

Baiting technology is typically applied on a single-structure basis. However, one house in a neighborhood with a baiting system could theoretically control termites that are also feeding on neighboring houses. Thus, there has been interest in using these systems to control populations over larger areas. Government-funded, area-wide projects in Mississippi and Louisiana have certainly decreased termite incidence (*65*, *66*), although long-term results are not conclusive given that each area, where treatment has been applied, is surrounded by neighborhoods with extensive termite populations and activity. Baiting was also used successfully to eliminate subterranean termites from a large area in Florida, although social and political factors impinged on the effort (*67*).

The successful commercial implementation of termite baiting has led many to seek improvements to the existing paradigm, both to augment existing baiting systems and to establish unique new products. The first augmentation that we will discuss relates to decreasing the time and/or increasing the probability of termite discovery of in-ground monitoring stations. This is especially significant given the results of Puche and Su (*68*), who found that subterranean termites were unable to detect sound wood even within a distance as small as 2.5mm; i.e., termites have to tunnel directly to bait stations to locate them. So, if foraging workers cannot detect a piece of sound wood in close proximity, what can be done to the area around the monitoring or bait station to improve the odds and reduce the time until discovery?

One way to improve discovery might be to modify the soil around the monitor with a chemical that could cause termites to tunnel more readily in that area. One possibility is the use of an analog of termite trail following pheromone, 2-phenoxyethanol, isolated and identified from ball-point pen ink (*69*). Compounds associated with brown-rot fungi may also be attractive to termites. In laboratory studies, Cornelius et al. (*70*) recently confirmed that subterranean termite foragers will aggregate in the presence of brown-rot fungi-degraded wood when given the choice of sound or decayed wood. Su (*71*) further demonstrated that foraging subterranean termites will orient towards extracts of brown-rot fungi in the field, suggesting, as did Grace et al. (*72*) with trail pheromone, that these chemicals could be applied to the soil to direct termite foraging toward monitoring or bait stations. Practical difficulties in the application of such an approach to bait enhancement may be the stability of the extracted chemicals (*73*), and habituation by the termite foragers to their presence (*74*). It has also been demonstrated that subterranean termites tunnel more readily in damp sand as compared to low-moisture conditions (*75*). Carbon dioxide has also been suggested as a means of attracting termites to monitors (*76*).

Other possibilities to shorten discovery time involve modifying the physical environment around a monitor. For example, Ettershank et al. (*77*) found that *Gnathamitermes tubiformans* and *Amitermes wheeleri* discovered food items on the soil surface in desert environments far more readily than similar buried items, because termites seemed to detect the thermal shadow produced by items on the surface. More recently, Swoboda and Miller (*78*) found that *R. flavipes* aggregated in thermal shadows (cooler areas at 20°C) as compared to warm areas (25°C).

Other physical aspects of the subterranean environment that may be used to direct termite foraging are guidelines; termites naturally follow solid surfaces in the soil environment. Laboratory assays with various guidelines demonstrated that *Reticulitermes* spp. workers followed wood and wood thermoplastic composites in preference to plastic alone (*79*). Monitors connected with these materials were more likely to be encountered by termites than unconnected monitoring stations. However, bait consumption did not increase in connected monitors, presumably because termites initially fed upon the guideline materials themselves rather than the contents of the monitors.

Once termites are active in monitoring or bait stations, the next important parameter is feeding on the bait. If foragers do not feed on the bait, or it is not preferred, they may not accumulate a toxic dose. Various researchers are interested in chemicals that could be added to bait matrices to increase palatability and thus increase feeding. Chen and Henderson (*80*) investigated the feeding responses of termites to various amino acid additives. A number of these patented materials were also investigated by Cornelius (*81*), and only ergosterol proved to be a significant feeding stimulant. Various carbohydrates have also been investigated as termite phagostimulants, and papers treated with 1-3% glucose or sucrose have been identified as preferred in various studies (*82*, *83*). Other identified preferred sugars include fructose, rafinose, galactose and trehalose (*82*), and xylose (*83*). However, Cornelius (*84*) investigated the effect of a sports drink (Gatorade®) containing various carbohydrates on discovery times of *C. formosanus*, and did not find a significant effect over non-treated controls. In the same study (*84*), however, termites did appear to orient their tunnels towards an extract of a commercial bait enhancement product (Summon, FMC Corporation). Swoboda et al. (*82*) also observed that papers containing low concentrations of uric acid were preferred over untreated paper.

The Cutting Edge

As discussed previously, pathogens have been investigated with little success in developing a viable biological control agent for subterranean termites, due to their ability to detect and respond behaviorally or physiologically to these threats. Also discussed previously, baiting strategies have proven to be effective in the commercial setting. Researchers are currently investigating a novel hybrid of these two strategies. In this scenario, a bacterial "Trojan Horse" would be introduced into a termite colony via conventional "toxicant" baiting (*85*). The advantage of this variation is that the bacterial species is a naturally occurring termite symbiont, that would not be rejected by foragers. This bacterial species would carry within its genome a gene which would express a lethal toxin that could be turned-on (expressed) via some environmental or introduced factor. This might seem to be outside of the realm of possibility, if not for the fact that the tools for its implementation have already been developed.

Husseneder et al. (*86*) successfully introduced a model bacterium, *Escherichia coli*, transformed with a fluorescent marker, into groups of termites and demonstrated that the bacteria were rapidly spread throughout the group through trophallaxis. However, the transformed bacteria did not persist more than a week in a termite colony without constant feeding. Husseneder et al. (*87*) then isolated a termite gut bacterium, *Enterobacter cloacae,* and transformed it, as had been done with *E. coli*. These transformed bacteria were quickly established in the laboratory colony and persisted for at least 11 weeks, unlike the *E. coli* model. To assuage concerns about releasing a genetically transformed organism into the environment, the transformed bacterium was also introduced into soil, with no evidence that it persisted or that the marker genes were transferred into other soil bacteria. Clearly, more research is needed to develop this concept to commercial fruition, but the use of such genetically modified microbial control agents may represent the future of termite baiting technology.

Acknowledgement

Preparation of this review, and University of Hawaii research described herein, was supported by McIntire-Stennis funds for forestry research and USDA-ARS Specific Cooperative Agreement 58-6615-4-237.

References

1. Borror, D. J., C. A. Triplehorn, C. A.; Johnson, N. F. *An Introduction to the Study of Insects, 6th Ed.* Saunders College Publishing: Philadelphia, PA, 1989; 875 pp.
2. Krishna, K.; Weesner, F. M. *Biology of Termites*; Academic Press: N.Y., NY 1970; Vol. 1, pp. vi-vii.
3. Wilson, E. O. *The Insect Societies*. Belkamp Press of Harvard University Press: Cambridge, MA, 1970; 548pp.
4. Su, N. -Y.; Scheffrahn, R. H. 1988 *Sociobiol.* **1988**, *14*, 353-359.
5. Grace, J. K.; Abdallay, A; Karr, K. R.. *Can. Entomol.* **1989**, *121*, 551-556.
6. Grace, J. K. In *Proceedings of the National Conference on Urban Entomology*; Robinson, W. H., ed.; College Park, MD, 1992; pp. 67-76

7. Rust, M. K.; Grace, J. K.; Wood, D. L.; Reierson, D. A. *Calif. Agric.* **1998**, *42*, 15-18.
8. Su, N. -Y. *J. Econ. Entomol.* **1994**, *87*, 389-397.
9. Su, N. -Y.; Ban, P.; Scheffrahn, R. H. *Sociobiol.* **1990**, *17*, 77-94.
10. Weesner, F. M. *The Termites of the United States – A Handbook.* National Pest Control Association: Elizabeth, NJ, 1965; 69 pp.
11. Tamashiro, M; Yates, J. R.; Yamamoto, R. T.; Ebesu, R. H. In *Pest control in the 90s: problems and* challenges; P.K.S. Lam, P. K. S; O'Toole, D. K., eds.; Applied Science: City Polytechnic, Hong Kong, 1990; Pp.77-84.
12. Grace, J. K.; Yates, J. R.; Tamashiro, M.; Yamamoto, R. T. *J. Econ. Entomol.* **1993**, *86*, 761-766.
13. Su, N. -Y.; Tamashiro, M.; Yates, J. R.; Haverty, M. I. *J. Econ. Entomol.* **1982**, *75*, 188-193.
14. Grace, J.K.; Yates, J. R.; Tamashiro, M. *Pest Control.***1993**, *61*, 60-61.
15. Hu, X. P. *J. Econ. Entomol.* **2005**, *98*, 509-517.
16. Gahlhoff, J. E.; Koehler, P. G. *J. Econ. Entomol.* **2001**, *94*, 486-491.
17. Remmen, L. R.; Su, N. -Y. *J. Econ. Entomol.* **2005**, *98*, 906-910.
18. Shelton, T. G.; Grace, J. K. *J. Econ. Entomol.* **2003**, *96*, 456-460.
19. Ibrahim, S.A.; Henderson, G.; Fei, H. *J. Econ. Entomol.* **2003**, *96*, 461-467.
20. Su, N. -Y. *J. Econ. Entomol.* **2005**, *98*, 471-475.
21. Lenz, M.; Morrow, P. and Runko, S. In *International Research Group (IRG) on Wood Preservation, 29th Annual Meeting*; IRG Secretariat: Stockholm, Sweden. IRG/WP 98-10264, 1998; 8pp.
22. Su, N. -Y.; Ban, P.; Scheffrahn, R. H. *J. Econ. Entomol.* **2004**, *97*, 570-574.
23. Ebeling, W.; Pence, R. J. *J. Econom. Entomol.* **1957**, 50, 690-692.
24. Tamashiro, M; Yates, J. R.; Ebesu, R. H. In *Biology and Control of the Formosan Subterranean Termite*; Tamashiro, M.; Su, N. -Y., eds.; College of Tropical Agriculture and Human Resources, University of Hawaii: Honolulu, HI,1987; pp. 16-20.
25. Grace, J. K.; Yates III, J. R.; Tome, C. H. M. *Sociobiol.* **1996**, *28*, 365-372.
26. French, J. R.; Ahmed, J. B.; Trajstman, A. *Sociobiol.* **2003**, 42, 129-149.
27. Lenz, M.; Runko, S. *Sociobiol.* **1994**, *24*, 1-16.
28. Grace, J. K. *J. Agric. Entomol.* **1997**, *14*, 281-289.
29. Culliney, T. W.; Grace, J. K.; Ford, P. W.; Davidson, B. W. *Bull. Entomol. Res.* **2000**, 90, 9-21.
30. Grace, J. K. *Sociobiol.* **2003**, *41*, 115-121.
31. Lenz, M. In *Proceedings of the Fifth International Conference on Urban Pests*; Lee, C. –Y.; Robinson, W. H., eds.; Executive Committee of the International Conference on Urban Pests, Singapore, 2005; pp. 47-52.
32. Myles, T. G. *Sociobiol.* **2002**, *40*, 257-262.
33. Zoberi, M. H.; Grace, J. K. *Mycologia* **1990**, *82*, 289-294.
34. Osbrink, W. L. A; Williams, K. S.; Connick, W. J. Jr.; Wright, M. S.; Lax, A. R. *Environ. Entomol.* **2001**, *30*, 443-448.

35. Leong, K. L. H. Master's thesis, University of Hawaii, Honolulu, HI, 1966.
36. Jones, W. E.; Grace, J. K.; Tamashiro, M. *Environ. Entomol.* **1996**, *25*, 481-487.
37. Wang, C.; Powell, J. E.; Nguyen, K. *Environ. Entomol.* **2002**, *31*, 381-387.
38. Mankowski, M. E.; Kaya, H. K.; Grace, J. K.; Sipes, B. *Biocontrol Sci. Technol.* **2005**, *15*, 367-377.
39. Grace, J. K. In *Proceedings, Hawaii Agriculture: Preparing for Growth*; College of Tropical Agriculture and Human Resources, University of Hawaii at Manoa: Honolulu, HI, 1995; Pp. 166-167.
40. Rosengaus, R. B.; Traniello, J. F. A.; Lefebvre, M L.; Brown, J. J. *Naturwissenschafzen* **1999**, *86*, 588-591.
41. Rosengaus, R. B.; Jordon, C.; Lefebvre, M. L.; Traniello, J. F. A. *Naturwissenschafzen* **1999**, *86*, 544-548.
42. Rosengaus, R. B.; Maxmen, A. B.; Coates, L. E.; Traniello, J. F. A. *Behav. Ecol. Sociobiol.* **1998**, *44*, 125-134.
43. Grace, J. K.; Zoberi, M. H. *Sociobiol.* **1992**, *20*, 23-28.
44. Chen, J.; Henderson, G.; Grimm, C. C.; Lloyd, S. W.; Laine, R. A. *Nature* **1998**, 392, 558.
45. Boucias, D. G.; Stokes, C; Storey, G.; Pendland, J. C. *Pflanzenschutz-Nachrichten.* **1996**, *49*, 103-144.
46. Woodrow, R. J.; Grace, J. K. *Sociobiol.* **1998**, *32*, 17-25.
47. Woodrow, R. J.; Grace, J. K.. *J. Econ. Entomol.* **1998**, *91*, 905-909.
48. Woodrow, R. J.; Grace, J. K. *Sociobiol.* **1998**, *32*, 27-49.
49. Lewis, V. R.; Haverty, M. I. *J. Econ. Entomol.* **1996**, *89*, 922-934.
50. Lai, P. Y. Ph.D. Dissertation, University of Hawaii, Honolulu, HI, 1977.
51. Su. N. –Y.; Scheffrahn, R. H. *Sociobiol.* **1996**, *28*, 521-530.
52. Grace, J.K.; Tome, C.H.M.; Shelton, T.G; Oshiro, R. J.; Yates, J. R. III. *Sociobiol.* **1996**, *28*, 511-520.
53. Esenther, G. R.; Beal, R. H. *J. Econ. Entomol.* **1974**, *67*, 85-88.
54. LaFage, J. P. *Proceed.Canadian Wood Preserv. Assoc.* **1984**, *5*, 77-81.
55. Delate, K.M.; Grace, J. K. *J. Appl. Entomol.* **1995**, *119*, 429-433.
56. Grace, J.K.; Yamamoto, R. T.; Tome, C. H. M. *Sociobiol.* **2000**, *35*, 457-466.
57. Grace, J. K.;Yamamoto, R. T. *J. Econ. Entomol.* **1994**, *87*, 1547-1554.
58. Forschler, B. T. *Sociobiol.* **1996**, *28*, 459-483.
59. Thorne, B. L.; Breisch, N. L. *J. Econ. Entomol.* **2001**, *94*, 492-498.
60. Jones, S. C. *Sociobiol.* **1989**, *15*, 33-41.
61. Rojas, M. G.; Morales-Ramos, J. A. *J. Econ. Entomol.* **2004**, *97*, 2015-2020.
62. Su, N. –Y.; Scheffrahn, R. H. In *Termites as pests of buildings*; Abe, T.; Bignell, D. E.; Higashi, M., eds.; Termites: Evolution, Sociality, Symbiosis, Ecology; Kluwer: Dordrecht, The Netherlands. 2000; pp. 437-453.
63. Grace, J. K.; Su, N. -Y. *Sociobiol.* **2001**, *37*, 301-310.

64. Campora, C.E.; Grace, J. K. *J. Econ. Entomol.* **2001**, *94*, 1193-1199.
65. Ring, D. R.; Morgan, A. L.; Woodson, W. D. *Sociobiol.* **2001**, *37*, 293-300.
66. Lax, A. R.; Osbrink, L. A. *Pest Manage. Sci.* **2003**, *59*, 788-800.
67. Su, N. -Y.; Ban, P.; Scheffrahn, R. H. *J. Econ. Entomol.* **2004**, *97*, 2029-2034.
68. Puche, H.; Su, N. -Y. *J. Econ. Entomol.* **2001**, *94*, 1398-1404.
69. Chen, J.; Henderson, G.; Laine, R. A. *J. Entomol. Sci.* **1998**, *33*, 97-105.
70. Cornelius, M. L.; Daigle, D. L.; Connick, W. J.; Williams, K. S.; Lovisa, M. P. *Sociobiol.* **2003**, *41*, 513-525.
71. Su, N. -Y. *J. Econ. Entomol.* **2005**, *78*, 471-475.
72. Grace, J. K.; Wood, D. L.; Frankie, G.W. *J. Chem. Ecol.* **1989**, *15*, 129-139.
73. Grace, J.K. *Proceed. Entomol. Soc. Wash.* **1990**, *92*, 773-777.
74. Grace, J.K. *Sociobiol.* **1989**, *16*, 165-182.
75. Su, N. -Y.; Puche, H. *J. Econ. Entomol.* **2003**, *96*, 88-93.
76. Bernklau, E. J.; Fromm, E. A.; Judd, T. M.; Bjostad, L. B. *J. Econ. Entomol.* **2005**, *98*, 476-484.
77. Ettershank, G., J.; Ettershank, J. A.; Whitford, W. G. *Environ. Entomol.* **1980**, *9*, 645-649.
78. Swoboda, L. E.; Miller, D. M. *Sociobiol.* **2005**, *45*, 811-828.
79. Swoboda, L. E.; Miller, D. M. *J. Econ. Entomol.* **2004**, *97*, 1404-1412.
80. Chen, J.; Henderson, G. *J. Chem. Ecol.* **1996**, *22*, 2359-2369.
81. Cornelius, M. L. *Sociobiol.* **2003**, *41*, 583-591.
82. Swoboda, L. E.; Miller, D. M.; Fell, R. J.; Mullins, D. E. *Sociobiol.* **2004**, *44*, 547-563.
83. Waller, D. A.; Curtis, A. D. *Ann. Entomol. Soc. Am.* **2003**, *96*, 81-85.
84. Cornelius, M. L. In *Proceedings of the Fifth International Conference on Urban Pests*; Lee, C. –Y.; Robinson, W. H., eds.; Executive Committee of the International Conference on Urban Pests, Singapore. 2005; pp. 159-163.
85. Husseneder, C.; Grace, J. K. *Clean Asia* **2004**, *11*, 20.
86. Husseneder, C.; Grace, J. K.; Oishi, D. E. *Current Microbiol.* **2005**, *50*, 1-5.
87. Husseneder, C.; Grace, J. K. *Appl. Genet. and Molecular Biotechnol.* **2005**, *68*, 360-367.

Chapter 16

Improving the Performance of Organic Biocides by Using Economical and Benign Additives

Tor P. Schultz and Darrel D. Nicholas

Forest Products Laboratory/FWRC, Box 9820, Mississippi State University, Mississippi State, MS 39762–9820

The copper systems currently used in North America to treat lumber for residential applications will likely be replaced by organic systems at some future time. However, most organic wood preservatives used or being considered for wood preservation are relatively expensive. Also, organic biocides can be depleted by various chemical, physical or biological mechanisms over the long service life expected from treated wood. One way to address these problems is to employ selected additives that increases an organic biocide's activity and/or minimizes depletion. This would mean less biocide is required, possibly lowering the cost of a wood preservative system. Additives that may achieve these goals include antioxidants, metal chelators and water repellents. Laboratory and exterior exposure data with additives will be described, and other possible additives briefly discussed.

Introduction

The preservative chromated copper arsenate, CCA, was used for about 80% of all wood products treated in the U.S. up to the end of 2003, but CCA is no longer permitted for treating residential lumber. New 2nd generation preservatives labeled for residential applications, the major use for treated wood in North America, contain high levels of copper(II) and an organic co-biocide to inhibit copper-tolerant fungi (*1, 2*). However, environmental concerns with copper and questions over future disposal of metallic-treated lumber are being raised. Three European countries already require 3rd generation totally-organic systems, and some minor usage restrictions will be placed on copper-treated lumber starting in 2010. Many professionals predict that at some future point the U.S. will require totally-organic systems for residential applications.

Development of effective, economical, and dependable totally-organic wood preservatives, especially in areas with high or severe deterioration hazards, will be difficult. Indeed, at the present time no totally-organic preservative for exterior above-ground or ground-contact residential applications is commercially available in the U.S., although several companies are developing, have submitted for standardization [obtaining regulatory approval of] or have recently obtained standardization of an organic system for non-millwork residential applications. Two major problems in developing organic preservatives for residential exterior exposure are: 1) the relatively high cost of the new organic biocides [often $ 30-60/kg] being examined as potential wood preservatives compared to copper(II) [$ 2-3/kg]; and 2) the biotic and abiotic depletion reactions that occur with organic biocides but not copper.

We have studied certain non-biocidal additives that could be employed in wood preservation, with the objective to improve the performance of organic biocides at protecting wood by addressing one or both of the above problems.

To address the first problem, the high cost of organic biocides, we are examining non-biocidal and low-cost additives that, when mixed with organic biocides, enhance the activity of organic biocides. This means that less of the relatively expensive biocide is needed, which may reduce the overall preservative system cost and lessen environmental and disposal concerns.

A second major concern with organic biocides is the greater potential for biocide depletion. Specifically, metallic biocides are only depleted by water leaching. However, organics not only leach but are also biodegraded by numerous microorganisms and depleted by various abiotic (chemical, vaporization and photo/sunlight) pathways.

It is not surprising that the agrochemicals being examined as potential wood preservatives are depleted over the long service life expected from treated wood, since they were originally developed for the agrochemical market (*3*). Specifically, the desired properties of a wood protection biocide differ greatly from that for an agrochemical. Agrochemicals are typically applied to a crop to control a specific pest for no more than a few weeks; after this short time it is desired that the chemical rapidly degrades. When the same agrochemical is examined as a potential wood preservative, however, the biocide is expected to protect wood for many years against a wide variety of organisms. Our hope is that if a low cost additive can reduce biodegradation of an organic biocide over the long service life expected from treated wood, less of the relatively expensive biocide would be required when the wood is first treated.

For several years we and other groups have examined various non-biocidal compounds that may address the above problems. Many of the additives examined are relatively benign; indeed, several have a GRAS (Generally Classified As Safe) classification by the U.S. FDA and, thus, are approved food additives. The compounds examined include free radical scavengers/ antioxidants, metal chelators, and water repellents. Other possible additives will be briefly discussed.

Antioxidants

Overview

About 15 years ago we decided that the 1st generation wood preservative systems, CCA, pentachlorophenol and creosote (*2*), would eventually be restricted. Consequently, a need existed for more environmentally-benign systems. One approach tp develop greener wood preservatives is to understand why the heartwood of certain tree species is naturally durable. Consequently, we have studied the role that extractives, particularly stilbenes (*4*), play in heartwood durability. Surprisingly, we found that extractives in highly durable heartwood have very poor fungicidal activities compared to commercial biocides (*5*). In further work (*6*) we hypothesized that extractives may protect heartwood against fungal degradation by having dual properties: The extractives have relatively poor fungicidal activity but are superb free radical scavengers (antioxidants). This dual defense hypothesis was based on the facts that white- and brown-rot fungi both employ free radicals to degrade wood (*7*) and that the phenolic extractives in durable woods are excellent antioxidants (*8*).

To test this hypothesis we ran several wood-containing laboratory decay tests and measured the activities of: 1) an antioxidant that had no fungicidal activity; 2) a biocide that had no antioxidant activity; and 3) a combination of the two. If our hypothesis was correct then we should observe enhanced efficacy with the biocide and antioxidant mixture; i.e., the combination would be synergistic.

Laboratory Decay Studies of Organic Biocides and Antioxidants

Laboratory decay tests (*8, 9*) did indeed show enhanced efficacy for all biocide:antioxidant mixtures tested compared to the organic biocide alone. Synergism was observed with all seven organic biocides examined for various brown- and white-rot fungi, and gymnosperm and angiosperm woods. The biocides studied have either been examined as potential wood preservatives or are in commercial wood preservatives (*1, 2*). Initial studies were conducted with the antioxidant butylated hydroxytoluene (BHT), a low cost antioxidant that has a U.S. FDA GRAS classification and, thus, is benign to humans. We also examined the antioxidant propyl gallate (PG), another economical food additive. Better results were obtained in laboratory decay tests with PG than BHT. This may be due to PG's good metal chelating ability, since we later found that the combination of an organic biocide:antioxidant:metal chelator was more effective than a biocide:antioxidant mixture (*9, 10)*.

While the initial laboratory decay tests were encouraging and enabled us to obtain several U.S. patents, we had several concerns. First, we initially ran standard laboratory decay tests with 8-12 weeks of incubation but observed no synergism. Generally, we only observed synergism in laboratory decay tests with short incubation times of 4-6 weeks. Secondly, relatively high and uneconomical BHT levels of 5% were used.

Recent laboratory decay studies in France examined 2-hydroxypyridine-N-oxide (2-HPNO) (*11*), which was shown to have both fungicidal and metal chelating properties. The combination of 2-HPNO and the antioxidant Irganox 1076 was found to be synergistic. However, a tebuconazole and Irganox 1076 mixture was not synergistic. The relatively long incubation time employed by these researchers, 12 or 16 weeks, may be one possible explanation for the lack of synergism. We observed synergism in our studies with tebuconazole and BHT against both brown-rot and white-rot fungi, but used a short incubation time of 4 to 6 weeks (*8-10*).

Exterior Exposure Results

Synergism in short-term laboratory decay tests does not ensure that a preservative will be effective for the many years expected from treated wood. Indeed, our initial laboratory results indicated that uneconomically high levels of BHT are necessary and the protective effect was only observed for relatively short incubation times. Thus, we judged that this concept would probably not work in exterior real-world applications. However, even with a high probability for likely failure we felt that the potential benefits that might be obtained if this concept did work in "real world" conditions compelled us to test our concept in outdoor exposure. Consequently, an exterior ground-contact study was installed at two test plots in Mississippi in high and severe deterioration zones (*9, 12*). The wood stakes were SYP sapwood, treated with the commercial biocide chlorothalonil (CTN) and BHT. We selected CTN since it has both the fungal and termite activity necessary for a ground-contact study, and CTN has been extensively examined as a potential wood preservative (*1, 2*). The objective was to verify the antioxidant concept in outdoor exposure, not to develop a commercial system. Thus, lower CTN levels were employed than would be used in commercial systems and an organic solvent, toluene, was used. Also, lower BHT levels were employed in this first outdoor exposure study, 2 and 4% in the treating solution, than the 5% level used in the laboratory decay studies.

The initial results were extremely encouraging (*9*), and the results after the relatively long exposure time of 74 months (Table I) continue to be very promising. Generally, co-addition of BHT resulted in greater efficacy against fungal and termite attack than observed with stakes treated with CTN alone. The exception was that one set at Dorman Lake, treated with the highest biocide (0.5%) and BHT(4%) levels, had no difference in the average fungal efficacy rating compared to samples treated with only 0.5% CTN. This was due to one of the seven samples failing in the 4% BHT/0.5% CTN set at Dorman Lake owing to fungal degradation. This sample had had the bottom 3 inches removed after 48 months of exposure to measure BHT depletion, and it is possible that decay fungi were able to colonize the stake through the freshly-cut surface. This set did show increased termite efficacy compared to the matched set with no BHT, and the same treatment set at Saucier had greatly enhanced efficacy against both decay fungi and termites with the co-addition of BHT. Interestingly, samples treated with 4% BHT alone performed about as well as samples treated with the lowest biocide level at both locations.

The initial outdoor exposure results were better than we had anticipated for several reasons. First, the antioxidant concept was based upon the free-radical mechanisms by which decay fungi attack wood (*7*). However, BHT's protective effect against termite attack was unexpected. Also, BHT was only effective for a short time in laboratory decay tests, as discussed above. Since decay fungi would attack ground-contact stakes almost continuously the possibility existed that BHT might only last for a relatively short time. Finally, lower BHT levels of 2 and 4% were employed in this initial outdoor exposure test than the 5% level used in the earlier short-term laboratory decay tests.

BHT depletions after 48 and 54 months of ground-contact exposure were generally about 30-50% (*12, 13*), which we considered to be relatively low. It is possible that the oxidized BHT radical, formed as BHT scavenges a fungal-generated radical, interfered with the reductive portion of the fungal redox cycle(s) with the BHT being regenerated and, concurrently, disrupting a fungal redox cycle (*10*). If this does occur then the BHT would be recycled, possibly explaining the relatively low depletion of BHT observed in outdoor long-term exposure and would be a second mechanism by which BHT inhibits fungi.

Table 1. Average decay and termite ratings after 74 months of exposure for SYP sapwood ground-contact field stakes at the Dorman Lake, MS (high deterioration zone) and Saucier, MS (severe deterioration zone) test plots. Decay and termite ratings were based on AWPA Standard E7-01, with a "10" rating sound to trace of degradation, a "9" rating trace to 3% degradation, etc., down to a "0" [failed] rating, and are the average of seven replicate stakes per treatment/site.

Treatment (wt. %)		Average Retention (kgm^3)	Dorman Lake		Saucier	
CTN	BHT		Decay	Termite	Decay	Termite
0	0	0/0	0	0	0	0
0	4.0	0/19.4	2.9	4.8	2.4	0.4
0.15	0.0	0.74/0	3.0	3.1	1.1	0.6
0.15	2.0	0.72/9.5	6.7	7.0	5.6	4.1
0.15	4.0	0.70/18.8	6.4	6.4	5.9	3.9
0.30	0.0	1.47/0	5.7	7.1	3.5	2.7
0.30	2.0	1.54/10.1	7.9	8.7	8.0	7.3
0.30	4.0	1.41/18.8	8.0	8.1	6.0	5.1
0.50	0.0	2.42/0	8.3	8.6	6.3	5.9
0.50	2.0	2.41/9.6	9.9	9.9	9.0	8.6
0.50	4.0	2.43/19.5	8.1	9.9	9.3	8.4

Samples with 2% BHT are performing about as well as samples with 4% BHT. This is likely explained by BHT not being a biocide. Thus, while increased biocide levels result in enhanced efficacy, increased antioxidant levels would not give greater decay and termite resistance; all that matters is that at least some BHT is present. Depletion measurements showed that quite a bit of BHT remains after 54 months (*13*) in the stakes treated with 2% BHT, as discussed above and, thus, it is not surprising that similar results are obtained for stakes treated with the two different BHT levels.

The antioxidant BHT also appears to protect the organic biocide CTN (*13*) in ground-contact wood. For example, stakes that were treated with only 0.3% CTN and exposed at the Saucier test plot for 54 months had 65 ± 4% CTN depletion in the ground-contact section, while the stakes treated with 0.3% CTN plus 4% BHT had only 18 ± 8% CTN depletion. For the same 0.3% CTN set exposed at Dorman Lake, the samples without co-added BHT had 41 ± 16% CTN depletion and the samples with 4% BHT co-added had 29 ± 19% CTN depletion.

While no other reports on an antioxidant protecting biocides in treated wood could be found, a patent has reported that the antioxidant PG stabilized 3-isothiazalones solutions (*14*) and that the stabilized biocide could be used for various applications including wood preservation. Further research is needed to definitively confirm, or disprove, the possible BHT protective effect using various organic biocides in ground-contact and above-ground applications.

Based on the encouraging results from the first exterior study (*9*) a few additional ground-contact and above-ground samples were installed. Since the antioxidant PG gave better results in the laboratory decay tests than BHT, we included PG in most of these later outdoor exposure experiments. Unfortunately, results with PG in a ground-contact study were disappointing and only modest results were obtained in above-ground tests with PG and two biocides. The best results obtained with PG were when a water repellent was co-added to above-ground samples, suggesting that in outdoor exposure PG may be slowly leaching.

Above-ground samples with the antioxidant BHT continue to show encouraging results. For example, SYP lap-joint samples were treated with only BHT at the relatively low retention of 0.94 kg/m^3. After five years of exposure in a severe deterioration zone, BHT-treated samples had an average rating of 7.3 compared to a 4.5 rating for the untreated controls. Another study examined lap-joint samples treated to three retentions with a quat formulation, with or without BHT co-added at a retention of 2.5 kg/m^3. After four years of exposure at CSI's Hilo test site, the average ratings for samples treated with only the quat biocide ranged from 5.8 (lowest quat retention) to 7.6 (highest quat retention, compared to 7.8 to 8.0 for samples co-treated with 2.5 kg/m^3 BHT and the quat formulation.

To summarize, exterior exposure results with BHT were more promising than was originally anticipated. Specifically, the data indicate that relatively low BHT levels appeared effective in outdoor exposure as compared to the uneconomically high levels employed in the laboratory decay tests. Furthermore, BHT remained effective for multiple years in ground-contact, and one study indicated that BHT reduced biodegradation of an organic biocide. Additional experiments need to be run and longer exposure times are required to definitively resolve if an antioxidant can reduce biodegradation of organic biocides, determine effective biocide/BHT levels for various biocides and applications, and decide if this concept is economical.

In contrast to the generally good results obtained with BHT, exterior studies with the antioxidant PG gave poor results in ground contact and only modest results in above-ground exposure.

Metal Chelators

Overview

Our original antioxidant concept was based on the excellent antioxidant properties of extractives in naturally durable heartwood. In continuing this reasoning, it is well known that many heartwood extractives complex with metals (*9*) and that fungal degradation often involve various metal co-factors in the fungal redox cycles and/or as part of an enzyme (*15*). Consequently, metal chelation may be an additional means by which extractives protect heartwood (*16*) and, furthermore, mixtures of metal chelators and organic biocides may synergistically protect wood.

Laboratory Decay Studies with Metal Chelators and Organic Biocides

We employed the same technique used earlier with the antioxidants to examine the metal chelation hypothesis. Specifically, we ran short-term (5-6 weeks incubation) laboratory decay tests with: 1) metal chelators that had no fungicidal activity; 2) various biocides that had no metal chelating ability; and 3) biocide:metal chelator combinations (*9*). These laboratory tests showed synergism; specifically, the blocks treated with both an organic biocide and a metal chelator had less fungal degradation than the blocks treated with only the organic biocide. Furthermore, the trinary combination of an organic biocide:metal chelator:antioxidant was usually more effective than binary mixtures of biocide:antioxidant or biocide:metal chelator.

We usually obtained the best laboratory results with metal chelators employing the agar block rather than soil block test. Agar medium has relatively low levels of metals compared to soil and, consequently, metal chelators may be more effective in agar-containing than soil-based decay tests.

The laboratory decay studies in France with 2-HPNO (*11*) also examined 2-HPNO and the metal chelator EDTA, and reported that the combination was synergistic. However, synergism with EDTA and tebuconazole was not observed with long incubation times.

Exterior Exposure Studies with Metal Chelators

Many of our laboratory decay tests with metal chelators employed sodium salts of EDTA, a metal chelator that is an approved food additive. However, in outdoor exposure it would be expected that EDTA, and other metal chelators with carboxylic groups, would leach relatively rapidly.

We examined one solventborne metal chelator, 1-10-phenanthroline, in outdoor exposure tests. Lap-joint samples were treated with the same three quat retentions as before, along with 0.4% BHT (2.5 kg/mg^3), 0.2% of the above metal chelator (1.1 kg/mg^3), and wax (about 4.8 kg/m^3) as a water repellent. After four years of exposure in the extremely severe deterioration conditions at CSI's test site in Hilo, HI, the samples with all the above additives and the three quat retentions had average ratings from 9.5 to 10.0. By contrast, samples treated with only the biocide had average ratings of 5.8 to 7.6. While 1,10-phenanthroline is relatively expensive and toxic and, thus, unsuitable for a commercial preservative system, the results support the laboratory data that metal chelators enhance the efficacy of organic biocides.

Many microorganisms employ metals as co-factors when they degrade organic compounds. Thus, co-addition of a metal chelator may also reduce biodegradation of an organic biocide. Unfortunately, at this time we do not have any samples in outdoor exposure to test this possibility.

Wood in soil contact can have increased levels of iron and other metals due to metal migration into the wood from the adjacent soil (*17*). It is possible that when a metal chelator is added to treated wood for a ground-con tact application the metal chelator may be relatively quickly saturated and rendered ineffective. However, the potentially high benefits that might be obtained from the metal chelator concept suggest that this assumption needs to be tested. At this time we believe that metal chelators may be most useful as part of an organic preservative system for above-ground residential exposure, as suggested from results from the one above-ground study discussed above.

We have tried to identify metal chelators that are relatively inexpensive and commercially available, safe to humans and other non-target animals, and can be formulated into a waterborne system for treating lumber for residential applications but once impregnated into wood will not leach. A chance remark by a colleague let us to examine the potential of resin acids. Resin acids, a commercial product mainly produced as a by-product of tall oil obtained by the kraft pulping of pine and called tall oil rosin but better known by the acronym TOR, are inexpensive (about $0.25 – 0.40/lb, bulk shipment). [Other sources of resin acids are isolated from old pine stumps (wood rosin) and oleoresin tapped from living trees (gum rosin)]. Resin acids have a GRAS classification for foodstuff use such as chewing gum base or in paper that comes in contact with foodstuffs and, thus, are also benign. (Many of us used a rosin bag to help our grip when playing various sports in our younger days and some of us have eaten potatoes cooked in rosin.) Finally, one of the commercial uses of resin acids are as metal complexing agents, and resin acids are a renewable resource. All the above suggest that resin acids/rosin may be a suitable metal complexing additive to combine with organic biocides for wood preservative systems.

When we ran laboratory decay tests with an organic biocide and a solventborne (toluene) resin acid, some limited synergism was observed. However, when the resin acid was formulated into a waterborne system by forming the carboxylate anion, very good results were obtained with the agar block decay test (*18*). Indeed, treatment with 4% of a resin acid alone was sufficient to protect wood against fungal degradation in the agar block test. Furthermore, the resin acid anion in the waterborne formulation acts as a surfactant, so that some organic biocide can often be directly added to the formulation. We are continuing to examine the potential of employing TOR as an additive, and hope to report additional results shortly.

Water Repellents

As noted by Preston (*3, 19*), water repellents based on waxes or oils and formulated into a waterborne emulsion offer many advantages to treated wood. Water repellents lower the amount of rainwater soaked up by above-ground lumber, reducing the fungal decay potential and biocide leaching. Another positive benefit is the enhanced dimensional stability which reduces undesired bowing, checking, splitting, etc. Lumber treated with the new copper-rich systems and a wax-based water repellent is available as premium decking in North America. Some work is also being conducted on treating wood with silicone compounds to give water repellency and perhaps other benefits. Two general silicone formulations, waterborne and oilborne, are apparently available.

We examined the potential of resin acids/TOR as metal complexing agents, as noted in the prior section. Resin acids are also hydrophobic and may impart water repellency to wood. In water swelling tests with a solventborne resin acid, only slight water repellency was observed. However, when wood blocks were treated with a waterborne system extremely good water repellency was observed (*18*). For example, small southern yellow pine wood wafers were treated with 1.0% wt/vol waterborne resin acid or 1.1% wt/vol paraffin wax in toluene. The samples were dried to 12% MC, then placed in water and the radial swelling measured. The Water Repellency Efficiency (WRE), the radial swelling of treated wafers relative to the swelling of the matched untreated wafers, was measured at 30 and 90 minutes. At 30 minutes, the average WRE of four replicates was 40.2% for the 1.0% resin acid-treated sample but only 21.9% for the 1.1% wax samples. At 90 minutes, the resin acid samples still had some water repellency with an average WRE of 22%, while the wax-treated samples had a negative WRE. We also treated SYP 5/4's decking boards with a waterborne treatment formulated using a commercial source of resin acids to give a solution with 3% TOR, and after 5 weeks of exterior above-ground exposure the treated and matched untreated boards were weighed after a thunderstorm. The TOR-treated boards had an average moisture content of 25.3% versus the 31.0% of the control/untreated boards.

We have previously examined the wide variability inherent in the natural durability of above-ground untreated SYP sapwood samples (*20*), where some samples show extensive decay fairly quickly while other untreated samples only slowly degrade. SYP sapwood is reported to contain a variable and sometimes high level of resin acids (*21*). [The high temperatures used in kiln drying SYP lumber will degrade some of the resin acids naturally present, and may also disrupt the microdistribution of resin acids and so negatively impact the metal complexing capability.] It is interesting to speculate on the possibility that the amount of resin acids naturally present in a particular SYP sapwood sample may impart some inherent decay resistance (*22*) to above-ground SYP lumber.

Other Potential Additives

Many other non-biocidal additives might enhance biocide efficacy and/or reduce depletion. Possibilities include:

1. Additives that reduce vaporization and/or water leaching of biocides. One possibility might be to add a small amount of a heavy oil as an emulsion in a waterborne formulation. An organic biocide should be soluble in the small amount of heavy oil and, thus, have reduced likelihood to leach or vaporize.
2. Sunlight degrades organic biocides that are on or near to the wood surface, leaving the lumber surface susceptible to microorganism colonization. A compound that reduces wood photodegradation may also help protect organic biocides.
3. Preston (*3*) noted that future wood preservative systems will employ relatively low biocide retentions and, consequently, nonuniform biocide distribution within wood may be a future concern. Additives might help improve biocide distribution to obtain a more homogeneous micro- and macrodistribution in treated wood.
4. A major concern with organic biocides is that most do not fix to wood, unlike metallic biocides. Consequently, organic biocides could be susceptible to leaching. Additives that would fix with, or anchor the biocide to the lignocellulosic substance, should reduce leaching. However, the biocide also needs to be available to the decay fungi to be effective, so this approach needs to ensure that the biocide is not so tightly fixed so that it is no longer effective.

Acknowledgements

Funding for the authors' work was provided by the USDA-Cooperative State Research, Education and Extension Service - Wood Utilization Research, National Research Initiative and McIntire-Stennis Programs, Lonza Corp., and Mississippi taxpayers. Approved as Journal Article FP 351 of the Forest and Wildlife Research Center, Mississippi State University. The authors thank the many industrial professionals who gave suggestions and encouragement for our research.

References

1. Schultz, T.P.; Nicholas, D.D. Wood Deterioration and Preservation, Amer. Chem. Soc. Symp. Series 845, Washington, D.C., 2003, chp. 26.
2. Freeman, M.H.; Nicholas, D.D.; Schultz, T.P. *Environmental Impacts of Treated Wood,* CRC Press, chp. 2, (in press)
3. Preston, A.F. *Wood Deterioration and Preservation,* Amer. Chem. Soc. Symp. Series 845, Washington, D.C., 2003, chp. 22.
4. Schultz, T.P.; Boldin, W.D.; Fisher, T.H.; Nicholas, D.D.; McMurtrey, K.D.; Pobanz, K. *Phytochemistry* **1992**, *31*, 3801-3806, and references therein.
5. Schultz, T.P.; Harmes, W.B.; Fisher, T.H.; McMurtrey, K.D.; Minn, J.; Nicholas, D.D. *Holzforschung* **1995**, *49*, 29-34, and references therein.
6. Schultz, T.P.; Nicholas, D.D.; Minn, J.; McMurtrey, K.D.; Fisher, T.H. Synergistic combination of an antioxidant and wood preservative: A preliminary study. Internat. Research Group on Wood Preservation paper 98-30172, Stockholm, Sweden.
7. *Wood Deterioration and Preservation*; Goodell, B.; Nicholas, D.D.; Schultz, T.P. Eds., American Chemical Society Symp. Series 845, Washington, D.D., 2003.
8. Schultz, T.P.; Nicholas, D.D. *Phytochemistry* **2000**, *54*, 47-52, and references therein.
9. Schultz, T.P.; Nicholas, D.D. *Phytochemistry* **2002**, *61,* 555-560, and references therein.
10. Schultz, T.P.; Nicholas, D.D.; Henry, W.P.; Pittman, C.U.; Wipf, D.O.; Goodell, B. *Wood Fiber Sci.* **2005**, *37*, 175-184.
11. Mabicka, A.; Dumarcay, S.; Rouhier, N.; Linder, M.; Jacquot, J.P.; Gerardin, P.; Gelhaye, E. *Internat. Biodeter. Biodegrad.,* **2005**, *55*, 203-211.
12. Schultz, T.P.; Nicholas, D.D.; Prewitt, ML. *Holzforschung,* **2004**, *58*, 300-304.
13. Schultz, T.P.; Nicholas, D.D.; Prewitt, M.L.; Kirker, G.T.; Diehl, S.V. *Inter. Biodeter. and Biodegrad.* **2006**, *57*, 45-50.
14. Gironda, K.F. *U.S. Patent* 5,534,487, 1996.
15. Henry, W.P. *Wood Deterioration and Preservation,* Amer. Chem. Soc. Symp. Series 845, Washington, D.C., 2003, chp. 10.
16. Binbuga, N.; Chambers, K.; Henry, W.P.; Schultz, T.P. *Holzforschung* **2005**, *59*, 205-209.
17. Schultz, T.P.; Nicholas, D..; Lebow, S. *Forest Prod J.* **2003**, *53(9)*, 77-80.
18. Schultz, T.P.; Nicholas, D.D.; Kelley, S. **2006**. A nonleachable waterborne composition of resin acids and wood preserving organic biocides. U.S. Provisional Patent 60/743,669, filed March 22, 2006.
19. Preston, A.F. *Forest Prod J.* **2000**, *50(9)*, 12-19.

20. Nicholas, D.D.; Schultz, T.P.; Sites, K.; Buckner, D. **2005**, Internat. Research Group on Wood Preservation Paper IRG/WP 05-20310.
21. Ingram, L. L., Jr. 2005. Personnal communication, Miss. State Univ.
22. Harju, A.M.; Kainulainen, P.; Venalainen, M.; Titta, M.; Viitanen, H. *Holzforschung* **2002**, *56*, 479-486.

Chapter 17

Biocide Depletion: Chemical, Physical, and Photodegradation

John N. R. Ruddick

Department of Wood Science, Faculty of Forestry, University of British Columbia, Vancouver, British Columbia V6T 1Z4, Canada

Introduction

The performance of preservative treated wood is directly dependent on the efficacy and amount of biocide present in the wood. Consequently, it is important to understand the impact of different environmental conditions on the permanence of the active ingredients. There are a number of ways that the biocide content can decrease and Cockroft and Laidlaw, (*1*) have reviewed some of these. These include evaporation, water leaching, chemical or biological degradation of either the biocide or the biocide-wood complex formed, and UV degradation. Due to dominance of the inorganic copper based preservatives the most important mode of depletion for pressure treated wood is likely to be leaching. However, organic biocides are widely used for non-pressure treatment of wooden components in structures and as their use expands to include wood products currently protected with copper based preservatives then alternative degradation pathways will become more important and standard testing methodologies will need to be modified to reflect this. For example, when stability towards UV light is considered to be suspect, then a pre-exposure to UV light prior to assessment of the performance should take place. Thermal degradation is unlikley to be a factor unless the temperature at which it occurs is low and could be reached during natural exposure. Long term hydrolysis could also impact on performance. However, identifying such changes based on field experiments must be done with care, since during field exposure, potential for biological degradation of the biocide by bacteria or molds may difficult to eliminate.. An additional factor that also needs to be considered is the movement of biocide within timber while it is in service. This is particularly important for wood species which have significant amounts of heartwood so that the treated wood may be limited to a thin outer shell. This review will focus on strategies that have been used to assess depletion of wood preservatives from treated timber under both laboratory and field conditions. While some data are presented in the review, this is used for illustrative purposes. It is not intended to review the leaching characteristics of different preservative systems.

Evaporative Aging

For organic biocides long term instability to hydrolysis or photodegradation, and the potential for evaporation are likely to be primary causes for their depletion rather than leaching. Consequently, testing of both the biocide and treated wood using evaporative aging prior to efficacy testing will be important for organic preservatives.

Van Acker et al. (*2*) reported on the importance of an evaporative aging procedure prior to leaching during decay testing of creosote treated wood. They noted that the diluting solvent and also the type of creosote impacted on evaporative losses. They concluded that a combination of aging and leaching is recommended as part of assessment protocols and that different ageing and leaching regimes should be employed for testing treated wood against Basidiomycete and soft rot fungi. Edlund et al. (*3*) examined several different aging methods and concluded that they gave comparable results to natural aging for the biocides studied. The methods examined included a wind tunnel, a forced air oven with temperatures ranging from 40 to 70 °C, as well as leaching. It should be noted that the use of a small sample size, 50 x 25 x 15 mm, will accentuate the effect since losses will directly depend on the surface area being exposed. In 1991 Valcke and Stevens *(4)* examined the stability and distribution of propiconazole following either 18 months storage, an evaporative aging (wind tunnel) or a thermal stress exposure (forced air at 50 °C). No changes in the distribution of the biocide were noted. The use of small blocks in these tests would magnify any loss of biocide suggesting that thermal stresses will not impact on biocide performance for propiconazole..

Suttie, Orsler and Dearling (*5*) reported the impact of evaporative aging using lindane as the biocide. In considering their approach they noted that the use of elevated temperatures to determine the evaporative losses, is suspect, since raising the temperature will increase the vapor pressure for different molecules by amounts reflected by their temperature-vapour pressure relationship. Since this relatiuonship is different for different biocides the use of a "standard" elevated temperature will bias the results. Suttie and coworkers used natural exposure experiments which they then attempted to replicate in controlled laboratory weathering cycles.

Schoknecht et al. (*6*) examined the use of chambers developed for measuring volatile organiuc compound (VOC) emissions. They based theirresearch approach on the EN 1341-1 (*7*) and EN 13419-2 (*8*) standards which have been developed to assess the emission of volatile organic compounds. They noted that most of the biocides being used in wood preservation are either not volatile or are only semivolatile. Consequently the detection of emissions is very difficult unless the chambers are small and loaded with high amounts of biocide. In studying the emission data there were distinct differences between volatile chemicals and the biocides examined. For the latter, the emission rate is low so that concentration- time curves show a much slower initial increase before eventually reaching a steady state. Two chemicals dichlofuanid and 3-iodo-2-propynyl butylcarbamate (IPBC) in film coated wood, showed emissions more comparable to the rates for VOCs and during the 130 day test lost 2 and 1.6 % chemical, respectively. The same study showed the important roles that other components present in the formulation may have. A second formulation containing IPBC developed for masonary, showed losses after 120 days of only 0.001%, although a steady state had not been reached. Propiconazole recorded relatively low llosses to air, between 0.01 and 0.02 % over the 120 days. The authors noted that because of the relatively low emissions of the wood preservative actives, the influence of other chemical "sinks" can be important. These "sinks" can be important – particularly for indoor applications - because they can take up the biocide and reduce the concentration present in the air. They concluded that further studies are required to develop a better protocol for evaluating the evaporative loss of wood preservatives. A current European standard has been published relating to assessing evaporative loss of biocides (*9*) although few reports of its use have been published.

The role of additives to wood preservatives can also be important in ensuring long term performance. This can be readily understood if one compares the performance of pentachlorophenol solubilized in petroleum oil with that of light solvent treatment or liquefied gas treatment. When solubilized in oil it is able to provide utility poles with a service life in excess of 30 years, whereas without this barrier, loss of biocide by evaporation leads to a reduced performance (*10*). Even in oil, loss of biocide by evaporation can be detected (*11*). One method of assessing the volatilization of a preservative in the laboratory is through the use of bioassays on agar plates. The influence of a positive vapor pressure can be observed during bioassays with fungal cultures on plates when zones of inhibition are created even though no movement of biocide into the agar takes place. Alternatively, evaporation under evacuated conditions can indicate relative ease of loss from timber in service. Such simple laboratory tests can indicate whether evaporative loss from timber in service could compromise long term performance .

Degradation of the Biocide

A second factor which may influence the long term performance of wood treated with an organic biocide is the breakdown of the preservative. This may occur as a result of both biological and chemical action. An obvious example is the photolytically induced breakdown of natural pyrethroids used to protect timber against insects. Studies of chlorophenates have suggested that fungi can chemically alter the biocide through methylation to change its leaching characteristics *(12)*. Other studies of quaternary ammonium compounds have suggested that mold fungi are able to alter the long alkyl chains thereby reducing their performance (*13*) and enhancing the leaching of the quaternary compound (*14*). The biodegradation of preservatives is discussed in a separate chapter and therefore will not be reviewed further here. Suttie, Orsler and Dearling (*5*) examined the hydrolysis reactions of trihexylene glycol biborate to determine what suitable pretest environment could be used during preservative assessment. They found that one cycle of freezing and high humidity hydrolised all of the glycol borate. They also concluded that a single weathering cycle which included a 15 minute spray on 12 mm thick samples, would create an adequate environment for hydrolysis to occur. Field experiments suggested hydrolysis losses approached 50 % after two months exposure.

Chemical degradation of the chrome and copper based preservatives is not common since the active ingredients used are relatively stable to chemical reactions. However, the fixation of these components is complex and incomplete fixation can lead to loss of biocide. This has resulted in a large number of leaching studies where the focus is on examining the impact of different fixation conditions of the resulting permanence of the treated wood. Much of this research has employed small sapwood blocks fully impregnated with preservative. Guo et al. (*15*) reported the leaching characteristics of acid copper chromate (ACC) and compared it to those of CCA. Ruddick (*16*) has reviewed several studies on the fixation chemistry of ammoniacal copper based-preservatives, using leaching of small blocks to identify the degree of fixation. Subsequent research identified factors affecting the mobility of copper in amine copper preservative treated wood, including solvent and fixation temperature (*17*). The formation of stable diamminecopper wood complexes in wood was confirmed by crystallographic studies. Similarly, Ung and Cooper, (*18*) used leaching to examine the unbound chemical to monitor the fixation reaction in alkaline copper quat (ACQ) treated wood. They concluded that the copper stabilized much faster at lower copper concentrations. This confirmed earlier studies using small blocks under laboratory conditions reported by Lucas and Ruddick (*19*) which showed that a 0.5 % solution produced more complete fixation of copper than 1 and 5% solutions. They also determined that during exposure to water, both amine and copper were leached from the treated wood resulting in a stable copper-amine-wood complex was formed, with a 1:1 ratio of amine:copper.

Jiang and Ruddick (*20* and *21*) compared the influence of two amine solvents on the leaching characteristics of copper. The stronger copper-solvent complexes formed by ethylenediamine resulted in significantly greater mobility and leachability. Earlier research by these same authors (*22*) used a citrate buffered leaching solution to try to distinguish the copper complexed to wood from the precipitated copper. This strategy of using different media for leaching treated wood, to differentiate between precipitated and bound biocide is important as they may have different susceptibilities to solvolysis or degradation. Kamdem and Zhang (*23*) also used laboratory leaching of small blocks to explore the influence of different solvent types and ratios of copper to solvent on the fixation of alkaline copper systems. They found that increased solvent to copper ratio enhanced leaching. Earlier these same authors had reported that the source of copper can influence the pH of the resulting solution and thus the fixation or leaching. The use of carbonate or hydroxide gave a higher solution pH which enhanced the penetration into the wood compared to the solutions produced from the more acidic copper nitrate or sulphate. However the leaching was inversely proportional to the pH of the solution, with greater leaching from the hydroxide and carbonate treatments (*24*). The study also confirmed that the use of secondary and tertiary amines as solvents was not recommended due to less stable copper-amine complexes. From all of the studies it seems clear that due to the limited ion exchange capacity of wood the amount of copper bound to the wood by that mechanism is limited and that high loadings of copper will need to be complexed with other preservative components to form insoluble compounds.

Chemical stability of copper based preservatives in wood has also been compromised by the presence of acids such as humic acid and citric acid, which form very stable complexes with copper. This will be discussed in more detail in a later section when the impact of soil chemistry is considered on preservative leaching. It is also worth noting that Hager (*25*) has suggested that during service CCA in treated wood undergoes a slow hydrolysis reaction resulting in decomposition of fixation products.

Not all reactions lead to increased mobility of copper. For example, corrosion caused by the galvanic interaction of iron in metal fasteners in contact with copper preservative treated wood, results in the reduction of mobile copper. This can impair the performance of the treated wood, since many checks in wood surfaces are associated with fasteners. These checks allow water to penetrate and collect, which enhances the decay potential at these locations. As has been shown by Choi et al. (*26*) mobile copper plays a critical role in copper treated wood, by migrating and protecting the wood exposed in these checks. This will be discussed later under the importance of field assessments.

UV Degradation of Wood and Biocide

The depletion of wood preservative from treated wood by interaction with ultraviolet light is limited to the surface of the timber as the wavelength of the light is such that penetration into wood is very limited. In addition, the presence of any coatings on wood above ground limits the effect of the weathering. Alternatively, metals such as chromium or copper in preservative treated wood are able to stabilize the lignin, greatly slowing down the weathering process. This has been attributed to the copper or chromium being chelated to the guaiacyl functionality, thereby preventing the breakdown of the lignin polymer via the formation of phenoxy radicals. Nevertheless, the gradual formation of smaller water soluble phenolic chemicals will lead to loss of preservative over time. However, such losses do require exposure to UV light over a lengthy period, and would not be significant for wood in ground contact. An interesting study by Militz (*27*) also revealed that UV light was beneficial to the fixation of chromated copper preservatives and reduced subsequent leaching, while infrared radiation treatment retarded fixation. A 60 sec. radiation reduced leaching of chromium and copper by 70%.

The role of ultraviolet light in reducing preservative efficacy may be expected to play a greater role in organic biocides, since they are not able to prevent lignin breakdown. The addition of free radical inhibitors will be needed to reduce the impact of UV light. The UV degradation in naturally occurring pyrethroids has been documented (*28*). The authors reported that at depths greater than 2 mm there was virtually no change in the biocide during the first two years of exposure. They also suggested that after 4 years their was some evidence of permethrin breakdown in the outer 0.5 mm. However, Lloyd et al. (*29*) observed a dramatic losses of pyrethroid (200 g/m^3 to 50 g/m^3) in approximately 4 weeks in natural sunlight. The authors cautioned about the impact of the degradation on achieving the desired performance. The role of quaternary ammonium compounds as free radical initiators (*30*) would imply that preservatives based on quaternary ammonium compounds will need components to reduce the impact of UV degradation on treated wood. The increasing importance of surface disfiguring fungi on the performance of treated wood must also be considered, when evaluating the impact of surface depletion of organic biocides. Methylenedithiocyanate (MTC) and thiocyanomethylthiobenzothiazol (TCMTB) have both been studied for controlling mold and staining fungi on wood surfaces. Perrau and Paazian (*31*) have described the decomposition of MTC and TCMTB in aqueous environments. They reported that the decomposition of MTC and TCMTB is highly pH dependent and considerable decomposition was recorded for both at pH>11. However, when both MTC and TCMTB were exposed to sunlight, while the former was stable, TCMTB gradually decomposed. with a first order reaction.

Suttie, Orsler and Dearling, (*5*) have reported the impact of photolysis on the degradation of tri-n-butyltin oxide. They found that 1 week of continuous UV irradiation reduced TBTO concentrations to 55 %. Based on this they suggested that 7 day weatherometer cycles should provide a suitable pretest environment, to ensure any effects of UV on biocide degradation can be identified during preservative assessment.

Currently none of the protocols used to assess preservative performance require data on the impact of ultraviolet light. As organic biocides are used more widely there will be a greater need to establish standard methods to assess this characteristic. Clearly for many organic molecules exposure to ultraviolet light can raise the potential for biocide degradation.

Impact of Leached Chemical in Optimizing Preservative Formulations

Considering the factors impacting preservative depletion from treated wood in service, the most important factor by far is loss due to leaching. Although, as we will see later, much of the emphasis is now on environmental impact, this was not considered important when the leaching tests were first devised, since the objective was to simply measure long-term protection of treated wood.. It was reasoned that any preservative that was not bound to the wood would be lost when the product became wet, and so could not contribute to the long term performance. Hence the role of leaching was generally to remove non-fixed chemical so that the efficacy of the fixed chemical could be determined. This is illustrated by the classic laboratory studies reported by Henry and Jeroski (*32*) and Fahlstrom et al., (*33*). For their study they examined the leaching of chemical from wood blocks impregnated with several formulations to determine which formulation was the most leach resistant. This study was therefore very influential in moving the industry from CCA type-A or CCA type-B to the more balanced CCA type-C. Therefore, their study had achieved its purpose. But the authors clearly recognized that leaching from such small blocks would not relate to that observed in practice. More recently, the simple laboratory leaching trial and subsequent fungal efficacy evaluation of three novel wood preservatives reportedly confirmed the usefulness of this approach (*34*).

Impact of Soil Chemistry on Leaching of Wood Preservatives

When assessing the potential for loss of preservative from treated wood in service, it is important to distinguish between wood used in ground contact and that exposed above ground. For wood in ground contact, the surrounding soil or water will provide a continuous aquatic phase which creates a severe leaching environment. In addition it may well contain metal ions and complexing agents such as humic acid which can greatly impact on the depletion of chemical from treated wood. Plackett (*35*) has reported the use of inorganic salt solutions to examine their influence on preservative loss from CCA-treated wood under laboratory conditions. As might be expected, copper losses increased with increasing ion concentration of the solutions. The author suggested that the losses could account for problems of premature failure of treated posts in horticultural soils.

The loss of copper from treated wood in soils with a high humic acid content has been noted (*36*) and suggested as a possible cause of premature failure of preservative treated wood. Cooper and Ung (*37*) examined the leaching of CCA from small cubes placed in compost and noted high losses after a 12 month exposure. A study was also made of chemical losses from a compost bin built from CCA-treated wood (*38*). Copper was readily detected as being lost from the treated wood. A subsequent more comprehensive study (*39*) used humic acid solutions (swamp water) to leach CCA treated wood resulting in elevated losses of copper. The results were compared to losses from the outer 5mm surfaces of CCA-treated utility poles where both arsenic and copper were found in leach water adjacent to the poles. Evans (*40*) has reported leaching of CCA into food, drinking water and silage. Similar to results of Cooper and Ung, he found high levels of CCA components in the grass in the silo, and noted that copper and arsenic were still observed in the rainwater runoff after two years.

Ruddick (*41*) investigated the premature failure of CCA-treated stakes in a test site with a high water table and the presence of sulphate reducing bacteria. Laboratory studies confirmed the enhanced loss of copper and the subsequent reduction in performance when the samples were exposed to decay fungi. These studies are supported by research on the potential of organic acids to remove copper preservative from treated wood as a strategy for removing preservatives from treated wood waste, Stephan et al. (*42*). In these studies of chromated-copper biocides, copper was readily removed except when oxalic acid was the solvent, since in this case insoluble copper oxalate was formed. Kartal and Hwang (*36*) examined hard water, sea water and humic acid on the removal of CCA components from treated wood, and reported that a 1% humic acid solution was the most effective. Thus any attempt to determine long term service in ground contact needs to consider the variety of soil types and physical conditions present in different gfeographic regions, and the impact this will have on biocide degradation, when designing preservative assessment programs.

While organic acids clearly do impact on the loss of copper from wood due to their ability to form water soluble chelated metal complexes, acid rain also has been suggested as a cause for the reduced performance of inorganic preservative treated wood (*43*). Warner and Solomon (*44*) attempted to investigate the impact of acid rain on CCA-treated wood but used an incorrect synthetic acid rain solution. The inclusion of sodium citrate naturally led to chelation with copper and enhanced copper loss. At a pH of 3.5 they were able to remove 100% of the CCA components. The impact of the organic citric acid buffer was subsequently confirmed by Cooper (*45*) who showed that when inorganic acids were used losses were more consistent with other studies. A comprehensive study of the influence of leaching water temperature and pH on depletion of CCA from treated wood was described by Eetvelde et al. (*46*). These researchers concluded that increasing water acidity led to increased copper loss, and that stirring the water during the study further enhanced leaching. The authors also pointed out that laboratory tests can provide only a broad relative assessment of likely leaching under natural service conditions. In a second phase of the study, a Dutch pre-standard was used to assess the leaching of the preservative (*47*). The researchers found that there was a significant increase in loss of the actives when the temperature and water acidity were increased. They also commented on the remarkably low level of emitted quantities of CCA components from profiled pine boards compared to spruce posts.

A natural extension of this research is to consider the pre-exposure of treated wood to soil prior to carrying out efficacy studies. Gray (*48*) reported the pre-exposure of CCA-treated pine sapwood blocks prior to decay evaluation, but failed to find any significant effect. More recently, the American Wood Preservers' Association has developed a standard method for the leaching of preservative from treated wood during exposure to different soils. (*49*). This was based on extensive collaboration of several researchers. The results of the collaborative study have been reported by Lebow et al. (*50*). An important finding of the collaborative study was that soil contact leaching of samples generally generated more copper losses than that from water alone. They also concluded that the amount of copper and arsenic leached was influenced by both the initial loading of preservative and the soil properties. Venkatasamy and Okwara (*51*) discussed the loss of CCA from tropical hardwoods using soil-extract water and concluded that leaching of CCA components was enhanced. Ruddick and Kundzewicz (*52*) reported the influence of a soil pre-exposure on the decay of untreated pine stakes in a laboratory test. The uptake of iron from waterlogged soils did not in itself result in enhanced decay. Thus, observed acceleration of decay in CCA-treated stakes associated with iron uptake must be a result of interaction of the iron from sulphate reducing bacteria and the CCA preservative components. The development of leaching data based on both water and water/soil mixtures will become more important, particularly as more organic based preservatives are proposed for ground contact enduses. It will be important to indicate the conditions used, and also the emission rates in terms of weight of chemical, wood surface area, water volume of exposure and duration of exposure.

Impact of Marine Exposure on Preservative Mobility

Another environment where leaching can be exaggerated is the marine ecosystem. Albuquerque et al. (*53*) examined the leaching of CCA using natural marine versus laboratory salt water environments. They concluded that there was a time related loss of CCA from the cubes with losses of chromium and arsenic being negligible in the marine environment. However, copper losses in the marine environment were related to those measured in the laboratory, although at a slower rate of loss. Other studies have reported migration of copper within timber submerged in the marine environment (*54* and *55*). The development of data for loss of preservative from treated wood in the marine environment will continue to be important both to ensure adequate performance, but also due to the increasingly sensitive nature of contamination of aquatic systems. Consequently the future development of testing methodology, while providing valuable information on the leaching of biocides, will almost certainly focus on the rates of loss into aquatic environments and assessment of environmental impact and fate.

Loss of Preservatives in Landfill Environments

The more observant readers will have noted that there has been a subtle shift from using leaching studies to assess the impact of the preservative loss on the performance of treated wood to the impact that the chemical which leaches from the treated wood, creates on the environment. This may be either when the wood is in service or when the treated wood is placed in a landfill during disposal at the end of its life. For example, Dubey et al., (*56*) described the metal leaching from pressure treated wood in a sanitary landfill. This study identified metal losses from the treated wood after soaking with leachate collected from six landfills. Two other sample sets were leached according to the toxicity characteristic leaching procedure (TCLP) or the synthetic precipitation leaching procedure (SPLP). In this way the authors were attempting to determine the impact of the leached metals on the landfill leachate. In a second study, Dubey et al. (*57*) compared the variation in copper leaching from blocks cut from commercially treated lumber and leached using natural and de-ionized water. Copper concentrations and toxicity were lower in the natural water compared to the de-ionized water. The effects of the physical and chemical characteristics of the soil on the adsorption of leached CCA and ACQ have been studied by Stefanovic and Cooper (*58*). Based on their laboratory experiments the adsorption was linearly related to the concentration of the solution applied to the soil and saturation was not attained. They were not able to identify a soil chemical parameter that consistently explained the adsorption results.

Comparison of Laboratory and Field Exposures

Although the inorganic based wood preservatives are based on relatively stable compounds, their long term stability remains unexplored. Hager (*25*) suggested that hydrolysis reactions in CCA treated timber can lead to loss of preservative. However, the rate of loss will depend on the commodity and type of exposure. A number of studies have therefore focused on losses from full sized treated wood under natural conditions. Nurmi (*59*) examined the losses of chromium, copper and arsenic from CCA-type B and CCA-type C treated utility poles after 11 years. A greater amount of arsenic was detected from the type B formulation. The author also noted that only a small amount of the migrated chemical into the soil adjacent to the poles was in a soluble form. The presence of naturally occurring iron in the soil reduced the mobility of arsenic. This rapid immobilization of leached preservative at the soil wood interface can enhance the performance of the treated wood, since it prevents growth of the fungi in the soil close to the treated wood. However, the impact of this on the performance of treated wood is not addressed during standard laboratory testing of leachability.

Jin and Preston (*60*) compared the depletion of preservatives using laboratory, fungal cellar and field tests. They concluded that for simple alkaline copper formulations there was good agreement between the relative leaching of copper in the three test methods. However, when a quaternary ammonium biocide was in the formulation laboratory tests were inadequate for predicting the leaching from soil depletion studies. Evans et al. (*61*) assessed the distribution and leaching in field test stakes. They observed the greatest loss from the end grain and a re-distribution from the inner part towards the surface. However, there was no consistent pattern between the two preservatives evaluated. Lebow and Foster (*62*) examined the impact of chromium, copper and arsenic which leached from a treated boardwalk over a wetland. They noted that in an area of high rainfall, measurable losses of copper, chromium and arsenic were found in close proximity to the wood. The concentrations in the soil increased with time up to the 5 years of the testing. The corresponding measurements of the sediment showed a more rapid increase, reaching a maximum value within the first year. In addition, the CCA components were found as much as 3 m from the structure.

An important factor affecting leaching of wood preservative from wood in service is the wettability of the timber. It has been shown by several researchers that a water repellent can significantly reduce the preservative loss in above ground applications. For example, Cui and Walcheski (*63*) showed that loss of arsenic could be reduced by more than 35%, while studies of water-repellent treated ACQ decking boards showed similar reductions in copper losses (*64*). Thus when studying the performance of treated wood under laboratory conditions, the impact of field applied water repellent treatments is not easily identified. It may be much easier to measure the benefits when the water repellent additive is included as part of the formulation. Because of the impact of field applied water repellent treatments the development of field data is essential to better understand the biocide depletion rate, since this will directly impact the performance of the treated wood.

It is to be expected that the depletion rate will be strongly influenced by the ratio of the volume to surface area of the treated wood. Thus leaching of chemical from sawdust is much more aggressive than that from solid wood blocks. Ruddick and Sam (*65*) varied the surface area to volume ratio in a small block laboratory leaching study of didecyldimethylammonium chloride from 1.4 to 2.3, and reported that the leaching increased from 11 to 32 %. Yamamoto et al. (*66*) examined loss of preservatives from 10 x 10 x 25 cm (ratio 0.40), 5 x 10 x 25 cm (ratio 0.60) and 2 x 10 x 25 (ratio 1.2) specimens. The specimens with a ratio of 0.4 leached 76.2 g/m^3 copper from ACQ and 6.9 g/m^3 copper from CCA, while the samples with a ratio of 1.2 leached 153 g/m^3 from ACQ and 14.1 g/m^3 from CCA. Similarly, timber with a high surface area to volume ratio (such as plywood) loses chemical much more rapidly than wood that has a low surface area to volume ratio (e.g. a utility pole). This is important and leads to the obvious conclusion that preservative depletion measured in small samples in the laboratory will be much greater than that from typical products in service such as utility poles or fenceposts. Few studies have been reported in which commercial sized material is exposed to water leaching. However, Evans (*67*) reported that poles treated with CCA type B and stored in water showed no measurable loss of chromium or copper in the outer 5 mm but did lose about 20 % of the arsenic.

Yamamoto et al. (*68*) compared the leaching of CCA, ACQ and didecyldimethylammonium chloride/boric acid in natural exposure and laboratory leaching. The concluded that 10 day laboratory leaching was always higher than the natural exposure. They also noted the influence of the initial loss of mobile preservative during the initial leaching. The researchers also observed a wide variation in the field leaching of the chromium copper and arsenic, which is a limitation of field leaching experiments.

Relatively few studies have been reported on the leaching of organic preservatives under laboratory and field conditions. Schoknecht, Wegen and Melcher in 2004 described their application of the OECD guideline proposals for laboratory leaching of preservative treated wood using propiconazole *(69)*. They reported good repeatability of the data. The results were compared with filed data and in general were found to be higher than those for natural rain leaching. They also noted an influence of the surface roughness on the leaching of propiconazole. Schoknecht et al. also compared the data generated for propiconazole to results reported by other researchers. For example the field losses normalized for exposed area and rainfall were 0.6 to 1 mg/m^2/L propiconazole, while those reported by Morsing in 2003 (*70*) were 3 to 5 mg/m^2/L The difference was ascribed to the fact that the rainwater had a greater distance (760 mm) to travel in the Danish study, compared to only 100 and 150 mm in the field experiment of Schoknecht et al.

Kennedy and Collins (*71*) compared the losses of tebuconazole from copper azole treated decking material under laboratory and field exposure. They reported about 5% loeached during the AWPA E11-97 method compared to about 1% after 10 months field exposure where the rainfall was 600 L/m^2.

Simulated Rainfall and Shower Tests

While the focus of many simulated rainfall tests has been on the loss of chemical to the environment, the use of shower tests represents a better assessment of preservative depletion than small block leaching studies. Consequently such experiments can provide insights into how mobile chemical can be lost from wood in service, a key factor influencing performance. In one such study Walley et al. (*72*) compared a simulated rainfall approach in which they measured the amount leached from CCA-treated timber, with the amount of unconverted chromium in the wood using a chemical reagent. They observed that the chemical spot test gave results that fluctuated with time, whereas the simulated rain method provided a more realistic assessment of preservative leaching. Shower tests had been used previously to assess the impact of washing wood on the leaching of preservative. Walley (*73*) also confirmed the importance of the rate of water application in shower tests used to monitor leaching. They found greater chromium losses with higher volumes of simulated rain. Storage of the treated lumber prior to leaching reduced chromium losses. The authors recommended a standard simulated rain test be developed to monitor fixation. Warburton and Cornfield (*74*) also reported the successful use of a rainfall simulation method to assess CCA permanence in pine posts and spruce rails and boards. They concluded that an excellent correlation existed between the rainfall method and a laboratory leaching assessment of small chips leached for 14 days in cold de-ionized water. Havermans et al. (*75*) described a shower test for assessing preservative losses from treated wood under simulated open storage conditions. Clearly, the intent of this approach was to determine the impact of rainfall upon the losses of preservative from relatively large volumes of treated wood during storage and relate the results to preservative fixation.

However, the shower test does have limitations, such as large volume of timber needed to obtain a reasonable result. Homan and Militz (*76*) compared the shower test to an "on-site" method to assess fixation. They sampled wood at different storage times and the samples were subjected to a short period of agitation (2 min.) followed by a static leaching for 15 min followed by a further agitation for 2 min. Analysis of the leachate showed the results to be comparable to the shower test method. Recently, Esser et al. (*77*) compared the shower test described by Havermans et al. (*75*) with other test methodologies, including a simple spray of 500 ml of deionized water using a plant sprayer and a 6 hour submersion. They concluded that a test similar to that used by Homan and Militz was not suitable for ACQ treated wood, while simple spray tests showed good correlation with shower tests for ACQ.

Schoknecht et al. (*6*) examined the rate of biocide loss in laboratory text based on EN 84 and a simulated rain experiment. The authors reported that emissions were much higher during the initial stages of leaching tests than under simulated rain exposure. However, the losses during the later stages of water leaching were comparable to those observed with simulated rain. This lead the authors to conclude that leaching losses can be estimated from laboratory leaching studies. The authors recommended that the volumes of water and the emitting surface area for the leaching, need to be identified in any protocol used to measure leaching.

Field Assessment of Preservative Loss Above Ground

In 1976, Cymorek (*78*) showed that a light continuous rainfall with low precipitation exerted a much stronger leaching effect than short heavy showers. He also concluded that the time of water flowing on the wood surface was important on how much chemical was lost. These observations hold important clues in modeling preservative losses from wood exposed above ground. Since over 65 % of treated wood used in North America is used above ground, it is important to better understand how and why it performs with respect to preservative mobility. Consequently, there has been a concerted effort to develop data on the natural losses of preservative from decking and fencing samples exposed to natural rainfall and weathering.

In addition, the historical close link between the leaching of treated blocks and biological evaluation has meant that most of the data developed in the laboratory has been based on fully treated sapwood blocks. This was necessary in order that rapid decay would take place in the control or reference blocks and the impact of the preservative retention could be easily seen. However, this choice does not necessarily provide relevant information on the leachability of the preservative in a more practical application since, apart from the high levels of sapwood found in southern pine, most timber from Canada, the western USA and Europe contains large quantities of heartwood. Thus, leaching of preservative from such material will be much more complex. For example, the presence of extractives can lead to complex formation between the heartwood extractives and metals such as copper or chromium, resulting is less metal being leached.

In addition, whereas a southern pine decking board may be fully penetrated, those of hemlock, spruce, lodgepole pine, jack pine, western red cedar and Douglas fir will generally have only a thin outer shell of treatment with only the sapwood fully treated. During service, as the wood becomes wet the mobile chemical can migrate with the moisture to reactive sites in the heartwood where it is immobilized. Thus, it is clear that the performance of the preservative treated decking boards will not only depend on how much chemical is leached during service, but also on how well the chemical is able to migrate into areas of untreated wood.

Perhaps, the most important failure of the leaching studies reported for treated wood exposed above ground during the last decade is not recognizing the key influence of the drying of wood between periods of wetting. While the use of shower tests of solid wood can explore the diffusion of mobile chemical to the wood surface where it is leached, the role of wood drying on chemical movement is not often assessed. For that reason recent strategies have focused on studying the migration of chemical within wood as well as the loss of chemical during the seasonal wetting and drying of the timber above ground (*26* and *64*). Such studies are critical if projections of preservative loss are to be made, but are even more important in understanding the role of mobile chemical in *enhancing* the performance of shell treated wood such as decking. The migration of preservative onto check surfaces, the formation of which exposes untreated wood, is a key factor leading to enhanced performance. It is also responsible for the protection of freshly cut surfaces where no end protection has been applied. These key observations reveal the importance of field assessments when attempting to predict performance of the shell treated wood in service.

In the pioneering research of Choi et al. (*79*), they demonstrated that mobile chemical can migrate during weathering onto checked surfaces. Further, the research showed that spores were unable to geminate and grow on wood surfaces with levels of copper found on the surfaces of checks within 6 months of installation (*80*) and that spores of copper tolerant fungi did not exhibit copper tolerance. These observations revealed the importance of a more comprehensive assessment of preservative mobility when projecting performance of treated wood used above ground. For that reason it is important to establish above ground tests with a view to relating the climatic conditions to preservative movement, including loss. In one of the first such studies, Choi et al. (*26*) were able to correlate the arsenic loss from treated hem-fir decking to such climate factors as rainfall, sun hours, and temperature. Subsequent research also confirmed that preservative retention is also an important factor.

The importance of these field tests in establishing a better understanding of the factors which enable treated wood to provide extended service can not be overestimated. With such understanding, more appropriate laboratory experiments can be designed to enable better predictions of performance of wood preservatives. For example, if spore germination is the primary mode of infection of treated wood exposed above ground, can lower biocide retentions be successfully used, than those based on the standard soil or agar block methods? Would improving such characteristics as dimensional stability be beneficial to preservative performance in products like decking and fencing, by eliminating checking? Further, in shell treated wood, the long term performance may be limited by the ability to provide mobile copper to coat checks throughout the life of the product. To be able to make this kind of prediction of service life modeling of the losses of mobile copper are needed.

Modeling of Preservative Losses from Treated Wood

Perhaps the most comprehensive study on the modeling of leaching has been reported by Waldron et al. (*81* and *82*). These authors used diffusion measurements and dissociation constants for preservative components to model and predict long term leaching from treated wood. The authors concluded that when predicting emissions based on diffusion the exposure time to rain rather than rain intensity is more important, citing earlier studies by Cockroft and Laidlaw (*1*) and Klipp et al. (*83*) to validate this conclusion. This is consistent with Cymorek's conclusions in 1971 (*78*) that slow steady rainfall is more effective at leaching than heavy rainfalls. However, as shown by Choi et al. (*26*) for CCA and by Chung and Ruddick (*64*) for ACQ, intervals of drying have a profound effect on the loss of copper from treated wood above ground. In addition, it was possible to predict losses over extended periods that were consistent with available mobile copper present in the treated boards. However, related studies by Chung and Ruddick (*84*) have confirmed that other factors such as wood species and sample dimension can influence the loss of preservative when described on a unit area basis. In addition, a correlation has been observed between the initial losses of preservative and the long term depletion. This lends credence to "predictive" tests being developed by the AWPA to assess the leachability of wood shortly after treatment and fixation. In modeling the losses, time of exposure had the greatest effect on copper lost while rainfall had a second order effect, supporting the suggestion that simple rainfall is not a good indicator of potential for loss since a slow shower with short drying intervals has a greater impact than a heavy rainfall.

Taylor and Cooper (*85* and *86*) reported the influence of climatic variables on the leaching of CCA from decking and developed regression models to help explain the variation in component losses. The models were able to explain 44% of the variation in copper leaching and the authors concluded that long term studies are needed to establish seasonal and annual trends in leaching rates. They also revealed that post treatment washings were inconsistent in their effect on leaching while the application of commercial water repellents reduced preservative leaching. Waldron and Cooper (*87*) described the application of a physical model applicable to a first-order fixation reaction with wood. They used this model and laboratory experiments to determine the loss of CCA components from small unfixed wood samples. The modeling was based on the determination of the diffusion coefficients and reaction rate constant for Cr^{6+} in red pine.

From the modeling of preservative depletion under field conditions, over time, it is possible to identify key factors influencing the biocide performance. This is important as laboratory preservative assessments are generally done under quite artificial conditions. For example pine sapwood is the common substrate used in testing. However, in practice with a few geographical exceptions, wood with substantial quantities of heartwood needs to be protected. Similarly, the depletion of biocide under continuous water exposure will be different from an exposure to intermittent wetting. Often the fungi isolated from treated wood which has failed are not the standard fungi used in biocide evaluation. Such field isolations can provide important evidence of which fungi should be used when screening new classes of preservative. Preservative depletion from treated sapwood is often quite different from the corresponding heartwood, particularly where the heartwood contains abundant reactive molecules.

Standardization of Leaching Assessments

Modeling of chemical emissions from treated wood into surface and ground water is an important requirement in Europe for treated wood used in above ground and ground contact situations. Consequently, the development of testing protocols to evaluate leaching has received considerable attention. Such studies also can provide valuable insights into biocide depletion since the strategies of testing are generally more closely related to the end use of the treated wood. One of the earliest cooperative studies was reported by Morgan (*88*). The collaborators in this round robin concluded that even following the standard DIN test methodology resulted in variable data. The DIN 52 176-2 used here was later modified to form the EN 84 leaching protocol (*89*).

Eetvelde et al. (*90*) reported studies of modeling copper emissions according to two standard protocols, the NEN 7345 from the Netherlands and the European ENV 1250.2 (91). The authors noted that apart from the high correlation of the cumulative increase in copper with time, the model did not appear to be of value in predicting the long term impact of copper leaching. Subsequently, Eetvelde et al. (*92*) reported further studies, which suggested that in a comparison of European ENV 1250.2 with the European EN84 aging method, the latter was better suited as an early stage method to assess leaching of new preservatives. Modeling of chemical losses from treated wood was also explored by Orsler and Holland (*93*). The objective was to examine the possibility of adapting a simple laboratory method to monitor preservative loss from treated wood. Using disodium octaborate as the preservative, small treated blocks with selected faces sealed to prevent leaching were immersed in water and losses of boron via the end grain or the lateral surfaces were measured. The ability to link this information to identifying potential losses from treated timber in service was not identified.

An important consideration in comparing leaching information is the protocol used to develop it. Habicht *et al.* (*94*) compared three protocols, the European EN84 (*89*), the North American AWPA E11 (*95*) and the Japanese JIS K 1571 (*96*) for developing leaching information on N-cyclohexyldiazenium-dioxy-copper or CuHDO preservative (CX). They found that the three methods gave different assessments of the leaching. This may be expected since several key parameters varied between the methods. For example, the ratio of the cross section to the longitudinal areas varied from 1:5, 1:2 to 1:1 for the European, American and Japanese procedures, respectively. The increased cross section resulted in an increase of the preservative leached from 6 to 7 % for the EN 84, to 8% for the AWPA E11, and 15 to 16 % for the JIS K 1571.The same trend was found for ACQ where leaching increased from 7 to 9.9 % for the AWPA E11 to 12.5 to 13.9 % for the JIS K 1571 procedure. The authors also found that fast drying of the samples had the potential to lower fixation and, hence, increase the leached chemical. The authors also noted no clear concentration dependence on the copper leaching. Walley et al. (*97*) also reviewed the methodologies for assessing leaching of chemical from treated wood and concluded that there were such a large number of methods that a uniform leaching assessment could not be determined.

In a series of papers, Morsing and co-workers explored the leaching of active components from preservative treated timber used above ground. The results from this research were intended to support the requirements of the Biocides Products Directive, which were to identify leaching rates of chemical from treated wood exposed above ground. The design of the test assemblies was outlined by Morsing in 2003 (*70*). Test assemblies containing pressure treated and flow coated sapwood were constructed to allow vertical, sloped at 45 degrees, and horizontal exposure to rain. All samples were conditioned before installation in the leach test. The rainwater, after passing over the samples, was collected in a stainless steel tube which directed it into storage containers for collection for analysis. Organic preservatives studied were propiconazole, tebuconazole and IPBC. Subsequently, Morsing and Lindegaard (*98*) reported the leaching of active components from treated wood. The authors concluded that the method was useful in characterizing the leaching behavior from treated wood exposed above ground. The highest leaching rates were observed immediately after installation. They found no significant differences due to end grain sealing. However, enhanced penetration of preservative strongly influenced the leaching rate. Smooth surfaces leached less chemical than rough surfaces. The loss of preservative correlated directly with the preservative retention.

To complete the study, the final phase was a comparison of the semi-field method with a laboratory study (*99*). The method was based on the CEN/OECD,

Feb. 2003 "OECD draft guideline for testing of chemicals proposal for new guideline XXX. Estimation of emissions from preservative treated wood held in the storage yard after treatment and for wooden commodities exposed in Use Class 3 (not covered and not in ground contact)." Scots pine (*Pinus sylvestris*) samples (150 mm x 25 mm x 100 mm in cross section) treated with the candidate preservative is stored for periods of 12 days and 9 months at 20 °C. The samples were leached by immersing in 550 ml of demineralized water for 1 minute, twice each week over a 60 day period. The samples were immersed three times at 3 hour intervals for a total of 18 x 3 immersions. Results were calculated in terms of mg of active ingredient/m^2 wood surface/immersion day. For each subsequent week the solutions were bulked into one sample. The average leaching of preservative for the 1st to 10th day and 11th to 60th day is determined and presented as mg active ingredient/m^2 wood surface. The results were presented for propiconazole and tebuconazole. As may be expected, the leaching rate of the material stored for only 12 days fixation was slightly higher than that stored for 9 months. The leaching during the first 10 days was 15 % higher for propiconazole and 25 % higher for tebuconazole than the leaching for 11 to 60 days. The results showed that the laboratory data underestimated the field leaching. The authors concluded that it was not possible to estimate long term leaching in exterior exposure from short term laboratory data, there being no constant ratio between field and laboratory data. The authors believed that in-service testing is best able to characterize the leaching behavior of biocides. While much of this kind of testing is aimed at examining the environmental impact of the treated wood, it also provides a valuable insight into biocide depletion, a key parameter linked to performance.

In a similar study, Melcher et al. (*100*) examined the depletion of copper and boron from chromated copper borate (CCB) using the Draft guidelines developed by OECD for estimation of emissions from preservative treated wood for exposure in Class 3 and 4. These methodologies, as noted above, involve immersion of treated wood blocks for relatively short periods of 1 minute to 1 hour over 60 day duration. They also compared this procedure with the European standard for leaching, EN 84, using EN 113 sized blocks. Two laboratories collaborated in the project. They reported good reproducibility of the test methods. Boron leached about 90 % whereas copper losses were about 10 %. They suggested that such data would correspond to losses after 10 years in service. The authors noted that the use of Guideline #1 may be compromised by low levels of copper loss which might be below the detection limit and suggest that the longer immersion time used in Guideline #2 may be preferable, provided an agreement to factor the results for the lower hazard experienced in Use Class 3 can be agreed upon.

In the USA, the standard assessment of chemical efficacy as a wood preservative is based on exposure of small (19 mm cube) sapwood blocks to a leaching regime, prior to placing into the soil jars for exposure to selected decay fungi (*101*). There is also a standard leaching test which also uses small 19 mm sapwood blocks to determine losses (*95*). In 1977 Rak proposed a number of changes to the leaching protocol in use in North America at the time (*102*). He rationalized that the since the leaching experimental design did not simulate actual field exposure conditions, then the requirements for the standard should be limited to reproducibility and a relatively rapid test. He proposed that by raising the temperature of the leaching solution the time could be shortened. He suggested that the only question associated with higher temperature leaching was the quality of compounds leached at the room and higher temperature are different.

The most widely used method to assess the field performance of preservatives is the use of a field stake (*103*), which also contains a section describing the measurement of biocide depletion. However, almost all of the methods for calculating the average service life of preservative treated wood from field stake tests requires a lengthy exposure period. During this period the failure of the stakes is correlated with the service life (for example *104*, *105* and *106*). However, no attempt is made to correltate the failure of the stakes to biocide degradation or loss. If the field tests are ever to become predictive in their scope, correlation of performance with biocide loss or degradation will be required.

It has long been recognized that neither the laboratory testing nor the use of field stakes can adequately assess the performance of treated wood above ground. Consequently, several new test protocols have been introduced to assess the performance of treated wood in above-ground exposure. The earliest of these was the use of an L-Joint to simulate the lower corner joint of a wooden window (*107*). However, there is no requirement to monitor biocide depletion in this method. Recognizing the influence of water-trapping on decay potential, another method, the "lap-joint" has been standardized (*108*). This method has gained recognition as a useful test by some researchers, but it too does not measure preservative depletion. The most recent field test protocol for above ground exposure is the "ground proximity test" (*109*). In this procedure small samples are placed on cinder blocks and covered with a shade cloth. It has been widely used in tropical or semi-tropical environments when more rapid decay of the samples compared to more temperate climates has been reported. However, again, there is no requirement to monitor leaching of preservative. The most recent test added to the American standards is the measurement of leachability of wood preservatives from treated wood in contact with different soils (*49*). The intent of this method is to examine the influence of different soils in contact with the treated wood, on the loss of preservative. Consequently, it does not provide information on the depletion of wood in service above ground.

In summaryizing the extensive activity in researching the leaching behaviour of treated wood exposed under standardized laboratory and field test conditions, it is clear that a common approach needs to be considered. Within Europe progress has been made in this regard by trying to correlate research data using standardized laboratory exposure to that developed in the field. Comparisons with different international standard protocols are helpful in understanding when they may produce similar results and when different methods may show bias. However, the goal in most testing is to be able to predict long term performance, such as biocide depletion. If this predicitive capability of laboratory and field testing of wood preservatives is to be realized, a much greater understanding of the role of biocide degradation or loss on why treated wood fails is required.

Summary

The main factors resulting in preservative depletion from treated wood are reviewed with a focus on laboratory and field assessment methodologies. The factors resulting in depletion include: evaporation, degradation and depletion by weathering processes. Due to the dominance of the inorganic copper based wood preservatives, the most important of these is leaching. However, with the increasing use of organic biocides, the importance of ultraviolet, thermal and chemical degradation should not be neglected. Historically, the early studies, focused on preservative formulation optimization by accounting for chemical loss in small block laboratory leaching studies. It was recognized that soil conditions with humic acid or low pH offered a more challenging environment, and laboratory experiments were developed to understand their impact on performance. With the emerging focus on the environmental impact of treated wood, experimental strategies were developed to assess the loss of the preservative to the environment. Such studies are valuable in providing information on biocide depletion since the design protocol is usually more closely related to the typical exposure of the treated wood in service. Standard methodologies were also being developed to examine the leaching of preservatives after treatment but before being shipped from the facility. Other research developed methods for examining the impact of landfill environments on the leaching of chemicals from disposed treated wood waste. With the dominance of the waterborne preservatives, the opportunity to study the impact of treated commodities in service emerged and that could provide a more realistic assessment of the environmental impact. However, the focus of researchers was, and to some extent remains, on maximizing the fixation of the preservative components in wood. The fixation efficiency is most commonly assessed through laboratory leaching studies. However, shower tests of lumber have become widely accepted for measuring the possible losses of chemical from stored treated wood prior to shipment. Such studies are important as they contribute to an awareness of the impact of the fixation chemistry on biocide depletion.

The use of treated wood expanded rapidly with the acceptance in the residential market place. Consequently, it was important to understand the factors contributing to the efficacy of wood used above ground as decking, fencing or landscaping. Ground contact models are inappropriate. The use of water repellents could significantly affect preservative leaching. Recent research has revealed valuable information on which parameters strongly impact leaching of chemicals from above ground products like decking. Mini decking tests have demonstrated that drying intervals between periods of rain are very important in redistributing mobile preservative within the timber. This can be beneficial for shell treated wood, since it enables the checks to be coated. Parallel research has shown that the level of copper relocated to freshly exposed checks is sufficient to prevent spore germination, even of copper tolerant fungi, since at the spore stage thee fungi do not exhibit copper tolerance. Long term decking testing showed that long-term exterior depletion could be linked to the initial losses so that poorly fixed biocides could be identified through simple laboratory measurements. Comprehensive modeling of leaching has been initiated based on diffusion and dissociation of the preservative complexes formed in the wood, although as yet it is not possible to predict long term depletion by leaching. With an increased interest in predicting performance of treated wood in service above ground, a number of standard testing methodologies have been developed. However, at present not many of these new standard methodologies require monitoring of preservative depletion. This should be remedied as the depletion of the preservative with increased exposure is a key factor impacting on performance. Without any understanding of the rate which biocides degrade or deplete, and how this impacts on performance, the predictive capacity of the biocide assessment protocols will not be realized.

References

1. Cockroft, R. and R.A. Laidlaw. 1978. Internat. Res. Gp Wood Preserv. IRG/WP/Doc No. 3113
2. Van Acker, J., K. Ghekiere and M. Stevens. 2002. Internat. Res. Gp Wood Preserv. IRG/WP/Doc No. 02-20256.
3. Edlund, M.L., B. Henningsson, B. Jensen and C.E. Sundman. 1988. Internat. Res. Gp Wood Preserv. IRG/WP/Doc No. 3476.
4. Valcke, A.R. and M. Stevens. 199. Internat. Res. Gp Wood Preserv. IRG/WP/Doc No. 3647.
5. Suttie, E.D., R.J. Orsler and T. Dearling. 2000. Internat. Res. Gp Wood Preserv. IRG/WP/Doc No. 00-20193.
6. Schoknecht, U., R. Wegener, W. Horn and O.Jann. *Environ. Sci. & Pollut. Res.* **2002**, *10*(3):154-161.

7. EN 13419-1. 1999. Building products – Determination of the emission of volatile organic compounds – Part 1: Emission test chamber method. (European Prestandard).
8. EN 13419-2. 1999. Building products – Determination of the emission of volatile organic compounds – Part 2: Emission test cell method. (European Prestandard).
9. ENV 1250-1. 1994. Wood preservatives – Methods for measuring losses of active ingredients and other preservative ingredients from treated timber – Part 1: Laboratory method for obtaining samples for analysis to measure losses by evaporation to air.
10. Mortimer, W.P. Environmental persistence and migration of wood pole preservatives: Part One. *Proc. Can. Wood Preserv. Assoc.* **1989**, *10*:57-74.
11. Ruddick, J.N.R. *J. Wood Protect.* **1991**, *1*(2):77-83.
12. McBain, A., F. Cui, L. Herbert and J.N.R. Ruddick. The microbial degradation of chlorophenolic preservatives in spent pressure treated timber. **1995**, *Biodegradation 6*:47-55.
13. Burgel, J., J. Dubois and J.N.R. Ruddick. 1996, Internat. Res. Gp. Wood Preserv. IRG/WP/Doc No. 96-10166.
14. Nicholas, DD, A.D. Williams, A.F. Preston and S. Zhang. 1991. *For. Prod. J.* **1991**, *41*(1):41-45.
15. Guo, A., P.A. Cooper and T. Ung. 2005. *For. Prod. J.* **2005**, *55*(7/8):72-75.
16. Ruddick, J.N.R. *Proc. Amer. Wood Preserv. Assoc.* **1996**, 92:32-49.
17. Ruddick, J.N.R. *Proc. Can. Wood Preserv. Assoc.* **2003**, *24*:26-59.
18. Ung, T.Y. and P.A. Cooper. *Holz als Roh und Werkstoff.* **2005**, *63*:186-191.
19. Lucas, N. and J. N.R. Ruddick. 2002. Internat. Res. Gp Wood Preserv. Doc. No. IRG/WP/Doc No. 02-30285
20. Jiang, X. and J.N.R. Ruddick. 2000. Internat. Res. Gp Wood Preserv. Doc. No. IRG/WP/Doc No. 00-30233.
21. Jiang, X. and J.N.R Ruddick. *For. Prod. J.* **2004**, *54*(12):213-216.
22. Jiang, X. and J.N.R. Ruddick. 1997. Internat. Res. Gp Wood Preserv. Doc. No. IRG/WP/Doc No. 97-30158.
23. Kamdem, D.P. and J. Zhang. 2000. Internat. Res. Gp Wood Preserv. IRG/WP/Doc No. 00-30216.
24. Zhang, J. and D.P. Kamdem. 1999. Internat. Res. Gp Wood Preserv. IRG/WP/Doc No. 30203.
25. Hager, B. *For. Prod. J.* **1969**, *19*(10):21-26.
26. Choi, S., J.N.R. Ruddick and P.I. Morris. *For. Prod. J.* **2004**, *54*(3):33-37.
27. Militz, H. 1992. Internat. Res. Gp Wood Preserv. IRG/WP/Doc No. 3687-92.
28. Orsler, R.J and M.W.S. Stone. 1984. Internat. Res. Gp. Wood Preserv. IRG/WP/Doc No. 3288.
29. Lloyd, J.D., M.W. Schoeman and F . Brownsill. 1998. Internat. Res. Gp Wood Preserv. IRG/WP/Doc No. 98-30177.

30. Liu, R. and J.N.R. Ruddick. 1993. Internat Res. Gp Wood Preserv. Doc. No. IRG/WP/Doc No. 93-10020.
31. Perraud, R. and M. Papazian. 1993 2nd Internat. Symposium Wood Preservation. Internat. Res. Gp Wood Preserv. IRG/WP/Doc No. 5001:383-393.
32. Henry, W. and E.B. Jaroski. *Proc. Amer. Wood Preserve. Assoc.* **1967**, *63:*187-196.
33. Fahlstrom, G.B., P.E. Gunning and J.A. Carlson. *For. Prod. J.* **1967**, *17*(7):17-22.
34. Cowan, J. and S. Banerjee. *For. Prod. J.* **2005**, *55*(3):66-70.
35. Plackett, D.V. 1984. Internat. Res. Gp Wood Preserv. IRG/WP/Doc No. 3310.
36. Kartal, S. N. and W. J. Hwang. 2005. Internat. Res. Gp Wood Preserv. IRG/WP/Doc No. 05-50228.
37. Cooper, P.A. and Y.T. Ung. *For. Prod. J.* **1992**, *42*(9):57-59.
38. Cooper, P.A. and Y. T. Ung. 1995. Internat. Res. Gp Wood Preserv. IRG/WP/Doc No. 95-50048.
39. Cooper, P.A., D. Jeremic, J.L. Taylor and Y.T. Ung. 2000. Internat. Res. Gp Wood Preserv. IRG/WP/Doc No. 00-50151.
40. Evans, F.G. 1987. Internat. Res. Gp Wood Preserv. IRG/WP/Doc No. 3433
41. Ruddick, J.N.R. 1992. *Mat. u. Org.* **1992**, *27*(2):135-146.
42. Stephan I., H.H. Nimz and R.-D. Peek. 1993. Internat. Res. Gp Wood Preserv. IRG/WP/Doc No. 93-50012.
43. Murphy, R.J. and D.J. Dickinson. Internat. Res. Gp Wood Preserv. IRG/WP/Doc No. 93-50001:189-194.
44. Warner, J.E. and K. Solomon. *Environ. Toxicol. and Chem.* **1990**, *9*:1331-1337.
45. Cooper, P.A. 1991. *For. Prod. J.* **1991**, *41*(1):30-32.
46. Eetvelde, G. van, R. Orsler, G. Holland and M. Stevens. 1995a. Internat. Res. Gp Wood Preserv. IRG/WP/Doc No. 95-50046.
47. Eetvelde, G. van, W. Homan, H. Militz and M. Stevens. 1995b. Internat. Res. Gp Wood Preserv. IRG/WP/Doc No. 95-50040:195-208
48. Gray, S. M. 1991. Internat. Res. Gp Wood Preserv. IRG/WP/Doc No. 2385.
49. AWPA E20-04. 2005. Standard method for determining the leachability of wood preservatives in soil contact. Amer. Wood Preserv. Assoc. Book of Standards. Alabama, USA
50. Lebow, P., R. Ziobro, L. Sites. T. Schultz, D. Pettry, D. Nicholas, S. Lebow, P. Kamdem, R. Fox and D. Crawford*Wood and Fibre Sc.* **2006**, *38*(3):439-449.
51. Venkatasamy, R. and D.N. Okwara. 2003. Internat. Res. Gp Wood Preserv. IRG/WP/Doc No. 03-50203.
52. Ruddick, J.N.R. and A. W. Kundzewicz. *Mat. u Org.* **1991**, *26*(3):169-181.
53. Albuquerque, R.M., S. Cragg and J.D. Icely. 1996. Internat. Res. Gp Wood Preserv., IRG/WP/96-50080.

54. Gjovik. L.R. 1977. *Proc. Amer. Wood Preserv. Assoc.* **1977**, *73*:142-152.
55. Irvine, J. and S.-E. Dahlgren. *Holzforschung* **1976**, *30*(2):44-50.
56. Dubey, B., T. Townsend and H. Solo-Gabriele. 2004. Internat. Res. Gp Wood Preserv. IRG/WP/Doc No. 04-50220.
57. Dubey, B., T. Townsend, G. Bitton and H. Solo-Gabriele. 2005. Internat. Res. Gp Wood Preserv IRG/WP/Doc No. 05-50236.
58. Stefanovic, S. and P. Cooper. 2003. Internat. Res. Gp Wood Preserv. IRG/WP/Doc No. 03-50200.
59. Nurmi, A.J. 1993. Internat. Res. Gp Wood Preserv. IRG/WP/Doc No. 93-50001:91-99.
60. Jin, L. and A.F. Preston. 1993. Internat. Res. Gp Wood Preserv. IRG/WP/Doc No. 93-50001:109-119.
61. Evans, F.G., B. Nossen and M-L Edlund. 1994. Internat. Res. Gp Wood Preserv. IRG/WP/Doc No. 94-50026.
62. Lebow, S. and D. Foster. *For. Prod. J.* **2005**, *5*(2):62-70.
63. Cui, F. and P. Walcheski. 2000. Internat. Res. Gp Wood Preserv. IRG/WP/Doc No. 00-50158.
64. Chung, P.A. and J.N.R. Ruddick. 2004a. Internat. Res. Gp Wood Preserv. IRG/WP/Doc No. 04-50219.
65. Ruddick, J.N.R. and A.R.H. Sam. *Mat. u Org.* **1982**, *17*(4):299-313
66. Yamamoto, K., S. Motegi and A. Inai. 2000. Internat. Res. Gp Wood Preserv. IRG/WP/Doc No. 00-50160.
67. Evans, F. G. 1978. Internat. Res. Gp Wood Preserv. IRG/WP/Doc No. 3122.
68. Yamamoto, K., S. Motegi and A. Inai. 1999. Internat. Res. Gp Wood Preserv. IRG/WP/Doc No. 99-50134.
69. Schoknecht, U., R. Wegener and E. Melcher. 2004. Internat. Res. Gp Wood Preserv. IRG/WP/Doc No. 04-50209.
70. Morsing, N. 2003. Internat. Res. Gp Wood Preserv. IRG/WP/Doc No. 03-20276.
71. Kennedy, M.J. and P.A. Collins. 2001. Internat. Res. Gp Wood Preserv. IRG/WP/Doc No. 01-50171.
72. Walley, S., P. Cobham and P. Vinden. 1996a. Internat. Res. Gp Wood Preserv. IRG/WP/Doc No. 96-50075.
73. Walley, S., P. Cobham and P. Vinden. 1996b. Internat. Res. Gp Wood Preserv. IRG/WP/Doc No. 96-50074.
74. Warburton, P. and J.A. Cornfield. 1991. Internat. Res. Gp Wood Preserv. IRG/WP/Doc No. 2373.
75. Havermans, J.B.G.A., W. Homan and M.J Boonstra. 1993. Internat. Res. Gp Wood Preserv. IRG/WP/Doc No. 93-50001:77-90.
76. Homan, W.J. and H. Militz. 1993. Internat. Res. Gp Wood Preserv. IRG/WP/Doc No. 93-50001:189-194.
77. Esser, P, W. Suitela and H. Trompetter. 2000. Internat. Res. Gp Wood Preserv. IRG/WP/Doc No. 00-50150.

78. Cymorek, C. 1976. Internat. Res. Gp Wood Preserv. IRG/WP/Doc No. 264.

79. Choi, S. M., J.N.R. Ruddick, and P.I. Morris. 2001. Internat. Res. Gp Wood Preserv. IRG/WP/Doc No. 01-30263.
80. Choi, S.M., J.N.R. Ruddick and P.I. Morris. 2002. Internat. Res. Gp Wood Preserv. IRG/WP/Doc No. 02-10422.
81. Waldron, L., P.A. Cooper and T.Y. Ung. *Holzforschung* **2005**, *59*:581-588.
82. Waldron, L., T.Y. Ung and P.A. Cooper. 2003. Internat. Res. Gp Wood Preserv. IRG/WP/Doc No. 03-50199.
83. Klipp, H., H. Willeitner, K. Brandt and A. Muller-Grimm. 1990. Internat. Res. Gp Wood Preserv. IRG/WP/Doc. No. 3669.
84. Chung, P.A. and J.N.R. Ruddick. Proc. *Can. Wood Preserv. Assoc.* **2004b**, *25*:97-108.
85. Taylor, J.L. and P.A. Cooper. *For. Prod. J.* **2003**, *53*(9):81-86.
86. Taylor, J.L. and P.A. Cooper. *Holzforschung* **2005**, *69*:467-472.
87. Waldron, L. and P. Cooper. 2002. Internat. Res. Gp Wood Preserv. IRG/WP/Doc No. 02-50193.
88. Morgan, J.W.W. 1975. Internat. Res. Gp Wood Preserv. IRG/WP/Doc No. 246.
89. EN 84 1997. Wood Preservatives – Accelerated aging of treated wood prior to biological testing – Leaching procedure.
90. Eetvelde, G. van, S.L.G. Michon and M. Stevens. 1996. Internat. Res. Gp Wood Preserv. IRG/WP/Doc No. 96-20097.
91. ENV 1250-2 1994.Wood preservatives – Methods for measuring losses of active ingredients and other preservative ingredients from treated timber – Part 2: Laboratory method for obtaining samples for analysis to measure losses by leaching into water or synthetic sea water.
92. Eetvelde, G. van, M. Stevens, F. Mahieu, H.-W. Wegen and A. Platen. 1998. Internat. Res. Gp Wood Preserv. IRG/WP/Doc No. 98-50115.
93. Orsler, R.J. and G.E. Holland. 1993. Internat. Symposium Wood Preservation. Internat. Res. Gp Wood Preserv. IRG/WP/Doc No. 93-50001:195-214.
94. Habicht, J., D. Hantzshel and J. Wittenzellner. 2003. Internat. Res. Gp Wood Preserv IRG/WP/Doc No. 03-20264.
95. AWPA E11-97. 2005. Standard method of determining the leachability of wood preservatives. Amer. Wood Preserv. Assoc. Book of Standards. Alabama, USA
96. JIS K 1571 2004. Test methods for determining the effectiveness of wood preservatives and the performance requirements. Japan Standard Association. Japanese Industrial Standards, Chemical Engineering Division.
97. Walley, S, P. Cobham and P. Vinden. 1996c. Internat. Res. Gp Wood Preserv. IRG/WP/Doc No. 96-50076.

98. Morsing, N. and B. Lindegaard. 2004a. Internat. Res. Gp Wood Preserv. IRG/WP/Doc No. 04-20302.
99. Morsing, N. and B. Lindegaard. 2004b. Internat. Res. Gp Wood Preserv. IRG/WP/Doc No. 04-20303
100. Melcher, E., R.-D. Peek, U. Schoknecht and R. Wagner. 2004. Internat. Res. Gp Wood Preserv. IRG/WP/Doc No. 04-50208.
101. AWPA E10-01. 2005. Standard method of testing wood preservatives by laboratory soil block cultures. Amer. Wood Preserv. Assoc. Book of Standards. Alabama, USA
102. Rak, J. 1977. Internat. Res. Gp. Wood Preserv. IRG/WP/Doc No. 294.
103. AWPA E7-01. 2005. Standard method of evaluating wood preservatives with field stakes. Amer. Wood Preserv. Assoc. Book of Standards. Alabama, USA
104. Hartford, W.H. *Proc. Amer. Wood Preserv. Assoc.* **1972**, *68*:67-82.
105. Link, C.L. and R. D. DeGroot. *Wood and Fibre Sci.* **1990**, *22*(1):92-108.
106. Purslow, D.F. Build Res. Estab. Current Paper **1977**, CP 31/77.
107. AWPA E9-97. 2005. Standard field test for the evaluation of wood preservatives to be used in non-soil contact. Amer. Wood Preserv. Assoc. Book of Standards. Alabama, USA
108. AWPA E16-98. 2005. Standard field test for the evaluation of wood preservative to be used out of ground contact: horizontal lap-joint method. Amer. Wood Preserv. Assoc. Book of Standards. Alabama, USA
109. AWPA E18-04. 2005. Standard field test for evaluation of wood preservatives intended for use category 3B applications exposed, out of ground contact, uncoated ground proximity decay method. Amer. Wood Preserv. Assoc. Book of Standards. Alabama, USA

Chapter 18

The Role of Non-Decay Microorganisms in the Degradation of Organic Wood Preservatives

Derek F. Wallace, Steven R. Cook, and David J. Dickinson

Division of Biology, Faculty of Natural Sciences, Timber Technology, Sir Alexander Fleming Building, Imperial College, London SW7 2AZ, United Kingdom

Since the implementation of the Biocide Products Directive there has been increasing pressure on the wood preservation industry to utilise environmentally sensitive organic biocides. It has been reported that organic biocides are degraded in wood exposed to the full range of microorganisms resulting in the timber becoming susceptible to fungal decay by organisms otherwise controlled by the biocides. Biodetoxification was found to be mediated by a range of bacteria, with Gram negative proteobacteria often associated with biocide degradation. In particular, strains of *Pseudomonas* have been reported to degrade a number of QACs, IPBC, chlorothalonil and an oxathiazine derivative, although the mechanisms have not been elucidated. Despite the observed biodegradation of many organic wood preservatives, many of these compounds have been successfully used for a number of years. This chapter reviews the biodegradation of organic biocides that are used to protect wood.

Biocide Products Directive

In recent years the timber preservation industry has responded to environmental, legislative and economic pressures to utilise more environmentally sensitive formulations. The potential hazards posed by wood preservatives have long been highlighted by regulatory bodies, which culminated in the Biocidal Products Directive (BPD) being implemented by the European Parliament in 1998 (Directive 98/8/EC) to monitor placement of biocides and their formulations on the market. Quantification of the toxicity and persistence of these biocidal compounds was one of the principle aims of this legislation. The directive requires manufacturers to detail the environmental hazards of the active chemicals, with particular attention being paid to the ultimate fate of the compounds. Ideally, a candidate biocide should be completely biodegradable, leaving no toxic residues and having minimal environmental impact. It is clear that a number of the chemicals used prior to 1998 as wood-preservatives needed to be replaced and that this represented a significant challenge and opportunity for the industry.

The main problem associated with the development of environmentally friendly preservatives based on organic compounds, is the observed biotransformation of these biocides (*1, 2, 3, 4*). Activity against the causal decay organisms and resistance to physical losses are no longer the sole performance criteria that need to be understood when developing a new wood preservative. Despite these drawbacks, several modern organic biocides (examples of which are discussed to later) have found an important role in wood treatments. However, if they are to be utilised in the long term in high hazard situations, such as soil contact, it will be necessary to fully understand their biotransformation and attempt to control it.

Propiconazole

Propiconazole (*cis-trans*-1-[2-(2,4-dichlorophenyl)4-propyl-1,3-dioxolan-2-ylmethyl]-1*H*-1,2,4-triazole) is a triazole antifungal agent which inhibits ergosterol biosynthesis and causes electrolyte leakage and ultrastructural damage to cells (*5, 6*).

Developed for agricultural use, propiconazole has been shown to be effective in the control of sapstain, and both brown rot and white rot Basidiomycetes, with toxic values of between 0.2-0.5 kg m^{-3} (*7, 8*). These toxic values were not, however, found to be adequate to protect wood blocks from fungal attack in a soil burial test, where the wood showed decay after 8 weeks (*1, 9*). Although resistant fungi appeared to play a role in biocide depletion, it was clear that there was an additional biological factor in soil which contributed to propiconazole degradation, most likely members of the bacterial community (*1, 10, 11*). Subsequent studies isolated a number of bacterial strains which showed the ability to degrade propiconazole, illustrating the role of bacteria in the degradation of wood preservatives (*1*, *12*). However, studies of radiolabelled propiconazole revealed that the biocide is not utilized as a carbon source, with only 3-10% of the biocide being converted to CO_2 in a single year (*13*, *14*). Typically, propiconazole was found to degrade slowly in soil through hydroxylation of the propyl-side chain on the dioxilane ring; see Figure 1 (*13*, *14*).

Chlorothalonil

Chlorothalonil is a foliar, non-systemic, broad-spectrum, chlorinated fungicide, highly effective against pathogens that infect vegetables, fruits and ornamentals. Chlorothalonil mediates an anti-fungal effect by reacting with sulfhydryl groups and glutathione present in proteins or in cofactors (*15*). Most reports have shown that chlorothalonil was rapidly degraded by soil microorganisms in 5-36 days, under aerobic conditions (*16*, *17*).

Katayama *et al.* (*16*, *18*) demonstrated that bacteria of various taxonomic groups were capable of degrading chlorothalonil in pure culture. In total, 37 soil-inhabiting bacteria were tested for their ability to degrade the biocide, and it was found that strains of Gram negative bacteria, including *Acinetobacter*, *Agrobacterium*, *Azomonas*, *Pseudomonas* and *Xanthomonas,* could all degrade 0.5 mg L^{-1} chlorothalonil. A number of Gram positive bacteria, including *Bacillus*, *Corynebacterium* and *Nocardia* strains, were also found to be able to mediate clorothalonil's degradation, although these bacteria were less tolerant to chlorothalonil than the Gram negative isolates. It was considered that common enzymes ubiquitous in bacteria were involved in the degradation of chlorothalonil by cometabolism (*16*, *17*, *18*). There was also some evidence that fungi were capable of mediating chlorothalonil degradation (*19*, *20*).

There are reportedly two major metabolic pathways involved in chlorothalonil degradation. The first involves the displacement of one chlorine atom by a hydroxyl group, potentially via reductive dechlorination, to generate 4-hydroxy-2,5,6-trichloroisophthalonitrile (Mt-1), as shown in Figure 2 (*15*, *17*, *18*, *21*, *22*). A second mechanism suggested was the oxidation/hydration of one cyano group to a corresponding amide and organic acid, to generate Mt-2 and subsequently Mt-3 of Figure 2 (*15*).

Two months after chlorothalonil application to soil, Mt-1 was found to represent 37% of the applied chlorothalonil, whilst 22% had been degraded to 1,3-dicarbomoyl-2,4,5,6-tetrachlorobenzene, represented by the intermediate between Mt-2 and Mt-3 (*23*). Only 3-14% of the radiolabelled chlorothalonil applied to soil was degraded to $^{14}CO_2$ after 90 days, which illustrated that complete mineralisation was not the major metabolic pathway of chlorothalonil degradation (*21*).

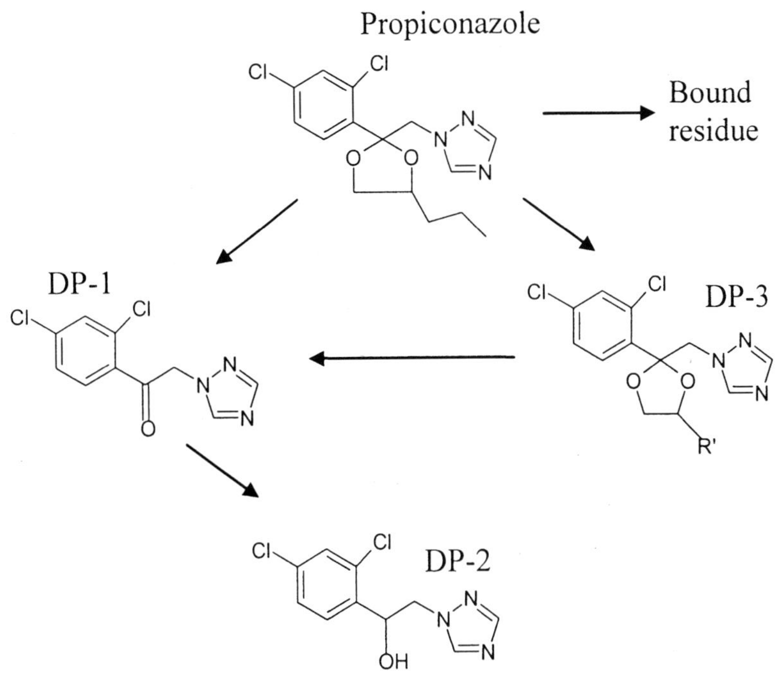

Figure 1. Proposed degradation pathway of propiconazole (reproduced with permission from reference 13. Copyright 2002 Kluwer Academic Publishers).

Kim *et al.* (24) recently reported the isolation of an *Ochrobacterium anthropi* strain capable of efficiently biotransforming the fungicide chlorothalonil from soil. A gene responsible for the chlorothalonil biotransformation was cloned into *E. coli* and subsequently identified as the open reading frame for glutathione-*S*-transferase (GST). It has been reported that GST catalyses the conjugation of the glutathione sulfur atom to a large variety of electrophillic compounds of both endobiotic and xenobiotic origin, resulting in detoxification (*25*, *26*). Glutathione-*S*-transferases are a very broad group of enzymes which have the cosubstrate glutathione in common. The enzymes' specificities are very broad, and the glutathione thiolate ion has been shown to attack the nitrogen of organic nitrite esters, the sulfur of organic thiocyanates and on the oxygen of organic peroxides (*26*). In a study of the mechanism of reaction with 4-substituted chlorobenzenes, it was concluded that the determining factor was the electrophilicity of the carbon atom attacked (*25*). This illustrates that there a number of different potential metabolic pathways by which a biocide can be microbially detoxified.

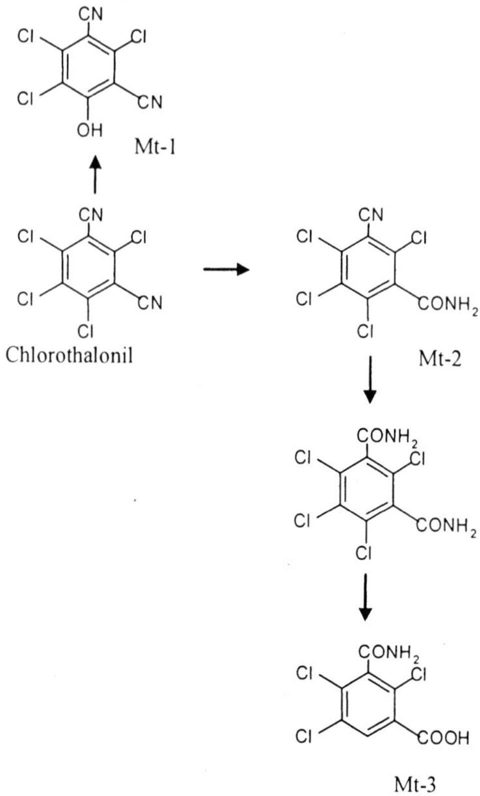

Figure 2. The degradation pathway of Chlorothalonil (scheme adapted from reference 15 by permission of The Royal Society of Chemistry).

Quaternary Ammonium Compounds

Quaternary ammonium compounds (QACs), defined by their quaternary nitrogen atom carrying hydrophobic alkyl chains, have been used extensively as disinfectants, fabric softeners, hair rinses and dispersion agents (27). QACs have also found use as wood preservatives, where they have been found to be highly effective at protecting timber from mould, wood-decaying and sap-staining fungi. Specifically, didecyldimethylammonium chloride (DDAC) was found to perform as well as copper chromic arsenate (CCA) in laboratory tests against wood decay fungi (*28*, *29*). Currently, products containing DDAC account for approximatley 95% of the Canadian sapstain control market (30).

The degradation of QACs in wood products has not been studied in detail, however QAC degradation in wastewater has been widely described. It has been reported that hexadecyltrimethyl ammonium bromide and decyltrimethyl ammonium bromide were both degraded by a co-culture of *Xanthamonas* and *Pseudomonas,* through the oxidation of the terminal carbon of the alkyl chain, resulting in the sequential cleavage of acetyl units by β-oxidation (*31*). Similarly, a strain of *Pseudomonas fluorescens,* isolated from activated sludge, showed the ability to utilise DDAC as a carbon source, with degradation mediated through two *N*-dealkylation steps. The initial dealkylation step was mediated by a mono-oxygenase, producing decyldimethylamine, and decanoic acid as the alkyl co-product, with the former compound being further degraded by a second inducible *N*-dealkylating enzyme (*27*).

Pseudomonas strains clearly have the ability to degrade a wide range of QACs, having been reported to also degrade laurypyridinium chloride, cetyltrimethyl ammonium bromide, hexadecyltrimethylammonium chloride, didecyldimethylammonium chloride and benzalkonium chloride (*27*, *32*, *33*, *34*).

A number of mould fungi have also been isolated which demonstrated the ability to tolerate and degrade QACs, including strains of *Penicillium*, *Trichoderma*, *Verticillium*/*Acremonium* and *Gliocladium* (*35, 36*). It was considered that the activity of these fungi may result in QAC degradation, through the hydroxylation of the QAC near the end of the alkyl chain, resulting in the susceptibility of the timber to decay (*37*). Notably these studies did not take into account bacterial degradation, and it is highly probable that a large bacterial QAC-degrading community would have been present in the wood.

IPBC

3-Iodoprop-2-ynyl-*N*-butylcarbamate (IPBC) is an organoiodine carbamate fungicide commonly used to control the defacement of softwood timber by sap-staining fungi, in place of the environmentally persistent pentachlorophenol (*38*).

It has been reported that IPBC is degraded *in vitro*, resulting in the loss of fungal toxicity (*2*, *3*). Nine strains of bacteria, belonging to the genera *Alcaligenes*, *Enterobacter*, *Microbacterium* and *Pseudomonas*, were isolated, and it was found that these isolates mediated the dehalogenation of IPBC to prop-2-ynyl-*N*-butylcarbamate (PBC) and iodine at a 1:1 stiochiometric ratio. The isolates were found to be unable to utilise IPBC as a carbon source and degradation was not mediated by either an extracellular moiety nor through acid hydrolysis (*39*). It was concluded that IPBC degradation occured via a reductive dehalogenation mediated by a cytosolic moiety which utilised NADH (and other reductive coenzymes) as a cofactor (*39*).

To overcome the limitations of using a readily degraded compound as a wood preservative, IPBC was combined with QACs such as DDAC (*40*). Since DDAC has a strong bacteriacidal activity, by disrupting the bacterial cell membrane (*41*), coformulation has proved to be an effective technique by which to prevent the bacterial-mediated degradation of an organic wood preservative.

The potential for bacterial degradation of this coformulation has not yet been ascertained.

Developmental Biocide

3-Benzo[b]thien-2-yl-5,6-dihydro-1,4,2-oxathiazine-4-oxide is a metal-free and halogen-free fungicide of a new chemical class. It is being developed for the control of stains, moulds and algae in paints, woodstains, wood preservatives and coatings. A study conducted by Forster *et al.* (42) revealed the effectiveness of this chemical against a variety of soft rot fungi and indicated its potential as a biocide to protect wood exposed in ground contact in the absence of heavy metals, which is likely to the next big challenge for the industry

A study undertaken to provide detailed information regarding the microbial degradation of the oxathiazine derivative resulted in the isolation of a number of proteobacteria that were able to mediate biocide detoxification (*43*, *44*). Using molecular techniques, the isolates were identified as *Pseudomonas*, *Alcaligenes*, *Achromobacter*, *Tetrathiobacter*, *Ralstonia*, *Stenotrophomonas*, *Variovorax* and *Serratia*. The isolates were found to show a range of tolerance to the biocide (see Figure 3 and Table I). It was found that the pseudomonads, *Ralstonia* and *Stenotrophomonas,* were able to grow rapidly in Luria Bertani medium containing even saturated quantities of the biocide, whilst the *Alcaligenes*, *Achromobacter* and *Serratia* isolates had an IC_{50} at a concentration just below that of the biocide's aqueous solubility (*45*). In contrast, *Tetrathiobacter*, *Variovorax* and *Alcaligenes defragrans* showed a low tolerance, and were inhibited by concentrations of 150-200 μM. However, the observation that these bacterial isolates degrade the oxathiazine derivative *in vitro* is not evidence, in itself, that the activity of these organisms will mediate its degradation in the timber product. It is therefore critical that these isolates are studied further to determine their modes of resistance and degradation mechanisms.

The potential for biocide degradation in treated timber is dependent on the particular application of the wood product (defined as hazard classes; *46*). In this study, treated woodblocks were exposed to soil (hazard class IV), which is the most severe environment in terms of potential for fungal degradation and bacterial activity. The potential for bacterial degradation of organic biocides in other hazard categories, such as interior timbers or external joinery, have not been assessed, but are likely to be much reduced.

Conclusions

It is clear that the proteobacteria are unparalleled in their ability to degrade organic wood preservatives, having been shown to degrade other biocides (*2*, *3*,

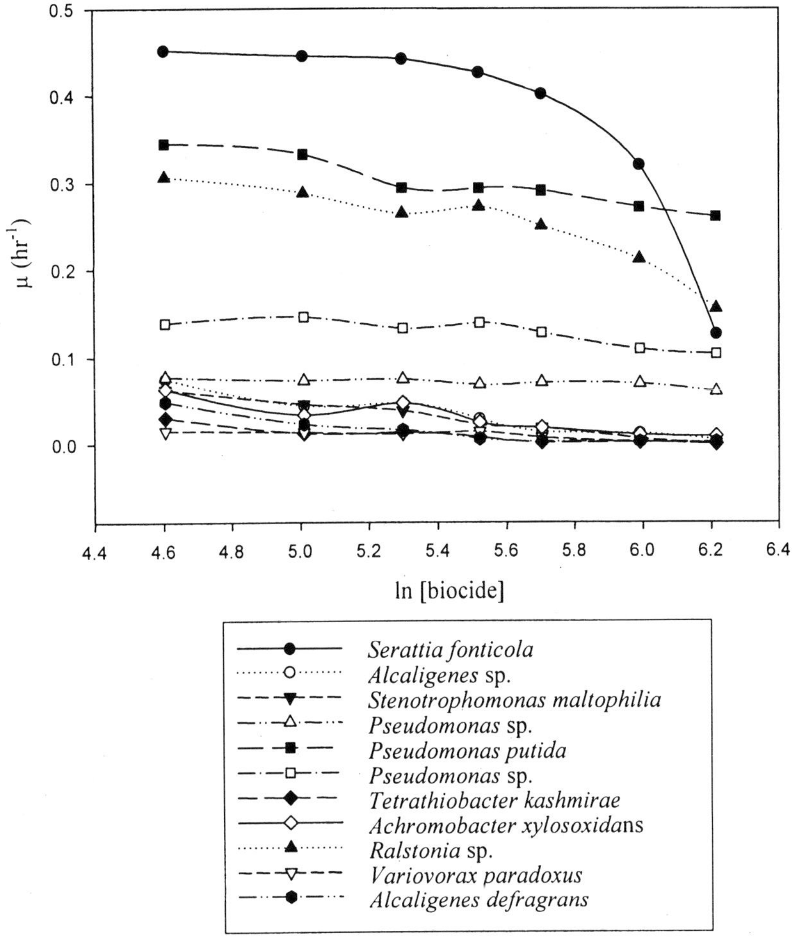

Figure 3. Determination of 50% inhibitory concentrations. Each bacterial isolate was grown in LB broth containing a range of biocide concentrations (0-500 μM), in triplicate. The specific exponential phase growth rate constant (μ) was then determined for each isolate at each biocide concentration. Each point is an average of three data points (46).

Table I. 50% inhibitory concentrations of the bacterial isolates. The 50% inhibitory concentrations were calculated from the data in Figure 2 using sigmoid non-linear regression (46).

Organism	IC_{50} *(μM)*	*Standard error (±)*
Tetrathiobacter kashmirae	177.7	1.05
Ralstonia sp.	-	-
Alcaligenes defragrans	158.9	1.05
Achromobacter xylosoxidans	428.4	1.02
Pseudomonas sp.	-	-
Pseudomonas putida	-	-
Variovorax paradoxus	159.2	1.04
Stenotrophomonas maltophilia	-	-
Serratia fonticola	448.1	1.01
Pseudomonas sp.	-	-
Alcaligenes sp.	502.7	6.7

16, *18*), along with other agrochemicals (47, 48, 49). In particular, bacteria of the *Pseudomonas* group have been reported as being able to degrade numerous QACs, IPBC, chlorothalonil and an oxathiazine derivative. However, with the introduction of the Biocidal Products Directive, it must be considered that the biodegradation of wood preservatives in soil, circumjacent to treated timber, is actually beneficial with regard to environmental persistence and potential bioaccumulation. What is more, many of the reviewed biocides have been successfully used in industry for many years, despite their observed microbial degradation.

References

1. Herring, I. J. PhD Thesis. Imperial College, London, UK, 1999.
2. Cook, S. R.; Dickinson, D. J. 2002. *The Intern. Res. Group on Wood Preserv.* Stockholm, Sweden, 2002. IRG/WP/02-10436, 1-5.
3. Cook, S. R.; Sullivan, J.; Dickinson, D. J. *The Intern. Res. Group on Wood Preserv.* Stockholm, Sweden, 2002. IRG/WP/02-10437, 1-9.
4. Cook, S. R.; Dickinson D. J. *The Intern. Res. Group on Wood Preserv.* Stockholm, Sweden, 2004. IRG/WP/04-10544, 1-20.
5. Tomlin, C. D. S. *The Pesticide Manual 12th Edition.* British Crop Protection Council, Surrey, UK, 2000.

6. Sun, G.; Thai, S.-F.; Tully, D. B.; Lambert, G. R.; Goetz, A. K.; Wolf, D. C.; Dix, D. J.; Nes Goetz, A. K.; Wolf, D. C.; Dix, D. J.; Nesnow, S. *Toxicol. Lett.* **2005**, *55*, 277-287.
7. Valcke, A. *The Intern. Res. Group on Wood Preserv.* Stockholm, Sweden, 1989. IRG/WP/3529, 1-18.
8. Wakeling, R. N.; Maynard, N. P.; Narayan, R.D. *The Intern. Res. Group on Wood Preserv.* Stockholm, Sweden, 1993. IRG/WP/93-30021, 1-31.
9. Herring, I. J.; Dickinson, D. J.; Gray, S. M.; Carey, J. K. *The Intern. Res. Group on Wood Preserv.* Stockholm, Sweden, 1997. IRG/WP/97-20116, 1-14.
10. Edlund, M. L.; Nilsson, T. *Materials and structures.* **1998**, *31*, 641-647.
11. Thorstensen, C. W.; Lode, O. *J. Environ. Qual.* **2001**, *30*, 947-953.
12. Molnar, S.; Dickinson, D. J.; Murphy, R. J. *The Intern. Res. Group on Wood Preserv.* Stockholm, Seden, 1996. IRG/WP/96-20089, 1-10.
13. Kim, I. S.; Beaudette, L. A.; Shim, J. H.; Trevors, J. T., Suh, Y. T. Plant Soil. **2002**, *239*, 321-331.
14. Kim, I. S.; Shim, J. H.; Suh, Y. T. *Pest Manag. Sci.* **2003**, *59*, 324-330.
15. Roberts, T.; Hutson, D. *Metabolic pathways of agrochemicals; Part 2: Insecticides and fungicides.* The Royal Society of Chemistry, Cambridge, UK, 1999; Vol. 1, pp 1380-1384.
16. Katayama, A.; Isemura, H.; Kuwatsuka, S. *J. Pesticide Sci.* **1991**, *16*, 239-245.
17. Motonaga, K.; Takagi, K.; Matumoto, S. *Biol. Fertil. Soils.* **1996**, *23*, 340-345.
18. Katayama, A.; Itou, T.; Ukai, T. *J. Pest. Sci.* **1997**, 22, 12-16.
19. Mori, T.; Fujie, K.; Kuwatsuka, S.; Katayama, A. *Soil Sci. Plant Nutr.* **1996**, *42*, 513-322.
20. Mori, T.; Fujie, K.; Katayama, A. *Soil Sci. Plant Nutr.* **1998**, *44*, 297-304.
21. Regitano, J. B.; Tornisielo, V. L.; Lavorenti, A.; Pacovsky, R.S. *Arch. Environ. Contam. Toxicol.* **2001**, *40*, 295-302.
22. Sato, K.; Tanaka, H. *Biol. Fertil. Soils.* **1987**, *3*, 205-209.
23. Rouchaud, J.; Roucourt, P.; Vanachter, A. *Toxicol. Environ. Chem.* **1988**, *17*, 59-68.
24. Kim, Y.-M.; Park, K.; Joo, G.-J.; Jeong, E.-M.; Kim, J.-E.; Rhee, I.-K. *J. Agric. Food Chem.* **2004**, *52*, 4192-4196.
25. Jakoby, W. B.; Ziegler, D.M. *J. Biol. Chem.* **1990**, *265*, 20715-20718.
26. Keen, J. H.; Habig, H. W.; Jakoby, W. B. *J. Biol. Chem.* **1976**, *251*, 6183-6188.
27. Nishihara, T.; Okamoto, T.; Nishiyama, N. *J. Appl. Microbiol.* **2000**, *88*, 641-647.
28. Butcher, J.; Preston, A.; Drysdale, J. *For. Prod. J.* **1977**, *27*, 19-22.
29. Preston, A. F.; Chittenden, C. M. *N. Z. For. Sci.* **1982**, *12*, 102-106.
30. Chen, T.; Dwyre-Gygax, C.; Smith, R. S.; Breuil, C. *J. Agric. Food Chem.* **1995**, *43*, 1400-1406.

31. Dean-Raymond, D.; Alexander, M. *Appl. Environ. Microbiol.* **1977**, *33*, 1037-1041.
32. Desbordes, J.; Jourdan, R. *Dev. Biol. Stand.* **1974**, *24*, 77-90.
33. Janota-Bassalik, L.; Olczyk, C.; Kaczorowska, M. *Acta Microbiol. Pol. Ser.* **1969**, *1*, 31-34.
34. Van Ginkel, C.G.; Van Dijk, J. B.; Kroon, A. G. M. *Appl. Environ. Microbiol.* **1992**, *58*, 3083-3087.
35. Doyle, A. K; Ruddick, J. N. R. *The Intern. Res. Group on Wood Preserv.* Stockholm, Sweden, 1993. IRG/WP/93-10020, 1-7.
36. Zheng, Y.; Ruddick, J.N.R. *The Intern. Res. Group on Wood Preserv.* Stockholm, Sweden, 1995. IRG/WP/95-10105, 1-11.
37. Dubois, J. W.; Ruddick, J. N. R. *The Intern. Res. Group on Wood Preserv.* Stockholm, Sweden, 1998. IRG/WP/98-10263, 1-10.
38. Xiao, Y.; Krebber, B. *Holzforschung* **1999**, *53*, 237-243.
39. Cook, S. R. Phd thesis, Imperial College, London, UK, 2002.
40. Hong, L. T. *The Intern. Res. Group on Wood Preserv.* Stockholm, Seden, 1989. IRG/WP/3532, 1-9.
41. Takasaki, A.; Hashida, T.; Fujiwara, S.; Kato, K.; Nishihara, T. *Jpn. J. Tox. Environ. Health.* **1994**, *40*, 344-350.
42. Forster, S. C; Williams, G. R.; Van Der Flaas, M.; Bacon, M.; Gors, J. *The Intern. Res. Group on Wood Preserv.* Stockholm, Sweden, 2002. IRG/WP/02-30301.
43. Wallace, D. F.; Dickinson, D. J. 2004. COST E22 Proceeding.
44. Wallace, D. F.; Dickinson, D. J. *The Intern. Res. Group on Wood Preserv.* Stockholm, Sweden, 2004. IRG/WP/04-10543, 1-10.
45. Wallace, D. F.; Cook, S. R.; Dickinson, D. J. In preparation.
46. Jermer, J. *The Intern. Res. Group on Wood Preserv.* Stockholm, Seden, 1988. IRG/WP/2310, 1-6.
47. Radosevich, M.; Traina, S. J.; Hao, Y.-L.; Tuovinen, O. H. *Appl. Environ. Microbiol.* **1995**, *61*, 297-302.
48. Mandelbaum, R. T.; Allan, D. L.; Wackett, L. P. *Appl. Environ. Microbiol.* **1995**, *61*, 1451-1457.
49. Kilbane, J. J.; Chatterjee, D. K.; Karns, J. S.; Kellogg, S. T.; Chakrabarty, A. M. *Appl. Environ. Microbiol.* **1982**, *44*, 72-78.

Non-Biocidal Chemicals and Processes to Protect Wood

The top photograph is courtesy of Militz/Becker of Georg-August University of Güttingen and the bottom two photographs are courtesy of Firma Mitteram-skogler of Austria. (See page 6 of color inserts.)

Chapter 19

Acetylation of Wood in Lumber Thickness

W. J. Homan

SHR Timber Research, P.O. Box 497, NL–6700 AL Wageningen, The Netherlands

Recent changes in legislation and public image in Europe and North America have lead to an increasing interest in wood modification as alternative for preservative treated wood. Acetylation of wood is probably the best-studied modification process at lab scale. Acetylation of solid wood has been studied in a 2500 l reactor on commercial, full-size lumber. The main properties of interest are durability, dimensional stability, UV-stability, strength properties, gluability, and paintability. Very favourable results were obtained on the above properties with a commercial size process. Durability class 1, dimensional stability up to 70% ASE, good UV-stabilisation, equal stiffness and bending strength, good gluability and excellent paintability are technically feasible for lumber of commercial thickness. At present, Titan Wood is building a commercial plant for the acetylation of solid timber in the Netherlands, and other possible commercial ventures are being examined elsewhere.

Cellulose acetate was invented about 100 years ago, and the earliest reports of acetylated wood date to the 1920's. Since then research activities on wood acetylation have appeared at many laboratories worldwide. However, hitherto acetylation of solid wood has not been commercially successful. The main reasons for this are the costs of acetic anhydride, the costs of processing recycle streams of the process (mainly acetic acid), the large supply of tropical timbers with good durability, the fact that traditional wood preservation was not hindered by much legislation and that environmental issues such as sustainable forest management has only recently become important. However, times have changed and the interest in acetylation has greatly increased in the past 20 years.

Developments in the Last 20 Years

Many developments in the last 20 years have influenced the development of acetylating lumber. First, the availability of high quality wood has diminished, mainly caused by depletion of the tropical rain forests. Secondly, environmental issues have become important to the public. This has led to a myriad of laws that regulate wood preservation with biocidal systems. The present implementation of the biocidal products directive in the European Union and the changing criteria at the EPA in the USA are typical examples. Thirdly, there are issues on the relatively low strength properties of heat-treated timber, the main non-biocidal alternative to chemical modification (*1*, *2*). Fourthly, the demand for raw materials, including wood, increases as the economies of China and other Asian countries grow. Fifthly, the importance of sustainable forest management is increasingly important. There are a number of initiatives or certification systems leading to sustainable forestry system, of which FSC is probably the best known. A drawback of these systems is that the availability of these timbers is far below the demand. Finally, technical developments have led to significant cost reductions in producing acetylated wood. Especially important is a cost-effective process that regenerates acetic anhydride from the acetic acid by-product.

Developments Leading to Titan Wood

In the 1980's research on acetylated timber was mainly done in the USA by Rowell and in Sweden by Simonson. In 1991, this work led to a European Patent (*3*). Further work in scaling up laboratory acetylation to semi industrial scale was done in the Netherlands at the SHR laboratories. At the end of the 90's, developments had advanced sufficiently so that a consortium of interested companies was formed: AKBV. The aim of AKBV was to fill in the gaps and study the acetylation of commercially thick lumber. These developments led to the construction of a relatively large pilot plant in Arnhem, The Netherlands, including a 2500 l reactor with a length of 4 m (internally) that can treat about 0.5 m^3 of timber per batch. In 2002, the AKBV consortium decided that although the technological proof was shown, they did not want to continue commercialization. The main reason for this decision was that a wood acetylation plant has the character of chemical industry, an unknown concept to the companies within AKBV that were mainly wood industries. A neighbor, at the Industry Park where AKBV was situated, happened to be Accsys Chemicals PLC. This company had been working on new processes for the production of cellulose acetate and had developed a completely new process for acetic anhydride production and recycling. Accsys Chemicals PLC formed Titan Wood Ltd. on April 17th, 2003, and took over the intellectual property rights for acetylated wood production. Titan Wood Ltd. is now working to produce acetylated wood at a commercially viable cost. The pilot plant was remodeled and is now producing sample material for potential customers. Also, a full-scale acetylation factory is being built that will have initial production capacity of 25000 m^3 per year. The plant is expected to start production in 2006.

The Acetylation Process

The basic chemistry of wood acetylation is rather simple and has been described by many authors (*4*, *5*). The structural components of wood are cellulose, lignin and hemicellulose. All three contain a number of hydroxyl groups that are responsible for the interaction of wood with the ambient water. In most of the chemical modification processes, these hydroxyl groups are substituted with hydrophobic groups (*6*). Acetylation with uncatalysed acetic anhydride is one of the simplest ways to modify wood. The basic reaction equation is given in Figure 1.

Quite obvious from the reaction is that with every acetyl group added to the wood one molecule of by-product acetic acid is formed. The removal of acetic acid from the wood and re-forming more acetic anhydride is therefore an essential part of an economical treatment process.

Durability

Perhaps the most important property of wood that man wants to enhance is its durability. In wood technology durability is usually defined as the resistance

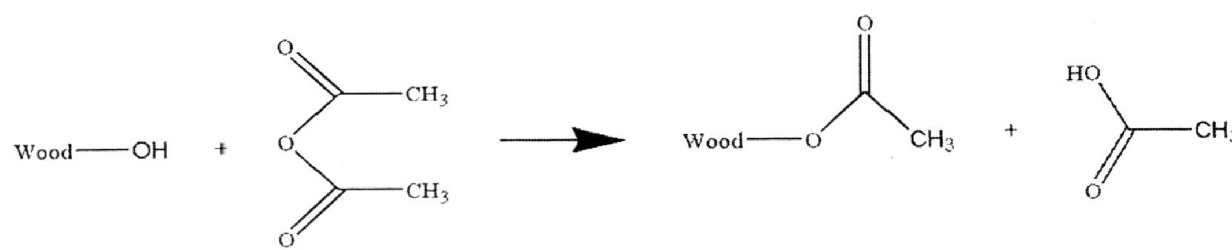

Figure 1. The wood acetylation reaction

of wood against biological decay. Fungi are by far the most important group of organisms with respect to wood durability. There are several explanations why acetylated wood is more durable compared to untreated wood. The most important ones are the lower equilibrium moisture content (EMC) that reduces the fungal decay potential, and the changed chemical structure that prevents fungus enzymes from degrading wood (*7*). Blocking of micropores may also play an important role (*8*). In any case, no biocidally active chemicals are responsible for the enhanced durability of acetylated wood. Whatever the mechanism(s), most authors obtain generally similar durability results with acetylated wood. There is a clear relation between the acetyl content of the wood and its resistance to fungi. Figure 2 shows SHR results of an EN 113 (*9*) test with 3 different fungi on acetylated beech.

Further brown rot data published by Ohkoshi (*10*) are shown in Figure 3. Figure 4 gives results of Imamura (*11*) with *T. palustris*.

In Figure 5, the field-test data (EN 252 stake test) of Larsson (*12*) are shown. In the last case not the weight loss but a decay index is given on the y-axis.

In Figures 2 and 5 acetyl contents are used, while in Figures 3 and 4 so-called Weight Percent Gain (WPG) is used. These variables are similar but can not be easily converted. The reason for this is twofold. The natural acetyl content in wood adds to the chemical analysis of the acetyl content, but does not influence the WPG. Also, some wood components are extracted from the wood

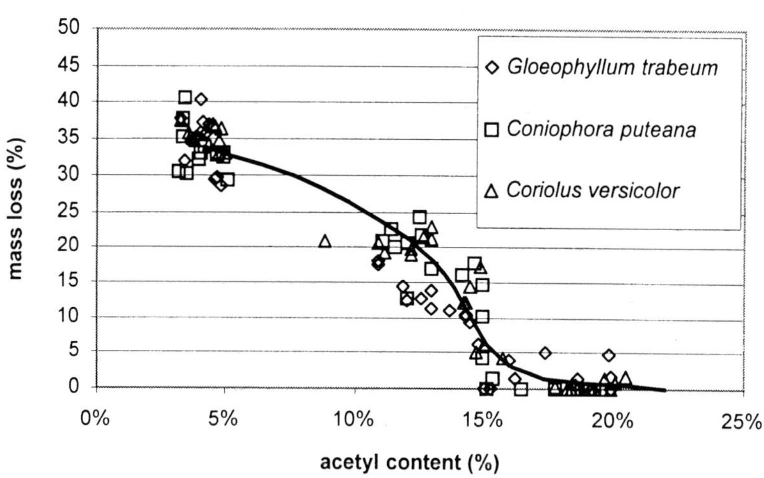

Figure 2. Resistance of liquid phase acetylated beech to basidiomycetes in laboratory decay tests.

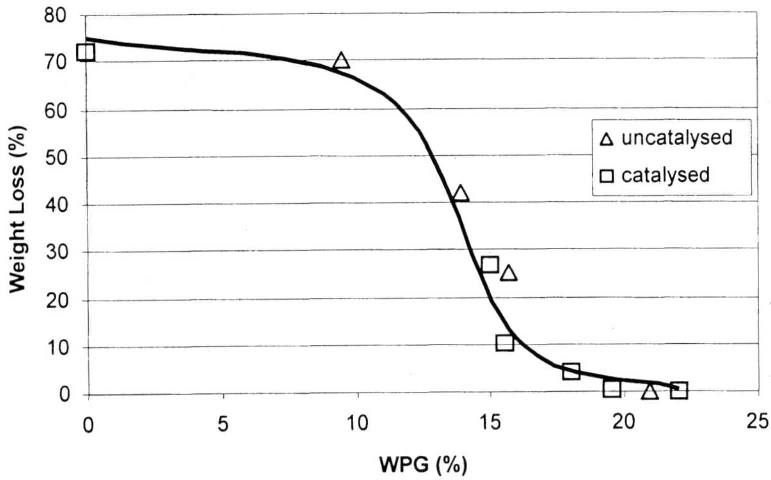

Figure 3. Resistance of liquid phase acetylated wood to brown rot fungi

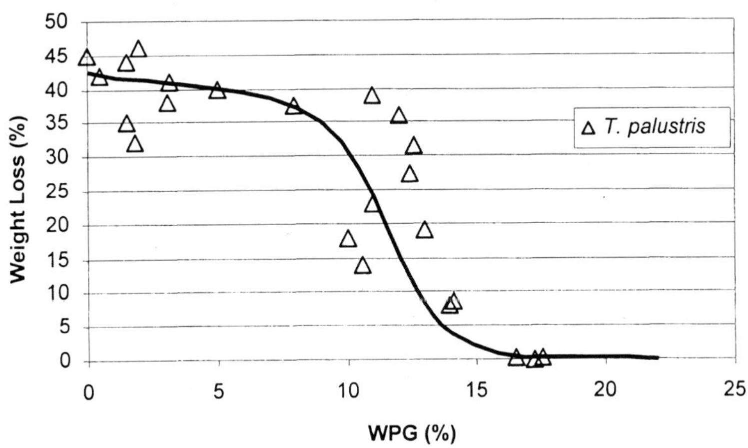

Figure 4. Resistance of acetylated wood to decay by T. palustris.

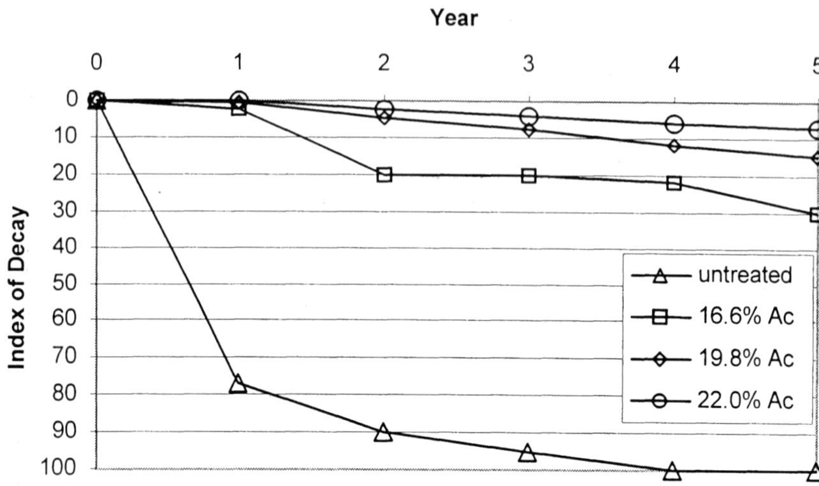

Figure 5. Results of a graveyard test in Simlångsdalen, Sweden, with acetylated Scots pine.

during the acetylation process. This leads to a reduced weight and, thus, a reduced WPG. However, taking all this into consideration, when looking at the bulk of literature - Figures 2-5 are just a small selection - it is very clear that at an acetyl content of about 20% wood becomes very durable.

Dimensional Stability

Much of the wood modification research examined biological durability. The durability of products made out of wood, however, is determined by many factors. The swelling and shrinkage behavior is also important. The poor dimensional stability of wood causes check formation, warping and twisting, and problems with paint and glue adhesion, and thus has a major impact on the service life and consumer satisfaction of the product.

Acetylation can reduce the swelling and shrinkage of wood to a great extent. The hydrophobic nature of the acetyl groups, in contrast to the original hydrophilic hydroxyl groups, reduces the EMC of the wood exposed to moisture in the surrounding air.

In Figures 6 and 7, the sorption lines for beech and Scots pine, respectively, are given. From these figures it is clear that the shift in EMC caused by a certain degree of acetylation is rather constant and not dependent on the wood species. In this research the equilibrium moisture content at the highest degree of acetylation was reduced by more than 50%. As a consequence the swelling and shrinkage of the wood was reduced. Figures 8 and 9 give the swelling (x-axis) in relation to the EMC (y-axis) for beech and Scots pine, respectively. These figures clearly show that the swelling as a function of the EMC does not change dramatically by acetylation; rather, the reduced EMC directly leads to reduced swelling. For beech a small effect in dimensional stabilization can be observed additional to the EMC effect. However, in Pine all deviations of the swelling curves of acetylated wood compared to untreated wood are within the experimental error. Since the swelling tends to accelerate in the higher EMCs, the dimensional stabilization is even stronger than would be expected on the basis of EMC reduction figures. Where the EMC is reduced by a little more than 50%, the swelling is reduced up to nearly 70% irrespective of wood species. This is true for swelling from 0% to 98% RH. In into the RH regions between 30% and 80% the effect is even stronger. Apparently, even though EMC is the most dominant factor determining the dimensional stabilization by acetylation, both the shapes of the sorption curves and the swelling curves determine the magnitude of this effect. Therefore the ASE values of the same material can vary in different RH segments.

Strength Properties

Much of the literature on acetylated wood report that the strength properties are not affected. In many cases this has been determined by measuring strength (MOR) and stiffness (MOE) of small pieces of clear wood. In construction, however, structural engineers need to work with more realistic values. In general for this so called characteristic 5% percentile figures are used which are determined on full-size lumber.

In a recent study (*13*), Radiata pine of structural sizes was tested in a 4-point bending test according to EN 408, to assess acetylation effect on MOR and MOE. First, the MOE was determined non-destructively for untreated timber. Then, the same pieces were acetylated to 20% acetyl content and the MOE and MOR determined after the acetylation treatment. The MOR and MOE of untreated Radiata pine were also determined using pieces from the same source. The density of all test specimens was recorded as well. Table I summarizes the results.

The data show that acetylation of commercially thick Radiata pine lumber did not significantly influence the bending MOE. Furthermore, the acetylation process did not alter the relationship between bending MOE and MOR. This is important since mechanical grading of acetylated lumber is thus still possible. The bending strength of the tested material was lower for the acetylated compared to untreated wood. The mean value of MOR is only slightly lower for

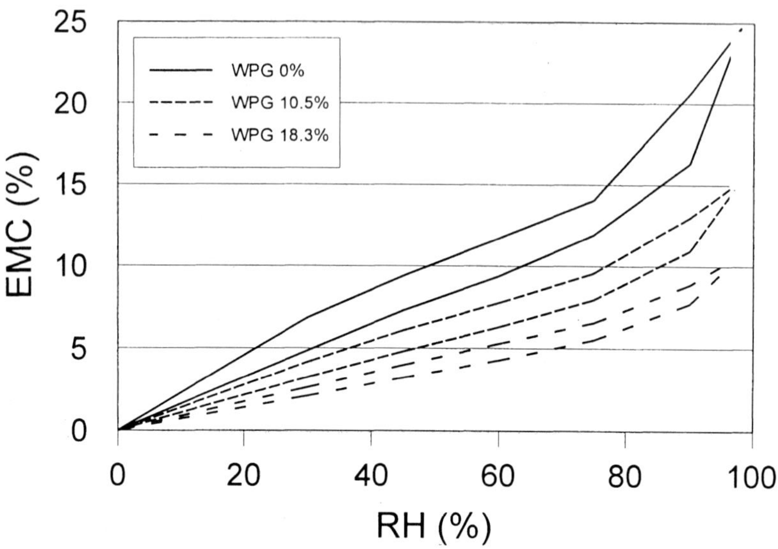

Figure 6. Equilibrium moisture content of acetylated beech as a function of relative humidity of the air.

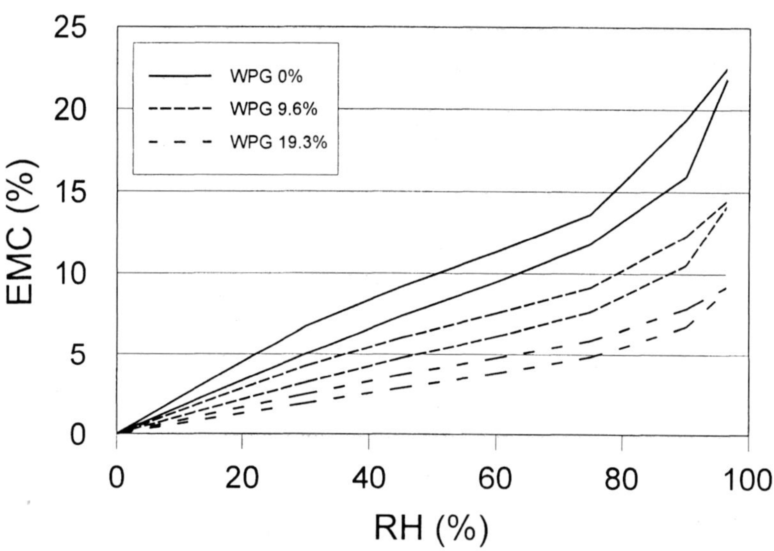

Figure 7. Equilibrium moisture content of acetylated Scots pine as a function of relative humidity of the air.

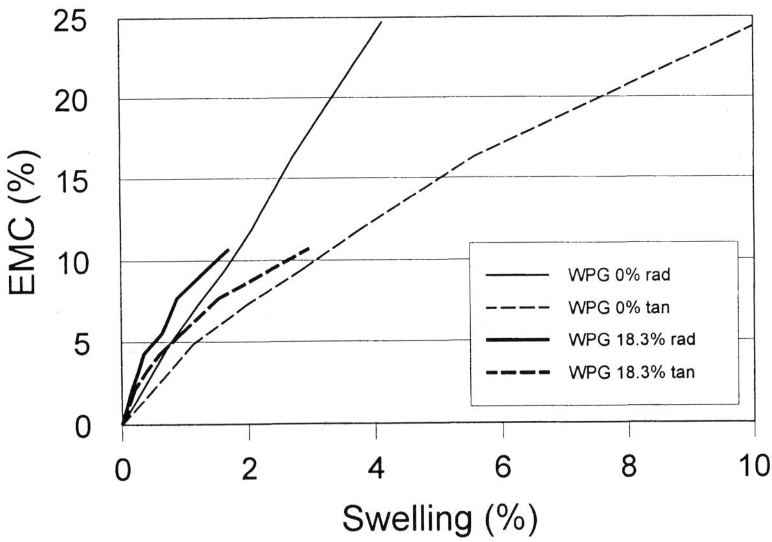

Figure 8. Swelling of acetylated beech as a function of EMC.

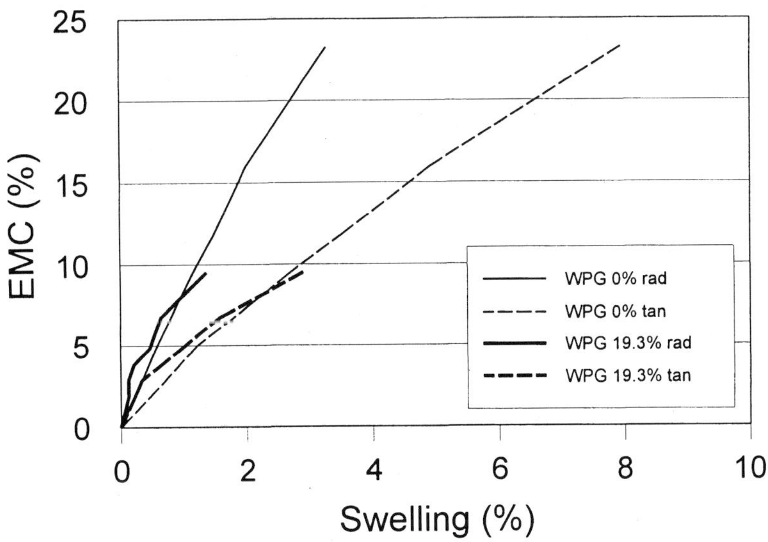

Figure 9. Swelling of acetylated Scots pine as a function of EMC.

Table I. Average strength, stiffness, density and moisture content of lumber thickness acetylated radiata pine

Treatment	Oven Dried Density [kg/m^3]	Moisture Content [%]	MOE [N/mm^2]	MOR [N/mm^2]
Series 1	417	12.2	9664	43
(untreated)	(25)	(1.3)	(2130)	(10)
Series 2 before	411	13.1	9064	
acetylation	(38)	(1.5)	(2268)	
Series 2	492	5.2	8788	39
acetylated	(48)	(3.1)	(2320)	(13)

NOTE: Values between brackets represent the standard deviation

the acetylated material, but due to an increase in variation the characteristic value (i.e. the lower limit of the 90% confidence interval) is significantly lower.

Paint Performance

The greatly reduced swelling and shrinkage behavior of acetylated wood benefits the durability of wood coatings. Many studies have shown this, but most used accelerated weathering systems. In 2005, (*14*) an outdoor evaluation test that lasted nearly 10 years was ended and the panels evaluated. In this study, acetylated and untreated Scots pine panels were coated with white, black opaque and teak pigmented transparent alkyd and acrylic coatings. The panels were evaluated after 5 and 10 years of exposure with regard to the degree of blistering, cracking, flaking and chalking. At the end of the study, the adhesion was determined under wet and conditioned circumstances and the coating layer thickness and fungal degradation was evaluated.

After 10 years of exposure, the coating of the control panels were severely cracked and flaked, or in some cases even completely eroded. Furthermore, all control panels had fungal degradation.

In comparison to the control panels, the coating on the acetylated panels showed remarkably good performance. In comparing the three coatings, the opaque coatings preformed better than the transparent ones and the white colors better than black coatings. The study showed that the expected lifetime of the coatings on acetylated wood is apparently only dependent on the intrinsic coating quality. The wood did not appear to affect the condition of the coating layer. This suggests that longer periods between repainting are possible depending mainly on the quality of the coating.

Use of Acetylated Wood

Titan Wood Ltd. has several intended areas of use for acetylated wood. Examples can be found on their web-site (*15*). Obviously the areas where the good properties of acetylated wood are best in place are outdoors and those indoors places where dimensional stability matters.

Although in general many properties of wood are enhanced by acetylation, there are still a number of possible problems that need to be addressed. First, acetylated wood is corrosive to ferrous metals. Experiments with hot dip zinc plated, electrolytical zinc plated and Cr^{6+} passivated material show that all of these will corrode within a short time. Only stainless steel (304 quality or better) fasteners are recommended. An other drawback of acetylated wood is that it does not provide any protection against blue staining fungi. Although these fungi do not affect the wood's integrity, they discolor the wood seriously. Research addressing the blue stain problem is still ongoing. Finally it should be mentioned, that acetylation processes, when performed at large scale, are not easy in terms of process technology. Careful process control is needed to ensure uniform and high quality products.

Conclusion

Owing to many recent changes in legislation and public perceptions, much research has been conducted on wood modification. It has become possible to produce, at commercially viable costs, acetylated timbers. Acetylated wood can have a durability class 1, reduced selling and shrinkage of up to 70%, is UV stable, is as strong as untreated wood, and performs well with glues and paints. Acetylated wood will likely be used for many high quality wood products within the near future.

References

1. Rapp, A.O.; *Review on Heat Treatments of Wood. Proceedings of Special Seminar.* COST Action E22: Environmental Optimization of Wood Protection; Antibes, France, **2001**
2. Militz, H.; *Heat Treatment Technologies in Europe: Scientific Background and Technological State-of Art.* IRG/WP 02-40241, **2002**
3. Rowell, R.M.; Simonson, R.; Tillman, A.M.; Eur. Patent EN 021352 B1, **1991**
4. Hill, C.A.S.; Jones, D.; Strickland, G.; Cetin, N.S. *Holzforschung,* **1998**, *52*, pp 623-629

5. Goldstein, I.S.; Jeroski, E.B.; Lund, A.E.; Nielson, J.F.; Weaver, J.W. *Forest Products Journal,* **1961**, pp 363-370
6. Homan, W.J.; *Wood Modification, State of the Art 2004.* COST Action E18: High Performance Wood Coating Exterior and Interior Performance, Paris, **2004**
7. Beckers, E.P.J.; Militz, H.; Stevens, M.; *Resistance of Acetylated Wood to Basidiomycetes, Soft Rot and Blue Stain.* IRG/WP 94-40021, **1994**
8. Hill, C.A.S.; Hale, M.; Forster, S.C.; *Investigations of the Role of Cell Wall Moisture Content and Micropore Blocking in the Decay Protection Mechanism of Anhydride Modified Wood.* COST Action E22: Environmental Optimization of Wood Protection, Estoril, Portugal, **2004**
9. European Standard: EN 113 *Wood preservatives – Test method for Determining the Protective Effectiveness Against Wood Destroying Basidiomycetes – Determination of the Toxic Values.* **1996**
10. Ohkoski, M.; Kato, A.; Suzuki, K.; Hayashi, N.; Ishihara, M. *J. Wood Sci,* **1999**, *45*, pp 69-75
11. Imamura, Y.; Nishimoto, K. *Wood Research,* **1987**, *74*, pp 33-44
12. Larsson Brelid, P.; Simonson, R.; *Resistance of Acetylated Wood of Biological Degradation. Evaluation of Field Test.* IRG/WP 97-30139, **1997**
13. Jorissen, A.J.M.; Bongers, H.P.M.; Kattenbroek, B.; Homan, W.J.; *The Influence of Acetylation of Radiata Pine in Structural Sizes on its Strength Properties.,* Presentation paper *in press.* 2nd European Conference on Wood Modification, Göttingen, Germany, **2005**
14. Bongers, H.P.M.; Creemers, J.G.M.; Kattenbroek, B.; Homan, W.J.; *Performance of Coatings on Acetylated Scots Pine after more than Nine Years Outdoor Exposure,* Poster *in press.* 2nd European Conference on Wood Modification, Göttingen, Germany, **2005**
15. *About Titan Wood,* URL http://www.titanwood.com

Chapter 20

Furfurylation of Wood: Chemistry, Properties, and Commercialization

Stig Lande[1], Morten Eikenes[2], Mats Westin[3], and Marc H. Schneider[4]

[1]Wood Polymer Technologies ASA, NO–1596 Moss, Norway
[2]Norwegian Forest Research Institute, NO–1432 Ås, Norway
[3]SP Swedish National Testing and Research Institute, Box 857, SE–50115 Borås, Sweden
[4]University of New Brunswick and Infinity Wood, Ltd., 989 Clements Drive, Fredericton, New Brunswick E3A 7J3, Canada

Furfurylated wood has proven to be a biological inert material and, thus, is an environmentally-benign alternative to wood treated with heavy metals or tropical hardwoods. The chapter will deal with the history, properties and uses of furfurylated wood to date.

Introduction

Due to the increasing awareness of the hazards of using toxic compounds for wood preservation, limitations have been introduced to the production, trade and use of the most utilized wood preservative, CCA (salts of copper, chromium and arsenic), in several European countries and the USA. Several alternative preservatives for wood are in use. Common for all these preservatives is their toxicity to fungi, bacteria, and insects along with non-target animals and humans. An alternative to employing toxic compounds is chemical modification, which changes the wood structure and wood chemistry so that the wood is less susceptible to bio-deterioration. One promising chemical modification method is the furfurylation process, which uses furfuryl alcohol that polymerizes in the wood structure and reacts with the wood substrate.

History of Furfurylation

Early research with furfuryl alcohol (FA) concerned FA-resins as a substitute to phenol formaldehyde resins (PF resins) for adhesives, molding compounds and inorganic composites. In the 1930's and 1940's several processes and patents concerning paper laminates, grinding wheels and metal casting moulds from FA-resin impregnated paper or inorganic compounds such as silicon carbide, were developed (*1*). Metal casting moulds and grinding wheels are still the main commercial applications for furfuryl alcohol. Research concerning modification of wood with furfuryl alcohol, hereafter referred to as "furfurylation" of wood, was initiated by the "pioneer of wood modification", Dr. Alfred Stamm, in the early 1950's. Additional early work on wood furfurylation was done by Irving Goldstein (*2-4*). Goldstein's process employed zinc chloride as a catalyst and mainly modified wood veneers. He reported that treatment with 90 % (weight/volume) solutions of FA resulted in products with high dimensional stability and increased resistance to fungal decay, alkali and acids (*2, 4*). This research lead to a small-scale production of furfurylated wood in the 1960's by Koppers Wood Inc. (USA). Among these products were laboratory bench tops, pulp mixer rotor-blades and knife handles (*5*). Anaya developed a process very similar to the Stamm/Goldstein process (*6, 7*). However, this process was never commercialized. The major problem with these processes was that zinc chloride used as a catalyst deploymerized the cellulose, which consequently reduced the strength properties of the modified wood.

Professor Marc Schneider and Dr. Mats Westin developed, separately and more or less simultaneously during the early 1990's, alternative catalysts for the furfurylation process. They both based their polymerization process on similar paths of chemistry using cyclic carboxylic anhydrides, mainly maleic anhydride, as key catalysts (*8-10*). These novel systems lead to solutions that are chemically stable at room temperature and have good properties with regard to the impregnation process. The final furfurylated wood products, shown to be of high quality, are described in this chapter. The properties of this second generation of furfurylated wood are generally superior to those wood modified by the first-generation process developed Stamm and Goldstein (*8*).

Chemistry of Furfurylation

The acid catalyzed polymerization of FA in wood has a very complex chemistry. The resulting polymer is a highly branched and cross-linked furan polymer that is chemically bonded to the wood structural components. However, the reaction parameters (i.e., type of and concentration of catalyst, the pH of the solution, processing temperature and time, and the presence of water) highly affects the final product by the degree and type of chemical bonds to wood, the major type of polymer units, and the degradation of wood components (mainly depolymerization of cellulose and hemicelluloses by acid hydrolysis).

The reaction types can be divided into several categories: a) homo-polymerization of FA, b) co-polymerization of FA and additives or wood extractive substances, and c) grafting of FA or polymerized FA to wood cell wall polymers (*11*, *12*).

Homo-polymerization of FA

Early studies stated that the reaction rate is a simple function of the pH inaqueous furfuryl alcohol solutions (*1*). However, there are several competing reactions and the rate of the dominating reaction mechanism is affected by factors other than pH. These factors include temperature, FA/water-ratio, and the presence of oxygen and/or weak organic bases. In the presence of a powerful acid catalyst, such as para-toluene sulphonic acid (PTSA), the dominating reaction is a very rapid radical polymerization which results in a high degree of ring-opening reactions (*13*). However, these kinds of systems are not suitable to use for impregnation and modification of wood since they are not stable at room temperature. Solutions used for impregnation of wood must be stable during storage and during impregnation of wood, but should react readily when the impregnated wood is heated.

In the initial phase of the polymerization of these stable systems, there are two competing condensation reactions as are illustrated in Figure 1: Reactions 1 and 2a (*14*). The reaction products according to reaction 1 normally dominate. High reaction temperature and high concentration of furfuryl alcohol further suppress the formation of the ether bridges in reaction 2a (*14*).

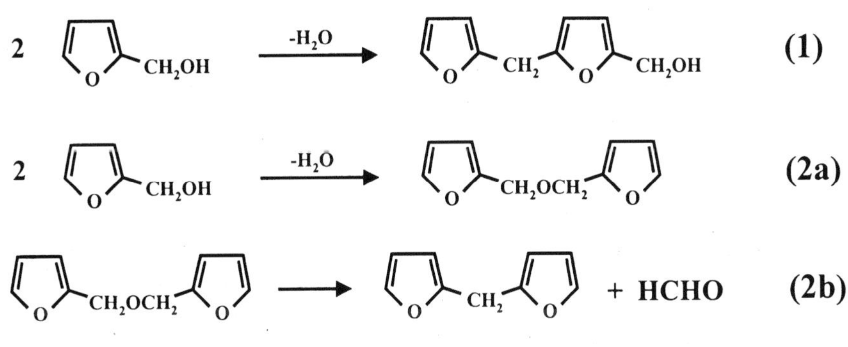

Figure 1. Initial polymerization of FA (acid catalyzed condensation reaction).

Since the ether-bridge in reaction 2a is an unstable intermediate, it is believed that it undergoes reaction 2b to form formaldehyde as an intermediate product. The terminal methylol group could later split off as formaldehyde. The latter reaction is highly temperature dependant. Initially, a linear polymer is formed. But as the polymerization progresses, cross-linking via reactions 3 and 4 occurs more frequently and the polymer becomes infusible. It is believed that formaldehyde, formed according to reaction 2b and by splitting off of terminal methylol groups, acts as a cross-linking agent according to reaction 4. However, the dominating cross-linking reaction is by condensation reaction 3, shown in Figure 2 (*1,14*).

Furthermore, it is commonly assumed that some cross-linking in the late stages of polymerization occurs through ring-opening reactions. A competing reaction to the FA ring-opening reaction is formation of levulic acid. This reaction, and oxidation reactions, can be suppressed by the presence of a weak organic base, e.g. triethanolamine (*1,15*). However, no significant difference in the polymer structure was found between polymerization carried out in the presence or absence of oxygen (*14*).

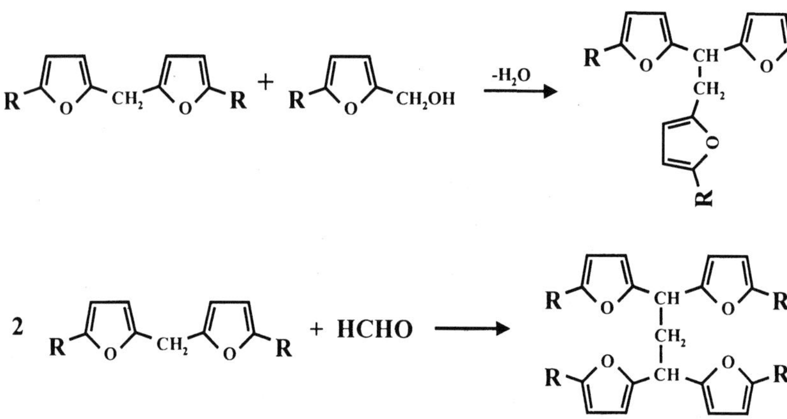

(3)

(4)

Figure 2. Branching (reaction 3) and cross-linking (reaction 4) of FA-polymer chains.

Reaction of FA to the Wood Cell Wall Components

The high degree of permanent bulking of the furfurylated wood cell wall is considered evidence of reactions between polymer and wood at an early reaction stage. Similarly to low-molecular weight PF-resins, FA impregnation leads to super-swelling of the cell wall (i.e. a slightly higher degree of swelling than for water). However, as PF-oligomers become more and more hydrophobic during polymerization, they begin to migrate out of the cell wall. This results in a maximum anti-shrink efficiency (ASE) of approximately 50 % for PF resins. Since FA-oligomers are even more hydrophobic than PF-oligomers at the same degree of polymerization (DP), one would expect anti-shrink efficiencies below 50 % for furfurylated wood. Instead, the ASE for FA-modified wood can be as high as 80 %. This is a strong indication that grafting reactions occur between the FA-polymer and wood during the early stages of polymerization.

Grafting of poly-FA onto cellulose was previously accomplished using ferric salts and peroxides as catalysts (*16*). However, these types of catalysts are too reactive to be used. Therefore, by using cyclic anhydrides, as described in this chapter, grafting reactions likely occur between the poly-FA and cellulose, hemicelluloses and, principally, the lignin in wood. Reaction 5 (Figure 3) is a possible reaction between FA and a guaiacyl unit of lignin (the predominant monomeric lignin unit in softwood lignin).

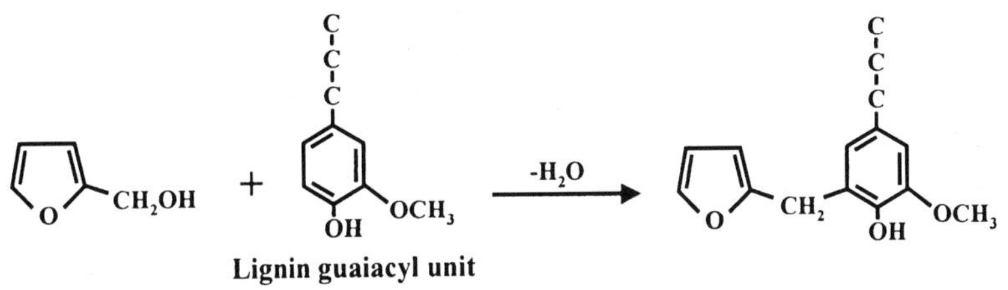

(5)

Figure 3. Suggested reaction between FA and a lignin unit

Biological Resistance

Furfurylation of wood can be done with different FA levels to obtain wood with varying degrees of modification, expressed as Weight Percent Gain (WPG). WPG is thereby an essential parameter when discussing the effects of furfurylation. A moderately low treatment level (WPG around 30%) will have only a minor effect on the physical properties of the modified wood. The major property improvement observed with a moderate WPG is increased resistance to biological degradation. Furfurylation has been shown to be a good method to increase the resistance against biological deterioration of wood using non-toxic chemicals. Various research groups have carried out laboratory and field tests on FA-modified wood. Irving Goldstein was the first to show that furfurylated wood has an increased resistance to biological decay. His formulation was based on high concentration of FA (90 %) that resulted in a high WPG. For commercial applications it is important to know the minimum treatment level required to achieve a desired property, to minimize the cost of the process. The studies reported in this chapter are published in international journals (*17, 18*) and some tests conducted for Wood Polymer Technologies ASA (Porsgrunn, Norway).

Fungal Decay Resistance

The results from laboratory testing (according to a modified version of AWPA E10, soil block test) of the decay resistance of furfurylated wood with basidiomycetes cultures of brown and white rot fungi are presented in Table 1 (mass loss caused by *Postia placenta* and *Phanerochaete chrysosporium*, respectively). The mass loss of furfurylated wood of medium and high WPG caused by brown rot decay are less than the mass loss of samples treated with CCA to a retention for Use Class 4, ground contact. Only for the furfurylated samples with a low treatment level have mass losses slightly higher than for CCA-treated samples. The conclusion is that the resistance to brown and white rot decay fungi is high for furfurylated wood treated to a WPG of 35 or more.

Laboratory tests with terrestrial microcosms (TMC) are more realistic test set-ups than typical laboratory decay tests, as long as ordinary soils are used. Accelerated conditions are gained by controlling the temperature and humidity. The results from TMC testing in three types of soils are presented in Table 2 (mass losses caused by fungal decay). Prior to the tests, the samples were subjected to artificial weathering by leaching in water. The mass losses caused by leaching are shown in the third column. No differences were found between the untreated samples of pine sapwood, and pine sapwood samples free of extracts (extracted with acetone). Only minor differences were found between the mass losses caused by leaching of the control samples and leaching of the FA-modified samples.

In-ground exterior field tests are the most realistic test setup that can be performed for testing decay properties. The climate and soil type will affect the rate of degradation. Therefore, outdoor exposure tests are normally run for at least several years using multiple test sites to obtain realistic results. A comparison of the performance of furfurylated mini-stakes in three different fields in Sweden, Simlångsdalen, Ultuna and Ingvallsbenning, are presented in Table 3. Although the type of decay is very different in these three fields, the performance of furfurylated wood is quite similar in all three fields. At the higher modification levels, furfurylation seems to provide a performance equal to or better than CCA in the retention level approved for Use Class 4.

Marine field tests (EN 275)

Wood used in marine applications is susceptible to marine borers. Laboratory tests exist for only some marine borers; the main option is to conduct

Table 1. Mass loss due to brown rot (*Postia placenta)* and white rot (*Phanarochaete chrysosporium)* decay in laboratory assays after 16 weeks.

		Mass loss caused by *Postia placenta*	Mass loss caused by *Phanerochaete chrysosporium*
Control I	Untreated	48.3	22.7
Furfurylated 1	WPG = 25	9.2	4.5
	WPG = 125	1.7	2.4
Control II	Untreated	65.6	24.2
Furfurylated II	WPG = 23	6.9	5.0
	WPG = 128	1.1	2.9
Control III	Untreated	60.0	20.0
Furfurylated III	WPG = 35	4.3	-
	WPG = 75	4.3	-
	WPG = 120	2.4	2.7
Control IV	Untreated	44.1	15.2
CCA	9 kg/m^3	4.0	1.1

Table 2. Mass loss for treated and untreated pine sapwood, after 12 months in 3 different TMCs (soil microcosms).

Modification type (/chemical)	*Modification level*	*Mass loss (%) due to leaching*	*Mass loss (%) in TMC 1*	*Mass loss (%) in TMC 2*	*Mass loss (%) in TMC 3*
Pine Control I	Untreated	2.4	78.0	62.9	20.1
	Acetone-extracted	2.1	76.7	58.5	14.8
Furfurylated pine	WPG = 22	2.4	5.7	5.2	8.5
	WPG = 41	1.7	2.1	2.8	5.0
	WPG = 60	0.6	1.0	1.8	1.9
Pine Control II	Untreated	Not leached	73	51	57
Furfurylated pine	WPG = 25	Not leached	7	3	4
Pine Control III	Untreated	Not leached	94.5	41.9	17.1
CCA (NWPC-Standard no.1)	4 kg/m³	Not leached	42.8	5.4	2.9
	9 kg/m³	Not leached	25.2	2.1	1.3
Birch Control	Untreated	1.5	85.8	30.0	28.2
Furfurylated birch	WPG = 40	1.6	5.6	6.6	7.1

Reprinted from Properties of Furfurylated Wood by Lande et.al. from Scandinavien Journal of Forest Research, www.tandf.no/forest, 2004, 19, 22-30, by permission of Taylor & Francis

Table 3. Condition of pine sapwood mini-stakes (8x20x200mm) after 8 years in three Swedish fields.

Treatment	*Treatment level*		*Index of Decay (0-100% decay)*		
	WPG	*kg/m³*	*Site A*	*Site B*	*Site C*
Furfurylation	15		92	71	92*
	33*		19*		
	50		4	17	25*
	>100		0	0	0*
CCA		2.1[b]	100	96	92*
(NWPC[a] Standard No.1)		9.0[b]	21	62	25*
Untreated controls	-	-	100	100	100*

a) Nordic Wood Preservation Council. b) CuO (19 wt-%); CrO3 (36 wt-%); As2O5 (45 wt-%) *Six years exposure. SiteA – Simlingsdalen, SiteB – Ultuna, SiteC - Ingvaldsbenning

tests in an actual marine site. In Europe, marine testing is done according to European Norm EN-275.

The furfurylated wood was found to perform well in marine tests. All untreated pine sapwood samples in the first set of controls were rejected when inspected after one year (Table 4, column to the right). Specifically, the controls were so severely attacked by Shipworms (*Teredo navalis)* that the teredo tunnels covered more than 90% of the X-ray pictures. The untreated samples were replaced with new controls and, at each annual assessment during the following years all controls failed due to teredo attack and were again replaced. Conversely, the furfurylated samples at medium to high modification levels were all rated sound after 5 years of marine exposure.

Termite resistance

Much deterioration is caused by termites. Although termites only exist in certain areas, the economical value of the wood destroyed is greater than the value of wood destroyed by decay fungi. Furthermore, some termites are capable of destroying dry wood in buildings that are typically not susceptible to fungal degradation.

Furfurylation of wood appears to reduce its susceptibility to termite attack; however, this is based on only a few tests. One laboratory termite test was done at the University of Hawaii, following AWPA E1 Standard. For untreated pine the average mass loss was 66 %, which gives an AWPA rating 0 (total failure).

Table 4. Condition of pine sapwood (25 x 75 x 200 mm samples) after 5 years of exposure on test rigs in the bay outside Kristineberg (Sweden) Marine Research Station.

	Chemical retention		*Replica*	*No. of samples classified as attacked*			*Overall rating*	
Chemical treatment	*wpg*	*kg/m*^{3}	*n*	*sound*	*Attack*	*reject*	*(0-4)*	
Furfurylation	11		5	-	-	5	4.0	Failed
	29		5	5	-	-	0.0	Sound
	50		5	5	-	-	0.0	Sound
	120		5	5	-	-	0.0	Sound
CCA (NWPC[a] Stand. No.1)		4[b]	6	-	-	6	4.0	Failed
		18[b]	6	5	1	-	0.2	Sound
Untreated controls	-	-	8+5+5 +7	-	-	8+5+ 5+7	4.0[c]	Failed[c]

a) Nordic Wood Preservation Council. b) CuO (19 wt-%); CrO3 (36 wt-%); As_2O_5 (45 wt-%). c) Controls; 4 sets, each set lasted only one year

For untreated beech the average mass loss was 38 % with an AWPA rating of 4.0 (severe attack). For furfurylated pine sapwood the average mass loss was 8 % and the AWPA rating 7.4. However, for furfurylated beech the average mass loss was only 2 % (AWPA rating 9.6). For approval of a treatment as effective against termite attack, a visual rating of at least 7 is required. However, the maximum single mass loss accepted is 5% with an average of 3%. Only the furfurylated beech fulfilled this criterion. Therefore, the general conclusion is that furfurylation provides good but not total termite resistance.

In-ground mini-stake tests in fields with high subterranean termite activity were used to test termite resistance in outoor exposure. Generally, highly furfurylated wood performed very well. All stakes were rated sound after one year in the Indonesian Bogor field, whereas very little remained of the control stakes (93-98% mass loss). In a supplementary test in the Indonesian Bandung field, medium-furfurylated pine (WPG=43) also seems to be resistant to termite degradation (Table 5).

Mechanical and Physical Properties of Furfurylated Wood

Chemical modification of wood can affect a wide range of properties. Cellulose, hemicelluloses, and lignin are intimately combined to form the

Table 5. Performance of mini-stakes (8 x 20 x 200 mm) in the Bogor and Bandung, two Indonesian termite fields

Material	*Mod. level*	*Bogor field, weight loss (%), month*			*Bandung field, weight loss (%), month*		
		3	*6*	*12*	*3*	*6*	*12*
Scots pine control		33	47	93	10	66	85
Furfurylated pine	WPG = 15	3	16	62	4	62	65
	WPG = 43	-	-	-	0	0	0
	WPG = 115	1	1	2	0	0	0
Agathis pine control		36	7	95			
Furfurylated Agathis	WPG = 15	2	2	61			
	WPG = 115	0	1	2			
Sengon control		16	85	98			
Furfurylated Sengon	WPG = 15	2	9	85			
	WPG = 115	0	1	1			

anatomical structure of wood. Chemical modification introduces chemicals that react with one or more of the three wood constituents to varying amounts.

Typically, the change in wood properties is related to the degree of the treatment. A slight chemical modification will have only a minor impact on wood properties, while high treatment levels will have a greater influence. The optimum treatment level is determined by the intended use and the manufacturing costs.. In furfurylation, the treatment level is conventionally expressed as WPG. By proper dilution of the impregnation solution, it is possible to obtain a WPG ranging from 10 to greater than 100% for low-density wood, with the WPG determined by the FA concentration and wood density. Low-density woods have a relatively large void volume, resulting in a high WPG compared to dense wood.

Furfurylation of wood results in polymer formation in wood cell lumens and cell walls. Polymers formed in the cell wall are fixed and will to some extent replace adsorbed moisture, resulting in a permanently swollen or bulked cell wall. This "bulking" effect reduces dimensional changes with changes in moisture content. This dimensional stabilization can be expressed as the anti-shrink efficiency (ASE). The permanent bulking and grafting of FA polymer to the cell structure will also affect the stiffness, strength and brittleness of the wood. Stiffness can be expressed as stiffness stabilization efficiency (SSE).

For furfurylated wood, the volumetric ASE has been demonstrated to be high, even at low WPG. For example, at WPG=32 the ASE is close to 50 %, and at WPG=47 the ASE is approximately 70 % (see striped bars in Figure 4). The stiffness stabilization efficiency (SSE, Grey bars in Figure 4) is also high with moderate WPG (*17*).

The increase in hardness is moderate at low WPGs but very high at high WPGs. The drawback is the decreased impact bending strength at medium to high WPGs (negative bars), although the drastic reduction shown by the bars in the diagram is to some extent an artifact caused by using much smaller samples than the standard specifies. In other impact bending tests of furfurylated wood with correct test specimen size, the results show a more modest reduction of impact strength (*19*). Furfurylated wood with WPG of about 30 % was also exposed to elevated temperature and humidity to investigate the long-term effect on impact bending strength. In general, this WPG level resulted in an impact bending strength loss of 25-35 % after exposure.

Environment and Eco-toxicity

Successful commercialization requires low environmental impacts with the use and disposal of furfurylated wood. The high decay resistance of furfurylated wood is believed to be caused by inertness of the modified wood to fungal enzymes, rather than to toxicity. In an attempt to understand the protective mechanism, several studies were performed. Information obtained included toxicity of airborne emissions, toxicity of water leachates and potential release of toxic chemicals upon disposal.

FA is a liquid that is classified as a hazardous chemical. FA is a strong solvent and a flammable liquid. Polymerized FA is non-toxic and safe to use and dispose. However, chemical reactions that take place during polymerization might leave unreacted byproducts, including monomeric FA, in the polymer. These could be phytotoxic and, therefore, contribute to the decay resistance.

Several studies of the furfurylated wood have helped to understand these potential side effects. Leach water from furfurylated wood was tested for FA content, and only extremely small concentrations of FA were found. Radial fungal mycelium growth test of basidiomycetes on FA-contaminated malt agar growth medium showed no growth inhibition, even with 1000 to 4000 times the concentration found in the leaching water. This suggests that any unreacted FA in the furfurylated wood does not contribute to the resistance against decay fungi.

Leached water from furfurylated wood has also been tested for eco-toxicity at Toxicon Laboratory in Sweden. Results were compared to leach waters from

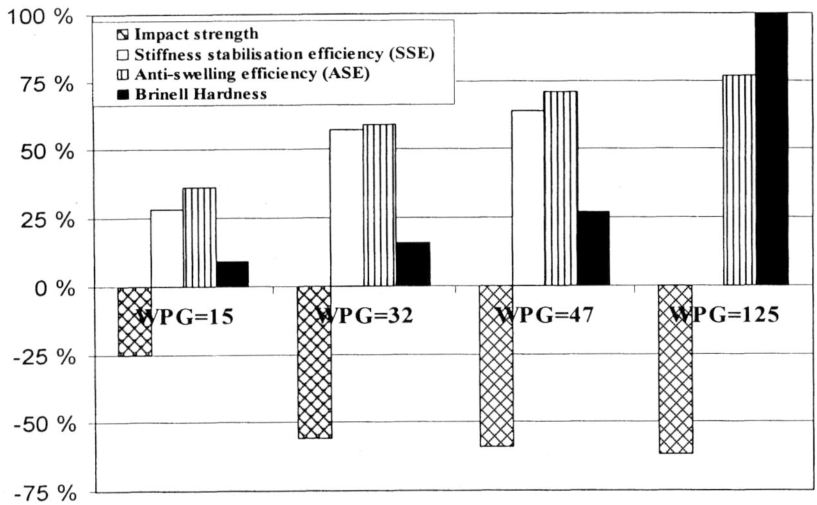

Figure 4. Impact strength, stiffness, dimensional stability and hardness of samples by treatment level (WPG) (Reprinted from Properties of Furfurylated Wood by Lande et.al. from Scandinavien Journal of Forest Research, www.tandf.no/forest, 2004, 19, 22-30, by permission of Taylor & Francis.)

other wood and building materials. The furfurylated pine wood had very low toxicity units, lower even than some untreated woods such as western red cedar, a tropical hardwood (azobé) and Scots pine (Table 6).

Air emissions from furfurylated wood exhibited low concentrations of volatile organic compounds (VOCs). Results from an emission test performed by an accredited laboratory are in Table 7 (*20*). The results are emissions of VOC reported as Toluene equivalents (TVOC) after 1 and 4 weeks of outgassing. Furthermore, burning furfurylated wood was found to release no more toxic gases than burning untreated wood (Table 8).

Biocide Directive

The European Commission has implemented a biocide directive to control and register all biocides in use. New biocides have to fulfill the requirements of this directive. However, since it has been demonstrated that furfurylated wood does not work by biocidal action, but by being biologically inert, the Commission ruled that furfurylated wood was not required to be registered under the biocide directive.

Table 6. Toxic Units (TU) of leaching water from structural materials.

Material	*Algaetox*	*Daphntox*	*Microtox*
Pine	16-32	<2	< 2
Pine – Heat Treated	8-16	2-4	2-4
Pine – Furfurylated	3.1	1.5	2.6*
Western Red Cedar	>32	4-8	16-32
Azobé	8-16	4-8	8-16
CCA – type C	>32	>32	8-16
Concrete	<2	<2	8-16
Recycled plastic	4-8	<2	<2

* Microtox 15 min

Reprinted from Chemistry and Ecotoxicology of Furfurylated Wood by Lande et.al. from Scandinavien Journal of Forest Research, www.tandf.no/forest, 2004, 19, 14-21, by permission of Taylor & Francis

Table 7. Total emissions of TVOC from furfurylated wood

Material	*WPG*	*TVOC 1 week ($\mu g/m^2h$)*	*TVOC 4 weeks ($\mu g/m^2h$)*
Furfurylated Beech	85%	60	50
Furfurylated Pine	30%	40	30
Furfurylated Maple	30%	90	60

Commercialization

There is growing interest in commercial production of furfurylated wood materials due to their useful properties and good environmental profile. Wood Polymer Technologies ASA (WPT) of Norway has been strongly involved in the commercialization process and holds several patents on wood furfurylation.

WPT has based its business model on being an R&D company which licenses the furfurylation technology to wood modification plants around the world.

WPT was established in 1996 to take the furfurylation wood modification process from the laboratory activity to commercialization. Before the start-up of the world's first production facility in Norway, it was necessary to document and test the properties of FA modified wood. WPT cooperated with leading universities and test facilities around the world to secure impartial and reliable test results on furfurylated wood.

These tests suggested attractive properties with furfurylated and, as a result a full scale production facility was constructed by Kebony Products in Porsgrunn, Norway. At Kebony Products the autoclave and curing chamber used

Table 8. Results from smoke gas analysis of furfurylated and untreated Scots pine.

Mode	*25 kW/m2*		*25 kW/m2pilot flame*		*50 kW/m2*	
	Untreated	*Furfurylated*	*Untreated*	*Furfurylated*	*Untreated*	*Furfurylated*
Optical dens.	664	352	402	70	131	70
CO, ppm	5900	6000	3200	2200	160	77
VOC, mg	590	370	260	120	51	28
PAH, mg	2.3	1.8	1.1	1.0	1.7	<1.0

to furfurylate wood was adapted from an oil-based wood impregnation plant. This demonstrated that it is possible to upgrade existing impregnation plants to utilize furfurylation technology.

As the commercialization of furfurylated wood was initiated, brand names were created for the products. The name VisorWood was selected as the brand name for furfurylated products based on pine (the sapwood is treated). Today, that market is dominated by heavy metal impregnated wood. A second brand, Kebony, is for products that will be alternatives to tropical wood from endangered rainforests. In these products, the whole cross section is treated. Today Kebony Products treats beech, birch and maple, but other wood species, such as rubber wood, is being tested.

Kebony Products' first year of full operation was 2004. Since then, it has gained valuable experience with optimization of production parameters and raw materials. This will be valuable information for future plants.

In 2005, there are two plants in operation using the WPT process for modification of wood. Boen Bruk is a Norwegian parquet flooring producer and WPT licensee making Kebony 100, which is used instead of the tropical hardwood Merbau in the flooring. WPT's daughter company, Kebony Products, produces VisorWood as an alternative to traditional heavy-metal impregnated wood and also making Kebony 30, an alternative to tropical hardwoods. Discussions are currently ongoing with potential licensees from Europe and Asia. In Europe, the focus of the possible licensees will be maintenance-free building siding and other construction materials with improved durability. The focus in Asia will be on furniture and flooring products that are currently produced from tropical timbers.

Examples of Uses of Furfurylated Wood

Furfurylated wood has been commercially available since 2004 from Kebony Products DA, Norway. Figures 5 through 8 illustrate some projects where furfurylated wood has been used.

References

1. Dunlop, A.P.; Peters, F.N. *TheFurans;* Reinhold Publishing Corp.: New York, NY, 1953.
2. Goldstein, I.S. *Forest Prod. J.* **1955**, *5*(4), 265-267.
3. Goldstein, I.S. Patent, GB846680, 1960.
4. Goldstein, I.S. and Dreher, W.A. *Ind. Eng. Chem.* **1960**, *52*(1), 57-58.
5. Stamm, A. J. In *Wood Technology; ChemicalAaspects*; Goldstein I., Ed.; ACS Symposium series 43; American Chemical Society: Washington, DC, 1977; 141-149.

Figure 5. Decking of furfurylated Scots pine on a café in Porsgrunn Norway. Photo: WPT Norway

Figure 6. Private residence in Sogndal, Norway showing the use of furfurylated wood siding material. Photo: WPT Norway

Figure 7. Canal lined with furfurylated wood, Norway. Photo: WPT Norway

Figure 8. Apartment buildings in Bergen, Norway, with furfurylated wood siding. Photo: WPT, Norway

6. Anaya, M.; Alvarez, A.; Novoa, J.; González, M.; Mora, M. *Revista sobre los Derivados de la Cana de Azucar*, **1984**, *18*(1), 49-53.
7. Anaya, M. Cuban patent CU21453, 1987.
8. Schneider, M.H. *J. of Wood Sci. and Tec.* **1995**, *29*(4), 135-158.
9. Westin, M.; *Mid-term status report to The Swedish Council for Forestry and Agricultural Research (SJFR).* Trätek: Stocholm, Sweden, 1995, 25.
10. Westin, M.; Ohlsson, B.; Simonson R.; Nilsson, T. In proceedings *212th ACS National Meeting*; Orlando, FL, Aug 25-29,1996.
11. Choura, M.; Belgacem, N.M.; Gandini, A. *Macromolecules* **1996**, *29*(11), 3839-3850.
12. Foo, L.Y.; Hemingway, R.W. *J. Wood Chem. Techn.* **1985**, *5*(1), 135-158.
13. Ekström, G.; In *Perstorpsboken – Plastteknisk handbok.* Maskinaktiebolaget Karlebo and Perstorp AB, Stockholm, Sweden, 1980.
14. González, R.; Martínez; Ortiz, P. *Makromol. Chem.*, **1992**, *193*(1), 1-9.
15. Delmonte, J.; U.S. Patent 2,462,054, 1949.
16. Philippou, J.L.; Zavarin, E. *Holzforschung*, **1984**, *38*, 119-126.
17. Lande, S.; Eikenes, M.; Westin, M. *Scand. J. of For. Res.* **2004**, *19*(5), 14-21.
18. Lande, S.; Westin, M.; Schneider, M.H. *Scand. J. of For. Res.* **2004**, *19*(5), 22-30.
19. Lande, S.; Eikenes, M; Westin, M.; Schneider, M.H. In Proceeding *Sec. Euro. Conf. on Wood Mod.* In press, 2005
20. SP Swedish National Testing and Research Institute; Report P4 01826-9-1, 2005.

Chapter 21

Wood Protection with Dimethyloldihydroxy-Ethyleneurea and Its Derivatives

Andreas Krause[1], F. Wepner[2], Y. Xie[1], and Holger Militz[1]

[1]Institute of Wood Biology and Wood Technology, Georg August University, Buesgenweg 4, D–37077 Goettingen, Germany
[2]Hamberger Industriewerke GmbH, Postfach 10 03 53, D–83003 Rosenheim, Germany

The cross-linking reagent DMDHEU (dimethyloldihydroxy-ethyleneurea) and its derivatives were used to chemically modify wood. The mode of action is based on DMDHEU cross-linking with the wood compounds and self-polymerization within the cell wall. The modified material is a polymer composite with the appearance and texture of solid wood. The impregnation causes a permanent bulking of the cell wall and reduces the swelling and shrinkage properties, with the dimensional stability thus considerably increased. This is reflected by an improved anti shrink efficiency (ASE) of up to 70 %. In addition, high durability against white, brown and soft rot fungi are obtained. The treatments also enhance the wood's surface hardness and weathering properties.

Introduction

The objective of wood modification is to improve some of the properties of wood, such as low dimensional stability, UV/sunlight degradation and biological durability. (*1*). The two basic methods to modify wood are thermal or chemical processes. Various chemical modification techniques have been investigated for many years, particularly acetylation (*2*) and treatments with melamine (*3*). One of the most promising chemicals was N-methylol based agents.

N-methylol compounds are widely used in the textile industry to improve cotton or other cellulose-based fabrics. They enhance wash- and wear-properties and help fix color or other agents to fibers (*4*). DMDHEU was the most widely used N-methylol compound in the textile industry, but due to formaldehyde emissions in the process and from the textiles, low-formaldehyde containing agents were developed (*5*).

Chemical wood modification with DMDHEU or its derivatives are applicable to both solid lumber and wood based composites. The mode of action is based on DMDHEU cross-linking with wood compounds and its self polycondensation within the cell wall. Technically, the modified material is a wood polymer composite with the appearance and texture of solid wood (*6*).

One of the main advantages of DMDHEU-based modification is an increase of dimensional stability. Impregnation causes a permanent bulking of the cell wall and, thus, reduces subsequent dimensional changes of the wood. Tests with DMEU (dimethylolethyleneurea) and DMDHEU resulted in anti-shrink efficiencies (ASE) of up to 70% (*6-13*). In addition, high durability against white, brown and soft rot fungi are obtained (*9,14-17*), but the mode of action against basidiomycetes is not fully understood (*18*). Also, only a slight increase in termite resistance was demonstrated (*19*). A final advantage is that DMDHEU is a non-toxic and environmental benign chemical (*20*). However, despite numerous advantages, the treatment of wood with DMDHEU poses some problems. The basic difficulty is the tendency for larger size wood to crack after treatment (*8,15*). Furthermore, some mechanical properties, such as bending strength or impact strength, are reduced (*21*).

Chemical Agents and Reactions

Chemical Agents

Various N-methylol compounds have been developed by the textile industry in the past 40 years (*22*), but only DMDHEU and it's derivatives were widely accepted. The reactive functional groups in the molecule are the two N-methylol groups (Figure 1A).

To reduce the formaldehyde emissions from DMDHEU, the molecule is also partially methylolated to mDMDHEU (Figure 1B). The reactivity of methylolated DMDHEU is lower than that of DMDHEU, however.

By an N,O-acetalization with an alcoholic compound, DMDHEU can be modified. This reaction prevents the hydrolytic release of formaldehyde (*23*). Formaldehyde emissions can be further reduced by adding formaldehyde scavengers, such as citric acid, chitosan or glyoxal (*24*).

Formaldehyde free finishing agents, such as dihydroxydimethyl-imidazolidinone (DHDMI), are also used for finishing of textiles. However, it has a low reactivity and is thus not suitable for wood modification (*6*).

Various catalysts are used to enhance the reactivity of cross-linking agents (*6,13*). One of the best catalysts was magnesium chloride ($MgCl_2$), which is used in the reported results below.

Mechanism of Reaction and Treatment of Wood

The chemical reaction mechanism has been extensively investigated by textile researchers (*4*). The N-methylol group reacts with hydroxyl groups to form acetal bonds. The following reactions can occur:

- Cross-linking with hydroxyl groups of wood
- Hydrolysis of N-methylol groups to formaldehyde and NH-groups
- Condensation with NH groups to form methylene bonds
- Condensation with hydroxyl groups of alcohols to form ether bonds

These reactions of N-alkoxymethyl compounds are subjected to a general acid catalysis (*4*).

The main goal in modifying wood with N-methylol compounds is to achieve both a high extent of cross-linking with wood components and for it to self-polymerize in the wood cell wall.

The treatment procedures of textiles and of solid wood are different. Wood tends to form cracks after treatment due to drying stresses. Also, since wood will undergo structural changes at temperatures above 130°C, relative mild reaction processes are necessary. The typical treatment consist of following steps:

- Impregnation of wood with an aqueous solution containing agent and catalyst
- Drying the wood to below fiber saturation point (optional)
- Curing at temperatures above 90°C and below 130°C

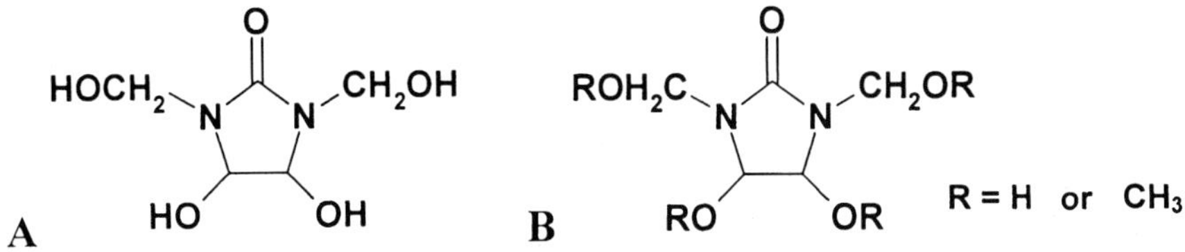

Figure 1. A = DMDHEU (dimethyloldihydroxy-ethyleneurea). B = modified (methylolated) DMDHEU

- Conditioning the modified wood to a final equilibrium moisture content (optional)

Dry wood (Figure 2A) is normally impregnated. During impregnation, the agent is incorporated into the wood cell wall (Figure 2B). The curing at high temperatures leads to the formation of cross-links (cl) between wood hydroxyl groups and N-methylol groups, and to polycondensation (pc) between N-methylol groups and NH groups (Figure 2C).

When treating large size wood, it is necessary to obtain an uniform distribution of DMDHEU within the wood. An uneven distribution will lead to heavy cracking of the treated wood when it is dried after treatment. Therefore, a novel curing process, which uses superheated steam, was developed (*25*). Wood in any dimensions can be treated with this modification.

Properties of Treated Wood

Moisture Content and Dimensional Stability

The sorption behavior of wood treated with DMDHEU or mDMDHEU is significantly changed compared to untreated wood. The equilibrium moisture content (EMC) at a specific relative humidity is influenced by the amount of the N-methylol compound in the wood as well as the type and amount of catalyst employed. The hygroscopic nature of wood is mainly due to hydroxyl groups (*26*) of the cellulose and hemicelluloses. The N-methylol compounds further contain two to four hydroxyls and, consequently, are also hydroscopic as monomers (Figure 1). Consequently, investigations on Scots pine sapwood (*Pinus sylvestris*) and European beech (*Fagus sylvatica*) impregnated with DMDHEU but without curing showed an increased EMC compared to untreated wood at the same relative humidity. However, upon further reaction of the N-methylol compounds so that polymerization and cross-linking has occurred, the

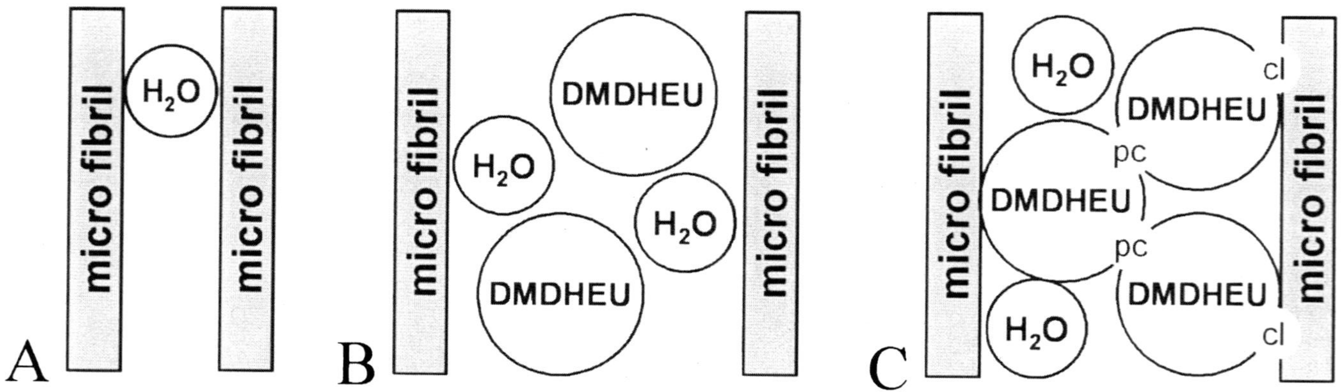

Figure 2. Scheme of modification of wood with DMDHEU. A = dry wood before treatment; B= impregnated wood; C = wood after reaction; cl = cross-linking; pc = polycondensation.

number of free hydroxyl groups was reduced with a concurrent decrease in hygroscopicity.

Beech wood was impregnated with an aqueous solution containing 22.5% DMDHEU and 1.5% $MgCl_2 * 6H_2O$ and cured at 120°C for 24h. At 20°C and 65% relative humidity (rh), the treated beech wood showed an EMC of 9.3%, significantly lower than the EMC of untreated wood (13%).

In addition to the effect of hydroxyl groups of wood, the pore structure in wood affects the hygroscopic behavior of wood. At humid climate conditions of 80% to 100% rh, water vapor condenses in the capillaries of the cell wall and is partly responsible for the EMC (*27*). The reduced water uptake for the modified wood is partially explained by reduced capillary condensation, which is induced by the DMDHEU deposited within the cell wall and the cross-linking effect (*6*).

The sorption behavior of treated wood is also dependent on the type and the amount of catalyst employed. Hygroscopic substances such as $MgCl_2$ lead to an increased moisture content compared to untreated wood (*6,28*).

Along with the changed EMC, the swelling behavior of treated wood differs from untreated wood. Monomeric N-methylol compounds are able to penetrate into the cell wall and fix the wood cell wall in a permanently swollen state. This bulking effect can increase the volume of treated beech up to 10% compared to the volume of untreated wood. Consequently, the swelling and shrinking of wood is significantly reduced. The complementary effects of bulking and cross-linking result in an anti-shrink efficiency (ASE) of up to 70%. The correlation between EMC and the swelling/shrinking of treated beech wood is not linear, however, unlike untreated wood (*29*). With increasing relative humidity, EMC increases more than swelling (*30,31*).

Durability Against Biological Decay

Protection of wood against biological decay is one of the main objectives of N-methylol modification. While DMDHEU-modified wood has enhanced fungal resistance, the mechanism of protection against biological decay has not been completely clarified. It is generally assumed that the protective property of DMDHEU is not based on a biocidal effect but on wood modification (*18,32*).

Brown and White Rot Decay Resistance

Durability tests against brown and white rot fungi, done according to EN113, studied treated beech and pine sapwood. The impregnation solution was based on mDMDHEU and diethyleneglycol (DEG) with magnesium chloride as the catalyst. Based on EN84, the wood was leached with water before incubation. Pine sapwood was exposed to brown rot fungi and beech wood to white rot fungi. Figure 3 shows a negative relationship between the chemical loading with DMDHEU/DEG and wood mass loss.

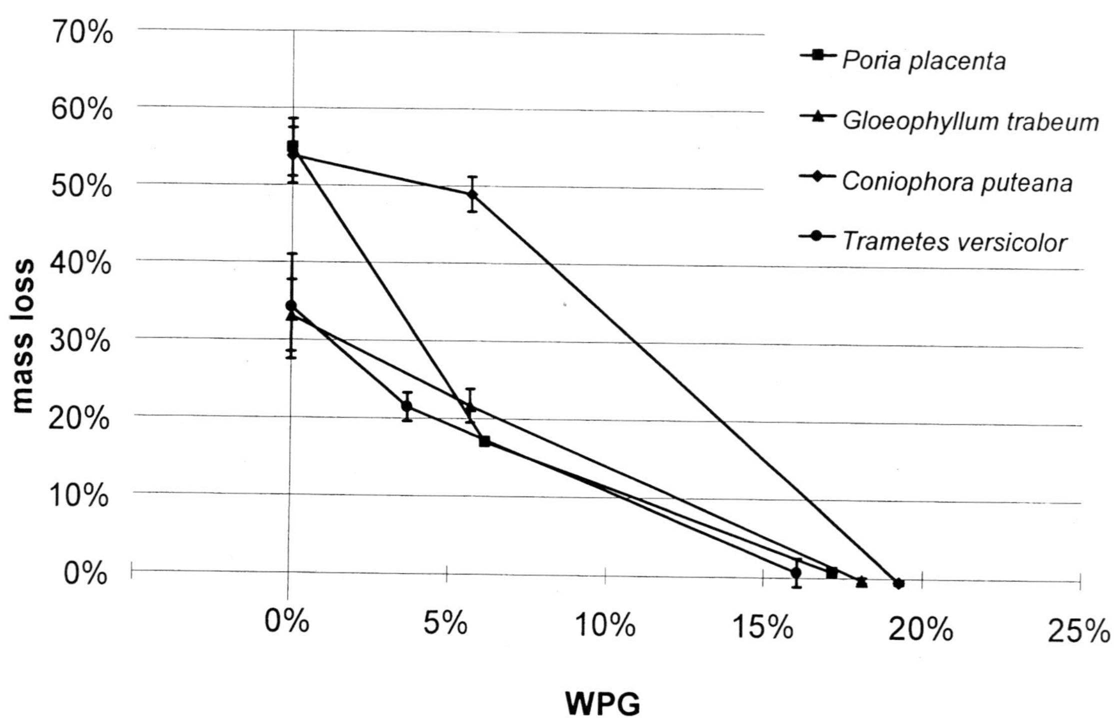

Figure 3. Mass loss of mDMDHEU/DEG-treated wood due to various fungi after 16 weeks incubation time, according to EN113.

The results clearly revealed that weight percent gains (WPG) of more than 15% to 20% assure complete protection against decay by the four fungi species. Consequently, this study confirmed that DMDHEU modification improves wood durability against basidiomycetes.

Soft-rot Decay Resistance

The durability of treated pine sapwood against soft rot decay was investigated in laboratory (ENv807) and field tests (EN252). As expected, the laboratory results showed that the resistance of modified beech wood in soil contact depends on chemical loading (WPG). The difference in decay resistance between wood treated with DMDHEU vs. mDMDHEU was minor. The laboratory tests indicated that beech wood treated with DMDHEU or mDMDHEU complies with durability class 1-2.

Reliable conclusions about the durability of modified wood can be only attained after ground-contact outdoor tests. Those tests, however, require considerable exposure time to obtain meaningful results. The results presented below have only been exposed for the relatively short period of three years and, thus, the results are only preliminary (Figure 4). The failure-rate of untreated pine sapwood samples in the test indicated a normal infestation by fungi in the test field. In contrast, only few of the DMDHEU/DEG treated samples exhibited minor evidence of decay, and DMDHEU-treated samples at a 24% WPG did not show any indications of decay.

Based on these results, modified pine sapwood at WPG's above 15% can possibly be classified at a high durability class, independent of the specific DMDHEU-agent employed.

Mechanical Properties

Wood is an excellent material for numerous engineering applications due to its good strength-to-weight properties. However, some wood modification processes, particularly heat treatment, can decrease the mechanical properties to a critical extent (*33*). Chemical treatment can also decrease the mechanical properties, although the negative impact is less than in the heat treatment process. For example, tensile strength tests (zero span method) with thin strips of DMDHEU-treated Scots pine sapwood yielded losses of approximately 40% (*28*). The treated wood showed an increased brittleness which is reflected by a reduced impact strength. Fortunately, DMDHEU enhances some of the mechanical properties of wood, such as hardness and modulus of elasticity.

The hardness of wood mainly depends on the species and on the density. In some applications, such as parquet flooring, hardness is a decisive property. Use of pine sapwood is limited in flooring applications, as it is naturally a soft

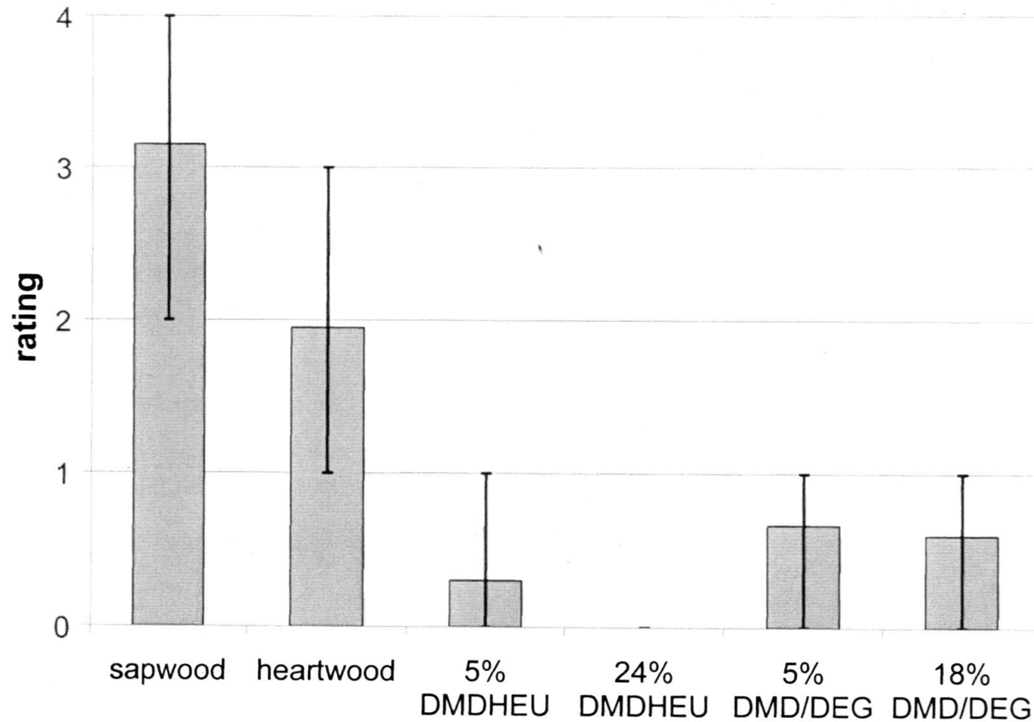

Figure 4. Rating of treated pine after 3 years ground contact in field according to EN252. DMDHEU and DMD/DEG content is expressed as WPG. DMD = DMDHEU. Column = mean value, line = maximal and minimal rating

material (approx. 15N/mm² Brinell hardness). Modification with DMDHEU showed that the hardness of untreated wood can be increased up to 4 times by a high-level DMDHEU-treatment (Figure 5), but a disadvantage is a loss in flexibility.

Surface Properties

One goal of the chemical modification is to stabilize the surface of uncoated and coated wood and thus increase its weathering properties. For example, the acetylation of wood enhances the weathering resistance of wood compared to untreated controls (*34*). The acetylated wood is compatible with finishes and improves the coating properties, such as adhesion or drying rates of finishes (*35*). Improvement of weathering resistance or surface behavior of wood treated with DMDHEU has not yet been clearly shown (*17*).

Weathering Resistance

Thin veneers of pine sapwood were treated with DMDHEU to 48% weight gain and were artificially weathered for 72 h. After weathering, the losses in tensile strength of DMDHEU-treated veneers were lower than that of untreated veneers, likely due to reduced cellulose degradation. Scanning electron microscopy (SEM) also revealed that DMDHEU treatment is highly effective at reducing the degradation of wood cell wall during weathering. Specifically, the tracheids in untreated veneers become distorted within 48 hours of artificial weathering, whereas the tracheids of modified veneers retain their shape even after 144 hours of weathering (Figure 6). The stabilization effect increases with increasing the DMDHEU-content within the veneers (*28*).

Flat-sawn panels of pine sapwood were modified with DMDHEU to 22% weight gain and naturally weathered for 18 months. The treated wood was found to have reduced discoloration and cracking on the wood surface compared to untreated wood (Figure 7). The surface erosion caused by weathering, especially in the less dense earlywood, is lower in the treated wood. The modified panels also had less colonization by blue stain or molds. This latter observation may be due to a reduced hydroscopicity of modified wood, rather than a biocidal effect.

Investigation of DMDHEU treated wood showed that the wettability of surfaces with several waterborne acrylic and solvent-borne alkyd finishes is similar to untreated wood (*36*). The drying rates of various finishes were not affected by the treatment, and the wet adhesion was significantly improved. Finally, modified pine sapwood coated with waterborne stains or oils showed

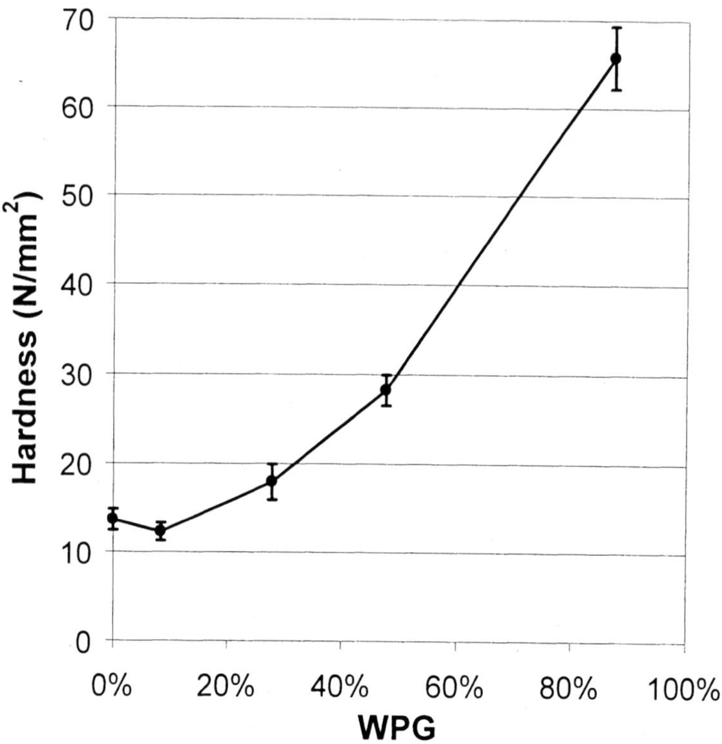

Figure 5. Brinell hardness perpendicular to grain of pine sapwood treated with DMDHEU.

significantly less cracking of coating and wood after weathering for 2 years than untreated coated wood (*37*).

Treatment Of Wood Based Panels

The treatment of solid wood with aqueous solutions at reaction temperatures of more than 100°C will likely be a complex process on commercial-size lumber, because of internal stresses within the wood. Also, many wood species, such as spruce (*Picea abies*) or oak (*Quercus spp.*), are difficult to impregnate. In contrast, small or thin wood parts, such as veneers or chips, can be easily impregnated for most wood species. Alternatively, these species could be modified with N-methylol compounds using veneers or as fibers (*38*) as the furnish, rather than solid wood. The curing and drying is relatively easy with veneers or fibers. Water evaporation is also very fast with

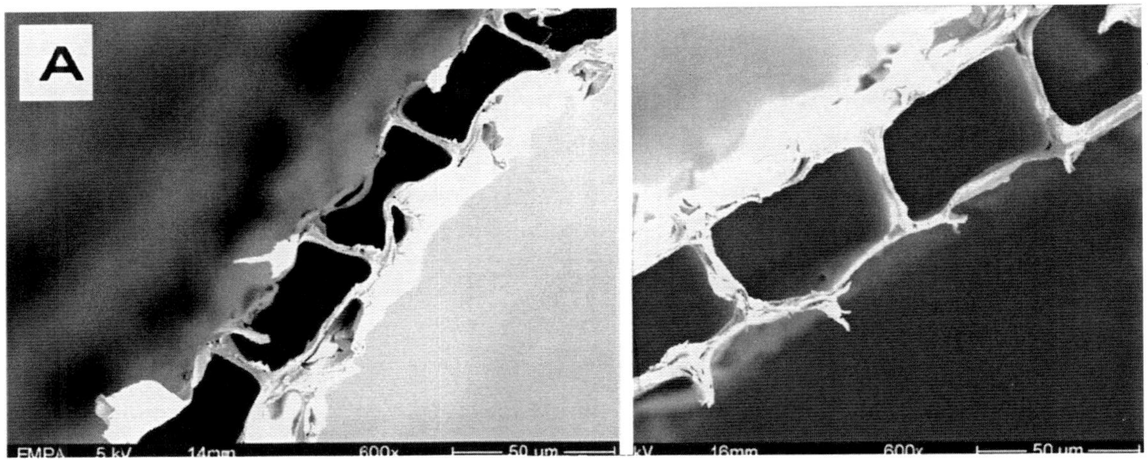

Figure 6.The cross sections of veneers after artificialy weathering of 144 h UV-light: A, untreated; B, treated with DMDHEU to 48% weight gain.

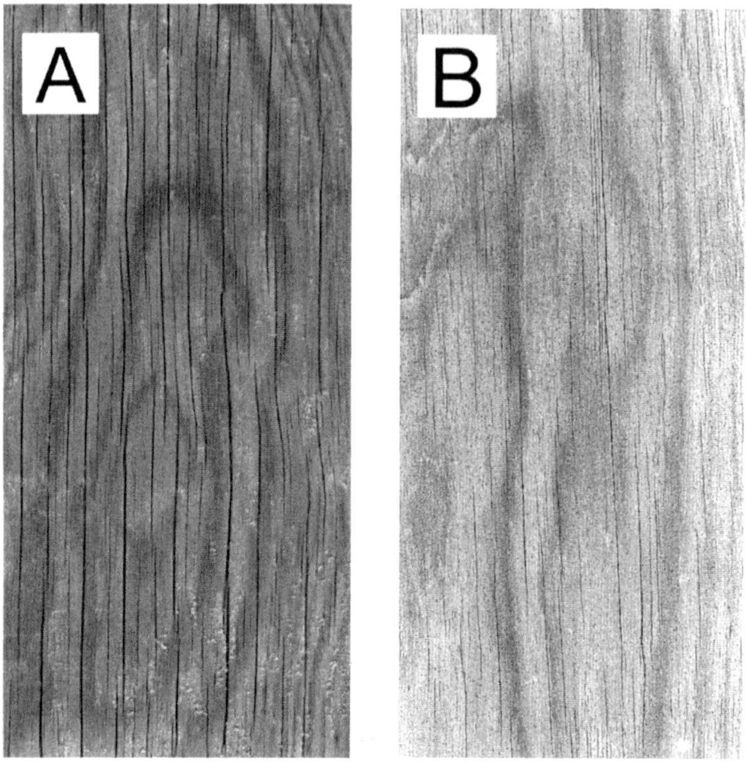

Figure 1. Pine sapwood after 18 month natural weathering. A: untreated; B: treated with DMDHEU (22% WPG). (See page 7 of color inserts.)

small wood pieces, so that the tendency to form cracks is reduced and high temperatures can thus be employed.

A number of possible products have been examined. These include parquet flooring, veneers to obtain plywood with enhanced stability and weathering properties (*31*), and particleboards or fiberboards with enhanced properties (*39*).

Conclusion And Outlook

New processes and chemicals enable wood to be treated with N-methylol compounds such as DMDHEU and its derivatives. Solid wood must be impregnated with aqueous solutions. Wood of small size, such as veneer or flakes, can also be treated.

Among the improved properties are dimensional stability, which depends on the concentration of the agent and the specific processing variables. The swelling and shrinking of wood can be reduced up to 70%. Mean ASE of 50% in an industrial process may be achievable. A durability which corresponds to the natural durability class 1 against fungal decay can be achieved. The treatment does not prevent the growth of molds and stains at surface, but it will reduce the growth of non-wood destroying molds to a high extent in exterior exposure because of the changed moisture behavior. The hardness of wood can be increased by several fold through the treatment, and may be useful as flooring material. The weathering behavior of treated wood is also improved. Less cracking and erosion of the wood surface, as well as greater dimensional stability in outdoor exposure, are also observed.

The possibility of usage N-methylol compounds for modification of various wood-based products, such as veneers, fiber boards or plywood, has good commercial potential.

References

1. Rowell, R. M. *Forest Products Abstracts* **1983**, *6*, 363-382.
2. Beckers, E. P. J.; Militz, H. In *Second Pacific Rim Bio-Based Composites Syposium* Vancouver, Canada, 1994, p 125-135.
3. Lukowsky, D. *Holz als Roh- und Werkstoff* **2002**, *60*, 349-355.
4. Petersen, H. In *Chemical Processing of Fibers and Fabrics. Funktional Finishes Part A*; Lewin, M., Sello, S. B., Eds.; Marcel Dekker, Inc.: New York and Basel, 1983, p 47-327.
5. Reeves, W. A.; Day, M. O. *Journal of Coated Fabrics.* **1983**, *13*, 50-58.
6. Krause, A.; Jones, D.; van der Zee, M.; Militz, H. In *European Conference on Wood Modification*; Van Acker, J., Hill, C., Eds. Ghent, Belgium, 2003, p 317-327.
7. Ashaari, Z.; Barnes, H. M.; Lyon, D. E.; Vasishth, R. C.; Nicholas, D. D. In *International Research Group on Wood Preservation*; IRG Secretary Stockholm, Sweden: Rotorua, New Zealand, 1990, p 9.
8. Ashaari, Z.; Barnes, H. M.; Vasishth, R. C.; Nicholas, D. D.; Lyon, D. E. In *International Research Group on Wood Preservation*; IRG Secretary Stockholm, Sweden: Rotorua, New Zealand, 1990, p 11.
9. Militz, H. *Wood Science and Technology* **1993**, *27*, 347-355.
10. Nicholas, D. D.; Williams, A. D. In *International Research Group on Wood Preservation*; IRG Secretary Stockholm, Sweden: Honey Harbour, Ontario, Canada, 1987, p 8.
11. Videlov, C. L. In *International Research Group on Wood Preservation*; IRG Secretary Stockholm, Sweden: Lappeenranta, Finland, 1989, p 7.
12. Weaver, J. W.; Nielson, J. F.; Goldstein, I. S. *Forest Products Journal* **1960**, 306-310.

13. Zee Van der, M.; Beckers, E. P. J.; Militz, H. In *International Research Group on Wood Preservation*; IRG Secretary Stockholm, Sweden: Maastricht, The Low Countries, 1998, p 10.
14. Acker Van, J.; Nurmi, A. J.; Gray, S. M.; Militz, H.; Hill, C.; Kokko, H.; Rapp, A. O. In *International Research Group on Wood Preservation*; IRG Secretary Stockholm, Sweden: Rosenheim, Germany, 1999, p 16.
15. Sudiyanni, Y.; Imamura, Y.; Takahashi, M. *Wood Research* **1996**, 55-59.
16. Yalinkilic, M. K.; Gezer, E. D.; Takahashi, M.; Demirci, Z.; et, a. *Holz als Roh- und Werkstoff* **1999**, *57*, 351-357.
17. Yusuf, S.; Imamura, Y.; Takahashi, M.; Minato, K. *Mokuzai Gakkaishi* **1995**, *41*, 785-793.
18. Ritschkoff, A.-C.; Rättö, M.; Nurmi, A. J.; Kokko, H.; Rapp, A. O.; Militz, H. In *International Research Group on Wood Preservation*; IRG Secretary Stockholm, Sweden: Rosenheim, Germany, 1999, p 8.
19. Yusuf, S.; Imamura, Y.; Takahashi, M.; Minato, K. *Mokuzai Gakkaishi* **1995**, *41*, 163-169.
20. Anonymous In *10th SIDS Initial Assessment Meeting*; UNEP-Publications: Tokyo, Japan, 2000.
21. Minato, K.; Yasuda, R. In *Chemical Modification of Lignocellulosics* Rotorua, New Zealand, 1992, p 97-106.
22. Petersen, H. *Textilveredlung* **1968**, *3*, 160-179.
23. Vieweg, R.; Becker, E. *Kunststoffhandbuch Band X*; Carl Hanser Verlag: Muenchen, 1968.
24. Bhattacharyya, N.; Doshi, B. A.; Sahasrabudhe, A. S.; Mistry, P. R. *American Dyestuff Reporter* **1993**, *82*, 96-103.
25. Schaffert, S.; Krause, A.; Militz, H. In *European Conference on Wood Modification*; Militz, H., Hill, C., Eds. Goettingen, Germany, 2005.
26. Kollmann, F. *Technologie des Holzes und der Holzwerkstoffe*; Springer Verlag: Berlin. Tokyo, 1951-1955.
27. Skaar, C. *Wood-Water Relations*; Springer-Verlag: Berlin. Tokyo, 1988.
28. Xie, Y.; Krause, A.; Mai, C.; Militz, H.; Richter, K.; Urban, K.; Evans, P. D. *Polymer Degradation And Stability* **2005**, *89*, 189-199.
29. Niemz, P. *Physik des Holzes und der Holzwerkstoffe.*; DRW-Verlag Weinbrenner DmbH & Co.: Leinfelden-Echterdingen, 1993.
30. Krause, A.; Xie, Y.; Militz, H.; paper submitted: 2005.
31. Wepner, F.; Militz, H. In *European Conference on Wood Modification* Goetingen, Germany, 2005, p 169-177.
32. Verma, P.; Mai, C.; Krause, A.; Militz, H. In *International Research Group on Wood Protection*; IRG Secretary Stockholm, Sweden: Bangalore, India, 2005.
33. Militz, H. In *Enhancing the Durability of Lumber and Engineered Wood Products*; Forest Products Society: Kissimmee (Orlando), Florida, 2002, p 239-249.

34. Feist, W. C.; Rowell, R. M.; Ellis, W. D. *Wood and Fiber Science* **1991**, *23*, 128-136.
35. Beckers, E. P. J.; de Meijer, M.; Militz, H.; Stevens, M. *Journal of Coatings Technology* **1998**, *70*, 59-67.
36. Tomazic, M.; Kricej, B.; Pavlic, M.; Petric, M.; Krause, A.; Militz, H. In *Woodcoatings-Developments for a Sustainable Future* The Hague, The Netherlands, 2004.
37. Xie, Y.; Krause, A.; Militz, H.; unpublished results: 2005.
38. Wepner, F. Master, Georg August Uníversity, 2002.
39. Bartholme, M. Master thesis, Georg-August-University, 2005.

Chapter 22

Processes and Properties of Thermally Modified Wood Manufactured in Europe

Holger Militz

Institute of Wood Biology and Wood Technology, Georg-August-University Goettingen, Buesgenweg 4, 37077 Goettingen, Germany

Several processes to thermally treat wood have been commercialised in Europe in the past decade. Due to the high temperatures with most processes using 180 – 220 °C, the chemical structure of the wood components are greatly changed. Heat treated lumber has altered biological and physical properties. The wood is more resistant against basidiomycetes and soft rot fungi, and has a lower equilibrium moisture content and fibre saturation point. Consequently, the dimensional stability is improved. Because of the increased brittleness of the wood, some strength properties are greatly decreased. Due to the enhanced durability, dimensional stability, and good appearance, thermally-treated wood is currently used in Europe in many indoor and outdoor applications.

Introduction

Research efforts have long examined processes to chemically modify wood. Because of the availability of tropical timbers with high natural quality and cheap and effective wood preservatives, however, only a few wood modification processes were commercialized in the past. This has changed in the last few years, with increased interest to find alternatives for tropical timbers and preservative-treated wood leading to several new wood treatments that have recently been commercialized in Europe. Acetylation with acetic anhydride, furfurylation with furfural alcohol, or treating wood with modifying resins are examples of processes that have been, or will shortly be, commercialized in Europe (*1*). Most of these new treatments are non-biocidal alternatives to conventional wood treatment with biocidal preservatives, and provide wood with improved dimensional stability and a pleasant appearance for interior or exterior use.

It has long been known that wood properties can be altered when wood is heated at elevated temperatures. However, only recently has this knowledge led to the development of commercial processes. Today, many production units with varying production capacities exist in Finland, Sweden, Germany, France, Switzerland, Austria, The Netherlands and possibly other East European countries. This article will give an overview of the existing technology and the material properties of thermally treated wood.

Treatment Processes

Initial attempts to use the scientific knowledge of Stamm et al. (*2*) and Burmester (*3*) to develop a commercial heat-treatment process for mid-European wood species were made by Giebeler in Germany (*4*). For more than 20 years knife handles were produced in a small scale production plant. The original goal, introduction of a large scale process for exterior wood, was not reached because of a lack of interest from the wood industry in the 1980's. Only about 10 years later was the idea of thermally treating wood taken up by several research groups and industry in Europe. More or less independently from each other, several processes were developed and taken from the laboratory to commercial production. All of these processes have in common a thermal treatment at elevated temperatures (160 – 240 °C) than that normally used to dry lumber (50 – 120 °C). The main differences between the various processes are the process conditions and treatment technology. To produce wood with good decay resistance and physical properties, the temperature, wood moisture content and minimizing oxygen during the elevated temperatures are the key paramaters.

The main characteristics of some of the processes are (*5*):

Plato-Process (PLATO BV, the Netherlands)

The PLATO-process (*6, 7*) involves three treatment steps and combines the hydrothermolysis step with a dry curing step. The process time depends on the wood species, thickness, shape of wood, etc., and uses a thermolysis step (1-2 hours at 160 - 190 °C) followed by an intermediate drying step (3-5 days) and a final curing step (8-12 hours at 170 - 190 °C). In some cases a conditioning step (2-3 days) is needed. Depending on wood species and thickness of the material, these times can be shorter. The heating medium can be steam or heated air (*8*).

Retification Process (NOW New Option Wood, France)

This is a one-step process that starts with relatively dry wood (approx. 12 % MC). The material is heated under oxygen-poor conditions (less than 2% oxygen) to 200 – 240 °C. A nitrogen atmosphere is used to minimize the oxygen present. The total duration of the process depends on the wood dimensions and wood species, and is approx. 9-12 hours. There are different production sites in France (*9*). The usual energy source is electricity.

OHT Process (oil-heat treatment, Menz Holz, Germany)

The main characteristic of the OHT process is the use of linseed oil as a drying medium and to improve heat flow into the timber. At the same time, the oxygen level in the vessel is low due to the oil. Fresh or pre-dried timber can be used in this process. The process is performed at 180 to 220°C for 2-4 hours in a closed vacuum-pressure process vessel. Additional time involves heating up and cooling down, with this time dependent on the wood dimension. Typical process duration for a whole treatment cycle (including heating up and cooling down) for logs with a cross section of 100 mm x 100 mm and length of 4 meters is 18 hours (*10*).

Thermo Wood Process (Stora, Finnforest, Finland)

An industrial scale wood heat treatment process, under the trade name of ThermoWood was developed in Finland. Today the process is licensed to members of the Finnish ThermoWood Association. The ThermoWood process consists of three steps. In a first step, the wood is dried in a high temperature kiln. The temperature is increased steadily to 130 °C, during which time high temperature drying takes place. In a second step, the temperature is raised to 185 - 230°C. The temperature is held for 2-3 hours, depending on the end-use application. The third step is a cooling and conditioning step. This final stage lowers the temperature using a water spray system, and at a temperature of 80 - 90°C re-moisturising and conditioning takes place to bring the wood moisture content to 4-6 %. The temperature inside the wood is used to regulate the temperature rise in the kiln. The wood employed can be freshly sawn or kiln dried (*11*).

Stellac Treatment (Finland)

The Stellac treatment is similar to the ThermoWood treatment described above, and is run under atmospheric conditions. The process takes place in an air tight stainless steel kiln. During the first step, the temperature is raised up to 100 °C, followed by a conditioning phase. The actual heat treatment takes place at temperatures up to 250 °C for several hours. The total process duration is about 24 hours.

Further Processes

An Austrian joint-venture (Mühlböck/ Mitteramskogler) produces thermally modified wood for mainly interior and exterior water resistant products. The technology uses kiln drying chambers where the wood is thermally modified at approx. 160 – 220 °C. New processes were recently introduced in Switzerland (Balz) and by a Russian-German cooperation (Barkett).

Products and Production of Heat Treated Wood

Depending on the wood species and the production process, several biological and technological properties of the wood are changed by the treatment. The colour of the wood turns brownish, which is used by some companies to give local wood the appearance of expensive tropical wood. Because no chemicals are used, heat treated wood can be used in both exterior (with increased resistance against wood degrading organisms) and interior applications. At the present time thermally treated wood is used in many applications, including windows, claddings, play ground equipment, sauna interiors, bath rooms, parquet flooring, decking, etc.

Official data on the real production of heat treated wood and the production capacity is not available. Militz (*12*) estimated the production in 2001 at approx. 165,000 m^3. The capacity could easily be increased, because the equipment to thermally treat the wood is relatively simple and has a low capital cost.

Chemical and Anatomical Changes

Chemical changes to the wood structural polymers caused by the high treatment temperatures, leads to altered wood properties, such as biological resistance against wood degrading organisms, altered physical/strength properties, darker colour, etc. Intensive research (*13-22*) has shown that many different chemical transformations occur during the thermal treatment.

Kotilainen (*13*) studied changes in the chemical composition of different softwood and hardwood species that were heated to 150 - 260 °C for several hours under steam, air or nitrogen atmospheres. He found that the reaction conditions influenced the extent and type of chemical changes in the wood components. Hardwood species tended to decompose more than softwood species. The main volatile compounds, beside water, were acids (formic acid, and acetic acid) liberated from the hemicelluloses. Sivonen et al. (*23*) used Electron Spin Resonance (ESR) and CP-MAS ^{13}C-NMR to show increased cellulose crystalinity and the presence of stable free radicals.

Relatively mild PLATO-treated wood was investigated with solid phase CP-MAS ^{13}C-NMR and Fourier transform infrared spectroscopy (FTIR) to understand at the molecular level the reasons for the enhanced properties improvements (*14- 16*). Acetic acid is liberated from the hemicelluloses, which leads to further acid-catalyzed carbohydrate hydrolysis causing a reduction in the degree of polymerisation of the carbohydrates. Acid catalysed dehydration and other reactions result in formation of formaldehyde, furfural and other aldehydes, as well as some lignin cleavage at the Cα and O4 interunit bonds. Some aldehyde groups may be produced from the lignin Cγ carbon. Lignin autocondensation through the cleaved, positively charged benzylic Cα may lead to formation of some methylene bridges in this first phase. An increase in the number of free reactive sites on the aromatic ring of some lignin units also occurs in the early phase, and continues into the next phase. In the second treatment step, autocondensation of lignin is believed to continue with formation of methylene bridges that connect adjacent aromatic rings. The aromatic nuclei sites are formed by demethoxylation and then react with the cleaved, positively charged benzylic Cα. Reactions occur with some of the aldehyde groups that were formed in the first step phase with lignin aromatic nuclei sites to form additional methylene bridges. This leads to an increase in cross-linking with consequent improvement in a dimensional stability and decreased hygroscopicity of wood. Esterification reactions were found to occur under dry conditions at elevated temperature in the curing step, indicated by an increase in the ester carbonyl peak at 1740 cm^{-1} in the FTIR spectrum. The formed esters were apparently mainly linked to the lignin complex, based on the observation that the newly formed carbonyl groups were present in heat-treated wood but absent in the isolated holocellulose. Finally, Weiland and Guyonnet (*24*) performed DRIFT spectroscopy on thermally modified *Pinus pinaster* and *Fagus sylvatica* and reported the formation of new ether linkages, along with the previously-reported acid hydrolysis.

The importance of hemicellulose degradation in combination with lignin changes on the swelling behaviour of thermally treated pine (*Pinus pinaster*) and beech (*Fagus sylvatica*) was demonstrated by Repellin and Guyonnet (*18*). Furthermore, Wikberg and Maunu (*19*) studied the chemical changes caused at temperatures between 160 and 195 °C in spruce (*Picea abies*), birch (*Betula pendula*), aspen (*Populus tremula*) and oak (*Quercus robur*) by ^{13}C CPMAS NMR spectroscopy. Spectra revealed the degradation of amorphous cellulose and hemicelluloses, which increased the relative amount of crystalline cellulose. Furthermore they reported changes in the lignin structure by cleavage of the β-O-4 bonds. In softwood lignin, a decrease in methoxyl groups was measured leading to a more condensed lignin structure.

An increase in cellulose crystallinity due to heat treatment was reported by studies of Bhuyian et al. (*25*) and Bhuyian and Hirai (*17*), based on X-ray diffractometry. This may be important for further process optimisation, in that the process conditions (oven drying versus high-moisture conditions) likely affects the final crystallinity. Specifically, when heating under dry conditions the crystallinity change was observed to be much less than with moist process conditions.

Nuopponen et al. (*21*) used FT-IR and UV Resonance Raman spectroscopy to analyse thermally modified scots pine (*Pinus sylvestris*). Analysis of acetone extracts of the modified wood showed that that the acetone solubility of the lignin increased with process temperatures above 180 °C. Increased levels of free phenolic hydroxyl groups were found in the lignin, probably due to the cleavage of β-O-aryl ether linkages. The amount of extractable lignin increased with increasing temperature, while the resin content in the extract decreased.

Sander and Koch (*22*), based on UV-spectroscopy studies, reported an increase in the 280 nm maximum in the S2 layer of sprucewood (*Picea abies*) caused by hydrothermal treatment, and concluded that the change in the lignin chromophoric behaviour could be due to hydrolysis reactions of carbohydrates.

Wood treated by the PLATO treatment was evaluated by light and scanning electron microscopy (*26*). This study suggested that, depending on the process and wood species, damage to the ray tissue and the vessels will occur. With optimised processes, however, no major tissue changes were seen.

Properties

In the last decade many publications have studied the material properties of heat treated timber. Overview articles are given by Rapp (*5*), Militz (*12*), Ewert and Scheiding (*27*). In general, as was earlier shown by Stamm et al. (*2*), Burmester (*3*) and Giebeler (*4*), the durability, sorption, shrinkage and swelling,

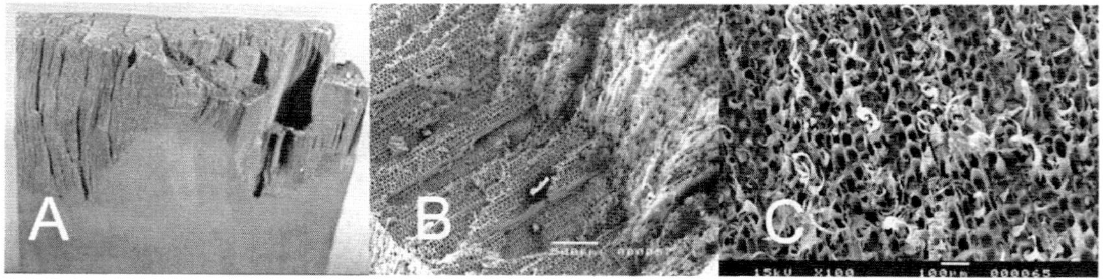

Figure 1. The typical fracture of a heat treated Norway spruce specimen after a bending test (A). Microscopical photo of a heat treated (B) and non-treated (C) Radiata pine, fracture surface after bending test (26)

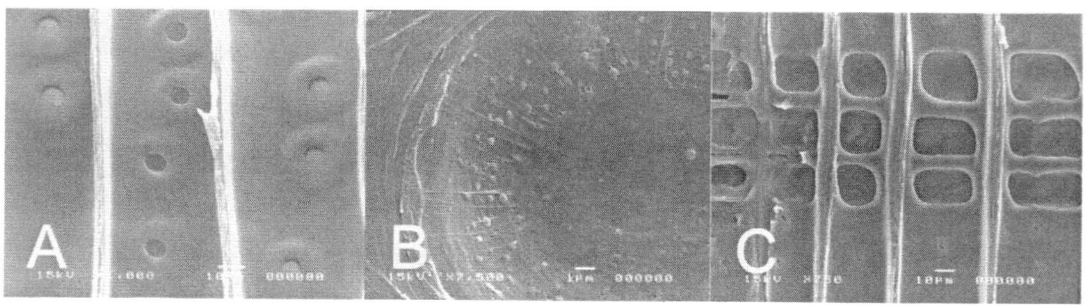

Figure 1. Two stage heat treated Scots pine: radial section of tracheids (A), bordered pit with opened pit chamber(B) radial section, crossing field (C). No changes of the tissue are visible (26)

and strength properties are changed by a heat treatment. The level of change depends on the wood species and process conditions, in which the temperature, the duration of treatment, the wood moisture content and the absence of oxygen are the critical process factors.

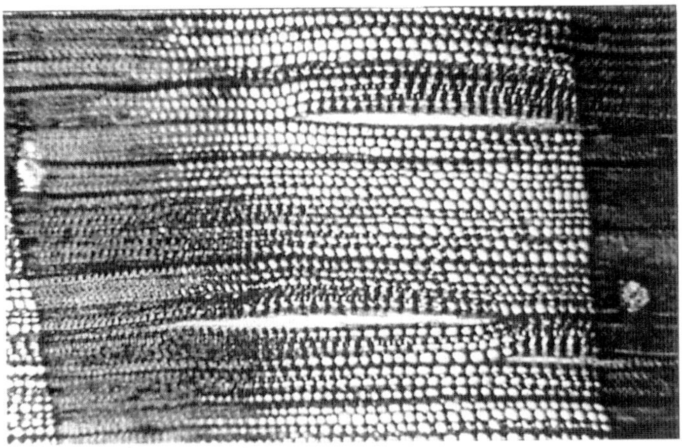

Figure 3. Heat treated Scots pine (cross section). Radial cracks in the earlywood occur due to harsh conditions (26)

Resistance Against Fungi and Insects

Many authors have shown that the durability of wood against decay fungi can be improved considerably by a thermal treatment of wood (*8, 10, 11, 14, 27-32*). The efficacy depends on the wood species and treatment conditions.

The durability of non-durable softwood species, like Norway spruce (*Picea abies)*, Scots Pine (*Pinus sylvestris)* and Maritima Pine (*Pinus maritime)* sapwood, can be increased, with the durability obtained dependent on the treatment temperature and process duration (*10, 11, 33*). Tjeerdsma et al. (*8*) studied the influence of wood moisture during the hydrothermal step of PLATO-treated wood. The resistance against all of the examined fungi, especially soft rot fungi, improved considerably. Decay resistance was found to be dependent on the applied process conditions. In the research of Tjeerdsma et al. (*8*), the treatment effectiveness was improved by employing a hydrothermal step prior to the dry heat-treatment step. The process conditions in the curing step appear to have the largest effect on the resistance against soft and brown rot decay. White rot decay was less dependent on the curing conditions and more affected by the hydrothermolysis step. The enhanced resistance to brown and soft rot degradation was partly assigned to the reduced hygroscopicity of the wood.

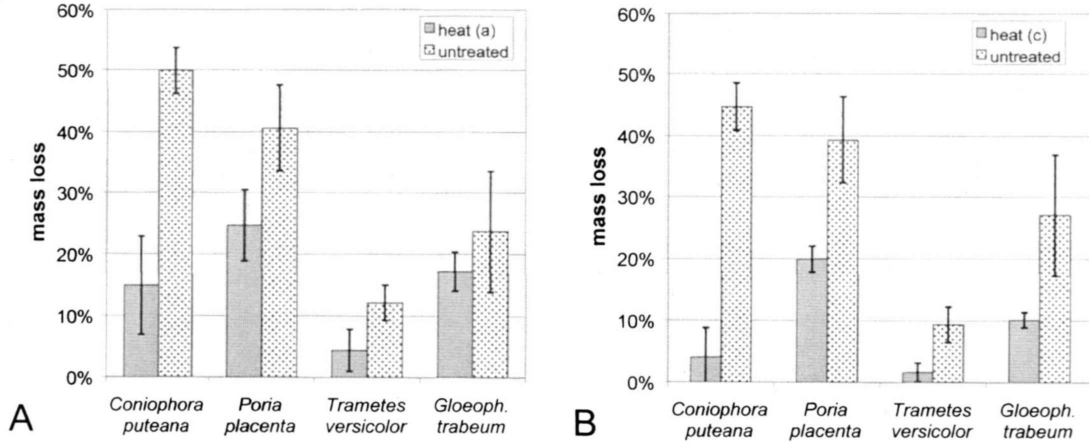

Figure 4. Mass loss of untreated and heat treated Scots pine (Pinus sylvestris) from two commercially available processes (heat a, c) in a 6 week monoculture test following a modified EN 113 (34).

For the OHT-process, Rapp and Sailer (*10*) and Sailer et al. (*35*) studied the resistance of heat treated wood to *Coniophora puteana* with different oil loadings. With increasing temperatures in the range of 180 – 220 °C, the resistance of heat-treated spruce and pine to the brown rot fungus *C. puteana* improved considerably. Mass loss of less than 2% was found in pine sapwood treated in oil at 200°C. With spruce, a decisive increase in resistance was only obtained at 220°C. Matsuoka et al. (*36*) treated Sugi (*Cryptomeria japonica*) in liquid paraffin at temperatures between 90 and 150 °C, but did not find much durability against brown rot fungi or termites, suggesting that temperatures greater than 150 °C are necessary.

Militz and Krause (*37*) and Ewert and Scheiding (*27*) tested the resistance of beechwood (*Fagus sylvatica*) and Pine sapwood (*Pinus sylvestris*) from several commercially available heat treatment processes against soft rot fungi and basiodiomycetes. The durability of both wood species (natural durability class 5) was considerably improved; however, the durability varied. Depending on the process and test fungi, some wood species were very resistant (class 1-2), while others showed only a slight improvement (class 3-4). Also, resistance against soft rot fungi was lower than the resistance against most basiodiomycetes. As shown earlier by Tjeerdsma et al. (*30, 33*), a treatment temperature of at least 180-200 °C is needed to improve the durability to class 1-2.

Ewert and Scheiding (*27*) tested resistance against blue stain fungi using EN 152-1 method with *Aureobasidium pullulan* and *Sclerophoma pithyophila*, and reported no difference between treated wood and controls in the colonisation at the surface of the samples. However, penetration of the hyphae into the wood was only seen with non treated (control) pine, whereas the heat treated wood was only superficially colonised.

Welzbacher and Rapp (*31*) took material from several commercial treatment batches of pine sapwood (*Pinus sylvestris*) and spruce wood (*Picea abies*). For most of the material, an improvement to durability class 3 was obtained. Mayes and Oksanen (*11*) and Viitanen et al. (*38*) also reported a higher resistance of heat treated timber (Thermowood, Finnforest) against several fungi. Furthermore, an improved durability of several bamboo species against basidiomycetes was demonstrated by Leithoff and Peek (*39*).

Research performed at the University of Kuopio (Finland) and at the French institute CTBA showed a higher resistance of thermally treated wood against longhorn beetles, *Annobium punctatum* and *Lyctus brunneus*. However, preliminary trials with termites showed no improved resistance (*11*).

Sorption and Dimensional Stability

Because of the chemical alteration of the wood cell wall structural polymers, the sorption behaviour of the thermally treated wood is altered (*40*). Tjeerdsma et al. (*8*) measured the hygroscopicity of Plato-treated wood. The strong impact of the treatment on reducing hygroscopicity with softwood and hardwood was illustrated by the reduced sorption curves of the treated versus control wood. Reduced hygroscopicity was most pronounced at higher relative humidities (R.H. > 70 %). The hysteresis effect between sorption and desorption was found to be unchanged by the heat treatment of wood. From corresponding research, it is known that the hygroscopicity of heat-treated wood can vary considerably with varying process time and temperature in the second treatment step (*8*).

Popper et al. (*41*) investigated the influence of temperatures between 100 and 200 °C on sorption properties and swelling properties of several wood species (*Pinus radiata, Pseudotsuga menziesii, Laurelia sempervirens, Castanea sativa, Quercus robur*). They noticed that even low temperatures resulted in a lower equilibrium moisture content, with the effect greater with increasing temperature. The sorption analysis, according to the Hailwood-Horrobin model, suggested that changes in the void volume and cross linking of the holocellulose could be responsible for this effect.

By a heat treatment of *Fagus orientalis* at temperatures above 180 °C, Yildiz (*42*) reduced tangential swelling and the rate of water absorption.

Goroyias and Hale (*43*) treated wood strands for OSB production and reported on the effects of temperature and treatment time on mechanical properties, dimensional stability and water absorption. High temperature treatments resulted in significant reductions in thickness swelling of wood strands but reduced MOR and MOE by up to 20%.

In a joint research programme with the German window industry, the physical properties of several commercial heat treatment processes were compared (*37*). All evaluated processes lowered the equilibrium moisture content in the range of relative humidities examined, as well as the volumetric swelling. At higher humidity, the volumetric swelling was reduced to approx. 50-60 % of its original values (Fig. 6). Liquid water uptake usually is affected much less by a heat treatment (Fig. 7) than bound water absorption. As shown by microscopical analysis, the high increase in free water uptake of heat-treated material may have been caused by blue stain fungi that had colonised the wood before the heat treatment.

Mechanical Properties

The changes in the cell wall chemistry (changes in the hemicellulose and lignin structures, cellulose depolymerisation and increased crystallinity, etc.) affect the mechanical properties of heat treated wood. (*8, 37, 44- 49*).

For PLATO-treated wood, the bending strength measurements of several wood species, non-treated versus heat-treated, showed an average strength loss of 5 to 18 % for heat-treated planks (40 mm x 150 mm x 2200 mm). Earlier

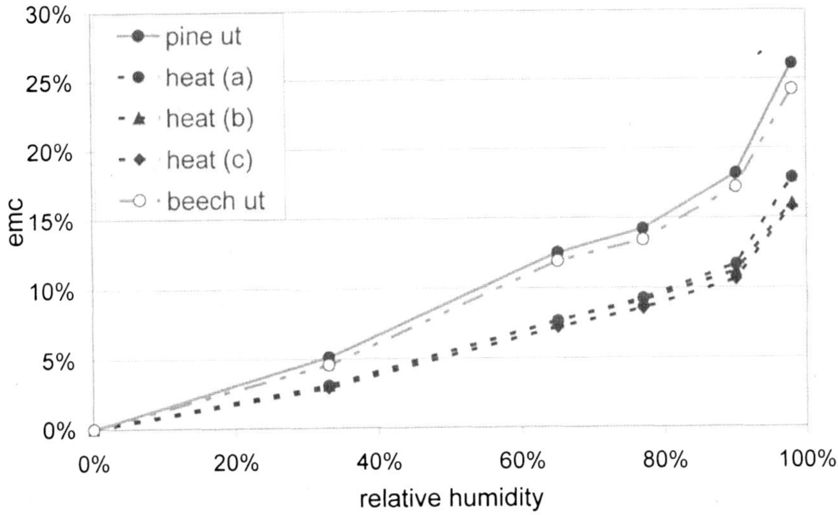

Figure 5. Equilibrium moisture content of untreated (ut) and heat treated Scots pine (Pinus sylvestris) and beech (Fagus sylvatica) from three commercially available processes (heat a-c)

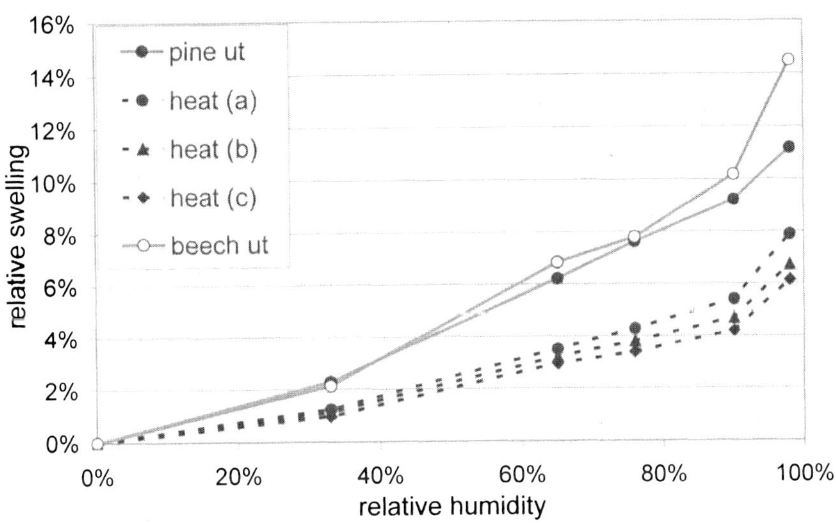

Figure 6. Maximal volumetric swelling of untreated Scots pine (ut) (Pinus sylvestris) and beech (Fagus sylvatica) and heat treated Scots pine from three commercially available processes (heat a-c)

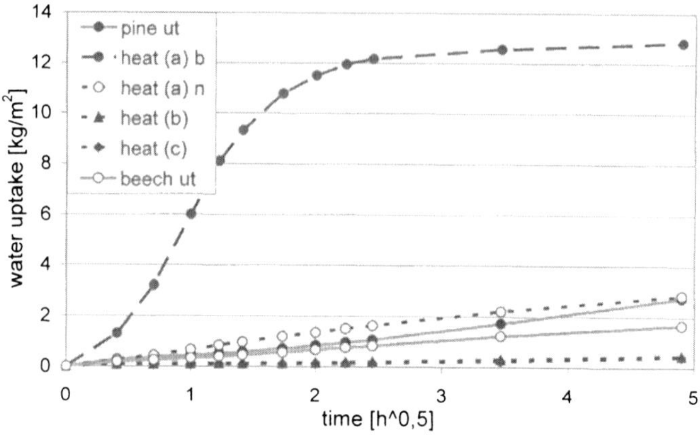

Figure 7. Capillary water uptake in tangential direction of untreated Scots pine (ut) (Pinus sylvestris) and beech (Fagus sylvatica) and heat treated Scots pine from three commercially available processes (heat a-c). Microscopical studies revealed that some samples from process a showed blue stain attack (code(a)b) wheras others were free of attack (code (a)n).

studies on this subject showed in general a strength loss to approximately 50% or more (*50*).

Strength results often are based on small wood samples free of defects and planks treated under mild conditions. During the process, high tension can occur in the wood as it is exposed to high temperatures and rapid evaporation of water. Some of the wood species were found difficult to treat and showed a number of defects, mainly cracks, if not treated carefully. Several softwood species are known to have a high resistance against liquid impregnation. These wood species were found to be difficult to heat treat and showed a relatively high strength loss. Altogether, the final strength was dependent the process conditions and affected predominantly by the process temperature in combination with wood species (*26*).

The MOE and MOR of OHT-treated wood (*10*) was determined in a three point bending test with medium force applied on 150x10x10 mm^3 samples parallel to the grain. There was no reduction in the MOE of coniferous wood with either heat treatment process. The MOR of wood that was oil-heat-treated at 220°C was reduced to about 70% of the value of untreated controls. Bengtsson et al. (*51*) tested the strength of heat treated beams (45 x 145 mm) of spruce (*Picea abies*) and pine (*Pinus sylvestris*) from higher temperature ranges (200 – 220 °C) and found reduced bending strengths of up to 50% but only minor MOE changes. Schmid et al. (*46*) compared several mechanical properties of material from three commercial heat treatment processes (Table 1). The compression strength remained relatively unchanged and the Brinell hardness was slightly reduced. Because of the increased brittleness, the impact bending strength decreased considerably. This was reported earlier by other authors (*10, 11, 52, 53*), who showed that the impact bending strength is the strength property most reduced by all heat treatment processes. Under non optimum process conditions, the impact bending strength can be reduced to about 50 % of its original values.

Kubojima et al. (*48*) examined the influence of temperature and oxygen with sitka spruce (*Picea sitchensis*) on static bending strength and impact bending. They found that at lower temperatures the static Young´s modulus increased, but it decreased with longer treatment times and higher treatment temperatures. The decrease was less in nitrogen than an oxygen rich atmosphere.

Table 1. Bending strength (MOR), Modulus of elasticity (MOE), Compression strength [all in N/mm^2], impact bending strength [kJ/m^2] and Brinell hardness of untreated and heat treated Scots pine (*Pinus sylvestris*) from 3 commercially available processes (heat a-c)

	Scots pine	*Heat (a)*	*Heat (b)*	*Heat (c)*
MOR	80	71	91	93
MOE	10589	9944	13144	11729
Compression strength	37	47	57	54
Impact bending	56	17	21	19
Brinell hardness	15	14	17	13

Colour and Odor

Due to the high temperatures employed, all heat-treated wood species show a characteristic brownish colour (comparable to the natural colour of *Thuja plicata*, Western Red Cedar). The colour is affected by the treatment temperature and the duration of the processes. The higher the temperature and the longer the duration the darker the colour. After treatment, the wood has a characteristic caramellish smell, likely due to to furfural formation. Measurements (*11, 13*) also showed that monoterpenes emission from treated pine is considerably reduced, but heat-treated wood does emit acetic acid.

Paintability and Coating Performance

The resistance of heat treated timber against weathering (UV-light, wetting) is not greatly changed compared to untreated wood, making a surface treatment with oils or paints necessary. No changes in paintability of heat treated wood with water borne acrylic or solvent borne alkyds were found (*54*). Due to its UV-degradability, opaque systems are recommended over priming oils and stains (*11*). After several years of outside exposure, heat-treated timber performs considerably better than non-treated wood. This is likely due to its higher dimensional stability, resulting in less flaking and cracking (*11, 30, 54, 55*).

Gluability

Several authors reported on the gluability of heat-treated wood (*11, 30, 55*). The glued wood was evaluated, following different German and European standards, for strength and moisture performance (*46*). Heat-treated timber can be glued with many industrial adhesives (polyvinyl alcohol and other polyvinylic glues, polyurethane, isocyanate, and resorcinol-phenolic glues). Due to the lower shear strength and tension strength perpendicular to grain with heat-treated lumber, a higher wood failure is found. Furthermore, the hydrophobic wood surface causes a slower penetration of the solvents from the glue to the surrounding wood, which makes it necessary to modify the gluing process.

References

1. Militz, H.; Hill, C. *Wood modification: Processes, Properties and Commercialisation.* Proceedings of the 2nd European Conference on Wood Modification, Göttingen, Germany. 2005. ISBN 3-00-017207-6.
2. Stamm, A. J.; Burr, H. K.; Kline, A. A.: Heat stabilized wood (staybwood). Rep. Nr. R. 1621. Madison: Forest Prod. Lab. 1946.
3. Burmester, A. *Holz als Roh- und Werkstoff* **1973**, *31,* 237-243.
4. Giebeler, E. *Holz als Roh- und Werkstoff* **1983**, *41,* 87-94.
5. Rapp, A. O. Review on heat treatments of wood. Proceedings of a special seminar held in Antibes, France. COST Action E22, Brussels. 2001.
6. Ruyter, H. P. European Patent Appl. No. 89-203170.9, 1989.
7. Boonstra, M. J.; Tjeerdsma B. F.; Groeneveld, H. A. C. *Intern. Res. Group on Wood Pres. Doc. no. IRG/WP 98-40123* **1998**.
8. Tjeerdsma, B. F.; Boonstra, M.; Militz, H. *Intern. Res. Group on Wood Pres. Doc. no. IRG/WP 98-40124* **1998**.
9. Duchez, L. New Option Wood. Personal communication. 2002.
10. Rapp, A. O.; Sailer, M.: Heat treatment in Germany. Proceedings of Seminar "Production and development of heat treated wood in Europe", Helsinki, Stockholm, Oslo. Nov. 2000.
11. Mayes, D.; Oksanen, O. *ThermoWood Handbook.* By: Thermowood, Finnforest, Stora. 2002
12. Militz, H. *Enhancing the Durability of Lumber and Engineered Wood Products*, Forest Products Society, Madison, WI, **2002**, 239-249.

13. Kotilainen, R.: Chemical changes in wood during heating at 150 – 260 °C. PhD thesis, Jyväskylä Universitiy, Research Report 80, Finland. 2000.
14. Tjeerdsma, B. F.; Boonstra, M.; Pizzi, A.; Tekely, P.; Militz, H. *Holz als Roh- und Werkstoff* **1998**, *56*, 149-153.
15. Tjeerdsma, B. F.; Boonstra, M.; Militz, H.: Chemical changes in hydro thermal treated wood; FTIR analysis of combined hydro thermal and dry heat-treated wood. Submitted for publication. **2005**.
16. Tjeerdsma, B. F.; Militz, H. *Holz als Roh- und Werkstoff* **2005**, *63*, 102-111.
17. Bhuiyan, M. T. R.; Hirai, N. *J. Wood Sci.* **2005**, *51*, 42-47.
18. Repellin, V.; Guyonnet, R. *Holzforschung* **2005**, *59*, 28-34.
19. Wikberg, H.; Maunu, S. L. *Carbohydrate Polymers* **2004**, *58*, 461-466.
20. Mitsui, K.; Murata, A.; Tolvaj, L. *Holz als Roh- und Werkstoff* **2004**, *62*, 164-168.
21. Nuopponen, M.; Vuorinen, T.; Jämsä, S.; Viitaniemi, P. *J. Wood Chem. Tech.* **2004**, *24*, 13-26.
22. Sander, C.; Koch, G. *Holzforschung* **2001**, *55*, 193-198.
23. Sivonen, H.; Maunu, S. L.; Sundholm, F.; Jämsä, S.; Viitaniemi, P. *Holzforschung* **2002**, *56*, 648-654.
24. Weiland, J. J.; Guyonnet, R. *Holz als Roh- und Werkstoff* **2003**, *61*, 216-220.
25. Bhuiyan, M. T. R.; Hirai, N.; Sobue, N. *J. Wood Sci.* **2000**, *46*, 431-436.
26. Boonstra, M. J.; Sander, C.; Tjeerdsma, B.; van Acker, J.; Stevens, M.; Militz, H.: The effect of a two-stage heat treatment on the anatomical structure of wood. Submitted to: *Wood Research*, 2005.
27. Ewert, M.; Scheiding, W. *Holztechnologie* **2005**, *46*, 22-29.
28. Buro, A. *Holz als Roh- und Werkstoff* **1954**, *12*, 297-304.
29. Scheffer, T. C.; Eslyn, W. E. *Forest Products J.l* **1961**, *11*, 485-490.
30. Tjeerdsma, B. F.; Stevens, M.; Militz, H.; van Acker, J. *Holzforschung und –verwertung* **2002**, *54*, 94-99.
31. Welzbacher, C. R.; Rapp, A. O. *Intern. Res. Group on Wood Pres. Doc. no. IRG/WP 02-40229* **2002**.
32. Kamdem, D. P.; Pizzi, A.; Jermannaud, A. *Holz als Roh- und Werkstoff* **2002**, *60*, 1-6.
33. Tjeerdsma, B. F.; Stevens, M.; Militz, H. *Intern. Res. Group on Wood Pres. Doc. no. IRG/WP 00-40160* **2000**.
34. DIN EN 113:1996 *DIN Taschenbuch 132 Holzschutz* Beuth Verlag GmbH –Berlin – Wien - Zürich, 1998, ISBN 3-410-14278-9.
35. Sailer, M.; Rapp, A. O.; Leithoff, H. *Intern. Res. Group on Wood Pres. Doc. no. IRG/WP 00-40162* **2000**.
36. Matsuoka, Y.; Ohmura, W.; Fujiwara, S., Kanagawa, Y. *Intern. Res. Group on Wood Pres. Doc. no. IRG/WP 02-40221* **2002**.
37. Militz, H.; Krause, A.: Modified wood for window and cladding products. Proceedings of the COST conference Florence. 2004.

38. Viitanen, H. A.; Jämsä, S.; Paajanen, L. M.; Nurmi, A. J.; Viitaniemi, P. *Intern. Res. Group on Wood Pres. Doc. no. IRG/WP 94-40032* **1994**.
39. Leithoff, H.; Peek, R. - D. *Intern. Res. Group on Wood Pres. Doc. no. IRG/WP 01-40216* **2001**.
40. Kollmann, F.; Schneider, A. *Holz als Roh- und Werkstoff* **1963**, *21*, 77-85.
41. Popper, R.; Niemz, P.; Eberle, G. *Holz als Roh- und Werkstoff* **2005**, *63*, 135-148.
42. Yildiz, S. *Intern. Res. Group on Wood Pres. Doc. no. IRG/WP 02-40223* **2002**.
43. Goroyias, G. J.; Hale, M. D. *Intern. Res. Group on Wood Pres. Doc. no. IRG/WP 02-40238* **2002**.
44. Garrote, G.; Dominguez, H.; Parajó, J. C. *Holz als Roh- und Werkstoff* **1999**, *57*, 191-202.
45. Schneider, A. *Holz als Roh- und Werkstoff* **1971**, *29*, 431-440.
46. Schmid, J.; Illner, M.; Schwarz, B.; Stetter, K.; Militz, H.: Einheimisches dimensionsstabilisiertes Holz für den Fenster- und Fassadenbau. Abschlussbericht, DGFH München, 2004. 170 p.
47. Epmeier, H.; Westin, M.; Rapp, A. *Scandinavian J. Forest Res.* **2004**, *19,* 31-37.
48. Kubojima, Y.; Okano, T.; Ohta, M. *J. Wood Sci.* **2000**, *46*, 8-15.
49. Yildiz, S.; Çolakoğlu, G.; Yildiz, Ü. C.; Gezer, E. D.; Temiz, A. *Intern. Res. Group on Wood Pres. Doc. no. IRG/WP 02-40222* **2002**.
50. Seborg, R. M.; Tarkow, H.; Stamm, A. J.: Effect of heat upon dimensional stabilisation of wood. *J. Forest Products Res. Soc.* **1953**, *3*, 59-67.
51. Bengtsson, C.; Jermer, J.; Brem, F. *Intern. Res. Group on Wood Pres. Doc. no. IRG/WP 02-40242* **2002**.
52. Militz, H.; Tjeerdsma, B.: Heat treatment of wood by the Plato-process: Proceedings of Seminar "Production and development of heat treated wood in Europe", Helsinki, Stockholm, Oslo. Nov. 2000.
53. Kamdem, D. P.; Pizzi, A.; Guyonnet, R.; Jermannaud, A. *Intern. Res. Group on Wood Pres. Doc. no. IRG/WP 99-40145* **1999**.
54. Krause, A.; Militz, H.: Properties of modified wood coated with different stains. Proceedings of the 2nd European Conference on Wood Modification, Göttingen, Germany. **2005**. ISBN 3-00-017207-6.
55. Rapp, A. O.; Sailer, M.; Westin, M.: Innovative Holzvergütung – neue Einsatzbereiche für Holz. In: Proceedings of the Dreiländer-Holztagung, Luzern, Switzerland. **2000**.
56. DIN EN 350 Teil 1+2:1994 *DIN Taschenbuch 132 Holzschutz* Beuth Verlag GmbH –Berlin – Wien - Zürich, **1998**, ISBN 3-410-14278-9.

Chemical Wood Preservative Systems

The first photograph is courtesy of Darrel D. Nicholas, the second and fourth photographs are courtesy of Tor P. Schultz, and the third photograph is courtesy of H. M. Barnes all of Mississippi State University. *(See page 8 of color inserts.)*

Chapter 23

The Process by Which a New Wood Preservative System for Residential Exterior Applications Is Developed: An Industrial Perspective

David B. Helmer

Director of Sales and Marketing, Janssen Pharmaceutica, 1125 Trenton–Harbourton Road, Titusville, NJ 08560

Factors influencing wood preservative development for residential exterior applications, the major use of treated wood, are numerous, complex and highly inter-related. Stakeholders in the development include consumers, wood treaters, environmental regulators, preservative standard and building code governing bodies, and commercial entities developing the active ingredients and treatment products. The use of treated wood in the residential environment increases the need for and conformation of a very low risk safety profile for humans, animals and fate in soil and water environments in and around homes. Performance standards are high, and expectations demand robust active ingredients with very low environmental impacts. These are extremely challenging requirements that take 5 to 10 years to substantiate. The step-by-step process involved in developing a new wood preservative system is described, and a short description of new or potential totally-organic systems is given.

Introduction

Products that claim protection of wood from decay and insect attack in the United States are required to be registered by the US Environmental Protection Agency (EPA). The EPA requirements for registering wood preservatives, as with all pesticides, have become increasingly complex and costly. Consequently, the time required to develop the necessary data to support a registration has increased considerably. The performance expectations, and the data to substantiate that a product will meet these expectations, also require significant investment of time and resources. Developing and marketing new wood preservative products is driven by the need for safer but still more effective and economical wood preservatives. The cost to develop a new wood preservative product must be monitored throughout the development process to ensure that the relative product performance and product attributes with respect to market acceptance are at a level that will result in a sufficient return on investment.

A wood preservative development timeline, beginning with the identification of a candidate active substance to a commercially available product, can span a decade or more and cost tens of millions of dollars. Maintaining a cohesive development plan and consistent financial commitment over a ten-year time span is a challenge. Priorities change, regulations change, and market preferences change. Strategies to minimize development timelines and costs are highly desirable, if not essential.

Consumer safety groups, health officials and the EPA have closely scrutinized the types of wood treatments that might be used in residential construction (*1*). Products intended to treat wood for residential use are required to exhibit minimal health and safety risk to those living in and around wood treated with these products. Further, the environmental impact of the treated wood in proximity to drinking water sources and residential soils needs to be well documented to meet current safety criteria.

In addition to these considerations, the consumer expects that the wood preservative treatments will impart a 10 to 25 year durability aspect to the wood under all environments and exposure conditions in the United States. The treatment compounds must be durable and robust, but not persistent in the environment. The active substances must be highly toxic to the target organisms, but have a high degree of safety for humans and wildlife. The compounds are preferably readily soluble in water for impregnation of the wood, but need to resist leaching out of the treated wood for 10 to 25 years of service life. Finally, the active substances cannot present any substantial risk to humans or the environment at the end of their service life, and should revert to relatively benign degradation products with little or no impact on the biology of the waste management stream.

Minimum Requirements

New active ingredients for wood preservation need to be cost effective, safe to man and the environment and fit for use. Failure to meet the minimum standards in any of these categories will prevent further development or limit the potential uses of the preservative.

Cost Effective

Wood preserving active ingredients, that is compounds that are fungicidal and/or insecticidal, must be cost effective. Effective control of the biological organisms that attack and degrade wood can be measured in laboratory and field test models to provide an indication of the relative performance of the active substances in comparison to well recognized standard compounds (e.g. pentachlorophenol or CCA) and at least one current commercial preservative. Consistently high levels of challenge from fungal decay or insects can be difficult to reproduce in field tests, although environmental factors may present additional challenges. Laboratory tests provide a useful indication of the relative preservative potential of the candidate substances. Using these results, some estimates of the cost in use of the test compounds can be made. Laboratory biological performance evaluations are skewed to favor the degradation of the wood samples. Cost effectiveness estimates based on laboratory data should be assessed as compared to performance of the standard products. Field trials give a more reliable dose response and, therefore, are more useful for determining the cost effectiveness of a candidate. Unfortunately, most field trials require at least five years of exposure to provide meaningful long-term performance data.

Safety to Humans and the Environment

The EPA reviews data provided by registrants. From this data approved use patterns may be established along with precautionary statements regarding handling and disposal of the product by means of a product label. The product label provides requirements for protecting the safety of workers who handle the treating products. EPA assess the potential hazards of a product to workers who work with the treated wood commodity and the potential hazard to those who come into casual contact with the leachates and the waste from treated wood. The nature of the potential hazards is different for each of these different exposure groups.

Some wood treaters come into contact with the concentrated wood treatment products in the course of their task to apply the treatments to the wood (*2*). Owners of wood treating operations expect the treatment products to be safe for workers during the daily handling of the concentrated wood treatments or the freshly treated wood.

Carpenters, construction workers and do-it yourself homeowners handle, saw, nail and drill treated wood. The wood treatment must not pose significant additional hazard to these users either by contact with the skin, or by inhaling the sawdust. The same safety standards apply to the consumers who use outdoor structures built with treated wood. They expect that the wood treatment not present any potential health hazards from skin contact or that the treatment not cause contamination due to the preservatives leaching out of the treated wood. Regulators have identified children as the most likely to be exposed (*2*).

Finally, candidate wood treatments can not impart significant environmental hazard while the treated wood is in service or at the end of its service life when it enters the waste stream. Treated wood may end up in municipal landfills, or be recycled as garden mulch. The preservative treatments need to be environmentally benign and capable of being disposed of without resulting in undesirable effects like leaching into ground water or bio-concentrating in the food chain. Degradation pathways of the treatment compounds in soil and water need to be well understood. Incineration may be a disposal option if it can be proven that hazardous byproducts will not result from combustion of the treated wood.

Fit for Use

The bioactive preservative compounds need to be amenable to incorporation into commercial treatment formulations that will adapt well to the current treatment practices for application to the wood, whether carried on an aqueous solution or a solvent based system using dip, spray or pressure treatment. The treatment compounds need to be robust and stable in transit, storage and in dilution, during application to the wood and finally in the wood while it is exposed outdoors. Failure to meet these requirements will result in a product that does not perform well and thus will not be commercially acceptable.

Additionally, the wood treatment product cannot impart negative aesthetic attributes to the wood. Wood treatment products that result in offensive or persistent odors, discoloration, non-uniform coloration or staining would be considered unacceptable candidates.

Meeting Expectations

A successful wood preservative treatment candidate must meet the expectations of three primary stakeholders; preservative developers, consumers and wood treaters. Many of the expectations are shared by all three, but each has their own focus of interest.

Preservative Developers

Prior to embarking on the long and costly path to develop an organic wood preservative system for residential use, a company needs to have confidence that they will likely reap economic return on their investment sufficient to justify that investment. Factors including time to market, cost of development and market value need to be weighed against the probability of success and profit potential (profit per unit, market size and projected market share versus competitor products). A critical factor is the length of time required to develop and commercialize a new wood preservative. Time is money and delays in development result in delays in financial returns on the initial investments. Also, sustainability of a product requires that a full cradle-to-grave life cycle analysis of the candidate product be conducted. Factors for evaluation include total energy demand, environmental loading, biodegradation and the overall safety profile of the products.

Consumers

Consumers expect treated wood used in and around their homes to be safe. Consumers also expect treated wood products to resist weathering, remain dimensionally stable and to have a low maintenance requirement. This is a big challenge for treated wood, and the plastic wood composite decking materials have taken a significant market share of the decking market (*3*), primarily on the perception that these materials retain their appearance better than treated wood. Consumers not only expect the wood preservative treatments to protect the wood from attack by termites and decay fungi, but to resist the effects from weathering such as cracking checking and warping.

Treaters

Wood treating companies expect the treatment products they apply to the wood to be safe and relatively easy to handle and apply, to pose no significant health risks to their workers, and have no significant environmental hazards to their property (*4*). The treatments need to be effective at an incremental cost increase over untreated wood that is acceptable to the consumer. Wood treating operations are subject to numerous health and safety regulations, so the less hazardous a treating product, the better. Creating polluted land sites as a consequence of environmentally persistent preservatives is not in the interest of the treating facility owner.

Efficacy Evaluations

Laboratory Decay Tests

Typically, preliminary performance evaluations are conducted in the laboratory. These tests are designed to identify candidates with likely sufficient activity that may meet or exceed the expected performance requirements. As a consequence, the laboratory evaluations are severe. The tests are designed to provide optimal conditions for organisms to degrade treated wood.

Soil block tests are a good indicator test for determining the efficacy of a compound on wood against the primary decay organisms (*5*). Small cubes of sapwood are treated to various concentrations with the test compounds. Untreated controls and a commercial preservative are included for comparison. The treated and untreated cubes are exposed to soil colonized with one of the major wood rotting fungi. The test samples are incubated at elevated temperatures and humidity for several months, or at least until the untreated blocks show significant weight loss as a result of fungal attack. Typically, the candidate preservatives are tested for efficacy against six primary decay fungi, three softwood rotting and three hardwood rotting fungi. The biocide candidate needs to have a high degree of activity against all test organisms. From these tests a threshold toxic value, expressed as pounds per cubic foot (pcf) or grams per cubic meter (g/m^3), is determined for each fungus. The minimum target retention is the amount per unit wood volume of a compound needed to prevent decay in wood by the fungus on which it has the **weakest** activity.

After treatment and before exposure to the test fungus, some of the blocks may be put through a leaching or an artificial ageing procedure. These procedures are intended to give some indication of the tendency for the candidate preservative to leach or volatize out of the wood.

One soil decay test method variation under evaluation employs thin wood wafers inoculated for shorter exposure times. After exposure to the test fungi the wafers are placed under a load cell and their relative compression strength is used to extrapolate the degree of fungal degradation in the wafers. This method has several advantages to the standard soil block method in that the compression strength provides a higher degree of sensitivity after a shorter exposure period compared to the weight loss measurement.

Once a compound or treating system has shown promise in the laboratory decay test, estimations of its relative suitability for outdoor exposure can be made. While laboratory tests are designed to encourage preservative failure these artifical tests only give a tentative indication of potential field performance, since the laboratory tests focus only on specific organisms in a controlled environment. Field exposure introduces biological and environmental factors in dynamic interaction.

Field Decay Tests

Field tests represent the most reliable but time-consuming way to evaluate the efficacy of a new wood preservative. The efficacy of the test compounds or products can be compared to benchmark industry standard preservatives. Also, the ability of the test compound to resist long-term depletion can be determined by decay rating and/or analysis of exposed samples for the active substance remaining. Further, any negative effects on appearance or wood strength are revealed in the field test.

Field Stake Trials

The most challenging field evaluation is the ground-contact field stake trial. Wooden stakes are treated with either the test preservative treatments, standard commercially available preservatives, or left untreated (*6*). The size of the stake can vary. Smaller stakes will degrade more rapidly than larger stakes. Several locations with high decay hazard are recommended. Stakes are removed from the ground annually and rated for the extent of decay (and termite) degradation at the soil line, above ground, and below ground. Some stakes may be sacrificied for chemical analysis after a specified exposure period to determine the depletion rate of the preservative(s). The test should be continued for at least the length of time for the untreated stakes to fail. Ultimately, the test data may be used to determine application rates necessary to achieve acceptable decay control compared to current commercially-accepted treatments.

L-joint Trials

Wood preservatives can be evaluated for above-ground decay resistance in a test that simulates the exposure of wooden windows or door. Wood pieces are machined with a mortise or a tenon and the pieces treated with the candidate preservative by pressure or vacuum impregnation, dipping, or brush application. Solvent based solutions, water based solutions, or both are evaluated (*7*). Each L-joint is assembled from separately treated mortise and tenon pieces. The treated assembly is coated with paint and the ends are sealed either with a water impermeable sealant or a corrosion resistant metal cap, or both. The L-joints are placed on racks and lined out for exposure. The joints are placed in the racks such that water can collect in the joint area, increasing the decay potential. The L-joints are inspected annually or bi-annually for decay. The joints are pulled apart and the joint area inspected and rated for decay. It is advisable to duplicate the L-joint test at several locations, since different climates will encourage different decay conditions. Hilo, Hawaii is a favored test site, since the average annual rainfall is in excess of 125 inches and the warm temperature results in an extremely high decay hazard. Florida, Georgia and southern Mississippi are also environments that encourage decay.

Lap Joint Trials

Lap joint trials are similar to L-joint trials in that they are designed to evaluate preservative efficacy in an above-ground exposure environment (8). The two pieces of a lap joint are overlapped and held together by a removable clip or clamp, enabling disassembly of the joint for inspection and evaluation. The lap joint wood components are left unpainted. Lap joints are also exposed in high rainfall, high decay environments. A criticism of the lap joint trial is that untreated control joints do not decay as rapidly as in other above ground tests.

Ground Proximity Trial

Ground proximity tests simulate the exposure conditions that ground level deck joists or a sill plate might encounter (*9*) . Wood pieces are placed on top of cinder or cement blocks and then covered with a shade cloth that will admit rain but reduce light intensity, thereby slowing the rate of evaporation from the enclosure. The porous nature of the cement blocks serves to wick up soil moisture and some nutrients, thus maintaining a high degree of decay potential.

Figure 1. L-Joint Samples.

Figure 2. Lap-Joint Samples.

Insects

There is an increasing need for insecticides for wood treatment. The use of organophosphates, like chlorpyrifos, for wood treatment have recently been restricted. Copper based wood treatments effectively protect wood outdoors from attack by insects, but as non-metallic systems are introduced protection from termite and other insects is needed.

Laboratory Tests

Laboratory tests have been developed for screening prospective termite control agents. As with the laboratory decay tests, they are designed to provide fast and reliable results. Depending on the test model, insecticide treated paper, wood or wood sawdust can be used, since termites don't typically distinguish between cellulose substrates. The treated substrate can be part of a choice or no-choice laboratory test.

The laboratory choice test offers the termites both treated and untreated cellulose sources (*10*). The termite can choose which to eat. If the termites make no distinction, i.e. eats equally from both, then the insecticide being tested is non-repellant. However, if the termites choose to eat only from the untreated substrate, then the termiticide being tested is repellant. Laboratory no-choice termite tests offer the termites only the insecticide treated food source (*10*). The termites must eat the treated food or die of starvation.

Termiticide Field Tests

Field Stake Test

The field stake test for termiticides may be conducted simultaneously with the field stake test for decay. The test methods are the same, except that separate evaluations are made for decay and for termite degradation. If the test is conducted solely for the purpose of evaluating termite protection, it is important to apply a decay protecting preservative that has no termiticidal affect and does not repell termites.

Lunchbox Test

Lunch box tests are a type of ground proximity test where the treated pieces of wood are contained within a weather protected box. Untreated feeder strips between the ground and the treated wood pieces in the box allow the termites to enter into the test chamber for feeding.

Ground Proximity Test

The ground proximity test for evaluating termiticides is very similar to the ground proximity test for decay control. Several key differences are that wood feeder strips, rising from the ground up into the enclosure, encourage termites to migrate up and feed within the treated wood samples. The cover over the test unit is solid, not porous. This provides the dark and humid environment preferred by termites.

Stake/plug Tests

This field termiticide evaluation employs treated wood plugs inserted into the top of untreated wooden stakes that are driven into the ground. As the termites feed on the wooden stakes, they eventually reach the treated plugs and feed on them. The degree of feeding on the treated plugs is then scored as

Figure 3. Lunch box termite field test.

Figure 4. Ground proximity termite field test.

termiticide efficacy. The degree of feeding on the untreated stakes indicates the level of termite pressure and will minimize instances where lack of termite feeding is presumed to be the result of the test treatment, when in fact there was no termite activity on that particular test unit. There is great degree variability in the field termite tests, since termites initially feed almost randomly. Eventually they will send chemical signals to the others in their colonies, but this takes time. If a test compound has repellant activity, there will be no chemical signals sent and feeding will be generally light, even in the untreated controls.

Non-Biological Characteristics

Once the biological activity of a candidate is confirmed in laboratory and/or field evaluations, the nature of certain physical-chemical attributes specific to wood treatment need to be ascertained. Preservative treatments may have some inherent characteristics that may make them unfavorable for treating wood, or that may cause subsequent problem to the treated wood.

Suitability for Use at a Treating Facility

Wood preservatives cannot degrade the treating equipment by being corrosive. Standardized corrosion tests for solutions to common metals and polymeric materials will identify if corrosion to equipment is likely. New wood preservatives also need to be shipped in concentrated form and then readily dilutable as appropriate for their application method, and these dilutions need to be stable over time. Treating plants prefer products that require minimal attention but remain effective.

Corrosivity of Treated Wood to Fasteners and Hardware

A wood treatment that causes treated wood to be corrosive to nails, screws, hanger brackets or other metal hardware is unacceptable. Mildly corrosive treatments may result in the slow degradation of fasteners over time (*11*). This could result in a hazardous structural integrity decline that may be unnoticed. Thus, corrosion rates of new preservative systems to fasteners and hardware are necessary. Efforts are being made to standardize corrosivity measurements.

Appearance of the Treated Wood

This is perhaps one of the most difficult aspects for new wood treatments. Most people are familiar green coloration from treatment with copper-based preservative systems. This color helps mask the gradual decline of the wood surface to a uniform gray color. Additionally, the chemical bonding of the metallic elements to the lignin in wood helps protect against photodegradation. Wood stains can also be used to impart a color to the treated wood. Only the most heavily pigmented treatments, high in solids, offer coloration that will last for more than a year or two. In fact, non-metal wood preservative systems have a tendency to cause wood surfaces to turn a non-uniform gray color relatively rapidly. Wood preservative developers need to keep in mind the expectations of the consumers for an attractive structure with the treated wood.

Development Time and Costs

In order to meet the requirements established by the consumer, the treating industry and the EPA, a considerable amount of data is required to support a new preservative candidate. It is important to distinguish between entirely new active ingredients and new preservative products comprised of already registered and recognized active ingredients. From a regulatory standpoint, new active ingredients for use as wood preservatives require a comprehensive set of human, animal and environmental safety tests. The composition and nature of the new chemical substance needs to be described. Once these are accepted by EPA and the product is registered as a wood preservative, developing new wood preservative products becomes much simpler and less risky from a business investment perspective.

It has been roughly estimated that the cost to register a new active ingredient as a wood preservative is $30 million and takes about 10 years from the time the candidate has been identified in the laboratory. The failure rate is high. Because of the complexity of the requirements a compound must fulfill, few satisfy all of them. One strategy to address the enormity of this challenge has been to identify compounds already registered with the EPA as agricultural pesticides and evaluate these biocides as wood preservatives. This strategy enables the developer to reference human, wildlife and environmental safety data that have been already approved by EPA to support the wood preservative use pattern. Product performance, suitability for use and worker and consumer exposure risk assessments still must be conducted. This strategy saves a considerable portion of the development expense; however, because of the nature of the long term field trials the time savings is not as great. Another benefit from this strategy is that one has a higher degree of certainty that the health, safety and environmental characteristics of the substance are acceptable to the regulators. The approximate cost to develop an already-registered agricultural compound as a wood preservative is $6 million, and the development time from laboratory to commercialization about 6 years.

This crossover strategy from the agricultural chemical arena has several inherent drawbacks. Generally, the agricultural pesticide markets command higher prices for their inputs. The wood industry is more commodity based and new products need to compete with older, relatively inexpensive chemistries. Also, most agricultural pesticides are not persistent enough, by design, to provide the multiple year service life of an effective wood preservative.

Because of the time and investment required and the uncertainty of the outcome and the low cost of the final product, developing new wood preservative compounds needs to be managed very carefully. Utilization of a stage-gate development process can help prevent potential investment in unsuitable candidates. The development stages need to be defined to align the key risk factors with the significant investment requirements. Confounding the stage-gate approach is the length of time needed to determine the most critical performance factors, the field exposure trials.

As described, the development of a new wood preservative system is lengthy and risky. Furthermore, developing a totally-organic system will be even more difficult than the prior metallic-based systems due to the difficulty of developing a waterborne treating solution since most organic biocides are not water soluble (but once in wood, it is desired that the biocide cannot be easily leached by water), also many organic biocides tend to be rapidly degraded by various microorganisms and/or environmental factors.

Organic Preservatives for U.S. Residential Applications

The current U.S. exterior residential preservatives are all based on the combination of copper with an organic co-biocide to control copper-tolerant fungi. However, as mentioned in other chapters in this book, the current 2nd-generation copper-rich systems have some concerns; principally, a relatively high level of copper leaching which can negatively impact aquatic ecosystems, and the question on the ultimate disposal of metallic-treated wood. For these reasons most professionals expect that totally-organic preservatives for U.S. residential applications will be required at some point in the future. Indeed, several European countries now mandate totally-organic systems. Organic systems have several inherent limitations, however, including: 1) most organic biocides are water insoluble, but treating solutions for residential applications will almost certainly be waterborne. Thus, emulsions or other technology will need to be developed, with the developer then needing to ensure that the waterborne system deposites the biocide uniformly and deeply into the wood product being treated, and that the biocide does not later readily leach out. 2) Metallic biocides are inert to biological, chemical or photodegradation, but organic biocides can be degraded by a variety of microorganisms, including non-wood degrading microorganisms such as bacteria and molds/stains, and are also degraded by sunlight and/or chemical pathways. 3) Most of the newer organic biocides are effective against only a few of the many organisms that degrade wood, and do not have the broad effectivness of the metallic biocides. Thus, most organic preservatives are/will be composed of a mixture of two or three biocides to ensure broad activity. 4) Copper is about $3/kg, while most organic biocides are about 10-fold more expensive. However, since many of these new organic biocides are highly effective compared to copper, the cost/effectivness ratios of copper vs. the new organic biocides are often comparable.

The organic systems that are currently available in Europe are composed of a mixture of biocides to ensure a broad activity, as mentioned above. The biocides were first employed as agrochemicals and then considered for their potential to protect wood once they were registered as discussed earlier. [One exception is the oligomeric preservative system PXTS, which is not discussed here.] A few of the European systems, or systems similar to commercial systems in Europe, are now becoming available in North America. Following are several examples of totally organic wood preservative systems for exterior residential applications currently available in North America, the biocides, and the approved applications. Additional systems are being considered or undergoing standardization [evaluation by an appropriate standard-setting organization]. Also, some of these systems that have been approved for limited applications, such as only millwork, are being considered for additional uses such as exterior above-ground applications.

Woodlife™ 111	propiconazole, IPBC, tebuconazole	Millwork
Wolman™ AG	propiconazole, tebuconazole, imidacloprid	aboveground decking, garden fences and furniture
Vacsol™ Azure	propiconazole, tebuconazole, permethrin	Millwork
Clearwood™ MW-2	tebuconazole, imidacloprid	Millwork

Because of the diverse chemical nature of the organic preservative active ingredients, the ability to bring complex mixtures of biocides with complementary biological activity into robust delivery systems remains a major challenge to formulation chemists. The formulation requirements of the organic active ingredients and their compatibility in concentrate and dilute conditions often precludes certain combinations. Quaternary ammonium compounds and iodine containing compounds are typically incompatible. Bethoxazin is insoluble in few solvents at concentrations above 3 to 5%. Insoluble compounds like bethoxazin or chlorothalonil may need to be delivered in a slurry or flowable formulation, but these can lead to penetrability problems with pressure impregnation application methods.

In contrast there are several examples where organic preservative delivery systems have enhanced the performance of the preservative by either physical of or biological mechanisms. For example, certain alkyl amine compounds when combined with triazoles have improved the efficacy of the preservative effect, purportedly by altering fungal cell wall permeability. Additionally these amines impart water repellency and anti corrosion to the treated wood. The potential for the use of antioxidants in organic wood preservative mixtures has shown to positively effect biological performance (*13*). The use of UV absorbers could be used to improve the biological efficacy of compounds subject to UV degradation. It seems reasonable that, as the practical experience with the various organic preservatives increases, the array of tools available to address the weaknesses of the individual components will increase.

Challenges for Organic Wood Preservatives

Traditionally, wood preservatives have been broadly adapted for all uses with little segmentation. A piece of treated wood could be used in ground contact, or aboveground, or coated or uncoated. Success of the organic wood preservatives is dependant on the use of different preservative systems for the different wood uses and corresponding exposure conditions. The market segmentation requires maintaining the identity of the treatment throughout the distribution chain. The use of the preservative treatment certainly adds value to the wood, but only if that value is retained through to the ultimate use of the treated wood.

Organic wood preservatives face challenges with the wood treaters who will need to adapt to treating wood with different products for residential and non-residential uses, or for ground contact and aboveground applications. The impact of UV degradation on metal free wood treatments may affect the performance of the preservative system, as well as the appearance of the wood in service. Organic preservative systems also require more sophisticated quality control and analytical techniques. The value of organic treatments with respect to their increased cost over metal based preservatives needs to be reconciled in the marketplace. Finally, all treated wood for use in outdoor residential exposures is being impacted by the steady increase in the market acceptance of plastic and wood-plastic composite building materials.

References

1. Consumer Products Safety Commission/EPA Post Interim Study on CCA-Treated Wood: News from Consumer Products Safety Commission. May 11, 2005.
2. Risk Assessment and Scientific Support Branch of the US EPA, OPP Antimicrobials Division, Human Exposure, Feb. 18, 2004, 10-16.
3. Nemhaueer, J.B. Arsenic and Old Wood: A medical toxicologists perspective. Canadian Wood Preservtion Association Proceedings, 2001, 294-324.
4. Morton, J. Wood Plastic Composites: Taking It to the Next Level. Proceedings of the WPC Conference 2004. Principia Partners, October 11, 2004.
5. American Wood-Preservers' Association Standard E10-01, American Wood-Preservers' Association Book of Standards, 2006.
6. American Wood-Preservers' Association Standard E7-01, American Wood-Preservers' Association Book of Standards, 2006.
7. American Wood-Preservers' Association Standard E9-97, American Wood-Preservers' Association Book of Standards, 2006.
8. American Wood-Preservers' Association Standard E16-98, American Wood-Preservers' Association Book of Standards, 2006.
9. American Wood-Preservers' Association Standard E18-04, American Wood-Preservers' Association Book of Standards, 2006.
10. American Wood-Preservers' Association Standard E1-97, American Wood-Preservers' Association Book of Standards, 2006.
11. Preservative Treated Wood, Technical Bulletin; Simpson Strong Tie, January 1, 2005, pgs 1-4.
12. Helmer, D.B., New Formulation Delivery Platform for Organic Wood Preservatives. Proceedings of the AWPA Annual Meeting, April 11, 2006.
13. Schultz, T., Nicholas, D. Environmentally Benign and Economical Totally Organic Wood Preservative Systems. Proceedings of the AWPA Annual Meeting, May 16, 2005

Chapter 24

Wood Preservative Formulation Development and Systems: Organic- and Inorganic-Based Systems

Mike H. Freeman

Independent Wood Scientist, 7421 Hunters Tree Cove, Memphis, TN 38125 (wooddoc@aol.com)

Wood preservatives lengthen the useful service life of valuable biologically-based building materials, thereby saving the consumers money and time, and minimizing the effects of deforestation and unnecessary consumption of perishable goods. The formulation of a wood preservative can greatly impact its performance, simplify application, and affect the materials employed to apply the system and construct the final treated wood product. This chapter discusses the formulation systems used for many wood preservation applications such as pressure treatment, anti-sapstain, general / DIY (homeowner applied) preservatives, millwork/joinery systems, and remedial treatments. Specific examples are given that have impacted the performance, efficacy, and ease of usage of both organic and inorganic-based systems. Use of inert ingredients as formulation aids is also briefly discussed.

Introduction

The efficacy of any pesticide, including wood preservatives, can be greatly affected by its formulation into a commercial product. Formulations play significant roles in how a biocide is available for both fungal and insect control as well as affecting the treated wood's aesthetic and physical properties including preservative leaching.

In general, wood preservative formulations contain actives that provide antifungal and/or insecticidal properties. In some cases blends of fungicides and insecticides or a single compound which is active against both fungi and insects may be employed, while in other cases activity against only decay fungi or insects is required.

Essentially all wood preservatives employ a liquid carrier as the solvent. For lumber and timbers destined for the exterior residential market, water is the preferred solvent for the actives. Most residential preservatives currently used in North America are water borne copper-rich systems, where additional amines or ammonia are used to dissolve the copper containing preservative and minimize corrosion in the treating plant. It is expected that at some point in the future totally organic systems that employ water insoluble agrochemicals will be required for exterior residential applications. For these systems the active will likely be formulated in water using sophisticated systems such as an emulsion or nanodispersion. Some millwork systems employ light organic solvents, but waterborne systems are growing more predominate. For treated wood products used in the industrial sector, except for the waterborne chromated copper arsenate (CCA) system the organic or organometallic actives are dissolved in petroleum oil solvents, where the oils often provide some additional protection to the treated wood. The few exceptions of nonliquid carrier preservatives include the liquid creosote where the active is the carrier, a few solid preservatives such as zinc borate employed for in-process treatment of wood composites, and solid rods for remedial treatment of in-service wood.

A variety of additives are also used in wood preservative formulations. The most widely used are paraffin based water repellents that are incorporated directly into water borne treatments as emulsions or dispersions. Corrosion inhibitors, surfactants, dispersants, stabilizers, colorants, and other performance enhancers are also used but typically these types of additives are in small amounts.

This chapter reviews formulation insights including current solvents and biocides, and formulation additives for many wood preservative systems. Included in this overview will be current policies, practices, biocides, additives and solvents employed, the author's personal experience in formulating many of these preservative systems and a historical review. Although all segments of the wood preservative market will be discussed, the sections are arranged in the approximate market size: pressure treating chemicals, anti-sapstain systems, millwork/joinery products, general preservative systems, and remedially treatment chemicals.

Pressure Treating Preservative Systems

Pressure Treating systems are the major wood preservatives, and have been employed to preserve wood since 1838 when John Bethel first patented the use of creosote for full cell treating of products such as railroad ties and timbers. Pressure treating chemical sales in the United States in 2006 alone exceeded $1 billion just for the water borne products, predominately inorganic copper carbonate and borax based biocides. Other pressure treating chemicals include organic systems such as pentachlorophenol in oil, creosote, and creosote-petroleum systems. Interestingly, the wood preservatives used in the pressure treating arena amount to 75% of the total poundage of pesticides used in North America.

The typical goal for pressure treating systems is to have an easy-to-employ system that will penetrate most wood species to protect the sapwood from decay and termites. Pressure treated wood preservatives must also be reasonably inexpensive or of good value for the money, easy to analyze in both solution and treated wood, and safe for all producers and users.

The typical commodities which are pressure treated include lumber, timbers, utility poles, cross ties, piling, fence posts, fence boards, and plywood. The American Wood-Preservers' Association (AWPA) Use Categories UC1 to UC5 list over 500 various commodities which may be commercially pressure treated. In addition to wood preservation, many fire retardant chemicals are pressure treated into wood to hinder burning, glowing or flaming/cindering.

Pressure treating biocides fall into two categories. The restricted use pesticides (RUPs) are generally biocides where use at the treatment plants presents certain risks that require special knowledge for addressing so the operator must be licensed. Among the RUPs are chromated (CCA) and ammoniacal zinc copper arsenate (ACZA), other inorganic arsenicals, pentachlorophenol and creosote. The general use pesticides such as alkaline copper quat (ACQ), copper azole (CA), oxine copper, borax or DOT, and oil or water borne copper naphthenate are not considered as risky and no special use precautions are made with these systems.

Pressure treating chemicals can be further broken down into three dominate sub-categories: organic, metallic and organo-metallic systems. Organic biocides and solvents used for pressure treating for industrial applications in the USA include pentachlorophenol dissolved in AWPA P9 Type A oil or in mineral spirits and creosote either used as a stand alone AWPA P1/P13 creosote or creosote dissolved 50/50 in petroleum oils. The oils range from slurry oils to #5/#6 oils to Bunker C oils/slurry oils or HEGO oils and the resulting treatment solutiuon systems is called creosote - petroleum solutions. The metallic or organometallic systems include CCA, ACZA, ACQ (alkaline copper quat), and CA (copper azole), which are all water borne, copper naphthenate which is both water and oil borne, the water borne inorganic borates (predominately DOT), oxine copper which is difficult to formulate but available in both oil and water borne systems, and various non-biocidal water borne fire retardant formulations that contain borates, or nitrogen- and/or phosphate compounds.

Solvent Differentiation in Pressure Treating

Water Borne Systems

Most of the pressure treated systems in the United States today are water borne systems, and essentially all treated lumber used for residential exterior applications, the largest market, are water borne. The two systems in largest use, ACQ and CA, are based on copper with a small amount of an organic co-biocide to control copper tolerant fungi. Copper is the cheapest biocide per pound available today with the copper solubilized by employing amines, ammonia or a combination of these. (In addition there are acidic systems such as CCA and Acid Copper Chromate.) However, the added nitrogen stimulates mold growth on the treated wood which has to be further reduced by including a moldicide, typically an isothiazolinone (ITA), to the water based amine or ammonia system. The author and Pascal Kamdem of Michigan State University have recently experimented with tertiary and secondary amines and other solubilizers to minimize the use of moldicides such as isothiazolinones which are known skin sensitizers.

The ammonia formulated copper systems have some advantages in that they will partially penetrate the heartwood of pines and the sapwood of refractory species such as Douglas fir. These systems are extremely effective penetrants. Ammonia formulated systems do have residual odor thus minimizing their interior use until the ammonia is thoroughly evaporated from the treated wood product, and some mold growth occurs on the treated wood since ammonia also contains high percentages of nitrogen. There is also some splotching of the treated wood, due to the d9 co-ordination complex turning from a bright blue to a light green after water/ammonia has evaporated from the system and is re-wetted either by rain water or liquid spillage onto the treated wood. Fixation is usually considered complete after the ammonia evaporates from the treated wood. These ammonia systems are usually needed for biocide penetration in almost all Douglas fir treatments with the exception of Copper Xyligen and water borne copper naphthenate, which appear to treat Douglas fir well without ammonia.

All the major wood preservative suppliers in North America also offer water borne formulation that consists of a copper-based preservative with a wax added for water repellency. The wax is emulsified and provides enhanced dimensional stability for lumber for exterior above ground applications such as decking. This benign additive is very economical, safe, and also lessens the decay potential of wood and reduces biocide leaching.

Nano-particles or "micronized" systems are actively being used in wood preservation today, especially for copper based systems where the copper, usually basic copper carbonate, is micronized or reduced to nano-particle size of 0.2 microns or less. These are usually coupled with a co-biocide such as an azole or quat. These micronized systems offer many advantages over the alkaline based systems which use monoethanolamine (MEA) or ammonia to solubilize/chelate the copper including reduced corrosion of treating plant equipment and metal fasteners for building the final product such as decks or fences, reduced mold growth on the treated wood and less copper leaching. For example, recent tests have shown that micronized copper leaches ten times less that amine solubilized systems and have at least one-tenth of the corrosion to mild steel fasteners and virtually no corrosion to hot dipped galvanized fasteners.

Totally organic systems that contain water insoluble organic biocides for exterior residential applications will likely be formulated using nano particles dispersed in water or emulsion systems. Some of these systems are already available in Europe and similar systems are being developed for the likely future requirement for total organic systems for the residential North America market.

Interestingly, micronized systems are not new to wood preservation. This author personally started investigating micronized systems in 1988 when the author found that oxine copper dissolved in dodecylbenzene sulphonic acid and then neutralized with borax or other borates would form a pearlescent non-settling compound. This compound, when investigated either by a Coulter Counter or a laser light scattering device, had sub-micron particle sizes and was able to penetrate wood using normal pressure treating cycles such as Rueping, Lowry, Bethell and modified full cell processes. In addition, the author found that the two component wood treating system known as CDDC or Kodiak formed a water insoluble brown complex from either copper sulfate or copper amines as the copper source reacted with sodium dimethyl dithiocarbamate (SDDC). This water insoluble complex also had a sub-micron particle size but would settle. The author investigated passing this material through devices such as tissue homogenizers and attrition mills (attritors) to further reduce the particle size. Adding anti-caking and anti-settling compounds resulted in an effective pressure wood preservative. Chlorothalonil was also investigated since it was commonly micronized for the agricultural industry to give a particle size ranging from less than 0.1 micron to approximately 100 microns. Treating wood with formulations of chlorothalonil ground to commercial agricultural size, however, only resulted in limited penetration in southern pine due to blockages caused by the larger particles.

Today roughly fourty-five percent of the industry utilizes micronized copper instead of amine based systems to produce residential use pressure treated wood. The predominate species is southern pine but other species include Ponderosa and Radiata pines and incised Hem-fir.

Oil Borne Systems

Typically, organic and organo-metallic systems use AWPA P9 Type a Oil (#2 Fuel Oil), AWPA P9 Type C solvent (mineral spirits or VM & P Naptha), Number Six Fuel Oil or Bunker C Oil, or specialty organic solvents. Oil or solvent borne systems are usually used for industrial products where exudates and odor are not as much a problem as in the residential market. In addition, heavy oils enhance the efficacy of the actives. Creosote is often a stand alone (no solvent carrier) wood preservative.

For organics, creosote is used mainly for railroad crossties and certain regional markets of utility poles and piling. However the dominant utility pole treatment is pentachlorophenol in oil. The author has spent over two decades doing specialty formulations of pentachlorophenol type systems for this market. These are usually based on carbonyl compounds such as ketones or ketone byproducts and water insoluble alcohols such as kettle bottoms or still bottoms from the production of 2-ethylhexanol. In addition, pentachlorophenol is readily soluble in still bottoms and kettle bottoms of cyclohexanone as well as MiBK so that 40% to 50% concentrates can be produced, shipped to the pressure treater and then further diluted with Number Two Fuel Oil, P9 Type a Oil, or equivalent.

Pressure Treating Problems

There are certain problems which must be addressed with all pressure treating chemicals including oil borne solutions. If not correctly formulated the solutions can entrain wood sawdust, wood sugars, and inorganic debris which can cause sludging and/or emulsion formation. Also, particles in the solution will hinder penetration, leaving the untreated sapwood prone to decay. Additionally, the solubility of the biocide in the carrier system must be maintained. If the biocide is a crystalline product such as pentachlorophenol, it can recrystallize out of solution leaving the treated wood susceptible to attack. Thus, complete solubility of any biocide in the system is essential even though most organic biocides can not penetrate the wood cell wall. In addition, treating solutions must be relatively stable with continued re-use or penetration problems may occur with the treated wood. Problems that have occurred with water borne copper systems, including those which utilize a secondary biocide as an emulsion additive, include selective absorption, solution stability, penetration problems, and sludging/particle growth problems.

Anti-Sapstain Systems

Anti-Sapstain systems have been around for three to five thousand years. There is historical evidence that Egyptians used powered Natrum to prevent mold growth and stain organisms from infecting sheets of papyrus. In the Shang Dynasty in China some thirty five hundred years ago workers employed wood ash dissolved in water to prevent mold and mildew growth on wood for chopsticks. The goal in any anti-sapstain and mold control system is to prevent sapstain and mold organisms (fungi) from discovering and infecting green or unseasoned wood. Anti-sapstain chemicals often called prophylactic coatings.

Almost all biocides employed are dissolved or dispersed in water and applied to green lumber before air- or kiln-drying. Depending upon the application method, many different avenues of formulating an anti-sapstain chemical exist. Briefly, application methods include use of dip tanks either as bulk or inline dip tanks, spray booths which may be sprayed either laterally or inline, flood coaters and electrostatic spray coating.

In dip tank application, the volume of water is anywhere from fifty to two hundred times the level of the anti-sapstain concentrate for the anti-sapstain biocide. This allows for solutions of efficacious molecules to be applied to a relatively high surface area of green wood. In bulk dip tanks, typically the logs have been cut into board length material or flitches or veneers and are stacked on an automated lowering and raising device like a forklift mechanism and dipped for anywhere from thirty seconds to three minutes into a large volume tank which has the overflow capacity to hold both the lumber and the treating solution. Inline dip tanks take the wood that is coming linearly or laterally from the sawmill and, before further machining, quickly dip the material into an aqueous solution of the biocide dispersed or solubilized in water.

Spray booths, either lateral or inline, are exactly as defined by the term. In these applications dispersions or true solutions of biocides in water are sprayed as an ultra-fine mist onto all six surfaces of the board before it is dried. Flood coaters, typically called rain shower booths, have been used for years but are now falling out of favor due to environmental exposure to workers. Flood coaters basically applied the diluted anti-sapstain chemical in a ribbon-like feed where the wood was passed under the flood coater and coated on all four large surface sides with some ingrain coating as well. Electrostatic coverage usually involves first coating the wood with an ionic compound or solution so that the wood surface could carry a charge, or if the wood has sufficient moisture content charging the surface of the wood and then putting a counter ionic charge on the spray solution to allow coverage of the wood surface regardless of the application angle from any spray head.

Today's anti-sapstain preservative systems are predominately water soluble or water dispersible concentrates. The pH of today's systems can range from neutral to acidic or alkaline. Alkaline systems have typically been employed in the past to minimize iron stain due to iron contacting the green wood and later forming iron tannate complexes, as well as minimizing corrosion in the mild steel used to manufacture anti-sapstain application equipment.

Solvents

Water is the dominant solvent in use today due to economic and safety reasons and ease of application. In addition to water, there can be water dispersible co-solvents, cationic or anionic emulsifiers, alcohols, glycols, esters, and dispersing agents.

Biocides

Historically mercuric compounds, toxic to both man and fungi, launched the "modern" anti-sapstain work some 200 years ago. These mercury formulations included water soluble salts, usually phenol mercuric acetate or phenol mercuric lactate, with a lignosulphonate emulsifier and a co-solvent.

In the 1930s the chlorophenate compounds developed for wood preservation gained prominence for mold and stain control. Furthermore, synergism was found when chlorophenates were combined with borates. These systems also minimized iron stain on the wood, prevented corrosion in mild steel dip tanks and resulted in greater efficacy against decay organisms. Additionally, the small amount of free chlorine in the chlorophenate systems acted as a wood surface bleaching agent making the wood surface "brighter" to enhance the appeal of the treated wood. The only problem was that some *Cepheloascus* fungi were tolerant, but it was found that adding a trace of a mercuric compound to the chlorophenate formulation controlled the complete spectrum of mold and stain organisms.

Chlorophenates, however, began to be reviewed in the 1980s due to the negative publicity concerning compounds such as agent orange, dioxins, and other phenoxy herbicides. New biocides for the anti-sapstain began to replace the chlorophenates which had been used for over half a century.

The first launch into non-chlorophenate anti-sapstain control was based on oxine copper (copper-8-quinolinolate, or copper-8 or Cu-8). Patents were granted worldwide for the use of oxine copper dissolved in dodecylbenzene sulphonic acid (DDBSA). Work by West and Nagel and others found that oxine copper dissolved in DDBSA was a very robust and effective system for anti-sapstain control. Formulations based on this technology still exist today.

Further work on non-metallic systems included the investigation of quaternary ammonium compounds (quats) and carbamates. Included in these is a formulation known as NP-1 that is a mixture of IPBC and DDAC (3-iodo-2-propynyl butyl carbamate and didodecyldimethylammonium chloride). Additional investigation into phenolic based compounds included uses of OPP (*ortho*-phenoxy phenol) and Sodium OPP, commonly referred to Dowicide 1 and Dowicide A. These phenolic formulations have certain side effects which limit their use even though they are effective biocides.

Additional work into the investigation of mold and mildew control for lumber, based on testing of fungicides for the leather and pulp and paper industries by Stanley Buckman, included formulations based on TCMTB and MBT (2-(thiocyanomethylthio) benzothiazole and methylene *bis*-thiocyanate). These formulations, although extremely robust, had certain negative implications concerning worker exposure such as dermal and eye irritation.

Studies in America, New Zealand and Australia also found that certain co-biocide combinations, including those of oxine copper and carbendizim, or oxine copper and chrolothalonil (CTL), proved to be synergistic and very effective anti-sapstain and mold control combinations. These products are currently marketed under the trade names Nexgen and Hylite Extra, respectively. Additionally, a formulation innovation by West, Freeman and Accampo dissolved certain biocides in high concentrations in amine phosphonate systems. These amine phosphate systems, currently called The StaBrite System, could dissolve biocides such as IPBC, and the azoles azaconazole, propiconazole, and tebuconazole. The largest commercial system is the StaBrite P System, a formulation of IPBC in the patented amine phosphonate carrier. It was later discovered that the amine phosphonate carrier system was also an excellent bleaching compound for the surface of the wood to give a brighter appearance and greater appeal as well as inhibited corrosion with the mild steel dip tanks and removed saw marks and iron tannate stains from the wood surface. Additional patents were given for the amine phosphonate system as a hard surface cleaning agent, very similar to the patents issued for quaternary ammonium compounds for the same application.

The leading biocide formulations in today's anti-sapstain market include NexGen (MBT and CTL), NP-1 (IPBC and DDAC), Busan 1009 (TCMTB (2-(thiocyanomethylthio) benzothiazole, also called Busan 30) and MBT (methylene *bis*-thiocyanate)), MicoStat P (propiconazole and DDAC), BriteWood XL, BriteWood S, BriteWood P, and a new entry called Bazooka. In addition, new environmentally formulations launched in the last 24 months include formulations of sterol inhibitor-based chemicals that usually employ propiconazole or propiconazole/tebuconazole/IPBC combinations such as Premier and Anti Blu XP. Additionally in North America, particularly Canada, formulations of sodium carbonate mixed with borax and marketed under the name F2 have market niches where environmental concern for water run off from the treated wood and worker exposure give it a semi-permanent market position.

To recap, the biocides used in anti-sapstains are:

- Oxine Copper
- IPBC
- DDAC
- Isothiazilones
- TCMTB/Busan 30
- MBT
- CTL
- Propiconazole
- Tebuconazole
- OPP
- NaOPP
- Borax
- Sodium Carbonate

Formulation Additives

Formulation additives are typically used to add a particular trait or physical property to the biocide system to mitigate effects of the primary biocide system on operating equipment, enhance the efficacy of the actives, or improve the wood properties such as brightness and color. Borax, the historical anti-sapstain additive, is typically used to make the system alkaline to prevent iron stain from occurring on the wood, darkening of the heartwood, particularly in Douglas fir, and minimize corrosion to the mild steel equipment used in the treating process.

Substituted phosphonic acids began to gain wide appeal in the early 1990s when it was found that these products were inexpensive and would chelate iron and iron compounds into colorless or white coordination complexes which would not stain or discolor the wood. These substituted phosphonic acids chelated the iron from the solution and, when the phosphate content was high enough, would actually act to brighten the surface.

Emulsifiers and dispersants keep active ingredients concentration uniform and dispersed in both the concentrate and the ready-to-use (RTU) solutions.

These emulsifiers and dispersants, historically based on both anionic and cationic compounds which lead to selective absorption of the biocide on the wood, are now sometime coupled with other chemical formulation additives which include distinct surface active ingredients to prevent hard cake settling in dip tanks and ensuring uniform distribution of the biocide on the wood surface.

Although not considered an anti-sapstain organism, insect attack, especially to hardwoods, is of growing concern. It is known that insect movement can lead to the spread of sap and mold fungi from lumber piece to lumber piece or stack to stack. Therefore insecticides have been added to anti-sapstain formulations to prevent attack and infestation by Lyctid, Anoobiid. and Ambrosia Beetles and termites for over 70 years. Historically, the largest insecticide used for this purpose was Lindane (*gamma* hexachlorobenzene). Today's market includes other insecticides such as Chloropyrifos, Imidicloprid, and Permethrin.

Water repellents have been added to Anti-Sapstain chemical formulation to minimize checking and splitting while the wood was drying or seasoning, which further made the anti-sapstain chemicals more effective by not opening up untreated wood to new exposure sites.

New systems now generally contain a diffusible biocide plus a biocide that stays on the wood surface. These new systems are compatible with currently available insecticides, are stable in solution and the biocide concentration is easy to analyze both on the wood and in solution. In addition, new biocides can be easily detected on the wood surface by employing a presence indicator that can be sprayed or applied to the wood surface. The products in today's market provide cost effective, efficient, and safe yet effective and efficacious protection against stain and mold organism's infestation to green wood prior to seasoning. Although formulation improvements continue in the anti-sapstain market, the biocides will likely continue to be agricultural fungicides which have broad spectrum effectiveness and an extensive environmental data package.

Millwork and Joinery Systems

The principal goal in millwork and joinery systems is to deliver an active fungicide, or active fungicide and insecticide combination along with a water repellent to give surface mold, decay and insect protection and provide excellent water repellency and dimensional stability to the product. The secondary goal of a millwork or joinery system is surface preparation for future coating, staining, or cladding.

Historically, millwork systems used 5% pentachlorophenol (also called penta) in mineral spirits with additional wax, or mercurial compounds which were oil or mineral spirits soluble, with these systems employed for 70 years. However, concern arose for use of pentachlorophenol in this application during the US EPA RPAR Hearings of the late 1970s and the early 1980s, and mercury derivatives were also restricted. Additional work led to tributyltin oxide (TBTO) and tributyltin naphthenate some three decades ago, and although use of these compounds has greatly reduced they still maintain some market share. Following the conversion of pentachlorophenol from a general use to a restricted use pesticide in the late 1980s, millwork and joinery manufacturers decided that they would prefer to employ a general use pesticide rather then to obtain Certified Applicator licenses through Certified Applicator Training. This forced the chemical formulators to look at non-chlorophenate systems.

Biocides

Millwork biocides currently employed include:

- IPBC
- Oxine Copper
- Propiconazole
- Tebuconazole
- TBTO
- TCMTB
- Chlorpyrifos
- Permethrin
- Imidicloprid
- Bifenthrin

Since many of today's fungicides are non-insecticidal, an insecticide must be added to millwork systems to prevent insect attack. These include Lindane, Bifenthrin, Chloropyrifos, Permethrin, Pyrethrums, and Imidicloprid. Historically, Dieldrin, Aldrin, Chlordane, and Esfenvalerate have also been used. Recent studies on the insecticide compound Bifenthrin continue to show promise as this molecule is active at dosage rates 100-1000 times less than other insecticides in the same generic class.

The fungicide and insecticide combinations currently used are soluble or easily dispersible in mineral based systems, or can easily be converted to a water dispersible or water emulsifiable system through the use of co-solvents and emulsifiers. Of the 64 plants in North America currently treating with millwork preservative systems, IPBC is the major biocide employed for millwork. Sold under a wide variety names such as Wood Life, Mitrol IPBC, and DAP Ready-to-Use, IPBC has replaced 90% of the market that was formally pentachlorophenol and new formulations of IPBC and co-biocides continue to emerge. For example, a new formulation containing IPBC coupled with propiconazole and tebuconazole in a mineral based system that contains paraffin wax is marketed under the name Wood Life 111. Wood Life 111 takes advantage of the synergism between propiconazole and tebuconazole by adding them to the broad spectrum of decay protection of IPBC. Although IPBC will break down into iodine-based compounds under high heat or alkaline conditions, such as those typically those found in cladding operations, studies have found that the azole compounds in these multi-biocide systems remain effective.

Solvents

The major solvent currently used for millwork and joinery systems continues to be mineral spirits. Although over half a dozen different versions of mineral spirits are available, by far the largest is Rule 66, a low odor mineral spirits where the aromatic content is 2% or less. For millwork manufactures who want to quickly evaporate the solvent from the system, the mineral spirits can be readily replaced by VM & P Naphtha (Lacolene). Some millwork preservative systems employ water as the carrier and include wax for dimensional stability and block co-polymers to further give the wood dimensional stability and minimal grain raising during the application and re-drying process.

Some additives are included in the millwork and joinery systems. The most common is approximately 0.75% w/w of a ready-to-use solution of paraffin wax. And although many paraffin waxes are used, most common are those with an approximately 104 – 115 °F melt point and a medium to moderate needle penetration value. Some paraffin waxes have been replaced by small quantities of microcrystalline wax or hydrophobic polymers. Other additives to interior systems include trace levels of aromatic solvents in which an effective insecticide has been added.

Additional Factors

Additional formulation developments generally involve enhancing dimensional stability and water repellency. Although water repellency may be initially thought to be quickly and easily determined by simple test methods such as contact angle of water droplets on the wood surface, other investigations found that long term swellometer tests are more informative for determining the long term dimensional stability of the treated wood product. Automated swellometers coupled with computer systems have also shown that if dimensional stability is maintained for 100 hours, typically the product will perform well in service.

Additional tests on millwork formulated systems include simple paintability tests such as the Hatch Mark Test and analytical procedures for active ingredient determination for both penetration and retention, typically in the endgrain of the treated commodity.

New innovations in the millwork and joinery systems include pressure treating the commodities with water based system that includes a water repellent, allowing the wood to dry and then machining the wood product to its final dimensions. Innovative work in this area has recently been incorporated by the Jeld-Wen Corporation using their Aura Last Preservative System. However, there have been some complaints of increased corrosion to metal fasteners and other metal parts used in window manufacturing. Additionally, new water based preservative compounds such as Wolman AG (or Wolman L-3), a mixture of tebuconazole, propiconazole, and imidacloprid, for above ground uses of pressure treated wood, may allow treated wood to be used in millwork and joinery applications if dried after treatment and machined into place, similar to the Aura Last System. Although millwork and joinery preservatives, often called the LOSP (Light Organic Solvent Preservative) systems in Australasia and Europe, have been around for over 100 years, research continues to improve these systems to provide longer lasting products for the homeowner in the future.

General Preservative Systems

General Preservative Systems are those wood preservatives, biocides, and insecticide formulations which are best described by their name. These products are used as 'do-it-yourself' products by home owners and small industry users. These products employ general use pesticides to ensure their availability to over-the-counter purchasers and applicators without a certified pesticide license. Typical applications include brush-on, soak, dip, and spray on systems. The goal of general wood preservative systems is to provide protection against mold, mildew, decay fungi organisms and possibly insect protection when pressure treated wood is not available or not desirable. The secondary goal is that the products be easy to apply, relatively safe to workers and homeowners, easy to clean up, and employ general use pesticides. The ability for the treated wood to be painted, i.e. have paintability and stainability, are also desirable properties.

Biocides

With the exception of inorganic borates such as DOT (disodium octaborate tetrahydrate), most of the general wood preservatives are organo-metallic complexes or organic chemicals. These include copper naphthenate, both oil and water borne, IPBC, chlorothalonil, oxine copper, Folpet, zinc naphthenate, and quaternary ammonium compounds (quats) such as DDAC.

Solvents

The most common solvent for general preservatives are mineral spirits, which is usually of the variety of Rule 66 Low Odor Mineral Spirits which has a low aromatic content to minimize worker exposure as well as odor of the treated wood product. In addition, mineral spirits plus paraffin or microcrystalline wax is often used when water repellency is desired. Other solvents include AWPA Type P9A Oil (#2 Fuel Oil or equiv.) and recently biodiesel (methyl soyate). When a drying oil is desired typically tung oil, linseed oil, or their combination is employed. Water plus either surfactants or emulsifiers are also used, sometimes with polymers or waxes to impart water repellency to the treated wood.

General preservative systems typically have a few formulation specifics which they must meet along with being efficacious. These include easy clean-up, relatively safe for any exposure, and non-phototoxic so terrestrial and non-terrestrial plants are not affected if a spill occurs. These systems should also be compatible with almost all container types and should be tested accordingly to make sure they are compatible with typical containers which a home owner would use to store these products. In addition, the product should be shelf stable and have a uniform distribution both in the concentrate and in the ready-to-use solution during storage and immediately before application.

Remedial Treatment Systems

Remedial Treatment Systems are used to arrest decay and insect attack in both untreated wood and in wood where the original preservative has been depleted to below threshold value permitting attack by insects and decay organisms. Thus, the goal of remedial treatment systems is to extend the useful service life of a treated or previously untreated wood member, with the secondary goal to stop or eliminate any current attack by fungi or insects.

Remedial treatment chemicals used previously include many historical biocides such as pentachlorophenol, creosote, sodium dichromate, arsenic acid, borax or boric acid, sodium fluoride, copper naphthenate, lindane, and chloropyrifos as well as fillers/carriers like clays and thixotropic (water thicking) agents. Biocides currently used include copper naphthenate, copper hydroxide, boric acid, borax, sodium fluoride, and DOT. Remedial treatment systems generally fall into four predominately large categories; Thixotropic Gels or Pastes which may be either oil or water borne, Fumigants, Solid Rods, or Liquid Injectable Systems.

By far the largest category is the Thixotropic Gels or Pastes. The most effective remedial gels/pastes contain one agent that closely adheres to the wood surface, such as the base of the utility pole where a groundline decay can occur most readily, and a second agent which diffuses deeply into the wood, possibly even into the previously untreated heartwood. Products sold in this market over the years include many of the products under the Pole Nu or the Osmoplastic name usually followed by a number or letter such as Pol Nu CuRAP 20 or Osmoplastic B, Osmoplastic F, or Osmoplastic CF.

Fumigants in this area include products such as vapam (Metam Sodium) usually employed as a 32.7% water borne active ingredient concentrate, chloropicrin, sulfuryl fluoride or MITC (methyl isothiocyanate). In addition, solid rods such as IMPEL rods, FluroRods, or Cobra rods are also used as both a preventative and also as a curative in the remedial treatment arena. Rods have had some complaints against them in that they sit as a reservoir of wood preservatives until the treated wood has a high enough moisture content to dissolve the rod but that means the high moisture content can also promote and sustain decay.

In addition to solid rods there are also solid fumigants which are a sub-category of both the Fumigant Category as well as the Solid Chemicals Category. These include such new fumigants as Basimid (Dazomet) which produces methylisothiocyanate upon reaction with wood acids, as does Vapam and Mitc-Fume. Some researchers have found that the addition of an acidic copper compound, like copper sulfate, dissolved in water and co-introduced into the same application site as Basimid increases its toxicity and stoichiometric yield.

Liquid Injectable Systems are used to treat wood voids. Historically the biggest product used in this market for this application has been a product produced by Osmose called Hollow Heart. However, Hollow Heart contained significant amounts of toxic chemicals and has been replaced by a new environmentally friendly formulation called Hollow Heart CF which is a mixture of Copper Naphthenate and Sodium Fluoride. These liquid injectable systems are usually placed into wood voids after drilling the wood to determine if a void exists in the interior and injected under low pressure to fill the void to arrest decay and insect attack.

Summary and Conclusions

Wood preservative use has grown in the last 170 years into a modern, well segmented, complex and sophisticated market which extends the usable service life of wood thus saving the homeowners or other consumers' money and time while reducing harvesting and deforestation in the world. One should remember that chemical treatment costs pennies but saves dollars if the product is formulated correctly, and that toxic biocides can be safely handled if adequately formulated.

Specific goals for new wood preservative systems include good efficacy against the wide variety of oranisms that can degreade the wood product in its intended application, ease of formulating, ease of analysis in both the solution and the treated wood, good penetration into the wood, and safe for the environment, human applicators and final users. They must also have a good cost-to-benefit ratio and be economical in the long run for the producer. New wood preservative formulations now being examined are mostly based on agriculture organic biocides for employment in the large residential lumber market. These biocides are not water soluble so will have to be formulated using new formulation technologies such as nano-particle dispersion and emulsion systems. Emulsion systems recently developed do not have the faults of older emulsions such as water dispersible pentachlorophenol.

These new nano-dispersion and emulsion systems are already in the marketplace or going through the AWPA and/or the Building Codes (ICC-ES) for listing. Finally, new systems are essentially "cocktails" because many of the new agricultural chemicals being considered have a limited spectrum of activity, so biocide mixtures are used to obtain the broad fungicide and insecticide protection necessary for the long service life of treated wood. This means that formulation complexity may be increased.

Formulation is extremely complex and this chapter only provides an introduction. It is recommended that any proposed formulation change involve input by both a formulation chemist and wood preservationist, along with other professionals, to ensure that any of the many possible negative impacts will be properly addressed.

Suggested Further Reading

Anonymous. STP 752.Vertebrate Pest Control and Management Materials. ASTM International, ASTM Press, West Coshocton, PA, 1981

Anonymous. .STP 764.Pesticide Tank Mix Applications
Editor(s): Wright; Lindsay; Sawyer. ASTM International, ASTM Press, West Coshocton, PA, 1982.

Anonymous. STP 795. Pesticide Formulations and Application Systems. Editor(s): K G Seymour. ASTM International, ASTM Press, West Coshocton, PA, 1983.

Anonymous. STP 828. Pesticide Formulations and Application Systems: 3rd Symposium. Editor(s): Kaneko; Akesson, ASTM International, ASTM Press, West Coshocton, PA, 1984

Anonymous. STP 875. Pesticide Formulations and Application Systems. Editor(s): Kaneko; Spicer, ASTM International, ASTM Press, West Coshocton, PA, 1981

Anonymous. STP 915. Pesticide Formulations and Application Systems: Fifth Symposium.Editor(s): Spicer; Kaneko, ASTM International,ASTM Press, West Coshocton, PA, 1986.

Anonymous. STP 943. Pesticide Formulations and Application Systems: Sixth Volume. Editor(s): Vander Hooven; Spicer, ASTM International,ASTM Press, West Coshocton, PA, 1987.

Anonymous. STP 968. Pesticide Formulations and Application Systems. Editor(s): Beestman; Vander Hooven, ASTM International, ASTM Press, West Coshocton, PA, 1987.

Anonymous. STP 980. Pesticide Formulations and Application Systems. Editor(s): Hovde; Beestman, ASTM International, ASTM Press, West Coshocton, PA, 1989.

Anonymous. STP 1036. Pesticide Formulations and Application Systems: International Aspects.Editor(s): Hazen; Hovde, ASTM International,ASTM Press, West Coshocton, PA, 1989.

Anonymous. STP 1078. Pesticide Formulations and Application Systems. Editor(s): Bode; Hazan; Chasin, ASTM International, ASTM Press, West Coshocton, PA, 1990.

Anonymous. STP 1112. Pesticide Formulations and Application Systems: 11th Volume. Editor(s): Bode; Chasin, ASTM International, ASTM Press, West Coshocton, PA, 1992.

Anonymous. STP 1146.Pesticide Formulation and Application Systems. Editor(s): Devisetty; Chasin; Berger, ASTM International, ASTM Press, West Coshocton, PA, 1993

Anonymous. STP 1183. Pesticide Formulations and Application Systems, 13th Volume.Editor(s): Berger, Devisetty, Hall, ASTM International, ASTM Press, West Coshocton, PA, 1993

Anonymous. STP 1234. Pesticide Formulations and Application Systems: 14th Vol. Editor(s): Hall, Berger, Collins, ASTM International, ASTM Press, West Coshocton, PA, 1995

Anonymous. STP 1268. Pesticide Formulations and Application Systems: 15th Volume. Editor(s): H.M. Collins, F.R. Hall, M. Hopkinson, ASTM International, ASTM Press, West Coshocton, PA, 1996

Anonymous. STP 1312. Pesticide Formulations and Application Systems: 16th Volume. Editor(s): Hopkinson, Collins, Goss, ASTM International, ASTM Press, West Coshocton, PA, 1997

Anonymous. STP 1328.Pesticide Formulations & Application Systems: 17th Volume.Editor(s): G.R. Goss, M J. Hopkinson, H.M. Collins, ASTM International, ASTM Press, West Coshocton, PA, 1997

Anonymous. STP 1347. Pesticide Formulations and Application Systems: 18th Volume. Editor(s): Nalewaja, Goss, Tann, ASTM International, ASTM Press, West Coshocton, PA, 1998

Anonymous. STP 1373. Pesticide Formulations and Application Systems: Global Pest Control Formulations for the Next Millennium, 19th Volume. Editor(s): R. Scott Tann; John D. Nalewaja; Alan K. Viets, ASTM International, ASTM Press, West Coshocton, PA, 1999

Anonymous. STP 1400. Pesticide Fomulations and Application Systems, 20th Volume. Editor(s): Alan K. Viets; R. Scott Tann; Jane C. Mueninghoff, ASTM International, ASTM Press, West Coshocton, PA, 2001

Anonymous. STP 1414. Pesticide Formulations and Application Systems: A New Century for Agricultural Formulations, 21st Volume. Editor(s): Jane C. Mueninghoff; Alan K. Viets; Roger A. Downer, ASTM International, ASTM Press, West Coshocton, PA, 2001

Anonymous. STP 1430. Pesticide Formulations and Delivery Systems: Meeting the Challenges of the Current Crop Protection Industry. Editor(s): Roger A. Downer; Jane C. Mueninghoff; Gregory C. Volgas, ASTM International, ASTM Press, West Coshocton, PA, 2003

Anonymous. STP 1449. Pesticide Formulations and Application Systems: 23rd Volume. Editor(s): Greg Volgas, Roger A. Downer, Humberto B. Lopez, ASTM International, ASTM Press, West Coshocton, PA, 2003

Anonymous. STP 1460. Pesticide Formulations and Delivery Systems: The Continued Evolution of Agrochemicals, 24th Volume. Editor(s): G. Robert Goss; Gregory C. Volgas; Masoud Salyani, ASTM International, ASTM Press, West Coshocton, PA, 2005

Anonymous. Velsicol Pesticide Formulators Handbook. Velsicol Chemical Corp., Memphis, TN. 196 pp., 1974.

Anonymous. WITCO Pesticide and Emulsifier Handbook. WITCO Chemicals. 146 pp., 1977.

UNIDO. Pesticide Formulation. United Nations Industrial Dev. Organization. 484.pp., 1998.

Chapter 25

Copper-Based Systems for Exterior Residential Applications

Donatien Pascal Kamdem

Department of Forestry, Michigan State University, East Lansing, MI 48824

Environmental concerns and health issues related to the use of arsenic- and chromium-containing preservatives for residential applications have prompted the use of copper-based systems which typically contain copper(II) with an organic co-biocide added to control copper-tolerant fungi for exterior residential applications. This chapter deals with copper-based systems that contain no other metals. Copper-based systems include ammoniacal, alkaline and amine copper quat (ACQ types B, C and D), copper azoles (CBA and CA-type B), oxine copper (Cu-8), copper HDO, oilborne and water dispersible copper naphthenate, copper dimethyl dithiocarbamate (CDDC), copper borate and copper citrate (CC). The retention level, penetration, corrosiveness and leaching of copper from treated wood varies with the formulation of the treating solution, which is dependent on the form of copper (sulfate, nitrate, hydroxides, carbonates, metal), the amine/ammonia to copper ratio, the addition of corrosion inhibitors, the pH of the formulation and the inactive ingredients. The copper level generally required for residential above ground and ground contact applications is about 1.5 $Kg.m^{-3}$ and 2.5 $kg.m^{-3}$, respectively, but much less copper is necessary with the organometallics Cu-8 and copper naphthenate.

Background

Copper (Cu) has been used since the fourth millennium by the Egyptians as a mordant in dyeing. Copper sulfate was also used during the Greek civilization at the pre-Christian era of Hippocrates (circa 400 BC) to cure pulmonary disease. Copper is used for architectural purposes such as in roofs and gutters, piping, vehicle brakes, cooling towers, electroplating, metal finishing and semi-conductor manufacture. Copper is also used as an algaecides in ponds, spas and fountains, and to control growth of plant and roots in sewers, algae in swimming pools, ponds and lakes and on boat hulls. More than nineteen copper-containing pesticides are registered in the US. Today, Cu is still consumed by some inhabitants of Africa and Asia for healing sores and skin diseases, and widely used in baby foods, tonic and pills. Extensive studies show that copper is not a mutagenic, carcinogenic, a sensitizer, or embryo toxic.

Copper is the first element of Group IB in periodic table with four known oxidation states: Cu(0), Cu(I), Cu(II), Cu(III). It is a noble metal present as an element in nature in sulfides, arsenide's, chlorides and carbonates forms. Cu(0) is insoluble in water and is only superficially oxidized in air. Cu(I) is not stable in aqueous solution; it is present as insoluble CuCl in water. Cu(I) derivatives oxidize to more stable Cu(II) states but further oxidation to Cu(III) is relatively difficult. Few Cu(III) complexes are reported in the literature, they are mostly diamagnetic with four or five-coordination number. Most Cu(II) complexes are found in blue or green color with typical absorption bands between 600 and 900 nm in the UV/visible region (*1*).

Copper and Human Health

Cu is essential to human health. A diet must contain sufficient copper for the human body to have 1.5 to 2.2 mg of copper per kilogram of body. The recommend daily allowance (RDA) for copper for infant 1 to 6 years-old is 200 to 300 microgram per day and up to 1300 microgram for woman in lactation. The daily estimated dietary copper intake is equivalent to 500 to 600 micrograms as recommended by US national Academy of Sciences' Food and Nutrition Board (*2*).

Clarkson (*3*) reported that Cu is used in the formation of hemoglobin, bones, carbohydrate metabolism, catecholamine biosynthesis, cross-linking of collagen, elastin, hair keratin, and in the nervous system. Cu is also needed in cytochrome oxidase, tyrosinase, ascorbic acid oxidase, uricase, catalase, and peroxidase Copper deficiency in human may result in anemia, neutropenia, osteoporosis, pallor, dermatitis, anorexia, diarrhea, hematomegaly and slow growth (*4*, *5*).

An excess of copper will cause liver damage, nausea, vomiting, abdominal and muscle pain. An early report by Lehmann (*6*) listed the lethal dose of $CuSO_4$ in humans to be 7.5 g for oral ingestion. A genetic disorder known as Wilson's disease is associated with excess copper exposure (*7*, *8*). Low copper levels were found in the hair of patients over exposed to copper and with Wilson's disease symptoms, with high copper levels present in the livers and brains. A remarkable symptom of over exposure to copper is the blue and brown color of fingernails (*8*). One of the other reported diseases associated with excess level of copper exposure is Parkinson's disease, which affects about 1% of the population over the age of 60 years in the United States (*9*). The main symptom is a pronounced tremor of the extremities: the hands, chin or lips, stiffness or slowness of movement, a shuffling walk, stooped posture, and difficulties in performing simple tasks. USEPA has not provided specific data on the reference (RfD) of copper exposure (*10*). RfD is a numerical estimate of a daily oral exposure to the human population that will not likely cause harmful effects during a lifetime (*11*). Copper has not been classified as human carcinogen due to the lack of human data and inadequacy of animal data.

Copper in the Environment

Copper is present in cabbage leaves, peony buds, young oat plants and roots of rice plants (*8*). Copper concentrations in plant tissues can reach 5 to 20 mg/kg on dry weight basis (*12*). The bioavailability of copper to plants depends on the form of copper. Copper-based compounds with low water solubility, such as copper complexes, can interact with humic acid, citric acid, fulvic acids, and lignin in wood or soil to reduce its availability for "uptake" by plants (*8*). The copper content of plants varies with the region. An average value of 4.2 μg/g is reported for wheat growing in Oregon, 6.2 μg/g in Oklahoma, 7.8 μg/g in Texas, and 8.7 μg/g in California (*8*). Copper is also found in oceans, rivers, lakes and well water in the form of cupric ions (*13*). Copper levels of 1 to 10 ppb were reported in unpolluted waters in the US. However background levels in some water sediments varied from 10 to 70 ppm. Copper concentrations in tap water vary with exposure to copper tubing and tanks. The copper content is reported at 0.26 μg/ml in the morning and 0.11μg/ml in the evening, with the difference due to the solubilization of copper from the tubes overnight (*8*).

Copper as a Wood Preservative

Copper sulfate, known as blue vitriol, was recommended for preserving wood as early as 1767 and patented in 1837 by Margary in England (*14*). Today, copper is one of the main ingredients of commercial wood preservatives.

The mechanism of copper control of fungi and insects that degrade wood is complex. It is hypothesized that copper acts as a cofactor for electron exchange of several enzymatic reactions which involve oxidation, hydroxylation and dismutation. However, the presence of excess copper may harm fungi and insects by oxidizing proteins and lipids and promoting or initiating the formation of toxic intra- and extra cellular free radicals (*15*). Several fungicide chemicals used as wood preservatives are known to interfere with respiration either by inhibiting the formation of acetyl coenzyme A (CoA) or by interrupting respiratory chain phosphorylation (*16*). In summary, the most probable modes of copper fungicidal activity are: copper-enzyme interaction, promotion of the generation of highly reactive free radical and probable DNA damage (*17–20*).

Modern Copper-Based Preservative Systems

It is estimated that homeowners spent about 5 billions dollars yearly to address wood biodeterioration due to decay fungi and insects. For decades, chromated copper arsenate (CCA) was used as a wood preservative to increase the service life of solid wood and composite products for residential and industrial applications. Residential use now constitutes about 70-75% of the total volume of treated wood in the US. Due to environmental concerns, CCA was no longer labeled for residential applications since January 2004. Knowing that about 80% of the treated wood in the US was treated with CCA and that over 95% of this CCA treated wood was used for residential applications, effective alternatives that were environmentally benign and economical to replace CCA were needed. Creosote and pentachlorophenol, preservatives with good track record, are essentially limited to non-residential applications.

The objective of paper is to discuss copper-based systems, with no arsenic and no chromium present that are labeled or may shortly be labeled for exterior residential use in the US.

Copper based systems for residential exterior applications can be classified in three main categories based on the carrier-solvent used. The first and most common types are the waterborne systems, which use water for solvent. This class requires water with specific additives to stabilize the treating solution. The salt content, pH, and the hardness of the water all impact the stability of the copper-based systems made by mixing copper carbonate and an organic amine or ammonia. The main advantages of waterborne formulation are the relatively low cost and safety of water, and the paintability and lack of odor of the treated products. Thus, the vast majority of residential preservative systems are waterborne. The second category is the oilborne systems, which used oil as solvent and carrier. Oilborne systems are limited to copper naphthenate and copper quinolineate (Oilborne (*Bis*) copper-8-quinolinolate, also called copper oxine or Cu-8). Their uses in residential applications are limited to fences for copper naphthenate and to wood with contact with food such as fruits and vegetables bins for copper quinolineate. Waterborne or oilborne copper naphthenate is also used as a brush-on preservatives to treat cut-off surfaces of lumber from western species that are difficult to treat. The third category consists of oilborne formulations that are partially water dispersible.

The main waterborne copper-based systems in the US that replaced CCA are alkaline copper quats (ACQ) and copper azoles (CA). Their main advantages are that they are water soluble, effectively protect wood against decay fungi and insects destroying wood (*21*, *22*), the surface of treated wood is paintable with no oily residues on the wood surface, the treated wood is cost effective compared to alternatives such as steel, plastic lumber, and the absence of arsenic and chromium may make future disposal easier than CCA-treated wood. Copper-based systems are used to pressure treat lumber for numerous residential applications such as decks, outdoor furniture, play structures, raised beds and planters boxes, picnic tables and benches, fences, sill plates, structure members, and members for permanent wood foundations. Another copper system used in Europe, copper bis-(N-cyclohexyldiazeniumdioxy) (Cu-HDO, also abbreviated CX), consists of Cu-HDO, additional uncomplexed copper and borates, is used for above and ground contact applications. In the US, EPA registration of Cu-HDO is still pending as of the end of 2005 and thus not yet available.

ACQ

Waterborne alkaline copper quats are a combination of copper amine/ammonia and quaternary ammonium compounds with a typical ratio of copper oxide to quat of 2 to 1. The quats used in ACQ formulation are also in many home cleaning products and are thus an environmentally benign and safe biocide. Amine and /or ammonia are used to solubilize the copper carbonate in water. Ammonia is preferred for treatment of refractory species, although this increases the corrosiveness to metal fasteners. The different formulations of ACQ listed in the American Wood Preservers' Association (AWPA) book of standards are in Table I (*23*). Due to the metal corrosion concerns in the treating facility, the chlorine anion in quats has recently been substituted with carbonates. The difference between type A and the other three ACQ formulations is the ratio of CuO to quat. The most common formulations used in the US are ACQ types C and D. The minimum retention of the active ingredients is listed in Table III

The retention levels for residential applications are 4.0 and 6.4 kg/m3 for above and ground contact, respectively. Water-repellent, corrosion and UV inhibitors/stabilizers are sometimes added to improve the properties of the treated wood. Nature Wood™ and Preserve™ are the trade name for ACQ treated wood in the US. About 60 percent of the total plants that formerly used CCA to treat wood currently treat with ACQ.

CA-Type B

Today, about 40% of the treating plants which previously treated with CCA have switched to copper azoles (CA, both the older CBA formulation which contained boron and the newer formulation that does not contain boron, CA-type B). CA-B consists primary of copper as the main biocide and an azole, tebuconazole, to protect against copper tolerant fungi (Table II) . Copper carbonate is mixed with monoethanolamine and water at a ratio of 3.8 ethanolamine to copper, similar to ACQ and Cu-HDO. Tebuconazole is an organic azole used as a fungicide for various food and other agricultural crops. The ratio of copper to azoles and/or copper to amine is described in Table 2 (*23*). CA-B is used for above-ground, ground-contact and permanent wood foundations for residential applications CA has a lower copper retention, oxide basis, (1.3 kg/m3) than ACQ (1.68 kg/m3) for above ground applications, with same trend true for ground

Table I. Formulations and compositions of alkaline copper quat (ACQ)

Type		CuO, %	Quat, %	EA/CuO[3]	pH	
A	Ammoniacal	50 (45-55)	50	2.75	8-11	DDAC[1]
B	Ammoniacal	66.7 (62-71)	33.3	2.75		DDAC
C	Alkaline	66.7	33.3	2.75	8-11	BAC[2]
D	Ammoniacal	66.7	33.3	2.75	8-11	DDAC

[1]DDAC: Didecyl Dimethyl Ammonium Chloride or Carbonate

[2] BAC: Benzyl dimethyl Ammonium Chloride or Carbonate

[3] EA/CuO: Ethanol amine (EA) to copper oxide weight/weight

Table II. Formulations and compositions of copper azoles (CA)

Type	Cu,%	Borates, %	Azoles, %	EA/Cu*?
CBA-A	49 (44-54)	49	2	3.8
CA-B	96.1(95.4-96.8)	0	3.9	3.8

*EA/Cu: Ethanol amine (EA) to copper weight/weight

contact and permanent wood foundation (Table III). A copper oxide retention level of 0.66 and 1.0 kg/m3 were employed in CCA treated wood for above and ground contact exposure, respectively. Both of the copper-based systems in the US, ACQ and CA, require almost 2 to 3-fold more copper than was in CCA-treated wood, which results in relatively high copper leaching, discussed later.

Other Possible Copper Systems

A waterborne Cu-HDO, or CX, system is commercially available in Europe and is undergoing standardization [approval] process in the US. This system may be available for above-ground use once the active ingredients are registered by the US EPA. A minor fixed-copper waterborne system is copper citrate (CC), which is effective in above-ground applications but weak against copper-tolerant fungi in ground-contact applications. Only very small amounts of CC-treated wood are available in North America, and the system is no longer listed in the AWPA Book of Standards. Another complexed system is copper bis-(dimentyldithiocarbamate, or CDDC). This system requires a two-step treating process which is relatively costly, and CDDC is not commercially available. Acid copper chromate (ACC) was listed for years in the AWPA Standards, but was not available commercially. There has been some recent interest on re-introducing this system, but there is some concern over the presence of chromium.

Several copper-complexed systems have had some commercial success. Oxine copper, or Cu-8, has been used for over 30 years, but the difficulty in formulating it has minimized its commercial use. Research is on-going to develop a waterborne system based on phosphate. Copper naphthenate is made by complexing copper with naphthenic acids. Oilborne copper naphthenate has been used for over 50 years for treating wood, and starting in the 1990's some utility poles have been treated with this system. A waterborne formulation is available for brush-on applications, and a similar waterborne system has been submitted for standardization which, if approved, may greatly increase its market. These complexed systems, Cu-8 and copper naphthenate, require relatively low levels of copper compared to the wood treated with the ACQ and CA systems and, thus, may be preferred if disposal of lumber treated with high levels of copper or other metallics becomes a future issue.

Table IV lists some of these copper based preservatives their retention for above and ground contact applications.

Issues with Copper Based Systems

Photostability

Inorganic waterborne CCA is known to confer some photostability to treated wood in outdoor exposure, reportedly due to the presence of copper and chromium

Table III. Minimum retention (kg/m3) Of ACQ, CA-B and CBA-A

	ACQ (alkaline/ammoniacal copper qu			CA-B (Copper azole-Type B)		CBA-A (copper azole-type A)			
	Elementa Copper Cu	Copper oxi CuO	ACQ	Elemental Copper Cu	CA-B	Elemental copper Cu	Boron as i H3BO3 B	Azole Az	CBA
Above ground	1.68	2.1	4	1.3	1.7	1.28	1.28	0.05	3.3
Ground contact	2.72	3.4	6.4	2.6	3.3	2.56	2.56	0.10	6.5
PWF*	4.08	5.1	9.6	3.8	5	3.8	3.8	0.16	9.8

Table IV. Retetention (Kg/m3) of some alternative copper based systems for full protection for above and ground contact applications

Retention of active ingredients, Kg/m3		Above ground	Ground contact
Cu-HDO	61.5 % Cu (CuO+Amine) 14% CuHDO 24.5% Boric acid	2.4	
ACC acid copper chromate	CuO	4 (CuO:1.02)	8 (CuO: 2.03)
CDDC	CuAmine+SDDC	1.6 (Cu: 1.6)	3.2
Ammoniacal Copper citrta		4 (CuO:2.24)	6.4 (CuO: 3.6)
Oilborne Cu Naphthenate		0.64	0.94
Cu-8	Copper quinolineate	0.32	
CCA	Chomated copper arsenic	4 (CuO: 0.66)	6.4 (CuO: 1.0)
ACZA	Ammoniacal copper zinc arsenate	4 (CuO: 1.6)	6.4 (CuO: 2.56)
ACA	Ammoniacal copper arsenate	4 (CuO: 1.6)	6.4 (CuO: 2.55)

(*24–26*). Unfortunately, the photostability of wood treated with the newer of copper based systems is thus less than CCA-treated lumber (*24, 25*).

With the appearance of wood in residential applications an important factor desired by consumers, the color stability of copper based systems is an important factor in marketing solid treated wood to consumers. The utilization or incorporation of ingredients that will improve the color stability, such as pigments or inorganic light stabilizers such as talc or micronized iron, may help address this concern.

Currently, commercial formulations of copper based systems use an organic amine or ammonia to complex the copper in the pH range of 8 to 10 (*27*). Mono ethanolamine and ammonia are preferred due to their relatively low cost and ease of handling. Unfortunately, the presence of nitrogen and absence of arsenic encourage the growth of mold on the surface of freshly treated wet lumber. Mold growth also occurs on the surface of CCA treated lumber but was easily controlled with the addition of a mildewcide such as an isothiazolone. A much higher application rate of an mildewcide is needed for copper amine systems, probably because of the high pH which may degrade the mildewcide, the high content of nitrogen and the absence of arsenic. The mold species most common on the surface of ACQ-treated lumber was identified as *fusarium* species (*28*). Kiln drying the lumber after treatment is another method to prevent surface mold (*28*).

Corrosion

Steel, carbon steel, brass, aluminum, stainless steel, copper and other alloys are used to fabricate fasteners, plates, connectors, gutters, window frames, etc. which come in contact with wood treated with copper-based preservatives. Any free copper ions in the treated wood can interact with the metal fastener to give galvanic corrosion. [The presence of chromium minimized such corrosion with CCA-treated wood, and many construction personnel and homeowners did not fully appreciate the high corrosion potential of lumber treated with the new copper-rich systems, with the resulting problems.] Due mainly to economic reasons, steel is used to manufacture metal fasteners for residential use. The corrosion of stainless steel metals or fasteners in contact with treated wood is negligible. However, its relatively high cost has limited its uses for residential applications in the past. typically, lower-cost materials are coated onto the surface of steel or other inexpensive metal alloys to reduce corrosion, such as zinc or ceramic coatings (*29*, *30*). Due to some initial corrosion problems, the use of relatively high-cost stainless-steel fasteners, which only adds a relatively small amount to the overall cost of a finished product such as a deck, is becoming more common.

Leaching of Copper from Treated Wood

The leaching of copper from wood treated with the new systems is a concern that may limit the future use of copper preservatives. Copper is well known to negatively affect aquatic environments. Several studies have been examined the leaching of copper-treated wood in the laboratory and outdoor exposures (*31–36*). It is well established that the amount of copper leaching from wood in service is related to the initial retention of copper. The levels of copper in wood treated with the new copper-based systems, ACQ and CA-B, are two to three fold of that in CCA, and these systems do not have an oxidant to help fix the copper, as does CCA. Thus, the amount of copper leached from wood treated with the new systems is higher than that from CCA-treated wood. Specifically, the amount of copper leached from CCA lumber is reported to vary between 4 and 8 percent depending on the protocol used to collect data in the laboratory or the field, the wood species and the retention of CCA. Several studies reported a percentage of copper loss from wood treated with ACQ or CA-B to vary between 5 and 35 percent (*27*, *33*, *37–43*).

Several factors can influence the retention and the leaching of copper in wood treated with copper based systems: the wood species, the treatment parameters and the formulations (*43*, *44*). The strong fixation sites of copper in wood are the carboxyl groups in hemicelluloses and the phenolic groups in lignin, as confirmed by FTIR spectra (*43*). Aromatic esters groups and/or ketone groups present in wood can change to carboxylic acid groups through alkaline hydrolysis or oxidative cleavage and become potential bonding sites for copper. EPR was also used to suggest that the stereo-structure of copper complexes formed in copper amine treated wood is a tetragonal-based octahedral or square-based pyramidal complex with Cu interacting with two oxygen (*45*).

Conclusions

Copper-based systems for residential applications are reviewed. Copper is one of the cost effective ingredients for wood preservatives. Future challenges that may limit its use include: copper leaching with the regulating negative impact on the surrounding environment, especially aquatic ecosystems; future restrictions on disposal of metallic-treated lumber; the high potential for corrosion of metal fasteners in contact with treated wood; and the poor photostability of treated lumber exposed to sunlight. Several suppliers of copper based systems are working on new formulations that will reduce copper leaching and corrosion of metals fasteners, and are also educating the general public on the proper use of lumber treated with the new systems to reduce subsequent problems. In addition, waterborne formulations of several new copper-based systems are being examined or submitted for standardization. The future of the copper based systems will depend on how theses issues are resolved and also the performance and cost of substitute materials for residential applications such as wood-plastic decking.

References

1. Cotton, F. A.; Wilkinson, G. Advanced Inorganic Chemistry, 5th Ed. John Wiley and Sons, Inc. New York, 1988, 755-775.
2. US National Academy of Sciences' Food and Nutrition Board. Dietary reference intakes for vitamin a, vitamin k, arsenic, boron, chromium, copper, iodine, iron, manganese, molybdenum, nickel, silicon, vanadium, and zinc. NATIONAL ACADEMY PRESS. Washington, D.C.2000.
3. Clarkson, T.W. Inorganic and organometal pesticides. Handbook of Pesticide Toxicology. Academic Press. New York. 1991. Pp 497-583.
4. Montorsi, W.; Annoni, F.; Doldi, S.B.; Germiniani, R; Longoni, F. 1975. Concentration plasmatique du zinc et du cuivre après court-circuit intestinal. Nouv. Presse. Med, 1975, Vol 4: 1734.
5. Hambidge, K.M.D. *Ped. Clin. N. Amer.* **1977**, *24*, 95-106.
6. Lehmann, K.B.W. *Arch. F. Hygiene* **1897**, *31*, 279-309.
7. Lal, S.; Sourkes, T.L. *Toxicol. Appl. Pharmacol.* **1971**, *20*, 269-283.
8. Owen, J.C.A. Noyes Publications. Park Ridge, New Jersey. 1982, 1-23.
9. Gorell, J.M.; Johnson, C.C.; Rybicki, B.A.; Peterson, E.L.; Kortsha, G.X.; Brown, G.G.; Richardson. R.J. *Neurology* **1997**, *48*(3), 650-658.
10. Integrated Risk Information system (IRIS). 2003. Arsenic, inorganic (CASRN 7440-38-2). Copper (CASRN 7440-50-8). U.S. Environmental Protection Agency. 1998.
11. Klaassen, C.D; Watkins III, J.B. (eds. Essentials of toxicology. McGraw-Hill Medical Publishing Division. New York.). 2003.
12. Pätsikkä, E.; Aro, E.-M.; Tyystjarvi E. *Plant Physiology* **1998**, *117*, 619-62.
13. Agency for Toxic Substances and Disease Registry (ATSDR). Toxicological profile for copper. Draft for public comments, US department of health and services. Public health services: 2002. 286pp
14. Hunt G.; Garratt G. Wood Preservation. 1st Ed. McGraw-Hill Inc. New York NY.1938.
15. Goyer, R. A.; Cherian M. G. Toxicology of Metals, Biochemical Aspects. Springer-Verlag. 1997
16. Eaton, R. A.; Hale. M. D.C Wood: Decay, Pests and Protection. Chapman & Hall. London, 1993.
17. Hunt, J.V.; Simpson, J.A.; Dean, R.T. *Biochem. J.* **1988**, *50*, 87-93
18. Kobayashi, S.; Ueda,K.; Komano, T. *Agric. Biol. Chem.* **1990**, *54*(1), 69-76.
19. Hertzberg, R.P.; Dervan, P.B. *Biochemistry* **1984**, *23*, 3934.
20. Inoue, S.; Kawanishi, S. *Cancer Res.* **1987**, *47*, 6522.
21. Evans, F. G. *The International Research Group on Wood Preservation*, 2003. Paper IRG/WP 03-30303.
22. Evans, F. G. *The International Research Group on Wood Preservation*, 1999. Paper 99-30208.
23. American Wood-Preservers' Association Standards 2005. Selma, Alabama USA

24. Grelier, S.; Castellan, A.; Kamdem, D.P. *Wood Fiber Sci.* **2000**, *32*(2),196-202.
25. Fowlie, D.A.; Preston, A.F.; Zahora, A.R. *Proc. of Am. Wood –Preservers' Assoc.* **1990**, *86*, 148-159.
26. Liu, R.; Ruddick, J.N.R.; Jin, L. *The International Research Group on Wood Preservation Paper 1994,* IRG/WP/94-30040.
27. Zhang J.; Kamdem D.P. *Wood Fiber Sci.* **2000**, *32*(3), 332-339.
28. Zabel, R.A.; Morell. J.J. Wood Microbiology: Decay and its Prevention. Academic Press, San Diego, CA, 1992
29. Schweitzer P. A. Corrosion-Resistant Linings and Coatings. Marcel Dekker, Inc. New York NY 2001., pp 427.
30. Marcus P. Corrosion Mechanisms in Theory and Practise. 2nd ed. Revised and Expanded. Marcel Dekker Inc. New York NY 2002., pp 736
31. Jin, L.; A.F. Preston A. F. 1993. *The International Research Group on Wood Preservation Paper.* IRG/WP93-50001-07, 1993
33. Kennedy, M.J.; Collins. P.A. *The International Research Group on Wood Preservation Paper.* IRG/WP 01-50171, 2001.
34. Cao, J.; Kamdem, P. *Holzforschung* **2004**, *58*, 569-571.
35. Waldron, Y. T.; Ung, Y. T.; Cooper, P. A. *The International Research Group on Wood Preservation*: IRG/WP 03-50199, 2003.
36. Lebow S.; Foster, D.; Lebow P. *Forest Prod. J.* **2004**, *54*(2), 81-88.
37. Chung P.A.; Ruddick J.N.R. *The International Research Group on Wood Preservation Paper* IRG/WP/04-50219, 2004.
38. Waldron, L.; Cooper, P. A.; Ung, T.Y. *Holzforschung* **2005**, *59*, 581-588.
39. Lucas, N.; Ruddick, J. N. R. *The International Research Group on Wood Preservation Paper* IRG/WP 02-30285. 2002.
40. Jiang, X.; Ruddick, J. N. R. 2000. *The International Research Group on Wood Preservation Paper* IRG/WP/00-30233. 2000.
41. Pasek, E. A. *Proceeding, American Wood Preservers Association* Annual Meeting, Boston, MA. **2003**, *99*, 100-142.
42. Esser, P.; Suitela, W.; Trompetter, H. *The International Research Group on Wood Preservation Paper* IRG/WP 00-50150. 2000.
43. Yamamoto, K.; Motegi, S.; Inai, A. *The International Research Group on Wood Preservation Paper* IRG/WP 99-50134, 1999.
43. Zhang, J.; Kamdem D.P. *Holzforschung* **2000**, *54*, 119-122.
44. Jiang, X.; Ruddick, J. N. R. *The International Research Group on Wood Preservation Paper* IRG/WP/00-30233. 2000.
45. Zhang J. Ph D. Dissertation. Michigan State University, East Lansing, MI, 2000.

Chapter 26

Borate Wood Preservatives: The Current Landscape

Mark J. Manning

U.S. Borax, Inc., 26877 Tourney Road, Valencia, CA 91355

Borates have been used as wood preservatives for over 70 years - providing excellent control of wood destroying organisms such as decay fungi and termites. Because of their water soluble nature, the use of borates have been limited to applications where the wood is used in protected, above-ground applications such as lumber for residential construction. In recent years, less soluble borates such as Zinc Borate ($2ZnO{\cdot}3B_2O_3{\cdot}3.5H_2O$) have been used as preservatives to treat wood- and wood-plastic composites which are used in less protected applications such as exterior siding and above-ground decking. This paper will also highlight the current situation with the use of Zinc Borate as a preservative treatment for wood composites, providing protection for these commodities when used in exterior above-ground applications.

Introduction

Boron-containing wood preservatives are all derived from naturally occurring borate minerals. The mineral borax, Na2B4O7·10H2O is the most significant commercial source of B_2O_3; there are major deposits in both the United States and Turkey. Probably the most commonly used form of boron for wood preservation is the compound Disodium Octaborate Tetrahydrate (DOT, Na2B8O13·4H2O), sold under the trademark *Tim-bor*® Industrial (U.S. Borax Inc.); it exhibits high water solubility and has a near neutral pH. Borates in general, and DOT in particular, possess a number of properties that help to make them unique wood preservatives:

- Inorganic salts (nonvolatile)
- Low mammalian toxicity
- High toxicity to insects and fungi
- No tolerance by wood-destroying insects
- Odorless
- Near neutral pH
- Noncorrosive
- Colorless
- Strength properties similar to untreated wood
- Compatible with colorants if visual marking is desired

When used correctly they can give effective long-term treatment that is also economical and environmentally sound. Borates possess a very favorable eco-tox profile (*1*). They occur naturally in seawater, fresh water, rocks, soils, and all plants. The earth consists of trace amounts of more than 200 minerals that contain boron. In trace amounts, they are essential micronutrients for plants and believed to be nutritionally important for humans. Like many trace elements, boron is both essential at low concentrations and toxic at high concentration – allowing the borates to be effective preservatives against wood destroying organisms.

Borates are members of a class of waterborne chemical preservatives which are diffusible in wood. Using the available moisture in unseasoned wood, the chemical redistributes itself after the treatment – diffusing from areas of high concentration (of chemical) to areas of lower concentration. The water-soluble chemical equilibrates in such a way as to reduce the concentration gradient. This capability for the chemical to diffuse after treatment makes it possible to completely penetrate unseasoned wood, thereby allowing effective treatment of refractory species. The subject of diffusible preservatives has been reviewed in the context of these systems offering highly effective and flexible options, for both standalone preservatives and as components in more specialized formulations (*2*).

Boron compounds in current use as wood preservatives are susceptible to leaching under certain conditions, as they are not chemically fixed within the wood. A common misconception is that small amounts of moisture will readily leach boron out of the wood – this is simply not the case. In order for leaching to occur there needs to be a situation where liquid water enters the wood at one point and then leaves at another point, also as liquid water. This topic has been reviewed (*3*).

Because of this potential for depletion when exposed to significant moisture, borates are normally recommended for general building use in a protected environment and are not recommended for use in ground contact. This particular end-use is delineated in the original borate treating standard that was published by the American Wood-Preservers' Association: AWPA Standard C31-02 (*4*), "Lumber Used Out of Contact with the Ground and Continuously Protected from Liquid Water – Treatment by Pressure Processes". More recently, the AWPA has moved to a Use Category System (UCS), utilizing a framework whereby the intended application (end-use) of the treated wood product is used to specify the appropriate standard and preservative retention. Borates are listed in AWPA Use Category UC1 and UC2. UC1 is for wood and wood based materials used in interior construction not in contact with the ground or foundations (e.g., interior furniture and millwork). UC2 is for wood and wood based materials used for interior construction that are not in contact with ground, but may be subject to dampness – products that are continuously protected from the weather but may be exposed to occasional sources of moisture (e.g., interior framing and sill plates).

A number of recent trends are driving the development of wood preservation activity over the last decade. Among those highlighted have been: changing building practices (wood losing market share to termite resistant building materials such as steel and concrete), increased need for value added production and the emerging area of preservative treatment of wood composites (*5*). These issues all have relevance to the current development of boron based wood preservatives in North America and will be discussed in the following section. There have been several reviews published on the use of these preservatives (*1,6,7*) and a review of efficacy data against decay fungi and termites has also been compiled (*8*).

Recent Developments in North America

As described earlier, borates are referenced in the AWPA Book of Standards. The Standard for Waterborne Preservatives (Standard P5) lists

Inorganic Boron (abbreviated as SBX) and the acceptable boron preservatives: DOT, Sodium Tetraborate, Sodium Pentaborate, Boric Acid, and FR-1 (a fire retardant system containing boron). The wood species listed in AWPA Standard C31-02 as well as UC1 and UC2 are: Southern Pine, Hem-fir (Western Hemlock and Amabilis fir), Ponderosa Pine, Red Pine, Spruce-Pine-fir (SPF), and Coastal Douglas fir. The listing of Coastal Douglas-fir includes a requirement for incising, while the listing for Hem-fir and SPF does not include an incising requirement. The absence of an incising requirement for these refractory species is due to the diffusible nature of the borate preservative. Borates are also listed for the same species in the Canadian wood preservation standard CSA O80.34 (Canadian Standards Association).

The desire to protect buildings from wood destroying organisms (WDO) such as decay fungi and termites is well established and evident in mandates by building codes in many regions of the United States, as well as in standards required by home mortgage lenders. The predominant method of protecting homes requires establishing a barrier against subterranean termites by treating the soil under and around homes prior to construction and treating in and around existing homes when termites are detected. Additionally, treated wood is used for selected building elements such as sill plates. This approach provides some relief, but even perfect barriers don't protect structures against aerial colony-forming termites such as Formosan subterranean (*Coptotermes(FST) formosanus* Shiraki) and drywood termites, carpenter ants, or from wood decay. Additionally, chemical barriers are often disturbed in the construction process or by simple homeowner activities such as landscaping, and degrade over time.

Historically, organochlorine pesticides such as chlordane, aldrin, and heptachlor were used to place a chemical barrier at the soil level. These materials were long lasting (30 to 50 year life) and the vapor pressure from these materials would "fill in" any missed or disturbed areas to provide a continuous layer of protection. These properties, while functionally desirable in a termiticide, were environmentally unacceptable. As a result, chlordane and other organochlorine pesticides were removed from the market in the late-1980s. The current termiticides are less persistent (effective working lives of about 5 years) and much less forgiving of applicator error and disturbances.

A recently described approach for mitigating damage caused by WDO is the Six "S" strategy: Suppression, Site management, Soil barriers, Slab and foundation details, Structural protection, Surveillance, and remediation (*9*). Structural protection is the most straightforward and important single operation among the six. It involves two basic elements: managing moisture within the building assemblies and choosing materials that are protected against WDO. Wood that has been pressure treated with borates has been successfully used for the latter objective in various parts of the world where termites, decay and other wood-destroying organisms pose significant problems. In New Zealand, for example, the use of borate-treated lumber and plywood for the entire structure has been standard practice for more than 50 years. Borate-treated wood has shown excellent performance as a building material in the wet environment encountered in New Zealand, effectively making obsolete the need for remedial treatment and repair for homes built with this material (*5*).

Hawaii is known to have the highest Formosan subterranean termite pressure in the United States. The use of treated wood for structural systems has been a requirement of the building codes there since the early 1980s. When borate-treated wood was first introduced into the Hawaiian market in 1992, it quickly became the most widely used structural material for building homes. Since its introduction, borate-treated lumber and plywood (sold locally as *Hi-bor*®, registered trademark of STN Holdings) have become the market leader in treated framing – providing a safe, effective and economical option for the residential home builder. This Hawaiian model has recently been transplanted to the Southeastern U.S. where it is currently being used for protecting homes (both single and multi-family) from damage caused by decay fungi and wood-destroying insects such as the Formosan subterranean termite. Borate-treated wood has performed extremely well - both in rigorous field tests and as structural systems for homes built in these very challenging environments. These successful efforts in New Zealand, Hawaii, and the Southeastern U.S. has been combined in a design concept called Borate Treated Structural System (BTSS) – wherein the structural elements of a home are industrially pre-treated with borate wood preservatives.

One of the most significant uses of borates in North America is as a preservative treatment for sill plate material in residential construction (UC2 application described above). Extensive data was presented to the AWPA supporting the use of borate-treated lumber for this application, showing that there is no loss of efficacy when used as sill plate material. This has led to borate-treated lumber being listed in the most recent edition of the International Residential Code – the most widely accepted building code in North America for residential construction (*10*).

Additional support for the use of borates as a preservative treatment for sill plate is provided by research being carried out by the Wood Research Institute at Kyoto University, Forintek Canada, and the University of Hawaii (*11*). Field tests were established to evaluate borate-treated Hem-fir lumber against the Formosan subterranean termite in a covered, above-ground exposure that was specifically designed to simulate the use of dodai (sill plate) material in Japanese residential construction (*12-14*). This test is still on-going with end-matched samples exposed to active FST colonies: one piece evaluated in Hawaii while the 'sister' piece is exposed to an FST colony in Japan. It is tentatively planned to carry out the test for a total of 10 years at each location. After six years in Hawaii and seven years in Japan, the borate-treated samples are exhibiting performance equivalent to that of the 0.25 pcf CCA (Copper Chrome Arsenate) which is being evaluated as a comparison control in the same test. The six-year results from the test in Hawaii are displayed in Figure 1. Results from these tests were instrumental in helping to establish an AWPA borate retention for exposure to the FST: 0.28 pcf B_2O_3 – the same retention approved by the Honolulu Building Department for use as a preservative treatment in Hawaii. This test is very severe in that the test units are located directly on top of active FST colonies. It should also be noted that because of the warmer temperatures, the termite pressure in Hawaii is thought to be ~3x as severe as that in Japan, so that a 6 year exposure in Hawaii is considered to be comparable to ca. 18 years exposure at the Japanese test site in Kagoshima Prefecture (Island of Kyushu) (*13*).

Additional tests are ongoing which are evaluating a variety of wood species (Douglas-fir, Southern Yellow Pine, Spruce-Pine-Fir, Hem-fir, and Ponderosa Pine) treated to different borate retentions. In this test, the wood samples have been assembled into small 'house-like' structures (~ 1 m^3 in size) with sloped, plastic roofs, such that the test specimens are placed in protected, above-ground exposures. This equates to AWPA Use Category 2, which is the exposure for framing lumber used in residential construction. In this test (and the extended duration Hem-fir Dodai test previously discussed), it should be noted that the termite hazard is exceptionally severe in that untreated feeder stakes are driven into the ground with the upper surface placed in contact with the borate-treated sample; this is done in an ongoing effort to bring the FST up into the test unit and in contact with the treated specimens. In typical residential construction the exact opposite occurs - the builder and subsequent homeowner do everything possible to inhibit future termite pressure (*eg*., soil treatment, minimal use of wood in ground contact, regular inspections by pest control operators, etc.) (*9*).

In addition to termite pressure, wooden sill plates are also exposed to a decay hazard. In the extended sill plate exposures described above, *there has been no evidence for decay in any of the borate-treated samples* after 6 years in Hawaii and 7 years in Japan. This excellent performance of borates against decay fungi is highlighted in some L-Joint data developed by Forintek Canada (*15*). Hem-fir lumber was dip-treated with Disodium Octaborate Tetrahydrate (*Tim-bor* Industrial) to an average cross-sectional retention of 0.25% Boric Acid Equivalent (BAE) w/w and was subjected to a conventional L-Joint study (used to evaluate preservatives for exterior joinery; the treated wood is protected with a three coat paint system and is exposed directly to the elements, with no protective overhang as is typically seen in normal construction). A set of 30 treated samples was installed along with a set of 30 untreated samples. The results for average visual ratings after 12 years exposure are displayed in Figure 2. The untreated samples had an average rating of 1.5 (23 of the 30 untreated samples had failed) while the treated samples had an average rating of 9.3.

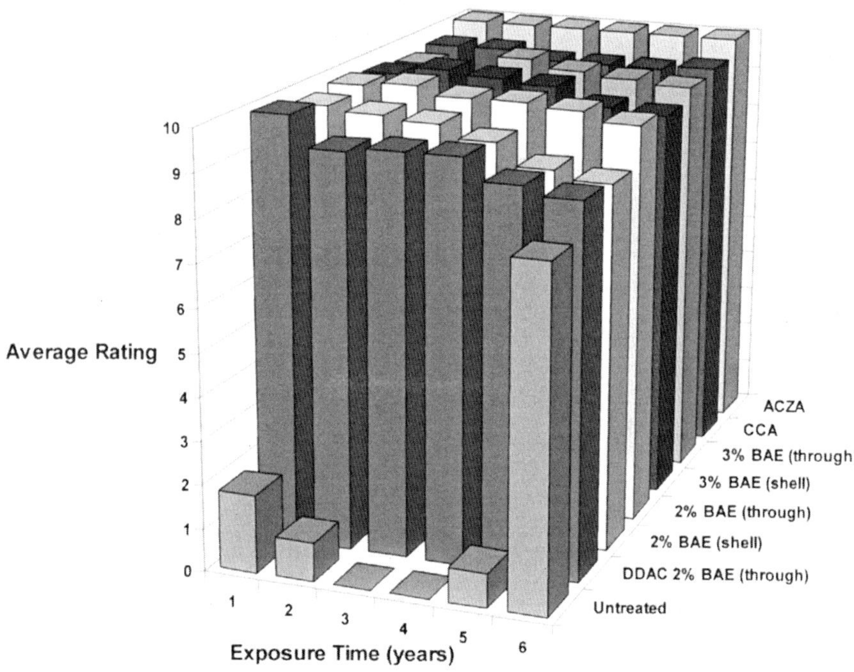

Figure 1. University of Hawaii / Forintek Canada / Wood Research Institute - Kyoto University. Six year data from Hem-fir Dodai samples exposed to the FST in a covered, above-ground test. Average visual ratings are on a 0 to 10 scale; rating of 10 indicates no attack while a rating of 0 indicates that the sample has been completely destroyed.

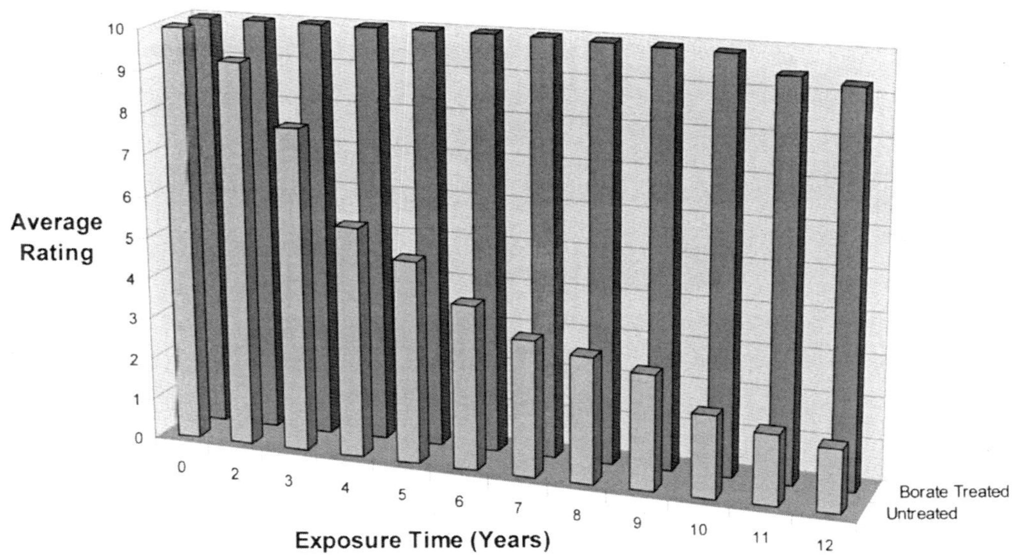

Figure 2. Average visual ratings for L-Joint samples exposed for 12 years in Vancouver, B.C. Average visual ratings are on a 0 to 10 scale; rating of 10 indicates no attack while a rating of 0 indicates that the sample has been completely destroyed.

The samples were exposed in Vancouver, B.C. where the average annual rainfall is ~40" (~1000 mm). After 5 years exposure (cumulative precipitation of ~200" (~5000 mm)) a set of five treated samples were removed and chemical assays were carried out; these results showed minimal boron levels (0.02% BAE) in the area of the joint with evidence for a 'reservoir effect' with boron being depleted in the area of the joint while being replenished from the ends. Of particular note here is the protection against decay being afforded by the extremely low levels of boron in the area of the joint. The boron is providing protection at retentions significantly below what is normally suggested as a toxic threshold - providing support that low levels of boron may inhibit spore germination. The excellent performance of the borate-treated samples after 12 years is even more noteworthy when one considers that the initial retention was approximately one-quarter of the non-Formosan AWPA Standard of 0.17 pcf B_2O_3.

Structural protection for WDO involves two basic elements: 1) managing moisture within the building assemblies and 2) choosing materials that are protected against WDO. Sometimes, non-wood building materials are chosen solely because they are resistant to decay and insect attack, when wood products would otherwise be the material of choice. However, common building materials such as steel and concrete can increase the water loading in the building envelopes, providing a more favorable environment for wood destroying organisms. Water can condense on the steel frame members, creating problems of corrosion, decay and insects. Concrete is a porous material and allows moisture to migrate and accumulate on the cool side of the block wall, creating a condition for WDO to prosper.

Wood that has been industrially pre-treated with borates has been successfully used for structural systems in various parts of the world where termites, decay, and other wood destroying organisms pose significant problems. This has resulted from the realization that it is far more cost effective and environmentally responsible to provide lasting, built-in protection to the very building components and systems that are susceptible to degradation than to provide remedial protection year after year.

In the United States, it has been shown that many new homes are suffering damage from WDO. At a national level, it has been estimated by the Wood Protection Council of the National Institute of Building Sciences that the annual cost to replace wood severely damaged by decay and termites increased from $750 million in 1988 to over $2 billion in 1993 (*16*). The economic loss continues to grow with some estimates indicating that Florida and Louisiana each spend $1 billion per year to repair and treat damage by WDO. It is widely believed that current costs are significantly greater.

The success in the use of borate wood preservatives for treating structural materials is due to its many performance attributes. Borates have a well-established record of performance against a broad spectrum of wood destroying organisms. Borates are cost effective and easy to use for the preservative treatment of solid sawn lumber, plywood, and wood composites such as oriented strand board (OSB). Lumber and plywood is pressure treated with waterborne borates such as DOT, whereas with OSB the borates are added to the wood composite during the manufacturing process. In treated wood, borates are colorless (although a dye is often added), non-volatile, and robust so they don't evaporate, degrade, or produce an odor during service; and are non-corrosive, requiring no special fasteners. The implication to builders and designers is that significant protection against termites and decay can be built into the structure without having to make drastic changes in design or to the construction process.

Preservative Treatment of Composites with Zinc Borate

The species, size, and quality of standing timber available for harvest is changing world-wide and this is promoting the development and extended use of wood composites in applications which require resistance to wood destroying insects and decay fungi. Traditionally, solid wood products are pressure treated with solutions of preservative chemicals. However, the nature of a composite makes it possible to incorporate a preservative into the product during its manufacture. This decreases total production costs and yields a superior product in which the composite can achieve a constant loading of preservative throughout its thickness.

Both DOT and Zinc Borate (ZB, $2ZnO{\cdot}3B_2O_3{\cdot}3.5H_2O$) are suitable for incorporation into wood composites, although ZB has been used almost exclusively for exterior composite products where there is a perceived risk of leaching. Zinc Borate is a white odorless powder (median particle size of 9 microns) and is typically mixed in the blender with the wood furnish, adhesive and wax. ZB is manufactured by U.S. Borax Inc. and sold under the trade name *Borogard*® ZB (registered trademark of U.S. Borax Inc.). ZB exhibits low water solubility at room temperature (<0.28%, w/w) and provides good efficacy, even after rigorous standard leaching tests (*17,18*). It is also relatively simple to incorporate ZB powder into a wood composite during blending (*19*).

Zinc Borate currently dominates the in-process treatment (preservative added during manufacture) of engineered wood products in the United States and Canada (*18, 20*). ZB is being used for the preservative treatment of commercial OSB sheathing and siding, MDF trim boards, and other exterior products. More recently there have been published reports (*21-24*) evaluating the use of ZB as a preservative for the treatment of Woodfiber-Plastic Composites (WPC). The AWPA has recently expanded its scope and now lists preservatives for nonpressure applications to wood products. ZB is listed in Standard P18 as a Nonpressure Preservative (*4*).

Zinc Borate has seen increased usage as a preservative additive introduced during the manufacture of WPC decking material. The properties that allow ZB to be used as a preservative treatment for wood composites are also important factors supporting its use as a preservative for WPC materials:

- Broad spectrum activity against wood destroying organisms
- Thermally stable during the manufacturing process
- Resistance to Leaching/Weathering
- Stable to ultra-violet radiation
- Excellent environmental profile
- Used for more than 10 years in wood composite plants
- Simple to introduce into the manufacturing process
- Relatively low cost

The production of WPC has seen growth averaging more than 25% a year since 1998 (*25*). This market is currently valued at greater than US$1.3 billion and climbing, with building products such as deckboards and railing accounting for just over US$1 billion in 2005. Current estimates have WPCs accounting for approximately 15% of the value of the residential deck market in North America, with some forecasts predicting growth to >35% by 2010 (*25*). When WPCs were introduced in the early 1990s, the common perception was that the 'wood fiber was encapsulated' by the plastic resin, thereby minimizing the potential for moisture absorption and protecting it against WDO. In reality, wood particles near the surface of commercial WPC products may be subject to levels of water absorption that can initiate and support fungal decay (it is accepted that this water absorption takes place in the wood component and not in the plastic phase).

Recently there have been reports which have examined the durability of these products in relation to artificial weathering and the impact on water absorption (*27*). This work has yielded data which shows that small samples (with high surface area to volume ratios) submerged in water can achieve the necessary % moisture content (MC) in the wood component to support decay fungal growth in less than 24 hours. Laboratory water absorption evaluations used by the WPC industry on ~300 mm lengths of conventional sized, freshly extruded samples which have not been weathered exhibited %MC values for the wood component which were less than 10%, well below the minimum threshold of 25% MC necessary for the onset of fungal decay.

This concept of the wood component being encapsulated by the plastic is supported by test methods that the WPC manufacturers utilize to evaluate moisture absorption. The ASTM D-1037 method utilizes a 24 hour water soak on large-sized specimens (12" long by 6" wide by product thickness) and is used by WPC manufacturers to generate water absorption values for their commercial products – typically yielding values of 1-2% weight gain. For a WPC material that is comprised of an approximate 50:50 wt. ratio of wood:plastic, a weight gain of 1-2% corresponds to a MC of 2-4% in the wood component (it is assumed that the plastic component does not absorb moisture).

If the %MC for the wood component is maintained at this low level, there would be insufficient moisture to support biological growth such as wood decay fungi. Zabel and Morrell (*28*) have outlined the growth needs for wood-inhabiting fungi:

- Water (minimum of 25-30% MC)
- Oxygen
- Favorable temperature range (15 – 45° C)
- Digestible substrate (wood)
- Favorable pH range (pH 3 – 6)
- Chemical growth factors (Nitrogen, vitamins)

As part of our work with these materials, commercial WPC samples were evaluated against decay fungi using the AWPA Soil Block test method (*3*). All of the samples were based on polyethylene/wood, with the wood content ranging from 50 to 70% of the product weight. Samples that were leached prior to exposure to the decay fungi exhibited weight losses ranging from 0.45% to 22.1% for the white rot organism *Trametes versicolor* and 0.23% to 7.99% for the brown rot organism *Gloeophyllum trabeum*. Unleached samples exhibited weight losses ranging from 0.20% to 18.1% for *T. versicolor* and 0.54% to 7.34% for *G. trabeum*. The higher weight losses in the *T. versicolor* tests were presumably due to the wood content in these samples being predominately hardwood. Samples with the highest weight loss had wood contents close to 50%; for samples with weight loss in excess of 20% (based on product weight), this yields a loss for the wood component of ca. 40%.

The image in Figure 3 was taken by an optical microscope and shows the surface of a WPC sample prior to being exposed to the decay fungi in the laboratory soil block test. Figure 4 shows the surface of the same WPC sample *after* exposure to the decay fungi – in this case the degradation of the wood component is clearly evident with the appearance of void spaces where the decay fungi have metabolized the wood. Other researchers have evaluated biological attack on WPC materials using a scanning electron microscope (*29*). In all of the samples that exhibited weight losses, the degradation was most severe on the surface that was in direct contact with the decay organism at the start of the test.

As part of the decay evaluation that was previously described, a series of extruded WPC samples (both commercial and toll produced) were treated with ZB at target retentions of 0.5, 1.0, or 2.0% ZB (w/w) during the manufacturing process and exposed in the AWPA laboratory soil block procedure. In all samples evaluated (both leached and unleached), the WPC materials treated with

Figure 3. Surface of untreated WPC sample (100x magnification) before AWPA E10 Decay Evaluation.

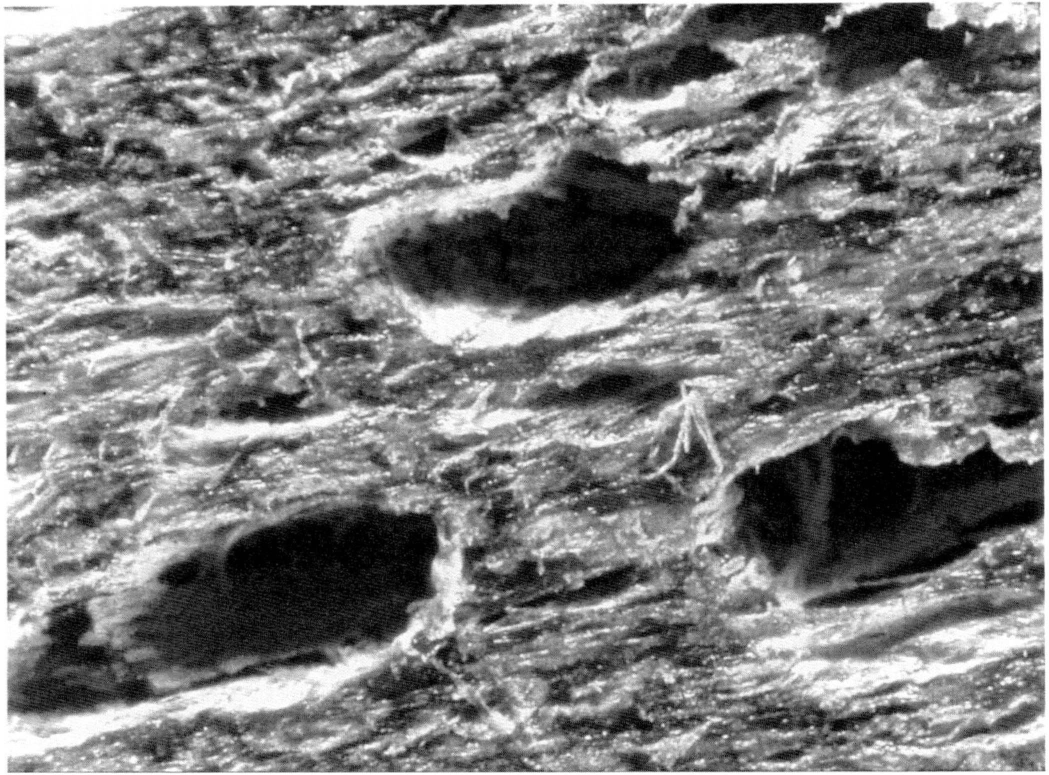

Figure 4. Surface of untreated WPC sample (at 100x magnification) after AWPA E10 Decay Evaluation.

ZB did not exhibit signs of fungal decay as evidenced by weight loss values that were all less than 1.1%. When compared with the earlier decay data on untreated WPC material, these results clearly demonstrate the significant improvement in protection against fungal decay afforded by the addition of low levels of ZB. In 2006, it is forecast that upwards of 35% of the WPC production in North America will be preservative treated with Zinc Borate.

Future Developments

There are still research efforts underway with the aim of producing a leach-resistant or "fixed" borate which could potentially expand the use of borates in the preservative treatment of solid wood. Many workers have been interested in this goal and some of the most interesting results to date have been achieved by the use of insoluble borates that are slowly solubilized to liberate boron in solution. Probably the best and most widely commercialized has been the use of the sparingly soluble zinc borate mentioned earlier. The low degree of water solubility exhibited by the zinc borate has allowed this preservative to pass currently utilized leaching tests and provide subsequent performance against decay fungi and wood destroying insects. The solid zinc borate slowly dissolves to yield low levels of zinc borate in the aqueous phase. This subsequently hydrolyzes to generate zinc hydroxide and boric acid. The zinc hydroxide precipitates in the wood while the low levels of boric acid can diffuse throughout the composite, providing protection where it is needed.

However, the use of zinc borate as a slow-release diffusible preservative is currently only being carried out as a treatment for wood composites. The ability to incorporate the "leach resistant" preservative during the manufacture of the wood composites is an obvious advantage over trying to generate such a complex *in situ*. Such a strategy, (*in situ* deposition of low solubility boron compounds which are also effective preservatives) has proven to be a challenging goal for the preservative treatment of solid timber, although work continues in laboratories around the world. When the balance between leach resistance through insolubility and enhanced diffusion through limited dissolution are fully understood, ways to optimize such treatments are possible. Judging by the degree of research activity in the area of diffusible preservatives, rational preservative design represents an achievable goal.

This update has touched on the wide range of technologies that can be used to treat wood with borates – from solid lumber to wood composites and, more recently, the emerging field of wood-plastic composites. Critical to the continued growth of boron-based wood preservatives will be their use in appropriate applications such as UC1, UC2, and more recently UC3A. By their very nature, diffusible borate preservatives are more forgiving of variations in treatment regimes than many of the more conventional preservative types.

Coupled with the broad spectrum of biological efficacy, this flexibility puts boron preservatives in a very strong position for the future.

References

1. Lloyd, J.D. 1997. International Status of Borate Preservative Systems. *In:* Proc. The Second International Conference on Wood Protection with Diffusible Preservatives and Pesticides. Forest Products Society. Madison, WI. pp. 45-54.
2. Manning, M.J., Lloyd, J.D. and M.W. Schoeman, 1997. The Future of Diffusible Preservative and Pesticide Systems. *In:* Proc. The Second International Conference on Wood Protection with Diffusible Preservatives and Pesticides. Forest Products Society. Madison, WI. pp. 157-168.
3. Lloyd, J.D. 1995. Leaching of boron wood preservatives: a reappraisal. *In:* Proc. Annual Conv. British Wood Pres. and Damp Proofing Association.
4. American Wood-Preservers' Association (AWPA). 2002. Standard C31-02. Lumber Used Out of Contact with the Ground and Continuously Protected from Liquid Water – Treatment by Pressure Processes. AWPA Book of Standards. AWPA, Granbury, Texas. pp. 187-189.
5. Vinden, P. 1990. Treatment with Boron in the 1990's. *In:* Proceedings of the First International Conference on Wood Protection with Diffusible Preservatives. Forest Products Research Society. Madison, WI. pp. 22-25.
6. Cockroft, R. and J.F. Levy. 1973 Bibliography on the Use of Boron Compounds in the Preservation of Wood. *J. of the Inst. of Wood Sci.* *6*(3): 28-37.
7. Barnes, H.M., Amburgey, T.L., Williams, L.H. and J.J. Morrell. 1989. Borates as wood preserving compounds: The status of research in the United States. Doc. No. IRG/WP/3542. International Research Group on Wood Preservation, Stockholm, Sweden.
8. Drysdale, J.A. 1994. Boron Treatments for the Preservation of Wood – A Review of Efficacy Data for Fungi and Termites. Document No. IRG/WP 94-30037. International Research Group on Wood Preservation, Stockholm, Sweden.
9. Morris, P.I. 2000. Integrated Control of Subterranean Termites: the Six S Approach. *In*: Proc. of the American Wood-Preservers' Association, 96. pp. 93-106.
10. International Residential Code for One- and Two-Family Dwellings, International Code Council, 2003.
11. Grace, J.K., Byrne, A., Morris, P.I. and K. Tsunoda. 2004. Six-year Report on the Performance of Borate-treated Lumber in an Above-ground Termite Field Test in Hawaii. Document No. IRG/WP 04-30343. International Group on Wood Preservation, Stockholm, Sweden.

12. Grace, J.K., Tsunoda, K., Byrne, T. and P.I. Morris. 1995. Field evaluation of borate-treated lumber under conditions of high termite hazard. *In*: Wood Preservation in the '90s and Beyond. Forest Products Society. Madison, WI. p. 240.
13. Tsunoda, K., J. K. Grace, T. Byrne, and P. I. Morris. 2002, *Mokuzai Gakkaishi 48*: 107-114 (in Japanese).
14. Grace, J.K. 1997. Review of recent research on the use of borates for termite prevention. *In*: Proc. The Second International Conference on Wood Protection with Diffusible Preservatives and Pesticides. Forest Products Society. Madison, WI. pp. 85-92.
15. Morris, P.I. 2000. Ten year Performance of L-joints Made From Borate Diffusion Treated Wood. Document No. IRG/WP 00-30225. International Group on Wood Preservation, Stockholm, Sweden.
16. Wood Protection Council. 1993. National Institute of Building Sciences, "Wood protection guidelines – protecting wood from decay, fungi, and termites", November 1993.
17. Laks, P.E. and M.J. Manning. 1995. Preservation of wood composites with zinc borate. IRG/WP 95-30074. International Research Group on Wood Preservation, Stockholm, Sweden.
18. Laks, P.E. 1999. The Past, Present and Future of Preservative –Containing Composites. In: 33rd International Particleboard/Composite Materials Symposium Proceedings. Wolcott, Tichy, Bender, Eds. Washington State University, Pullman, WA. pp. 151-158.
19. Manning, M.J. 2002. Wood Protection Processes for Engineered Wood Products. In: Enhancing the Durability of Lumber and Engineered Wood Products. Forest Products Society. Madison, WI. pp. 131-136.
20. Laks, P.E., D.L. Richter and G.M. Larkin. 2001. Preservative Systems for Wood Composites. In: Proc. 21st Annual Meeting of the Canadian Wood Preservation Association. pp. 177-187.
21. Silva, A., C. Freitag, J.J. Morrell, and B. Gartner. 2001. Effect of Fungal Attack on Creep Behavior and Strength of Wood Plastic Composites. In: Proc. The Sixth International Conference on Woodfiber-Plastic Composites. Forest Products Society. Madison, WI. pp. 73-77.
22. Simonsen J., C. M. Freitag and J.J. Morrell. 2001. Effect of Wood-Plastic Ratio on the Performance of Borate Biocides Against Brown Rot Fungi. In: Proc. The Sixth International Conference on Woodfiber-Plastic Composites. Forest Products Society. Madison, WI. pp. 69-72.
23. Verhey, S.A., P.E. Laks and D.L. Richter. 2001. The Effect of Composition on the Decay Resistance of Model Woodfiber-Thermoplastic Composites. In: Proc. The Sixth International Conference on Woodfiber-Plastic Composites. Forest Products Society. Madison, WI. pp. 79-86.
24. Manning, M.J. and F.M. Ascherl. 2003. Borates as Fungicides in Wood-Plastic Composites. In: Proc. The Seventh International Conference on Woodfiber-Plastic Composites. Forest Products Society. Madison, WI. pp. 69-78.

25. Morton, J. Current and Emerging Applications for Natural & Wood Fiber Composites. Paper presented at the 7th International Conference of Woodfiber-Plastic Composites, Madison, WI. May 19-20, 2003.
26. Pendleton, D.E., T.A. Hoffard, T. Adcock, B. Woodward, and M.P. Wolcott 2002. Durability of an Extruded HDPE/Wood Composite. *Forest Prod. J.*, *52*(6):21-27
27. Mankowski, M.E., Ascherl, F.M., and M.J. Manning. 2005. Durability of Wood Plastic Composites Relative to Natural Weathering and Preservative Treatment with Zinc Borate. IRG/WP 05-40316. International Research Group on Wood Protection, Stockholm, Sweden.
28. Zabell, R.A. and J.J. Morrell. 1992. Wood Microbiology: Decay and Its Prevention. Academic Press Inc. San Diego, CA. pp. 90-91.
29. Mankowski, M. and J.J. Morrell 2000 Patterns of Fungal Attackin Wood-Plastic Composites Following Exposure in a Soil Block Test. *Wood Fiber Sci.*, *32*(3):340-345.

Chapter 27

In-Process Protection of Wood Composites: An Industry Perspective

Glenn M. Larkin[1], Paul Merrick[2], Marek J. Gnatowski[3], and Peter E. Laks[1]

[1]School of Forest Resources and Environmental Science, Michigan Technological University, Houghton, MI 49931
[2]Level Research Center, Weyerhaeuser Company, P.O. Box 8449, Boise, ID 83707
[3]Polymer Engineering Company, Ltd., Burnaby, British Columbia V5B 3A6, Canada

Wood-based composites are commonly used as construction materials. In many of these applications, there is a potential for fungal and insect attack (e.g. millwork, sheathing, exterior siding, and decking). Traditional pressure, spray, and dip treatments suitable for solid lumber typically cannot be employed to treat wood-based composites. In-process application of preservative formulations, however, has proven to be a feasible option. Inorganic borate systems, particularly zinc borate, have an established commercial track record. Their success as an in-process preservative system can be attributed to meeting five basic attributes: they are relatively safe to use, have minimal environmental impact, have regulatory acceptance, are compatible with most wood composite manufacturing processes, and are economical. This chapter discusses the use of in-process preservative systems from a North American industrial perspective, and is aimed at formulation chemists who are developing biocides for in-process treatment by the wood composite industry. This chapter is based on the oral presentation, "In-Process Protection of Wood Composites: An Industrial Perspective" by authors Merrick and Gnatowski at the ACS 229th National Meeting In San Diego, CA, March 16, 2005.

Over the last thirty years there has been an increase in the use of wood-based composite building materials. They increasingly substitute for solid lumber in applications exposed to a risk of fungal and insect attack such as millwork, home framing products in termite prone areas, decking, and exterior siding (*1*). Wood-based composites used in these exposures commonly contain a wood preservative system. These uses of preservatives have sparked interest in new chemistries and technologies (*3, 4*). The Use Category System is a useful tool developed by the American Wood-Preserver's Association that correlates biodeterioration hazard to product application. Examples of the above applications include (*2*):

UC2 - sill plates;
UC3A - exterior siding, and coated millwork;

Solid wood products used in UC2 and UC3A exposures are frequently pressure treated with a waterborne preservative system, but most wood-based composites are not suited to this treatment approach due to permeability and dimensional instability issues. Gardner and Walinder (*5*) discussed the various strategies for manufacturing wood-based composites that are resistant to attack by fungi and insects. These strategies include in-process and post-process treatment, use of durable wood species and recycled preservative treated wood, and the use of chemically modified wood.

In-process treatment is the current, commercialized approach discussed by Gardner and Walinder (*5*). It is commercially used to treat particle-based composites including oriented strand board (OSB), laminated strand lumber (LSL) and medium density fiberboard (MDF). Figure 1 shows a typical OSB manufacturing process and the points that are most practical for preservative addition. During in-process treatment, a preservative can be applied to either wet or, more commonly, dry furnish, or the combined wood, adhesive, and water repellent that constitute the composite matrix. Wet flakes are treated after stranding before they are dried to allow time for preservative to diffuse into and fix (chemically bond) with the wood. The dry furnish is treated in the blender via biocide addition to either the wood, and/or adhesive, or combined with the water repellent. The preservative is distributed throughout the furnish during the blending process.

This chapter summarizes the selection criteria for biocides intended for wood composite use from the perspective of a wood composite manufacturer. Zinc borate is used to illustrate how these criteria can be met.

Key Attributes of an In-Process Biocide Additive

The key attributes of an in-process preservative additive can be summarized as follows:

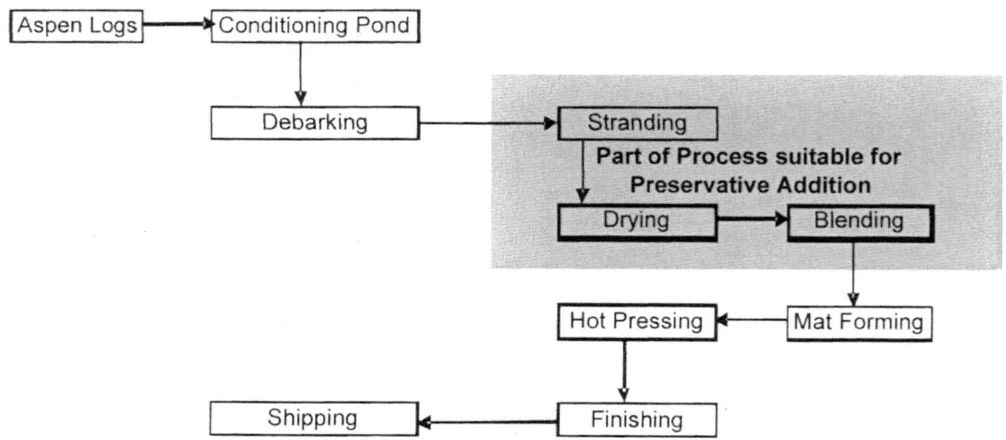

Figure 1. A schematic of the OSB manufacturing process showing the most practical stage for in-process wood preservative treatment. Adding the preservative to the flakes either prior to drying or to the furnish in the blender makes the most sense (gray shaded area). A detailed description of the OSB manufacturing process may be found elsewhere (6).

Is Safe
Low Environmental Impact
Meets Regulatory Requirements
Good Technical Performance
Favorable Economics

These attributes were first summarized by Laks (*7*), and are expanded upon here based on the authors' (Merrick and Gnatowski) industrial experience.

Safety

The toxicological profiles of the preservative system and of the treated composite are critical considerations for the manufacturer. Exposure of plant workers, supply chain personnel, secondary processors, and the end user must be considered. The manufacturer must:

- Plan for safe transportation, delivery, and storage of the preservative system;
- Provide for safe blending, dilution, or mixing of preservative systems by plant personnel. For this reason, a ready-to-use preservative formulation is usually preferred;
- Properly handle volatile off-gassing during the manufacturing process; and
- Confine dangerous process operations (when practical) to specific areas within the plant.

The treated wood composite will eventually be handled by construction workers and homeowners who may not use the same level of personal protective equipment. For this reason, the finished product must be safe for the secondary processors to machine and end users to handle.

Environmental Impact

Preservative systems used in wood-composite production may negatively impact existing manufacturing facility permits. Their use may also affect the manner in which waste streams are handled. A biocide that does not have a negative impact on existing permits will be more favorably received.

All plants generate waste streams that may be partially recycled. This may include re-use in the manufacturing process as furnish, by-product sale for other uses, and burning as boiler fuel. Some waste may be sold to other wood-composite manufacturers. For example, short strands from LSL production may be sold to OSB manufacturers. Secondary processor and end user waste disposal also needs to be considered. For example, window manufacturing plants may sell their sawdust as animal bedding, and construction site waste may be used as mulch. Waste is most frequently burned as fuel in the plant boiler. Therefore, the preservative system must not interfere with pollution control devices installed on the furnace, nor with ash disposal. The preservative system's contribution to air emissions must be considered against the current permitted levels.

Regulatory Acceptance

The safety and environmental impact of a wood preservative determine its ability to comply with legally required registrations and permits. There may also be commercially desirable listings for the preservative system and/or treated wood composite product.

Legal

Wood preservatives which claim pesticide attributes must have regulatory approval. Registrations are necessary with the appropriate agencies at the Federal, State, and local levels. In the United States the EPA, and in Canada the PMRA, manage the registration of pesticides at the Federal level. A partial list of preservative treatments for wood composites manufactured in the USA has been published by Laks (*1*).

In addition, if there is any emission related to biocide use, the manufacturing plant itself will have to obtain a permit at the state level for emission and waste disposal. For this reason, such emissions are not desirable and emission free biocides will receive preferential consideration.

Commercial

Secondary manufacturers and distributors of the treated composite may request that the preservative system and/or treated wood composite product be listed with various standards, codes or trade association bodies. This generally involves a substantial cost and effort in the evaluation of a preservative treated wood-based composite's efficacy and suitability for a particular end use. Pre-existing listings and support from the biocide supplier are required. Examples of trade association, standards bodies, and code agencies that recognize preservative actives and treated wood-based composite are given below:

- American Wood Preservers' Association (AWPA) - A respected association that writes treatment and performance standards for wood preservatives and treated wood products. AWPA Standards are referenced by the model building code (*2*).
- International Code Commission Evaluation Service (ICC-ES) – ICC-ES establishes standards for the model Building Codes. ICC Wood preservative listing requirements for the ICC are outlined in Acceptance Criteria (AC) 326 (*8*).
- Window and Door Manufacturer's Association (WDMA) - A well established trade association that writes performance specifications for manufacturers of doors, skylights, and windows in North America.

Technical Performance

The technical performance of a preservative for in-process incorporation into wood and wood-based composites ultimately needs to be evaluated by each manufacturer for their process, product, and intended end uses. There are, however, common elements for such assessments. Although each case is unique, a typical checklist might include the following:

Chemical Performance

- Is the preservative (formulation) stable while stored at the plant?
- Is the preservative compatible with other components of the composite, especially the adhesive? For example, amine and metal ions have the potential to catalyze the reaction of isocyanate adhesives which results in poor bonding between wood particles (*9*). Some borates react with phenols in phenol formaldehyde resins (PF) to preventi complete adhesive curing (*10*). Biocides may decompose in the strongly alkaline environment of the PF glue line rendering them ineffective (*11*).
- Does the preservative "bond" to the wood particle? Is there chemical fixation? Adsorption? Biocide fixation in the glue line may interfere with biological activity.
- Is the preservative corrosive to plant equipment?
- Will the preservative produce unsafe volatile by-products?

Manufacturing Performance

- Can the preservative be used with existing plant equipment? Is it easy to use?

- Will it arrive ready-to-use or require on-site tank mixing and/or dilution?
- Is it an emulsion, powder, or solution? What safety precautions are needed?
- Will the preservative tolerate the manufacturing process? How stable is it with respect to press temperatures, which may be in excess of 200°C?
- Is the preservative able to provide protection to multiple wood species?
- Will one formulation of the preservative allow use with two or more adhesives?

End Use Performance

- Does the preservative afford adequate protection for the intended end use?
- Is there an analytical method that allows for quick preservative assays in the plant setting? Are there clearly defined QA/QC criteria for the preservative and for the finished composite material?
- Do the preservative formulation and the treated wood composite have an acceptable environmental profile?
- Will the preservative supplier provide ongoing technical support?

These questions, while not exhaustive, provide a framework for the types of questions that need be asked when assessing the suitability of a preservative formulation for use within a wood composite manufacturing process. Wood composite manufacturers expect that preservative formulation suppliers actively assist in answering these questions.

Economics

The choice of which preservative system to incorporate into a wood composite is determined by whether the cost plus delivered performance is recognized by the customer as having value. Does the cost of the treated wood composite compare favorably with alternative building materials, such as treated lumber or steel?

In almost all cases, the treated wood composite will be more expensive to manufacture than the untreated version. Will the treatment enable the product to more effectively perform over the expected service life of the application? Will the addition of a preservative treatment enable the product to enter new markets, thus expanding the overall volume of product produced and perhaps lowering the manufacturing cost structure? These questions must be answered before any serious developmental work on new biocides or in-process treatment processes occurs.

Zinc Borate as an In-Process Composite Wood Preservative

Two forms of zinc borate are commonly used commercially (Table 1). One has been approved by the EPA for use as a wood preservative, and finds use as a biocide in the treatment of wood composites. Both are used as fire retardants in plastics. The number designations refer to chemical structure.

Table 1. Examples Commercially Available Zinc Borate Compounds

Commercial Reference	Chemical Formula	ZnO (%)	B_2O_3 (%)	H_2O (%)	Water Solubility (%)
ZB2335[a]	$2ZnO{\cdot}3B_2O_3{\cdot}3.5H_2O$	38.2	48.2	13.6	0.28
ZB223[b]	$2ZnO{\cdot}2B_2O_3{\cdot}3H_2O$	43.5	34.2	19.2	0.04

[a] AWPA Standard P18-04, Nonpressure Preservatives (Allows In-Proccess Treatment of Wood Composites)

[b] 223 & 2335 forms used as a fire retardant & smoke suppressant in plastics industry

Zinc Borate Safety, Regulatory Issues, and Environmental Impact

Zinc borate is currently the most commonly used in-process preservative for the protection of wood-based composites and has been in use for over a decade. It (ZB) is commercially available in several grades, and from several manufacturers in the United States. Type ZB2335 is registered with the United States Environmental Protection Agency (EPA) and listed in AWPA Standard P18-04, *Non Pressure Preservatives* (*2*) for use in wood composites. Commercial products differ including particle size (Figure 2) which may affect handling during processing. For example, fine particle size may cause excessive dust which would not be desirable.

Pure zinc borate is considered safe and environmentally friendly. It generally has a very low worker exposure hazard and relatively low environmental impact. However, all commercial preservative formulations, including zinc borate, may contain undesirable contaminants. An example of contaminants found in commercial samples of zinc borate is shown in Table 2.

Zinc Borate Technical Performance

The technical performance properties of zinc borate are well established and its use as a preservative for in-process treatment of wood composites is supported in the literature (*9, 12*).

Zinc borate arrives at the manufacturing plant as an easy to store and use powder, most commonly in palletized super sacks. At its current use rate, ZB is

Zinc Borate 1

Zinc Borate 2

Zinc Borate 3

Zinc Borate 4

Figure 2. Electron photomicrographs of 4 types of zinc borate (5000x). Note the differences in particle geometry and size. Photos from Marek Gnatowski.

Table 2. Select Elemental Analysis in four Commercial Samples of ZB2335[a]

	Zinc Borate (ppm)				
Element	A	B	C	D	Detection Limit
Antimony	---	---	1660	---	0.2
Arsenic	---	---	16	6	0.2
Barium	---	---	2	234	0.2
Cadmium	0.5	0.7	5	9	0.04
Calcium	1060	1230	1330	1030	10
Chromium	0.6	2.3	---	---	0.2
Copper	12	6	8	7	0.2
Lead	3	4	45	434	0.2
Sodium	95	---	689	143	10
Strontium	0.5	---	19	34	0.2

[a] Data from Marek Gnatowski

compatible with other wood composite components including water repellants and adhesives such as polymeric diphenylmethyl diisocyanate (pMDI) and phenol formaldehyde (PF). Discrete zinc borate particles (in OSB) physically adhere to the wood-adhesive interface during panel manufacture. It is considered a relatively "leach resistant" wood preservative with low water solubility. When exposed to moisture, a slow hydrolysis reaction follows (Equation 1), yielding zinc hydroxide and boric acid (*13*). Boric acid can then diffuse away from the glue line and into the wood fiber where it provides protection.

$$2ZnO \cdot 3B_2O_3 \cdot 3.5H_2O + 7.5H_2O \rightarrow 2Zn(OH)_2\downarrow + 6H_3BO_3 \quad (1)$$

This limited leaching in contact with environmental moisture is an important attribute of biocides intended for wood composites used in the construction industry. It makes it well suited for many applications which are out of ground contact and protected from direct weathering (AWPA UC1–UC3A) (*2*). Examples include trusses, sill plates, door and window millwork, exterior trim, and siding. Zinc borate is not considered corrosive to standard manufacturing plant equipment, or to metal fasteners used in construction.

From a dry-process composite manufacturing perspective, zinc borate is relatively easy to use. Typically, powdered ZB is added to the wood furnish, adhesive, and other additives in a blender on a percent weight basis. Alternatively, it may be suspended in a slurry with wax emulsions and spray applied (*14*). Zinc borate does not chemically degrade during wood composite manufacturing processes.

Manufacturing quality assurance/quality control (QA/QC) is relatively fast, with the potential to assay treated materials in a matter of minutes using X-Ray fluorescence (for zinc, with indirect calculation of boron content).

Economics

Zinc borate currently is low cost. It is not a formulated product, so there are no additional costs for co-solvents and diluents. It provides cost effective protection to wood composites against a wide variety of wood decay fungi and insects, including termites. The overall use cost and protection efficiency should be analyzed when considering a new biocide for in-process treatment of a wood composite.

Conclusion

In-process preservative treatment of wood-based composites has proven to be commercially viable for more than a decade. In light of the current technologies available, it is an economical approach to the protection of particle-based wood composites including OSB, MDF, LSL and particleboard. Important criteria in evaluating a candidate in-process preservative system are safety, environmental impact, regulatory acceptance, technical performance, and economics. Each manufacturer will place a different value and emphasis on these criteria. With continued research formulation chemists may develop new in-process treatments for wood-based composites which may become successfully commercialized. Zinc borate serves as an example.

Zinc borate is currently the most commonly used in-process preservative system for OSB and LSL. Research and commercial use have established that ZB has good efficacy as a wood preservative against wood decay fungi and termites when used in the appropriate applications. Successful commercial use of zinc borate demonstrates it has an acceptable, and minimal impact on production processes and satisfies all of the established criteria for use as an in-process preservative for wood-based composites outlined in this chapter.

References

1. Laks, P.E. Protection of Wood-based Composites. Proceedings: *American Wood Preservers Association* **2004**, *100*, 78-82.
2. *American Wood Preserver's Association Standards* **2005**: Selma, AL
3. Laks, P.E. In: *33rd International Particleboard/Composite Materials Symposium Proceedings*; Wolcott, M.P.; Tichy, R.; Bender, D.F.; Eds; Washington State University: Pullman, WA, **1999**, 151-158.
4. Smith, W.R.; Wu, Q. *Forest Products J.* **2005**, *45*[2], 8-17.
5. Gardner, D.J.; Wålinder, M.E.P. In: *Wood Deterioration and Preservation Advances in a Changing World* Goodell, B.; Nicholas, D.D.; Schultz, T., Eds; ACS Symposium Series 845; American Chemical Society: Washington, DC, **2003**; 399-419.
6. Yougquist, John A.; In Wood Handbook: Wood as an Engineering Material; FPR-GTR-113. Madison WI: USDA Forest Service, Forest Products Laboratory, **1999**, 10-13 – 10-14.
7. Laks, P.E.; Palardy, R.D.; *Protection of Wood-Based Composites*; Preston, A.F.; Ed. Forest Products Society: Madison, WI , **1993**, 12-17.
8. ICC-ES -AC326 - *Acceptance Criteria for Proprietary Wood Preservative Systems – Common Requirements for Treatment Process, Test Methods and Performance*; ICC Evaluation Service, Inc. **2006**, Whittier, CA
9. Tiele,L.; Becker, R. In: *Adv. In Urethane Science and Technology* Frisch,K.C.; Klempner, D., Eds.; Technical Publishing Company: Lancaster, PA, **1993**, 59-85
10. Laks, P.E.; Haataja, B.A.; Palardy, R.D.; Bianchini, R.J. *Forest Products J.* **1998**, *38*(11), 23-24

11. Collins, P.A.; Kennedy, M.J.; Vella, R.D., Stability of bifenthrin in a commercial plywood glue, *International Research Group on Wood Preservation* **2003** Doc. IRG/WP/03-30311
12. Laks, P.E. and M.J. Manning, *Proceedings: Second International Conference on Wood Protection with Diffusible Preservatives and Pesticides* Forest Products Society: **1997**, 62-88.
13. Schubert, D.M.; Alam, F.; Visi, M.Z.; Knobler, C.B. *Chem. Mater.* **2003**, *15*, 866-871
14. Fookes, D.; Gnatowski, M.J.; Pike, R.L.; Templeton, D.A.; 1999. US Patent 5,972,266.

Chapter 28

Organic Preservative Systems for the Protection of Wood Windows and Doors

Alan S. Ross

Kop-Coat, Inc., 3020 William Pitt Way, Pittsburgh, PA 15238

For over 70 years, the manufacturers of wood windows and doors have utilized preservative systems based on organic and organometallic chemicals to provide their products with protection against decay, mold, termites and water damage. For the first 50 of those years, pentachlorophenol (PCP) was the basic preservative of the industry. This was replaced with products having improved safety and environmental attributes such as tri-n-butyltin oxide (TBTO) and 3-iodo-2-propynyl butyl carbamate (IPBC). In recent years, a new generation of organic preservatives which focus on long-term performance has been introduced. These materials are designed to further increase the service life of wood windows and doors while maintaining excellent health, safety and environmental properties. This chapter reviews the evolution of treatments and treating methods and also examines several of the new generation treatments for wood windows and doors.

Introduction

Wood has long been the favored building material for residential windows and doors. It imparts a warm, high quality look, is easy to fabricate, has excellent strength properties and is a renewable resource. In spite of these advantages, in recent years wood has been losing market share to plastic – particularly in the residential window market. Wood's growth, which was 4.6% per year in the 1999-2004 period, is expected to slow to just 1.1% per year between 2004 and 2009 (*1*). Plastic windows, however, are projected to grow at an annual rate of 7.9% over the same period (*1*). What accounts for this disparity? Much of the increased use of plastic is due to its lower cost, but there is also a perceived advantage of superior long-term durability. Wood, being a natural material, faces the stigma of ultimately being susceptible to attack from decay, mold and termites, while plastic is considered to be generally impervious to these challenges. Wood windows and doors have been protected from decay and other organisms through treatment with organic preservatives since the 1930's, and they have an excellent record of performance in service. The use of plastic, however, has raised the bar on long-term durability expectations. The industry is responding to this potential market threat by improving on window design and incorporating new higher-performance preservative systems in their products. This is not the first time the wood window and door industry responded to a threat from an alternate building material. Over 70 years ago, metal windows posed a similar challenge.

Historical Perspective

Preservatives

Originally, wood windows and doors contained no preservatives. However, by the 1930's it was recognized that these materials were susceptible to decay and insect attack, particularly in areas of high moisture contact. The metal industries of the 1930's, in search of new markets, were promoting metal windows and using as their principal sales argument the fact that wood rots, warps, shrinks and swells (*2*). In response to this threat, the manufacturers of wood windows and doors upgraded designs and began evaluating the use of preservative chemicals and water repellents to protect the wood. Early materials evaluated included zinc chloride, beta naphthol and several chlorinated phenol isomers (*2*).

Dr. E. E. Hubert of the Western Pine Association evaluated over 25 candidate preservatives and in 1936 concluded that pentachlorophenol (PCP) provided the best performance in controlling wood decay and staining without imparting color to the wood (*3*). In 1938, pentachlorophenol was adopted as the standard preservative treatment for wood windows and doors, and it remained the workhorse of the industry for nearly 50 years.

In the mid 1980's, the U.S. EPA, acting in response to concerns over safety and environmental issues, placed restrictions on the use of PCP in many applications including the treatment of wood windows and doors. This necessitated the development of a second generation of preservatives. Tri-n-butyltin oxide (TBTO) had been utilized since the 1960's as a preservative for external joinery (millwork) in the U.K. and in the U.S. in antifouling paints and as a mildewcide for coatings (*4*). Similarly, 3-iodo-2-propynyl butyl carbamate (IPBC) was developed by Troy Corporation in the late 1970's as a paint mildewcide (*5*). During the mid 1980's, both of these preservatives were adopted for use in the treatment of wood windows and doors in the U.S., typically as 0.5% (active ingredient) solutions in a mineral spirits carrier. To a lesser extent, 2-(thiocyanomethylthio)benzothiazole (TCMTB) was also used for this application. Today, TBTO, IPBC and TCMTB are still used as preservatives for wood windows and doors. However, their use is being supplanted by a new generation of preservative systems, as will be discussed below.

Substrates

Historically, white pine was the basic wood species utilized in the manufacture of wood windows and doors in the United States (*5*). In the late nineteenth and early twentieth centuries, window manufacturers located their facilities along the upper Mississippi River and its tributaries due to the abundant supply of white pine in that region. This area, encompassing the states of Minnesota, Iowa and Wisconsin, remains today as a major center for the production of wood windows and doors even though the white pine supply is long exhausted. As the supply of white pine began to dwindle in the 1920's, producers looked west for sources of sugar pine and ponderosa pine. These species were chosen for their uniformity, ease of treatment and relative abundance in the western states. A second concentration of manufacturers soon developed in close proximity to the supply of these species in the states of Oregon and Washington. By the turn of the twentieth century, supplies of high quality sugar pine and ponderosa pine began to diminish, as well. As a result, window and door manufacturers are now incorporating alternative wood species and wood composites in their products.

Treating Methods

Most of the wood used in the manufacture of windows and doors in the United States is treated by immersion in the preservative solution. This has been the case since treating was started in the 1930's. Up until about 20 years ago, virtually all components were bundled and dipped *en masse* in an immersion tank for three minutes. After being allowed to drain, the bundle was either air dried or placed in an oven or conditioning room to drive off the solvent. To improve production throughput, many manufacturers today employ in-line treating systems. In these processes, individual milled components are passed on a conveyer line through a trough or spray of treating solution and then into a drying oven. As the parts emerge from the drying oven, they are ready for painting or further processing. Typical immersion times on in-line treaters are on the order of 15-30 seconds. Because parts are treated individually rather than as components of large bundles, appropriate penetration and retention of the preservative solution can be attained using these shorter immersion times. Many manufacturers today employ solvent recovery systems to reduce air emissions and enhance cost efficiency.

In Europe, the double vacuum process, which had been widely used to treat millwork for many years, has recently fallen out of favor due to environmental concerns over solvent emissions. In double vacuum treatments, the wood components are placed in a treating cylinder. After an initial vacuum is applied, the cylinder is flooded with the treating solution. The solution is then drained, and a final vacuum is applied. This process is only suitable for the application of low viscosity organic-solvent preservatives such as mineral spirits, and in situations where only limited penetration is necessary. It was most often used for the treatment of Scots pine as the sapwood of this species is readily treated while the heartwood possesses reasonable natural durability (*6*).

Standards

Window and door manufacturers realized early on that standards would be essential in ensuring that proper treatments were utilized to preserve and protect their products. As early as 1936 a Preservative Standards Committee was established by the industry. With help from the chemical suppliers and public and private research laboratories, the committee developed a "Code of Minimum Standards for Millwork Preservatives" and set up a "Seal of Approval Program." The program's stated objectives were: To counteract the propaganda of metal competitors who were making capital of the fact that wood rots; to stimulate the preservative treatment of millwork; to provide a method of identifying treated millwork; to establish minimum standards for treating solutions; and to establish minimum standards for the method of treatment (*7*). Over the years, this organization evolved into what is today the Window & Door Manufacturers Association (WDMA). The basic standard governing the treatment of wood window and door components is WDMA I.S.4 – 2005: Water-Repellent Preservative Non-Pressure Treatment for Millwork (*8*). The American Wood-Preservers' Association (AWPA) has also developed a standard for the preservative treatment of millwork components, AWPA N1-04: All Millwork Products – Preservative Treatment by Nonpressure Processes (*9*).

Special Protection Needs

In addition to decay resistance, wood windows and doors may require special protection needs depending upon their expected service life and exposure conditions. Three important protection considerations are water repellency, termite resistance and mold resistance. Each of these is discussed below.

Water Repellency

Wood in the presence of moisture has a tendency to swell, warp and check. In addition, moisture contents of 20% and higher can promote growth of decay and mold fungi. For these reasons, water repellents are an integral component of most millwork treating systems. Both of the major treating standards, WDMA I.S.4 – 2005 and AWPA N1-04, include a requirement for water repellency of the treating solution. The standard test method cited by each is the WDMA Test Method TM-2 Swellometer Test, "Test Method to Determine the Water-Repellent Effectiveness of Treating Formulations"(*10*). In this test, treated and untreated matched ponderosa pine end grain wafers are immersed in water. The swelling of each wafer is measured in the tangential direction, and the swelling of the treated wafer is compared to that of its matched untreated partner. Performance is expressed as water-repellent effectiveness. Typically, a water-repellent effectiveness of not less than 55% is required to meet the standard.

Historically, water repellents were hydrocarbon waxes, although today a variety of waxes and polymers are employed. Since many parts are painted subsequent to treatment, it is important for the water repellents to not interfere with paint adhesion or drying properties of the paint. This can be particularly challenging when water-based paints are coated over water-repellent preservatives.

Termite Resistance

Wood windows and doors are susceptible to attack from termites. Pentachlorophenol functioned as both a fungicide and an insecticide when it was in use; however, the fungicides which replaced it in the mid-1980's were not effective insecticides. As a consequence, millwork treaters began, in the late 1980's, to incorporate chlorpyrifos as an insect control additive to their treating solutions. Chlorpyrifos is a broad spectrum insecticide which is widely used for many applications including the protection of crops. In recent years, its use has been restricted from some household and remedial applications by the U.S. EPA, although its use as a millwork treatment is still permitted. In the past five years, permethrin, a member of the synthetic pyrethroid family of insecticides, has begun to replace chlorpyrifos in millwork protection applications. Both chlorpyrifos and permethrin are used at a level of 0.1% as active in the treating solutions.

Termite resistance is not a requirement of the WDMA or AWPA millwork treating standards; however, both standards recognize the need for a termiticidal additive, especially when windows are in service in areas of high termite hazard. Although termites have been found in all states except Alaska, the main regions of concern are Hawaii and the southern and Gulf Coast states. It is estimated that over 80% of wood windows manufactured in the U.S. are protected with an insecticide treatment.

Mold Resistance

In just over five years, the topic of mold growth in indoor environments has evolved from a minor nuisance area to a major public health controversy (*11*). Reactions to mold concerns have been dramatic: regulatory agencies have made mold a top indoor air quality issue; thousands of mold-related lawsuits have been filed; and insurance companies have almost entirely eliminated homeowner coverage for mold-related claims. Most importantly, the public has made up its mind that mold growth indoors is no longer acceptable. This presents a special concern for suppliers of wood and wood composite-based building materials such as windows and doors. These substrates, if unprotected, provide excellent surfaces for mold growth. Fortunately, many of the same wood preservatives used to protect wood windows and doors from decay fungi are also effective against mold.

New Directions in Millwork Protection

Preservatives

A new generation of millwork preservatives has been introduced in recent years (see Table I for a list of biocides used in these formulas). Instead of relying upon one active ingredient, the new formulas contain multiple active ingredients. There are two advantages to this approach. First, the multicomponent systems offer a broader spectrum of protection against the wide variety of mold and decay fungi present in the environment. Secondly, combinations of active fungicides sometimes provide synergistic effects enhancing efficacy at lower required concentrations. For example, one such product is based upon three fungicides: IPBC, propiconazole and tebuconazole. These three ingredients are present at 0.21% each for a total preservative content of 0.63%. They have been shown in laboratory testing to provide seven times the efficacy against brown rot fungi to a system based upon 0.5% IPBC alone (*12*).

Although most millwork preservatives continue to rely upon solvent-based carriers such as mineral spirits and naphtha, there are several products which use water as the preservative carrier. Advantages of using water-based preservatives are lack of flammability concerns, lower odor, and lower air emissions. In addition, with the price of oil continuing to rise, water is a much more cost efficient carrier than petroleum-based solvents. However, there are some disadvantages to using water-based systems. Water requires much more energy to evaporate, increasing drying times and slowing production speeds. Water also has a tendency to cause wood to swell which can affect the dimensional stability of machined millwork components. In some cases, it can cause grain raising which requires an additional sanding step before cladding or painting. For these reasons, water-based treatments are not as widely used as solvent-based systems. However, as long as petroleum prices increase and air quality regulations become more stringent, there will be an incentive to develop a non-petroleum based carrier with low air emissions which can offset some of the disadvantages of water-based treatments.

Table I. Biocides Used in Millwork Protection in the U.S. - 2006

Common Name or Abbreviation	*Chemical Name*
Fungicides	
IPBC	3-iodo-2-propynyl butyl carbamate
TBTO	Bis(tri-n-butyltin)oxide
ZB	Zinc borate
Propiconazole	cis-trans-1-[2-(2,4-dichlorophenyl)-4-propyl-1,3-dioxolan-2-ylmethyl]-1H-1,2,4-triazole
Tebuconazole	(RS)-1-p-chlorophenyl-4,4-dimethyl-3-(1H-1,2,4-triazol-1-ylmethyl)pentan-3-ol
TCMTB	2-(Thiocyanomethylthio)benzothiazole
Insecticides	
Chlorpyrifos	O,O-diethyl O-3,5,6-trichloro-2-pyridyl phosphorothioate
Permethrin	3-phenoxybenzyl (1RS)-cis,trans-3-(2,2 dichlorvinyl)-2,2-dimethlycyclopropanecarboxylate
Imidacloprid	1-{(6-chloro-3-pyridinyl) methyl}-N-nitro-2-imidazolidinimine

Substrates

As noted above, the supply of high quality ponderosa pine, which has been the basic component of wood windows for over 50 years, is diminishing. As a consequence, manufacturers are looking at two approaches for their substrates. The first is the use of alternate species such as poplar, maple and radiata pine. Poplar and maple are available domestically while most radiata pine is being imported from

Chile and New Zealand. These species do not have the inherent decay resistance of ponderosa pine. Therefore, the newer multicomponent preservative systems with their improved efficacies are better suited to protect them.

The second approach involves the increasing use of wood composites in window and door manufacture. Wood composites are combinations of wood components, such as flakes, veneers and strands, with non-wood components such as resins and plastics. The advantages of wood composites are cost, uniformity and availability. However, if the composite contains more than about 40% of wood components, it is susceptible to attack from decay and mold fungi as well as termites. As a consequence, it must still be treated with an appropriate preservative and insecticide system. One method for treating wood composites is to combine the preservative and water repellent with the wood furnish (flakes, strands or chips) prior to processing. This type of integral treatment has the advantage of providing protection throughout the thickness of the entire part. Of course, the preservative and water repellent must be able to withstand the processing conditions which can involve exposure to pressure and elevated temperatures. One such preservative used to treat wood composites in this manner is zinc borate. The zinc borate is introduced as a powder, or it is sprayed on to the wood flakes as a water-based emulsion prior to final processing (*13*). At the proper retentions zinc borate offers protection against both decay and termite attack; however, it is not particularly effective against surface mold.

Treating Methods

As noted above, most window manufacturers today treat their wood components through immersion in a wood preservative solution. At least one window manufacturer uses pressure treatment. This process can result in more uniform and deeper penetration of the preservative into the treated part, but it is a much costlier and slower process than immersion treatment.

Millwork manufacturers are thus faced with a dilemma: to date, greater preservative penetration can only be achieved with processes having slower line speeds and greater treating costs (e.g., double vacuum or pressure treatment). In-line immersion treating systems are efficient, but they only result in an envelope or outer shell treatment of the preservative. With the right preservative, this can still result in many years of service life. However, in extreme exposure environments or in situations where the protective envelope is breached, a more complete penetration of the preservative may be desired.

One recent development which addresses the penetration vs. line speed dilemma is the Penetrating Barrier System. In this system, an organic solvent-compatible borate ester is combined with a traditional mineral spirits-based water-repellent wood preservative. Parts can be treated in existing immersion equipment alleviating the need for costly treating cylinders. Initially after treatment, the borate ester is present in the outer envelope along with the traditional preservative. However, when the part is exposed to moisture or humidity, the borate ester hydrolyzes to boric acid which is then capable of penetrating to the core of the treated part through diffusion. Testing has shown that leaching of the borate out of the treated part is minimal under AWPA Use Category 3A exposure conditions (*14*). The main advantage of the Penetrating Barrier System is that it can provide penetration to the core of treated parts without the need for special equipment such as pressure cylinders.

Special Needs

Although water repellents present some challenges for paintability of treated parts, it is likely that they will continue to be incorporated in millwork preservative systems. In addition to providing efficacy against fungal attack by keeping moisture out, water repellents also help to maintain dimensional stability of the wood in case of exposure to moisture or high humidity.

Insecticidal additives will also continue to be used in millwork treatment. Subterranean termites are a problem in many parts of the U.S., and Formosan termites present a special challenge in Hawaii, the Gulf Coast and the Southeast. In areas of termite activity, wood windows need to be protected since they are one of the first points of attack in the home. Several U.S. window manufacturers export products to Japan. Japanese standards require termite protection in all wood windows, and they do not allow the use of chlorpyrifos for this purpose. For this reason, permethrin is used to protect exported windows. In the future, other insecticides such as deltamethrin and imidacloprid will see broader use in millwork products.

The need for mold protection of wood windows and doors continues to increase. The consuming public is firmly convinced that any mold growth in the home is unacceptable. Moreover, recent hurricanes (Ivan, Jeanne, Francis and Charlie in 2004 and Dennis, Katrina, Rita and Wilma in 2005) and associated flooding in the Southeast and Midwest have greatly compounded problems with mold growth. Interior windowsills and jambs and patio doors are often exposed to moisture and are potential targets for mold growth. Treatment of these components with millwork preservatives at the point of manufacture helps resist mold growth. Active ingredients with particularly good mold resistance include IPBC and isothiazolones. Additionally, primers and topcoats may contain these additives to help resist mold growth.

Conclusions

The U.S. wood window and door industry has a long history of providing high value, long-lasting products. Although the use of plastic has increased in recent years, wood and wood-based composites will continue to be major substrates in window manufacture. As such, there will always be a need to protect these materials from biological and environmental hazards such as decay, mold, termites and water damage. This need will be met by modern wood preservative treatments which offer the attributes of long-term efficacy, cost-effectiveness, ease of use, and favorable health, safety and environmental profiles.

References

1. Anonymous, *Door & Window Manufacturer Magazine* **September 2005**, *6(6)*, 72.
2. Hubert, E. *Indus. Engineering Chem., Indus. Ed.* **1938**, *30*(11), 1241.
3. Hubert, E. *Indus. Engineering Chem., Analytical Ed.* **1940**, *12*(4), 197.
4. Ross, A. S. *Amer. Wood-Preservers' Assoc. Proc.* **1988**, *84*, 309.
5. Ross, A. S.; Cheeks, C. N.; Smith, L. L. *Amer. Wood-Preservers' Assoc. Proc.* **2001**, *97*, 79.
6. Richardson, B. A. *Wood Preservation;* The Construction Press: Lancaster, England, 1978; p 98.
7. Lance, O. C. *History and Development of Wood Preservation for Millwork in the United States.* National Woodwork Manufacturer's Association (now Window & Door Manufacturers Association): Des Plaines, IL, 1958.
8. WDMA I.S. 4-2005; Industry Standard for Water-Repellent Preservative Non-Pressure Treatment for Millwork; Window & Door Manufacturers Association: Des Plaines, IL, 2005.
9. *Amer. Wood-Preservers' Assoc. Book of Standards;* Standard N1-04; All Millwork Products-Preservative Treatment by Nonpressure Processes; 2005; pp 102-103.
10. WDMA TM 2; Test Method to Determine the Water-Repellent Effectiveness of Treating Formulations; Window & Door Manufacturers Association: Des Plaines, IL, 2004.
11. Ross, A. S. *Canadian Wood Preservation Assoc. Proc.* **2004**, Twenty-Fifth Annual Meeting, Vancouver, B.C., Canada: October 19-20.
12. Kop-Coat, Inc., Internal Communication, 2000.
13. Knudson, R. M., Gnatowski, M. J., U.S. Patent 4,879,083, 1989.
14. Ross, A. S. *New Directions in Millwork Protection;* 8th Annual WDMA Technical Conference: Chicago, IL, May 4-6, 2004.

Chapter 29

Biological Degradation of Wood–Plastic Composites (WPC) and Strategies for Improving the Resistance of WPC against Biological Decay

Anke Schirp[1], Rebecca E. Ibach[2], David E. Pendleton[3], and Michael P. Wolcott[4]

[1]Institute of Wood Biology and Wood Technology, University of Göttingen, Büsgenweg 4, 37077 Göttingen, Germany
[2]Forest Service, Forest Products Laboratory, U.S. Department of Agriculture, One Gifford Pinchot Drive, Madison, WI 53705–2398
[3]Naval Facilities Engineering Service Center, 1100 23rd Avenue, Port Hueneme, CA 93043
[4]Wood Materials and Engineering Laboratory, Washington State University, Pullman, WA 99164–1806

Much of the research on wood-plastic composites (WPC) has focused on formulation development and processing while high biological durability of the material was assumed. The gap between assumption and knowledge in biodeterioration of WPC needs to be reduced. Although some information on the short-term resistance of WPC against biological degradation is available, long-term data are required in order to adequately evaluate the performance of these relatively new materials and to enable use of WPC for structural applications. The influence of abiotic factors on material performance also has to be considered. Thus, further market growth and differentiation, especially in Europe, may be achieved while simultaneously standardization issues are dealt with.

Introduction

Wood–plastic composites (WPC) or natural fiber reinforced thermoplastic composites represent a relatively new class of hybrid materials which have been gaining rapid market share in North America, primarily as a substitute for wood decking (*1-3*). Applications for WPC also include siding, fencing, windows, doors, car interior parts, furniture, sinks, concrete formwork, crates, boxes, flower pots, dowels and tool shafts. The use of WPC for load-bearing purposes is currently being investigated. In 2003, 400,000 t of WPC were produced in North America (*4*) and approximately 30,000 t in Europe (*5*). In contrast to North America, the European market is still undifferentiated but significant market growth is predicted.

As commercial WPC formulations contain a nutrient source in the form of wood or other natural fibers, it can be assumed that this material is susceptible to microbial decay. However, some WPC manufacturers initially believed that these materials were resistant to biodegradation due to encapsulation of the wood particles by the plastic. It is well known that polyolefins are highly resistant to biodegradation, especially without prior abiotic oxidation (*6-8*), due to their backbone being solely built of carbon atoms. Scheffer and Morrell (*9*) applied this knowledge in the wood preservation area by using polyethylene films for below-ground decay protection of small ponderosa pine sapwood stakes over a two-year period. It must be emphasized, however, that biodeterioration of synthetic polymers can be initiated due to decay of additives included in plastics.

Evidence for the presence of fungal decay and discoloration on WPC decking material in service was first presented by Morris and Cooper (*10*). Morris and Cooper (*10*) isolated and identified brown and white rot fungi as well as one blue stain fungus growing on a WPC boardwalk in Florida which had been in service for four years. This boardwalk consisted of 50% recycled wood fiber and 50% recycled plastic bags and film.

Biological degradation of WPC may include attack by decay fungi, moulds, algae, termites, and marine borers. In addition, abiotic influences such as moisture, UV-light and temperature affect WPC performance, but these are not considered in this review. This chapter will provide a state-of-the-art review on:

- Biological degradation of WPC due to wood-decaying, mould and staining fungi, insects, and marine borers;
- Current test methods for biodegradation of WPC;
- Antimicrobial treatments for WPC;
- Strategies for improvement of WPC durability;
- WPC standardization initiatives in North America and Europe.

Biological Deterioration of Wood-Plastic Composites: State of the Art

Influence of moisture on development of biological decay

The availability of moisture is a prerequisite for biological decay in a material. Although moisture uptake in WPC occurs relatively slowly, moisture levels in the outer 5 mm of commercial products were shown to be sufficient for fungal attack (*11*, *12*). Once water has entered into the material, it will leave only very slowly since the plastic provides a barrier to gas evaporation and, as a result, biodegradation may occur.

Individual WPC processing methods have an influence on the moisture sorption. Clemons and Ibach (*13*) used a two-week soaking or cyclic boiling-drying procedure to infuse moisture into composites made from high-density polyethylene (HDPE) filled with 50% wood flour and processed by extrusion, compression molding, and injection molding. It was determined that extruded composites absorbed the most moisture, compression-molded composites absorbed less, and injection-molded composites absorbed the least.

It is well known that moisture alone greatly reduces the strength and stiffness of wood (*14*), however, the reduction in strength properties due to fungal decay generally outweigh the effects of moisture in wood. It was recently shown that this does not apply to WPC (*13*, *15*). See the section on "Mechanical Property Losses in WPC" for further information on this aspect.

Decay Fungi

Test methods

A multitude of methods are available to test for fungal degradation of wood and plastics, however, there is no laboratory standard currently available for testing the fungal durability of WPC. At present, in North America, the soil-block test for wood (*16*, *17*) has been adopted for fungal durability tests of WPC in which weight loss serves as an indicator of decay. In Europe, the agar-block test, according to EN 113 (*18*), is commonly used in fungal decay testing. While ASTM D2017-05 (*16*) aims at determining the natural decay resistance of woods, EN 113 (*18*) is intended to determine the efficiency threshold of wood preservatives against wood-decay fungi. A new European standard, prCEN/TS 15083-1 (*19*), will be used to identify the natural durability of woods and is currently under review. In prCEN/TS 15083-1, the test fungus (*Poria placenta* for softwoods and *Coriolus versicolor* for hardwoods) is grown on sterile malt-agar in a Kolle flask until the medium is well covered with mycelium. Then two sterile glass support rods are placed on the fungal mat, and two test or two reference specimens are placed on the rods. Following 16 weeks of incubation, specimens are withdrawn from the vessels, any adhering mycelium is removed, and each specimen is weighed and dried to constant mass. Mass loss of each specimen is calculated in percent based on its initial dry mass. The soil block test is principally very similar to the agar-block test, however, soil and a so-called feeder strip, made from a non-durable wood species, are used as the substrate for growing the fungus. In ASTM D 2017-05 (*16*), *Gloeophyllum trabeum* and *Poria placenta* are obligatory test fungi for testing both softwoods and hardwoods, and *Coriolus versicolor* is a test fungus for testing hardwoods. When information on strength properties is sought as part of a soil or agar block test, the test method is usually changed in two ways: longer specimens are used and incubation vessels are placed horizontally instead of vertically (*13*, *15*).

Schirp and Wolcott (*15*) compared the North American and European methods for WPC fungal durability testing. The agar-block test was modified such that no support rods were employed to accelerate moisture uptake by WPC specimens which had not been pretreated, only steam-sterilized in an autoclave. It was determined that modified agar- and soil-block tests are equally suited for determining weight loss in WPC, but that agar-block tests can be completed in a shorter time span.

Dynamic mechanical analysis (DMA) is a laboratory method which facilitates the investigation of the effects of wood decay fungi in WPC (*20*). Dynamic mechanical analysis can potentially provide valuable molecular and morphological information about a material in the solid state by subjecting it to dynamic loads over a broad range of temperature and frequency. A significant advantage of DMA as compared to commonly used static mechanical strength tests is that incubation times in fungal decay experiments can be drastically shortened due to the employment of very small specimens.

Field tests provide additional valuable information on the durability of a material. The objective is to determine the types and causes of failure not only from moisture, fungal and termite degradation but also from other environmental factors such as UV radiation, thermal cycles and freeze-thaw cycles. Verhey et al. (*21*) exposed field stakes at a site near Hilo, Hawaii, and samples were pulled for impact and flexural strength testing as well as fungal isolations at three-month intervals over the course of one year. Additional field tests with WPC, based on AWPA Standard E7 (*22*), were sponsored by the U.S. Navy and conducted by Michigan Technological University.

Scientists at the USDA Forest Products Laboratory are currently evaluating the field performance of extruded WPC (*23*). In-ground and above-ground specimens were installed in Madison, Wisconsin, and Saucier, Mississippi, in October 2000. Criteria for evaluating the in-ground stakes are decay, moisture, termites, cracking, checking and swelling. Decay and termite grades have been determined according to ASTM D1758-02 (*24*). The above-ground stakes have been evaluated for color change, mold/mildew, warping/twisting, checking/cracking and flexural strength and stiffness.

Specific adjustments will have to be made to existing wood durability standards so that they can be appropriately used in WPC fungal decay testing. For example, weathering of WPC prior to fungal testing may be an effective way of accelerating the laboratory test by simulating outdoor conditions (*25*). In addition, suitable reference(s) for WPC in fungal decay testing will have to be determined, for example, a durable wood species, wood treated with a preservative, or a material based on 100% plastic. Finally, the fungal species that colonize wood and wood composites outdoors may not be the fungi that are predominant on WPC during long-term exposure. Hence, research is required to establish the predominant microorganisms occurring on WPC.

Weight losses in WPC

Mankowski and Morrell (*26*) conducted scanning electron microscopy to determine patterns of fungal attack in WPC following exposure in a soil block test. It was demonstrated that a formulation containing a 70/30 wood-HDPE mixture was most susceptible to fungal attack while two different 50/50 wood-HDPE composites displayed little or no degradation. Decay fungi were prevalent in the voids between wood and HDPE. Verhey et al. (*27*) and Pendleton et al. (*28*) also determined that an increase in plastic reduced weight losses during laboratory soil block tests. Maximum weight loss due to *Poria placenta* after 12 weeks was 8.5% when the formulation contained 70% wood filler (*28*).

Weight loss of an extruded WPC formulation with 70% wood filler and incubated with *T. versicolor* was twice as high as that of redwood in a modified agar-block test (6% versus 3%), however, only 1% weight loss was obtained when the formulation contained 49% wood filler (*15*). These results indicate that WPC can be designed to provide high fungal durability by controlling the material composition of the formulation. Generally, reported weight losses of WPC obtained in laboratory testing are below 10%.

Verhey and Laks (*29*) determined that weight loss in WPC due to fungal decay increased as the wood particle size increased, probably due to a more effective encapsulation of smaller wood particles by the polymer matrix. A similar protective effect due substrate encapsulation has been reported for wood-cement composites (*30*).

Weight losses in WPC are usually calculated based on the wood content of a formulation because of the aforementioned resistance of plastic matrices to biodeterioration. Only one publication regarding polyethylene degradation by white-rot fungi has been published (*31*), and there is no information available on PE-degradation by brown-rot fungi. The white-rot fungi *Phanerochaete crysosporium*, *T. versicolor* and one unidentified fungal isolate degraded high-molecular-weight PE-membranes under nitrogen- or carbon-limited culture conditions (*31*).

Mechanical Property Losses in WPC

It is well known that strength properties are the most sensitive indicators of fungal decay in solid wood (*32-35*) but this concept does not necessarily apply to a composite material consisting primarily of a thermoplastic polymer matrix and wood filler. At low filler concentrations, the wood flour used in WPC does not greatly contribute to the strength of the composite unless coupling agents such as MAPP and MAPE are included in the formulations. However, the wood filler tends to increase the stiffness of the composite (*2*, *36*) and may therefore be useful as an indicator of fungal decay in WPC. It is desirable to compare potential losses in weight and stiffness in WPC formulations caused by decay fungi to determine which of the two methods is more sensitive as an indicator of fungal decay. Khavkine et al. (*37*) performed three-point bending tests with compression-molded samples which had been subjected to water boiling, boiling and freezing cycles, or fungal exposure. They found that the flexural properties of the composites were more affected by cyclic exposure than either by 2 hours water boiling or fungal exposure, however, no significant weight losses following 12 weeks incubation with decay fungi were observed.

Research conducted by Silva et al. (*38*), Verhey et al. (*39*, *21*) and Ibach and Clemons (*40*) determined strength properties of WPC following exposure to decay fungi. The effects of moisture sorption and fungal decay on strength and stiffness were separated by Clemons and Ibach (*13*). Clemons and Ibach (*13*) determined that water soaking and cyclic boiling resulted in large losses in strength of the extruded composites, however, strength losses caused by fungal attack were found for boiled composites but not water-soaked ones. In contrast, weight loss results showed that both water-soaked and boiled composites sustained weight loss.

Schirp and Wolcott (*15*) separated the effects of fungal decay and moisture absorption on flexural strength and stiffness of extruded WPC specimens which had not undergone any preconditioning treatment, i.e., water-soaking or boiling. Stiffness of WPC was affected more severely by moisture absorption than by fungal colonization (Table 1). Strength of WPC was not affected following incubation with decay fungi but significantly reduced by moisture absorption for a formulation containing 70 % wood filler. However, no strength reduction due to moisture was observed when the formulation contained 49 % wood filler. Weight loss was determined to be a more sensitive indicator of fungal decay in WPC than flexural strength tests for the formulations investigated.

In a parallel study (*20*), it was shown that after short-term (24 days) exposure, wood decay fungi may improve interfacial adhesion and reinforcement of WPC. Using dynamic mechanical analysis, an increase in the storage modulus (E'), which is closely related to stiffness as determined in static tests, was observed for fungal treated WPC. This increase in E' was attributed to a reinforcing effect of the fungal hyphae present in the interfacial gaps between wood filler and polymer matrix (Figure 1). It was supported with the observation

Table 1. (previously published in Wood and Fiber Science 37(4), 2005, 643-652; copyright: Wood and Fiber Science): Stiffness (MOE), flexural strength (MOR) and weight loss (based on wood fraction) of two WPC formulations, yellow-poplar and redwood, following three months of incubation with T. versicolor and G. trabeum. Formulation #3 consisted of 49% wood, 45% HDPE and 6% additives; formulation #7 contained 70% wood, 24% HDPE and 6% additives. Each value represents the average of 16 replicates, except for yellow-poplar in the soil block test (15 replicates). Figures in bold indicate significant ($p < 0.0001$) difference between samples and incubated controls.

Material, treatment	Type of test	Density* (g/cm^3)		MOE† (MPa)		MOR (kPa)		Weight loss (%)	
		Mean	SD	Mean	SD	Mean	SD	Mean	SD
#3, *T.v.*	agar	1.05	0.01	1497	212	26274	2344	1.17	1.25
#3, *G.t.*	agar	1.07	0.01	1345	94	28096	1490	-1.41	0.13
#3, incub. control	agar	1.07	0.01	1461	194	26920	2926	-1.26	0.09
#3, non-incub. control	n.a.	1.09	0.01	1979	130	26862	1101	n.a.	n.a.
#7, *T.v.*	agar	0.99	0.02	797	163	11975	2226	**6.32**	0.47
#7, *G.t.*	agar	0.98	0.01	839	109	10590	2166	0.38	0.74
#7, incub. control	agar	1.02	0.01	806	116	11171	1948	-0.46	0.21
#7, non-incub. control	n.a.	1.12	0.01	1004	216	14696	1139	n.a.	n.a.
YP, *T.v.*	agar	0.21	0.03	**2525**	479	**16355**	3500	**55.94**	5.59
YP, *G.t.*	agar	0.40	0.05	**6258**	2098	**43098**	21167	**25.21**	5.82
YP, incub. control	agar	0.48	0.02	9719	787	97598	7374	-1.19	0.70
YP, non-incub. control	n.a.	0.61	0.02	15366	623	145183	8190	n.a.	n.a.

RW, *T.v.*	agar	0.44	0.03	9380	1661	81643	14405	**3.03**	0.51
RW, *G.t.*	agar	0.45	0.01	8325	1733	68763	13596	**2.76**	1.56
RW, incub. control *T.v.*	agar	0.45	0.02	10096	1157	89621	10501	1.29	0.54
RW, incub. control *G.t.*	agar	0.47	0.03	9244	1327	91625	21824	-0.78	0.94
#7, *T.v.*	soil	0.91	0.04	**963**	164	7203	1915	**7.90**	2.97
#7, incub. control	soil	0.94	0.04	560	180	8092	2434	0.77	0.94
YP, *T.v.*	soil	0.22	0.03	**2758**	948	**13525**	4050	**61.85**	4.32
YP, incub. control	soil	0.54	0.04	13479	1518	126235	14101	2.03	0.19

T.v. = *Trametes versicolor*
G.t.= *Gloeophyllum trabeum*
YP = yellow-poplar
RW = redwood
n.a. = not applicable (equilibrated only).
* determined at time of flexural strength test.
† Calculation based on the values for 20-40 % of the maximum load in the stress-strain curves.

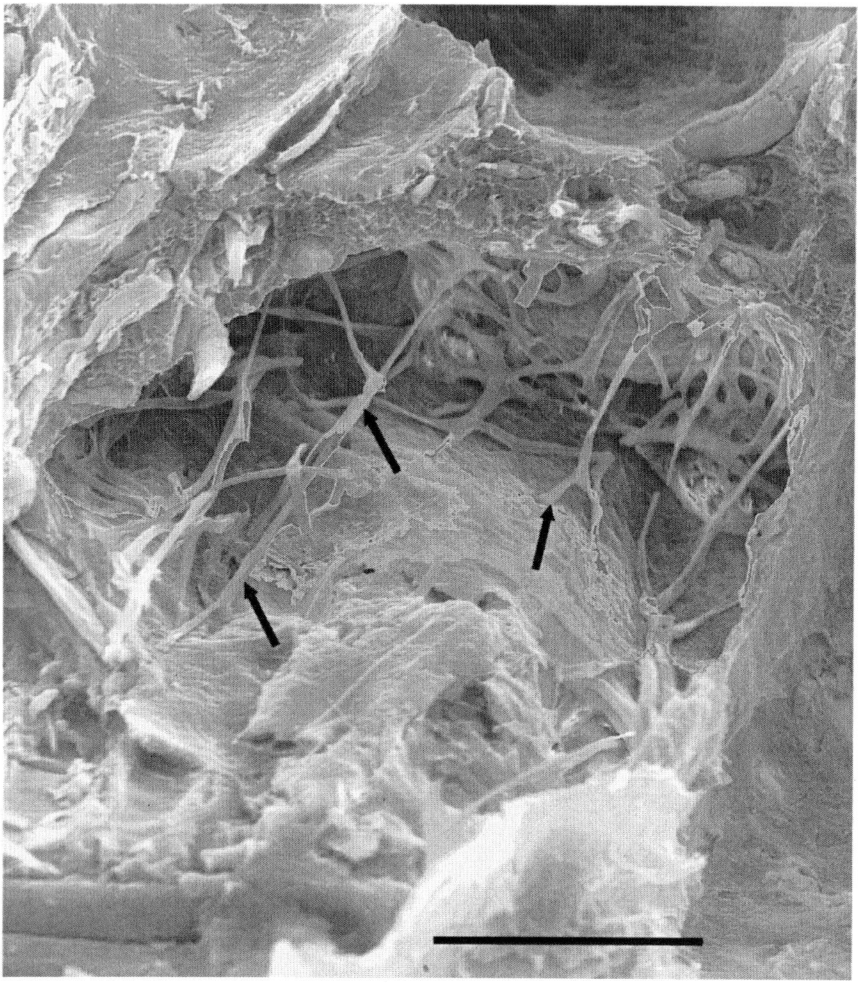

Figure 1. (previously published in Journal of Applied Polymer Science 99, 2006, 3138-3146; copyright: Journal of Applied Polymer Science): Reinforcement effect of fungal hyphae of T. versicolor (arrows) in WPC after 24 days of incubation (formulation #7, i.e, 70% wood, 24% HDPE and 6% additives). Bar represents 40 μm.

that higher activation energies were required for α-transition, i.e., chain rotation (*41*), in WPC samples incubated with the white-rot fungus *T. versicolor* as compared to untreated controls.

When using field tests, it is difficult if not impossible to separate the effects of moisture and fungal decay (*21*, *23*). In any case, a significant reduction in the flexural modulus of WPC stakes following three years of exposure above-ground in Mississippi and Wisconsin was observed (Figure 2, *23*). A significant loss in flexural strength was determined only for the above-gound WPC stakes placed in Mississippi (Figure 3, *23*). Examples of extruded WPC following three years of outdoor exposure are shown in Figure 4. Decay fungi could be identified in the wood flour component of the stakes only upon examination in the laboratory. The failure break of a WPC field stake after exposure in-ground for three years in Saucier, Mississippi, is depicted in Figure 5.

Field tests on WPC were also sponsored by the U.S. Navy and conducted by Michigan Technological University over a 54-month period near Hilo, Hawaii, ending in May 2004. In these tests, AWPA Standard E7 (*22*) was used. Test stakes, composed of HDPE with 2% and 5% zinc borate as well as PVC with 5% zinc borate, were placed in ground contact along with creosote-treated and untreated Southern yellow pine stakes as controls. Randomly selected stakes were pulled every six months and tested for physical and mechanical properties. At the conclusion of the study, there was no visual indication of decay in any of the WPC stakes whereas the untreated Southern yellow pine stakes were all destroyed by decay fungi within 30 months. Mechanical testing of the WPC stakes showed that there was a significant loss of strength and stiffness after six months but no further significant reduction in these properties occurred over the next 48 months.

At present, no published information on the durability of WPC against soft-rot and soil fungi is available.

Mould and Staining fungi

Test methods

Resistance of WPC against mould and staining fungi is currently tested either according to plastic, wood or paper standards. Some of the laboratory test methods for mould growth on WPC which may presently be used are:

- ASTM G21-96 (*42*);
- ASTM D3273-94 (*43*);
- ASTM D 4445-91, reapproved 1996 (*44*);
- AWPA Draft: Standard Method of Evaluating the Resistance of Wood Product Surfaces to Mold Growth (*45*);
- BS 3900: Part G6: 1989 (*46*);
- ISO 16869 (*47*);

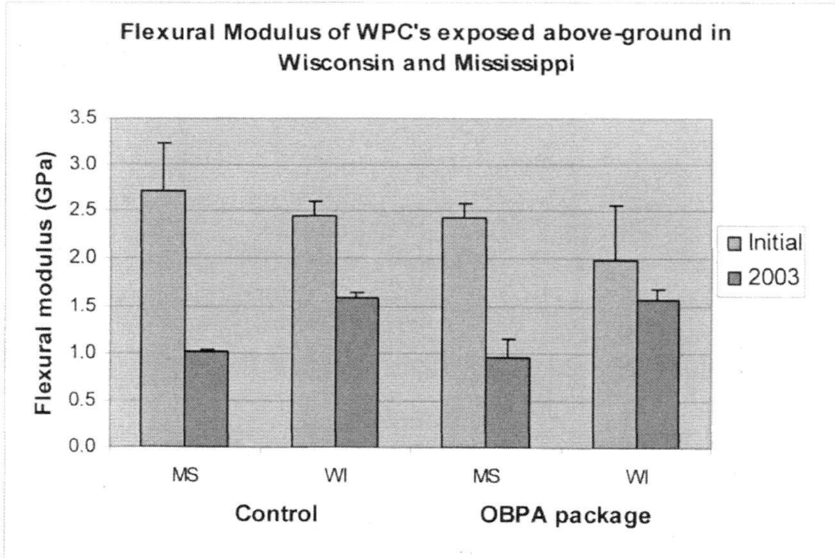

Figure 2. (previously published in Conference Proceedings, Progress in Woodfiber-Plastic Composites, Toronto, Canada, 2004, 1-13): "Control" consists of 48.5% HDPE (reprocessed milk bottles; H. Muehlstein and Co., Inc., Norwalk, CT), 50% Western pine wood flour (40 mesh, i.e., 420 μm, American Wood Fibers, Schofield, WI) and 1.5% light stabilizer package (Ciba Specialty Chemicals Co., Terrytown, NY). "OBPA package" contains 47.5% HDPE, 50% Western pine wood flour, 1.5% light stabilizer package plus a 1% OBPA package (Rohm and Haas, Philadelphia, PA). *(See page 9 of color inserts.)*

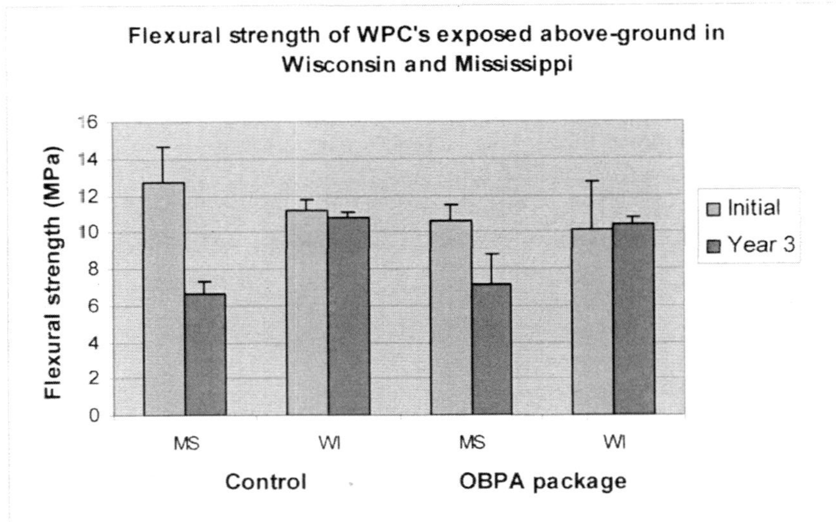

Figure 3. (previously published in Conference Proceedings, Progress in Woodfiber-Plastic Composites, Toronto, Canada, 2004, 1-13): "Control" consists of 48.5% HDPE (reprocessed milk bottles; H. Muehlstein and Co., Inc., Norwalk, CT), 50% Western pine wood flour (40 mesh, i.e., 420 μm, American Wood Fibers, Schofield, WI) and 1.5% light stabilizer package (Ciba Specialty Chemicals Co., Terrytown, NY). "OBPA package" contains 47.5% HDPE, 50% Western pine wood flour, 1.5% light stabilizer package plus a 1% OBPA package (Rohm and Haas, Philadelphia, PA). *(See page 9 of color inserts.)*

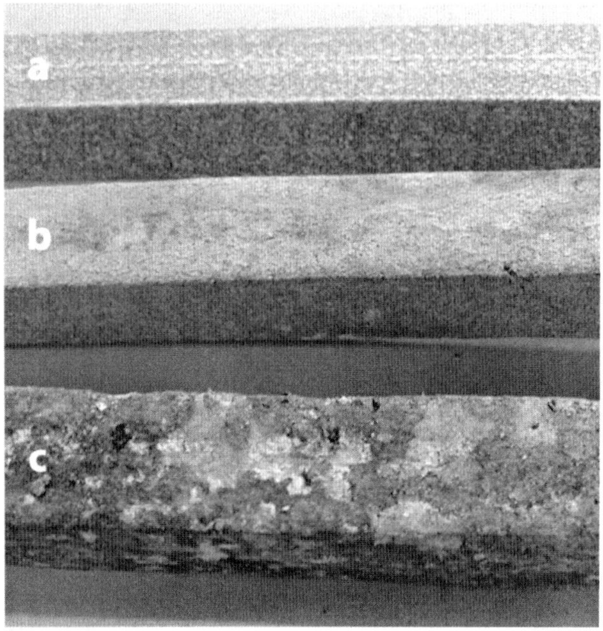

Figure 4. Extruded WPC: a) unexposed; b) above-ground exposure for 3 years in Saucier, Mississippi; c) in-ground exposure for 3 years in Saucier, Mississippi. (See page 10 of color inserts.)

Figure 5. Failure break of a WPC field stake after exposure in-ground for 3 years in Saucier, Mississippi. (See page 10 of color inserts.)

TAPPI T 487 pm-99 (*48*).

Fungal species used in mould testing are *Aspergillus niger*, *Aureobasidium pullulans*, *Penicillium purpurogenum*, *Stachybotrys chartarum*, *Chaetomium globosum* and others. Generally, a mixed fungal spore suspension containing a defined number of spores per mL is prepared and sprayed on the test specimens which are placed on an agar-based medium in a Petri dish. Specimens are incubated at a defined temperature (around 30°C) and high (around 85 %) relative humidity for a set amount of time, for example, 28 days (*42*).

All mould test methods listed above are based on visual quantification of fungal growth on the material surface. For example, in ASTM G21-96 (*42*), observed growth on specimens is rated between 0 (no growth) and 4 (heavy growth (60% to complete coverage).

Effect of mould and staining fungi on wood and the environment

Fungi which cause discoloration of wood in storage and service are generally described as staining fungi while those that grow superficially on wood are called moulds (*49*). Staining and mould fungi are members of the *Ascomycetes* and *Deuteromycetes* (*Fungi Imperfecti*) although a few of the moulds are *Zygomycetes* and are classified in the Mucorales. Both mould and staining fungi lower the aesthetic quality of wood-based products due to discoloration. It is well known that sapstaining fungi produce extracellular enzymes to utilize nonstructural components of sapwood, namely sugars, proteins, and extractives (*50-53*). Extracellular cellulase is not secreted by sapstaining fungi, however, selected species were shown to produce low amounts of xylanase and pectinase, and one species (*Ophiostoma piceae*) produces mannanase (*54*). The apparent inability of staining fungi to decay wood cell walls is not universal because some of these fungi are known to cause soft rot in hardwoods under optimal conditions (*49*).

Staining and mold fungi do not significantly affect the mechanical properties of wood (*55-58*), however, some species of sapstaining fungi may reduce impact bending strength or toughness of wood (*59-63*). Due to increased permeability in sapstained wood, the use of this material for appearance purposes is limited because it absorbs solutions excessively resulting in uneven finishes. Allegedly, some mould species such as *Stachybotrys chartarum* (syn. *Stachybotrys atra*) potentially produce health-related building issues. It was reported that 8,000 to 10,000 lawsuits were pending nationwide in the United States for mold litigation in the year 2002 alone (*64*). It must be stressed that mold and mold spores are always present in the air, indoors as well as outdoors, and that infections may occur only in immune-compromised human beings or following prolonged exposure to exceedingly high airborne mold concentrations, such as in an agricultural setting (*65*). The amount of mold that must be inhaled to cause an allergic response is unknown (*65*). In the United States, there are currently no regulations or exposure limits for molds or mycotoxins in homes and commercial or public settings. However, in the occupational setting, a general duty clause may apply to mold exposures, i.e., the rule that requires employers to provide workers with a safe and healthy work environment (*65*).

Effect of mould and staining fungi on WPC

In the short term, mould and staining fungi may represent a more relevant issue with WPC than decay fungi since these first two types of fungi develop more quickly on the material. Many of the mould fungi known to grow on wood have also been isolated from plastics, for example, *Aureobasidium*, *Aspergillus* and *Penicillium*, and therefore are of importance as screening organisms for both plastic and wood (*66*). According to anecdotal evidence (*10*), WPC are susceptible to mould and staining fungi, however, very little detailed information on this issue is available in the scientific literature (*66, 67*).

Laks et al. (*67*) examined the effect of several manufacturing variables (i.e., wood loading, extrusion lubricants, fungicides and extruder temperature) on mould susceptibility of wood-plastic composites. It was determined that an increase in wood content as well as surface roughness increased mould growth and that lubricants generally increased mould growth rates. Both in-process fungicides tested, chlorothalonil and zinc borate, significantly reduced mould growth on WPC, however, both fungicides were used in different concentrations. Thus, the efficiency of both biocides cannot be directly compared. The effect of extruder temperature on mould development could not be conclusively determined and needs to be investigated further.

In conclusion, so far the effect of staining and mould fungi on WPC has been evaluated only by visual quantification of microbial growth on the material. A drawback of this method is that it is based on subjective evaluation, unless this quantification could be achieved using a photometric colour coordinate system. In addition, measuring mould growth on a material cannot provide information on which components included in the plastic are degraded and if mould growth leads to a reduction in WPC strength properties. It must be emphasized again that synthetic polymers themselves are resistant to fungal growth but that generally, other components such as plasticizers, lubricants, stabilizers and colorants are responsible for fungal colonization and decay on plastic materials (*42*). Although mould fungi do not significantly affect the mechanical properties of wood, as previously mentioned, some fungi may be able to degrade certain additives in plastic and WPC, especially following oxidation (*8*), and thus reduce the WPC strength.

Termites

As previously mentioned, scientists at the USDA Forest Products Laboratory have been evaluating the field performance of extruded WPC, including deterioration by termites according to ASTM D1758 (*23*). Termite nibbling was visible on in-ground stakes after the third year of exposure in Saucier, Mississippi, but not on test material in Wisconsin due to the lack of termites in this state.

Results from termite field testing of WPC over a 27-month period were also obtained through work sponsored by the U.S. Navy and conducted by Michigan Technological University. Three WPC formulations were tested, HDPE with zinc borate, HDPE without zinc borate, and PVC without zinc borate. For the test, ten replicates were used for each of the formulations. The test specimens (50 x 125 x 25 mm or 50 x 90 mm) were randomized and placed horizontally on top of hollow concrete blocks four inches above the ground at a field site near Hilo, Hawaii. The test site is characterized by a naturally occurring, large Formosan termite population. Untreated aspen (*Populus tremuloides*) bait stakes (200 mm length x 25 mm thickness x 25 mm width) were driven vertically into the ground through holes in the concrete blocks to attract termites. In addition, untreated aspen feeder stakes (450 mm length x 25 mm thickness x 25 mm width) were placed between the rows of test blocks. The bait stakes were in direct contact with the feeder stakes which in turn were touching the test blocks. The assembled units were enclosed with a boxed lid to maintain dark conditions and provide shelter from rainfall. Termite-damaged bait and feeder stakes were replaced every six months during inspection of the test blocks, resulting in a permanent high termite hazard for the blocks. The rating system was according to AWPA E7 (*22*) in which a block rating of 10 equals perfect condition while a rating of 0 means failure. No termite activity on any of the WPC test blocks was observed throughout the test period whereas seven out of ten aspen control specimens were attacked by termites resulting in four failures.

There is no published information available in the scientific literature on WPC laboratory tests using termites.

Algae

No information is available in the scientific literature on algal growth on WPC although tests involving algae have been performed (*68*). Test method VDL RL 07 (*69*) was used to measure algal growth on WPC (*68*). This method is intended to determine the resistance of coating materials against algal growth by adding the coated test specimen to a suitable agar medium and inoculating the specimen with a standard algal solution followed by incubation for a specific duration. An alternative test method is IBRG/P98/03, however, it is available for members of the International Biodeterioration Research Group (www.ibrg.org) only.

Bacteria

Bacteria compete poorly with fungi on wood, however, where favourable conditions exist, they may represent a major form of decay of man-made structures (*70*). True wood degrading bacteria exist in all parts of the world, and have a tendency to be aerobic and exist in a diverse range of aquatic and terrestrial environments. Anaerobic and aerobic bacteria are known to cause a pronounced increase in the permeability of round wood, including refractory species like spruce, during ponding or water sprinkling.

Fungi are the most important microorganisms which colonize and degrade synthetic polymers, however, bacteria such as *Pseudomonas aeruginosa*, *Serratia marcescens* as well as several species of the genus *Micrococcus*, *Bacillus* and *Streptomyces* are also involved in deterioration of synthetic polymers (*71*). Lee et al. (*72*) provided the first report demonstrating degradation of oxidized, starch-containing polyethylene by *Streptomyces* species in pure culture. *Streptomyces* and other filamentous bacteria (actinomycetes) have also been reported as important colonizers of wood in ground contact although few reports showing degradation of lignified wood fibres are available (*70*). The effect of bacteria on WPC merits further research due to a lack of investigations carried out in this area.

Marine Organisms and Seawater

Marine test procedures with WPC according to ASTM D2481 (*73*) were conducted in Pearl Harbor, Hawaii (*74*). The test method is generally used to determine the relative efficacy of preservatives in small wood specimens exposed to a natural marine environment and was adapted to evaluate small WPC-specimens. A qualitative index of physical condition, determined visually during periodic inspections, was used to measure resistance to marine borer attack. In addition, flexural strength tests according to ASTM D790 (*75*) were performed periodically. The selected test site is known to have marine borer populations, including several *Teredo* spp. (shipworm), *Limnoria tripunctata* (gribble) and *Martesia* striata (pholad) that commonly cause failure of timber piling.

Two commercial WPC formulations were included in the study by Pendleton and Hoffard (*74*) consisting of a 70/30 mixture (by weight) of wood flour and HDPE and a 50/50 mixture of wood and LDPE. The LDPE specimens were added after the study had been in progress for two years. Untreated Southern yellow pine specimens were included as controls and as bait. All specimens were cut from larger pieces and measured 15.24 cm x 3.81 cm x 1.27 cm. Wood-plastic and wood specimens were attached to a test rack which was suspended from a pier about five feet below mean low tide. The wood specimens were replaced annually because of heavy borer attack by *Limnoria* and *Teredo*. Fouling organisms were removed annually to facilitate inspections.

The HDPE and LDPE specimens exhibited no marine borer attack after three years and one year of exposure, respectively. A loss in flexural strength and stiffness of the formulations was observed but attributed to wetting during exposure and not to marine borer degradation. Similar strength losses were found in the materials after simple wetting and drying.

Due to the high wood content and lack of any preservative in the HDPE formulation tested it may be concluded that current commercial WPC formulations with a low wood content are not at risk to *Limnoria* or *Teredo* attack. A remote risk exists that boring clams, such as *Martesia striata*, may present an occasional problem as these clams can bore into any material, including plastics and WPC, which are softer than the clam's shell. However, as no boring clams were observed in the wood specimens, no conclusions could be drawn in this regard.

Antimicrobial treatments for WPC

WPC manufacturers responded to the risk of fungal attack in their products by incorporating zinc borate into formulations prior to extrusion. Zinc borate has become the dominant preservative used for this purpose because of its relatively low water solubility, resistance to leaching, ability to withstand common extrusion temperature, low cost and very low environmental and work toxicity hazard (*76*, *77*). Zinc borate is effective against wood decay fungi and insects but, as mentioned previously, not highly effective against mould and staining fungi at low concentrations (*66*, *77*). It is suited for exterior applications with a low to moderate leaching hazard (American Wood Preservers Association H2 use level), e.g., siding, exterior trim and window components (*76*). Zinc borate at 2% concentration was shown to prevent any weight loss in extruded WPC and will take at least 20 years to completely dissolve and leach from the material (*28*). Simonsen et al. (*78*) determined that the addition of zinc borate or a blend of calcium and sodium borates before molding dramatically reduced weight losses of WPC due to decay fungi. Weight losses overall were below 2% when calculated on the basis of the wood content, however, the maximum weight loss of a WPC without any biocide was only 6%. For solid wood, according to ASTM D 2017-05 (*16*), a species displaying a mass loss due to Basidiomycete attack below 10 % is considered highly resistant. In European standard prCEN/TS 15083-1:2004 (*19*), a provisional durability rating scale is provided which classifies a timber as very durable at 5 % or less mass loss.

Dylingowski (*66*) compared the efficiency of an isothiazolone biocide, specifically, dichloro-octyl-isothiazolone (DCOIT), and of zinc borate against mould growth on WPC following testing according to a variety of methods and fungal isolates. DCOIT was shown to be more effective against mould growth than zinc borate. DCOIT is incorporated into a plastic pellet, added to the extrusion process, dissolved into the WPC matrix and slowly migrates to the WPC surface. It has a high thermal stability, very low water solubility and broad-spectrum activity against fungi, actinomycetes and gram positive bacteria (*66*).

Once mold growth has appeared on a WPC surface, several commercial WPC producers recommend cleaning the material with water-diluted sodium hypochlorite (bleach). However, bleach treatments do not completely eliminate fungal growth from wooden surfaces since they lack broad spectrum antifungal activity (*79*), and this may also apply to WPC. It may be necessary to use biocides to prevent mold growth or delay re-growth of mold on WPC. Periodic cleaning of WPC decking is advised to prevent the build-up of debris and pollen that can cause mold growth.

Some commercial WPC products, especially for use as decking material, contain broad spectrum treatments against bacteria, mold and mildew such as Microban® or Sanitized®. A bacteriostat, disinfectant (bacteriocide / germicide) and fungicide, 10, 10'-Oxybisphenoxarsine (OBPA), is currently being tested in field trials with extruded WPC formulations by scientists at the USDA Forest Products Laboratory (*23*).

Strategies for Improvement of WPC Durability

Introduction

All biological agents of deterioration have four basic requirements (*80*):

- Nutrient source;
- Moisture (generally greater than 30%);
- Temperature (generally between 5°C and 40°C, but optimally between 24°C and 30°C);
- Oxygen.

Since it is not feasible for WPC applications such as decking to limit oxygen supply and/or to adjust temperature to an out-of-optimum level for microorganisms, most strategies to inhibit biodegradation are either based on moisture or nutrient exclusion. Wood is rendered an unfavorable microbial nutrient if it is of high natural durability or if it is of low natural durability and treated with preservative(s), and synthetic polymers are relatively inert to biodegradation.

Nutrient exclusion or modification

WPC durability may be improved by minimizing the amount of nutrients available for microorganisms, i.e., wood and specific additives. As previously mentioned, laboratory tests have shown that no significant weight loss occurs in WPC containing 50% wood by weight. Similar baseline thresholds for additives (lubricants etc.) cannot be provided at present due to a lack of systematic investigations. If higher amounts of wood are incorporated into a formulation it may be advisable to also include zinc borate, but long-term field tests are required to clarify this aspect.

While it is recognized that fungal durability may be improved by reducing the amount and size of wood in a formulation, at the same time strength and especially stiffness of the WPC may be reduced and material costs will be increased because more plastic will be included. It was shown that the incorporation of particles with a high fiber aspect ratio may improve the mechanical properties of WPC (*36*).

Another strategy for nutrient exclusion would be to achieve inaccessibility of the degradable wood particles in WPC by encapsulating them with the plastic matrix.

The incorporation of wood originating from a naturally durable species may also improve biological durability of WPC. A promising alternative to the use of naturally durable wood is to chemically modify the wood filler prior to compounding and extrusion or molding. Chemical modification involves treating wood with chemicals that alter the water-sorption properties either by bulking the cell lumens to the point where water is excluded or by cross-linking the cellulose hydroxyls to reduce hygroscopicity (*81*). In order to improve compatibility of wood fibers with a thermoplastic matrix and improve material performance, chemical wood modification is aimed at changing fibers from a hydrophilic into a permanent hydrophobic state. The three best-investigated groups of modification agents for wood are anhydrides, isocyanates and epoxides (*82*). Acetylation has been the subject of most investigations due to the relatively low cost of the reagent and its relatively low toxicity (*82*). Liu et al. (*83*) investigated the potential of acetylation for modification of wood in WPC and determined that acetylation increased the interfacial shear strength between polystyrene and wood.

Compression-molded WPC based on wood flour modified with either acetic anhydride, butylene oxide or propylene oxide were prepared by Ibach and Clemons (*40*). A correlation between moisture and fungal decay was determined. Weight loss due to fungal decay increased with increasing specimen moisture content. Lowest weight losses were obtained with acetic anhydride modification, followed by butylene oxide modification, untreated control, and propylene oxide modification.

Despite the relatively low cost of acetic anhydride, the cost of modification includes the removal and recovery of acetic acid which may add complexity to the process. Thus, any composite based on an acetylated lignocellulosic substrate may have to be aimed at a value added market (*82*).

Moisture exclusion

Microorganisms depend on water in order to grow and cause deterioration. Thus, if moisture content of the wood filler can be kept out or at least below 20%, decay in WPC may be prevented. This could be achieved by complete encapsulation of wood particles by the plastic matrix, hydrophobation of the WPC surface, or by chemical modification of the wood substrate. It must be emphasized that no reasonable amount of zinc borate will prevent moisture entry into the composite. As a consequence, moisture exclusion is of paramount importance when designing and processing WPC.

Microbial pathway exclusion

Voids between wood and plastic represent entry points and proliferation pathways for microbes; thus, these should be eliminated or reduced. With regard to polymer-starch composites, biodegradation kinetics were examined based on scalar percolation theory (*84*). Scalar percolation theory deals with the connectivity of one component randomly dispersed in another (*84*). When the starch fraction exceeds a specific percolation threshold, significant pathways for microbial invasion are generated and degradation is accelerated.

It can be expected that the shape of the wood substrate in WPC, i.e., flour versus fibers, as well as wood density differences also have an influence on the degradation rate of WPC, however, this aspect has not been investigated.

WPC Standardization Initiatives in North America and Europe

With the improvement in WPC technology, it is likely that new product applications will result. Thus, standardization issues will become increasingly important on a world-wide level. The development of adequate WPC standardization is required to ensure market acceptance, growth and diversification, especially in Europe.

In North America, standards for plastic lumber as well as for WPC have existed for a number of years. Acceptance criteria based on ICC ES Acceptance Criteria AC174(3) are being used to address one of the significant WPC product applications in the United States, deckboards and guardrails (*85*). According to these criteria, a particular product is tested according to code-prescribed loads in the configuration for which the proponent has requested a code listing. Due to the unique nature of WPC in a specific application, acceptance criteria include test requirements such as temperature, moisture, UV and freeze-thaw effects. The important point is that in the United States, all materials for a specific application, must perform according to the same performance limits (*86*). For example, WPC may be compared with preservative-treated lumber when used as decking material. Currently, there is an on-going effort to eliminate termite and fungal decay testing in the ICC ES Acceptance Criteria for products with less than 70% wood content (*86*). The motivation for this effort is that no commercial WPC product with less than 70% wood has been shown to behave worse than preservative-treated lumber in tests according to ASTM D2017 (*16*), ASTM D1413 (*17*), ASTM D3345 (*87*), AWPA E1 (*88*) and AWPA E10 (*89*). At the same time, biological durability issues have become a top concern of the North American WPC industry due to a few early product failures (*86*). Hence, it remains to be clarified if the standards used for fungal durability testing of wood adequately address the long-term durability performance of WPC.

In Europe, several initiatives are currently working on WPC standardization (*4*):

- CEN/TC 249 Plastics, WG 13: Plastics – Wood-plastic composites (convenor: Claudine Bloyaert, Technical Marketing and Development, SolVin, Brussels, Belgium);
- CEN/TC 112 Wood-based composites, WG 12: Wood Plastic Composites (convenor: Dr. Alfred Teischinger, Institute of Wood Science and Technology, Department of Material Sciences and Process Engineering, University of Natural Resources and Applied Life Sciences, Vienna, Austria);
- ON (Österreichisches Normungsinstitut, Austrian Standardization Institute), Fachnormenausschuss 087 Holz, AG Holz-Kunststoff-Verbundwerkstoffe (convenor: Dr. Alfred Teischinger).

Standardization to be developed by CEN/TC 249/WG13 has the following objectives:

- To provide an exhaustive description of the test methods applicable to WPC;
- To define the common characteristics of the different applications;
- To provide characteristics and corresponding requirements for specified applications.

With regard to biological durability of WPC, publications and recommendations from experts in this field are currently compiled to serve as the basis for a durability standard to be issued by CEN/TC 249/WG13. CEN/TC 112 is currently working on the development of a scope for standardization, i.e., raw material characterisation and compounds, composites and products. The current main challenge in Europe will be to coordinate standardization initiatives such that the requirements of both the plastics and wood industries and research organizations will be met.

Conclusions

Despite a significant increase in WPC research in recent years, some fundamental questions on the biological durability of WPC still need to be addressed, for example:

- How durable is WPC under long-term outdoor exposure, including biological and abiotic factors?
- Can long-term outdoor exposure be adequately simulated by using an accelerated laboratory test?
- Which fungi colonize and degrade WPC under natural, i.e., outdoor conditions? Is there an ecological succession with regard to different types of fungi (moulds, decay fungi)?
- Which fungi and/or bacteria should be used as test organisms in WPC standards?
- How does the shape (flour versus fibers) and type (durable versus non-durable) of wood substrate influence degradation of WPC?
- Under which conditions can polymer matrices be biologically degraded?
- Are additives in WPC degradable and do they contribute to degradation?
- Which applications require fungicides in WPC formulations?
- Is there a possible interference between fungicides and coupling agents?

Acknowledgement

The authors thank Dr. Robert Tichy for supplying information with regard to standardization issues of WPC in North America.

References

1. Wolcott, M. P.; Englund, K. *Proceedings, 33rd International Particleboard/ Composite Materials Symposium*, 1999, Washington State University, Pullman, WA., 103-111.
2. Wolcott, M. P. *Wood-Plastic Composites*. Buschow, K. H. J.; Cahn, R. W.; Flemings, M. C.; Ilschner, B.; Kramer, E. J.; Mahajan, S.; Eds. Encyclopedia of Materials: Science and Technology; Elsevier: Amsterdam; New York, **2001**; 9759-9763.
3. Clemons, C. *Forest Prod. J.* **2002**, *52*(6), 10-18.

4. Teischinger, A. *Wood-Plastic Composites (WPC) – Technologie und Rahmenbedingungen für eine neue Werkstoffgeneration.* Symposium Proceedings, Holz Innovativ, Rosenheim, Germany, April 6 and 7, 2005.
5. Foster, C.; Hackwell, B.; Mann, D.; Pritchard, G. *The European Wood Plastics Composite Market 2003: Construction, Furniture and Automotive Applications.* The Hackwell Group: Tunbridge Wells, Great Britain, 2003.
6. Albertsson, A. C.; Andersson, S. O.; Karlsson, S. *Polym. Degrad. Stabil.* **1987**, *18*, 73-87.
7. Albertsson, A. C.; Karlsson, S. *J. Appl. Polym. Sci.***1988**, *35*, 1289-1302.
8. Hakkarainen, M.; Khabbaz, F.; Albertsson, A.-C. *Biopolymers*; Steinbüchel, A.; Ed. Wiley-VCH, Weinheim, **2003**; *Vol. 9*, 369-394.
9. Scheffer, T. C.; Morrell, J. J. *Forest Prod. J.* **1997**, *47*(5), 42-44.
10. Morris, P. I.; Cooper, P. *Forest Prod. J.* **1998**, *48*(1), 86-88.
11. Wang, W.; Morrell, J. J. *Forest Prod. J.* **2004**, *54*(12), 209-212.
12. Gnatowski, M. *Proceedings of the 8th International Conference on Woodfiber-Plastic Composites*, Forest Products Society, Madison, WI, **2005**, 249-256.
13. Clemons, C. M.; Ibach, R. E. *Forest Prod. J.* **2004**, *54*(4), 50-57.
14. Kollmann, F. F. P.; Côté, W. A. *Principles of Wood Science and Technology. I. Solid Wood.* Springer-Verlag, New York, 1968.
15. Schirp, A.; Wolcott, M. P. *Wood Fiber Sci.* **2005**, *37*(4), 643-652.
16. ASTM D 2017-05: *Standard method of accelerated laboratory test of natural decay resistance of woods.* American Society for Testing and Materials.
17. ASTM D 1413-05b: *Standard Test Method for Wood Preservatives by Laboratory Soil-Block Cultures.* American Society for Testing and Materials.
18. EN 113: *Wood Preservatives – Method of test for determining the protective effectiveness against wood destroying basidiomycetes – Determination of the toxic values.* Deutsches Institut für Normung e.V., 1996.
19. prCEN/TS 15083-1: *Durability of wood and wood-based products – Determination of the natural durability of solid wood against wood-destroying fungi, test methods – Part 1: Basidiomycetes.* Technical Committee CEN/TC 38, European Committee for Standardization (CEN), 2004.
20. Schirp, A.; Wolcott, M. P. *J. Appl. Polym. Sci.* **2006**, *99*, 3138-3146.
21. Verhey, S. A.; Laks, P. E.; Richter; D. L.; Keranen, E. D.; Larkin, G. M. *Forest Prod. J.* **2003**, *53*(5), 67-74.
22. AWPA E7: *Standard Method of Evaluating Wood Preservatives by Field Tests with Stakes.* American Wood Preservers' Association, 1996.
23. Ibach, R. E., Clemons, C. M. *Conference Proceedings, Progress in Woodfiber-Plastic Composites*, Toronto, Canada, **2004**, 1-13.

24. ASTM D 1758-02: *Standard Test Method of Evaluating Wood Preservatives by Field Tests with Stakes*. American Society for Testing and Materials.
25. Ibach, R. E.; Clemons, C. M.; Stark, N. M. *Proceedings of the 7th International Conference on Woodfiber-Plastic Composites*, Madison, WI, Forest Products Socity, **2004**, 61-67.
26. Mankowski, M. M.; Morrell, J. J. *Wood Fiber Sci.* **2000**, *32*(3), 340-345.
27. Verhey, S., Laks, P.; Richter, D. *Forest Prod. J.* **2001**, *51* (9), 44-49.
28. Pendleton, D. E.; Hoffard, T. A.; Adcock, T.; Woodward, B.; Wolcott, M. P. *Forest Prod. J.* **2002**, *52*(6), 21-27.
29. Verhey, S. A.; Laks, P. E. *Forest Prod. J.* **2002**, *52*(11/12), 78-81.
30. Goodell, B.; Daniel, G.; Liu, J.; Mott, L.; Frank, R. *Forest Prod. J.* **1997**, *47*(11/12), 75-80.
31. Iiyoshi, Y.; Tsutsumi, Y.; Nishida, T. *J. Wood Sci.* **1998**, *44*, 222-229.
32. Trendelenburg, R. *Holz als Roh- und Werkstoff* **1940**, *3* (12), 397-407.
33. Kennedy, R. W. *Forest Prod. J.* **1958**, *8*, 308-314.
34. Wilcox, W. W. *Wood Fiber* **1978**, *9*(4), 252-257.
35. Hardie, K. *Int. Biodet. Bull.* **1980**, *16*(1), 1-8.
36. Stark, N. M.; Rowlands, R. E. *Wood Fiber Sci.* **2003**, *35*(2), 167-174.
37. Khavkine, M.; Kazayawoko, M.; Law, S.; Balatinecz, J. J. *Int. J. Polym. Mater.* **2000**, *46*, 255-269.
38. Silva, A.; Freitag, C.; Morrell, J. J.; Gartner, B. *Proceedings of the 6th International Conference on Woodfiber-Plastic Composites*; Forest Products Society, Madison, WI, **2001**, 73-77.
39. Verhey, S. A., Laks, P. E. Richter. D. L. *Proceedings of the 6th International Conference on Woodfiber-Plastic Composites*; Forest Products Society, Madison, WI, **2001**, 79-86.
40. Ibach, R. E.; Clemons, C. M. *Proceedings of the 6th Pacific Rim Bio-Based Composites Symposium,* Portland, OR; Oregon State University, Corvallis, OR, **2002**, 574-583.
41. Saini, D. R.; Shenoy, A. V. *Polym. Commun.* **1985**, *26*, 50.
42. ASTM G21-96: *Standard Practice for Determining Resistance of Synthetic Polymeric Materials to Fungi*. American Society for Testing and Materials.
43. ASTM D 3273-94: *Standard Test Method for Resistance to Growth of Mold on the Surface of Interior Coatings in an Environmental Chamber.* American Society for Testing and Materials.
44. ASTM D 4445-91 (reapproved 1996): *Standard Test Method for Fungicides for Controlling Sapstain and Mold on Unseasoned Lumber (Laboratory Method).* American Society for Testing and Materials.
45. AWPA Draft: *Standard Method of Evaluating the Resistance of Wood Product Surfaces to Mold Growth.* American Wood Preservers' Association.
46. BS 3900: Part G6: 1989: *British Standard Methods of test for paints: Part G6: Assessment of resistance to fungal growth.* British Standards Institution.

47. ISO 16869: *Plastics – Assessment of the effectiveness of fungistatic compounds in plastics formulations.* International Organization for Standardization, 2001.
48. TAPPI T 487 pm-99: *Fungus resistance of paper and paperboard.* Technical Association for the Pulp and Paper Industry.
49. Eaton, R. A.; Hale, M. D. C. *Wood: decay, pests and protection.* Chapman and Hall, London, 1993.
50. Abraham, L. D.; Roth, A.; Saddler, J. N.; Breuil, C. *Can. J. Bot.* **1993**, *71*, 1224-1230.
51. Abraham, L. D.; Hoffman, B.; Gao, Y.; Breuil, C. *Can. J. Microbiol.* **1998**, *44*, 698-701.
52. Brush, T. S.; Farrell, R. L.; Ho., C. *Tappi J.* **1994**, *77* (1), 155-159.
53. Breuil, C.; Yagodnik, C.; Abraham, L. *Mat. Org.* **1995**, *29*, 15-25.
54. Schirp, A.; Farrell, R. L.; Kreber, B.; Singh, A. P. *Wood Fiber Sci.* **2003**, *35*(3), 434-444.
55. Chidester, M. S. *Proceedings of the American Wood Preservers' Association* **1942**, *38*, 134-140.
56. Wolf, F.; Liese, W. *Holz als Roh- und Werkstoff* **1977**, *35*, 53-57.
57. Seifert, K. A. *Ceratocystis* and *Ophiostoma*: Taxonomy, Ecology and Pathogenicity; Wingfield, M. J.; Seifert, K. A.; Webber, J. J.; Eds. American Phytopathological Society, Minneapolis, MI, 1993, 141-151.
58. Schirp, A.; Farrell, R. L.; Kreber, B. *Holz als Roh- und Werkstoff* **2003**, *61*(5), 369-376.
59. Findlay, W. P. K.; Pettifor, C. B. *Forestry* **1937**, *11*, 40-52.
60. Chapman, A. D.; Scheffer, T. C. *J. Agri. Res.* **1940**, *61*(2), 125-133.
61. Thunell, B. *Holz als Roh- und Werkstoff* **1952**, *10*(9), 362-365.
62. von Pechmann, H.; Graessle, E.; Wutz, A. *Forstwiss. Centralblatt* **1964**, *83*, 290-314.
63. Encinas, O.; Henningsson, B.; Daniel, G. *Holzforschung* **1998**, *52*, 82-88.
64. Smith, R. A. *The Wall Street Journal*, April 24, **2002**.
65. Robbins, C., Morrell, J. *Mold, Housing and Wood.* Western Wood Products Association, Portland, OR, 2002. www.wwpa.org.
66. Dylingowski, P. *Proceedings of the 7th International Conference on Woodfiber-Plastic Composites*, Forest Products Society, Madison, WI, 2003, 177-186.
67. Laks, P.; Vehring, J.; Verhey, S.; Richter, D. *Proceedings of the 8th International Conference on Woodfiber-Plastic Composites*, Forest Products Society, Madison, WI, 2005, 265-270.
68. Thometschek, M. *Proceedings of the 8th International Conference on Woodfiber-Plastic Composites*, Forest Products Society, Madison, WI, 2005, 279.
69. VdL-RL 07: *Richtlinie zur Bestimmung der Beständigkeit von Beschichtungsstoffen gegen Algenbefall* (Guideline for the determination of the

resistance of coating materials against algal growth). Verband der Lackindustrie e.V., Frankfurt am Main, Germany. www.lackindustrie.de.

70. Daniel, G. *Wood Deterioration and Preservation: Advances in Our Changing World.* Goodell, B.; Nicholas, D. D.; Schultz, T. P.; Eds. ACS Symposium Series 845, American Chemical Society, Washington, DC, 2003, 34-72.
71. Tschech, A.; Zeyer, J.; Kastien, H.; Sutter, H.-P. *Mat. Org.* **1992**, *27*(3), 203-233.
72. Lee, B.; Pometto III, A. L.; Fratzke, A.; Bailey, T. B. *Appl. Environ. Microbiol.*, **1991**, *57*(3), 678-685.
73. ASTM D 2481-05: *Standard Test Method for Accelerated Evaluation of Wood Preservatives for Marine Services by Means of Small Size Specimens.* American Society for Testing and Materials.
74. Pendleton, D. E.; Hoffard, T. A. *Phase II Engineered Wood Biodegradation Studies*; SP-2078-SHR, Naval Facilities Engineering Service Center, Port Hueneme, CA, 2000.
75. ASTM D 790: *Standard Test Methods for Flexural Properties of Unreinforced and Reinforced Plastics and Electrical Insulating Materials.* American Society for Testing and Materials.
76. Laks, P. *Proceedings, 33rd International Particleboard/Composite Materials Symposium*; Washington State University, Pullman, WA, 1999, 151-158.
77. Morrell, J. J. *Proceedings, 35th International Particleboard/Composite Materials Symposium*; Washington State University, Pullman, WA, **2001**, 17-24.
78. Simonsen, J.; Freitag, C. M.; Silva, A.; Morrell, J. J. *Holzforschung* **2004**, *58*, 205-208.
79. Taylor, A. M.; Freitag, C. M.; Morrell, J. J. *Forest Prod. J.* **2004**, *54*(4), 45-49.
80. Zabel, R. A.; Morrell, J. J. *Wood Microbiology: Decay and Its Prevention.* Academic Press, 1992.
81. Rowell, R. R. *Forest Prod. Abstracts* **1983**, *6*(12), 363-382.
82. Hill, A. S. *J. Inst. Wood Sci.* **2000**, *15*(3), 140-146.
83. Liu, F. P.; Wolcott, M. P.; Gardner, D. J.; Rials, T. G. *Compos. Interfaces* **1994**, *2*(6), 419-432.
84. Wool, R. P.; Raghavan, D.; Wagner, G. C.; Billieux, S. *J. Appl. Polym. Sci.* **2000**, *77*, 1643-1657.
85. Tichy, R. J. *Proceedings of the 7th International Conference on Woodfiber-Plastic Composites*, Forest Products Society, Madison, WI, 2003, 173-176.
86. Tichy, R. J. Pers. comm., 2005.
87. ASTM D 3345-74: *Standard Test Method for Laboratory Evaluation of Wood and Other Cellulosic Materials for Resistance to Termites.*

88. AWPA E1: *Standard Method for Laboratory Evaluation to Determine Resistance to Subterranean Termites*. American Wood Preservers' Association, 1997.
89. AWPA E10: *Standard Method of Testing Wood Preservatives by Laboratory Soil-Block Cultures*. American Wood Preservers' Association, 2001.

Registration and Approval, and Environmental Disposal Issues of Wood Preservatives

The first photograph is courtesy of Darrel D. Nicholas and the second and third photographs are courtesy of Tor P. Schultz, both of Mississippi State University. *(See page 11 of color inserts.)*

Chapter 30

The Federal Insecticide, Fungicide, and Rodenticide Act and Its Impact on the Development of Wood Preservatives: Registration Requirements

Henry M. Jacoby[1] and Mike H. Freeman[2]

[1]Henry Jacoby Regulatory Consultant, 6709 Ilex Court, New Market, MD 21774
[2]Independent Wood Scientist, 7421 Hubters Tree Cove, Memphis, TN 38125

An overview of FIFRA and its recent amendments and their impact on the registration requirements for wood preservatives. New amendments and statues have been passed in response to societal concerns and technological advances that have impacted on how EPA reviews pesticide products. This chapter will discuss some of these amendments/statutes and their impacts on the registration of wood preservative products.

The Environmental Protection Agency (EPA or The Agency) regulates the use of pesticides under the authority of two federal statutes: the Federal Insecticide, Fungicide and Rodenticide Act (FIFRA) and the Federal Food, Drug, and Cosmetic Act (FFDCA). These statutes are the basis of the Agency's regulatory authority as it applies to pesticides. The authority to establish tolerance or exemption from tolerance originates from the FFDCA. And the authority to regulatory pesticides is given to the Agency by FIFRA. In carrying out its responsibility, the Agency has promulgated rules that define its process and the responsibilities of the Agency as well as those of the regulated community. In these regulations, the requirements for issuing a registration, for restricting a registration, and for canceling as registration are defined. The regulations are further clarified by the use of Pesticide Registration Notices. These statues have been amended through the years to reflect the concerns of society regarding the use of pesticide and the exposure to pesticide residues from registered uses.

The Federal Food, Drug, and Cosmetic Act (FFDCA) authorizes EPA to set maximum residue levels or tolerance (the legal amount of residue contamination in or on foods and feeds as a result of a pesticide use) for each pesticide and its toxicologically significant residues which may be in or on foods or animal feeds as a result of the registered uses. The Federal Food, Drug, and Cosmetic Act also allows EPA to exempt pesticides from the requirement of a tolerance. Exemptions are allowed with those pesticidal chemicals or formulation chemicals that demonstrate little or no risk (reasonable certainty of no harm) from the residue levels associated with the pesticidal use. [(1)]This Act defines the rule-making process required in set tolerances or exemptions from tolerance. Compared with FIFRA, FFDCA normally does not take into consideration the benefits received from the use of the pesticide. However, it can in limited extreme circumstances. In order to establish a tolerance, or to exempt from a tolerance, for a pesticidal chemical the FFDCA requires the Agency to reach a finding that there is reasonable certainty that the tolerance or exemption from tolerance will result in "no harm". In reaching its decision, extra consideration must be given to protect infants and children. The FFDCA requires that a tolerance or exemption from tolerance be established prior to registration.

The Federal Insecticide, Fungicide and Rodenticide Act (FIFRA) provides EPA the authority for regulation, sale, distribution and use of pesticides in the U.S. The Act authorizes EPA to require the appropriate data necessity to reach its regulatory decision on a use-by-use basis. The Federal Insecticide, Fungicide and Rodenticide Act requires that all pesticide products must be registered before they can be manufactured, distributed for sale or sold. In registering pesticides, the Environmental Protection Agency (EPA) is required to take into consideration the risks posed by the pesticide to society and the environment compared to the societal benefits achieved from its use. That is, any risk to society and the environment as a result of exposure from the use of a registered pesticidal chemical and/or its residues or metabolites must be out-weighed by the benefits achieved from the use of the pesticidal chemical. In order to minimize risk from a pesticide, the Agency can regulate the pesticide product through use restrictions, labeling, packaging, composition and disposal. The Act also allows the EPA to ask for additional data (i.e., Call-in) in order to better understand the risks and/or exposure associated with a pesticidal chemical's use.*(1)*

Over the years, FIFRA has been amended in order to address new scientific technology or societal issues. "Passage of the 1972 amendments to FIFRA enacted through the Federal Environmental Pesticide Control Act *(2)* (FEPCA) was part of a wave of environmental legislation which completely overhauled Federal environmental regulatory authority."*(3)* "The Endangered Species Act (ESA) of 1973 prohibits any action that can adversely affect an endangered or threatened species or its habitat. In compliance with this law, EPA must ensure that use of the pesticides it registers will not harm these species."*(4)* "The Food Quality Protection Act (FQPA) of 1996 amended FIFRA and FFDCA. These amendments fundamentally changed the way EPA regulated pesticides whose uses may result in residue of the pesticide chemical or its toxicologically significant metabolites in or on food or feed. The requirements included a new safety standard of "reasonable certainty of no harm" that must be applied to all pesticides used on foods.*(5)* "The Pesticide Registration Improvement Act (PRIA) establishes pesticide registration service fees for registration actions carried out in the three pesticide program divisions: Antimicrobials, Biopesticides and Pollution Prevention, and the Registration Divisions."*(6)*

These amendments have had varying impacts on the registration process and data requirements. The Endangered Species Act mainly altered the way the supporting fish and wildlife toxicity data submitted to support registration of a technical chemical is used. The Food Quality Protection Act significantly altered the Agency's tolerance setting process with its impact on the data requirements in order to make a finding of "reasonable certainty of no harm" prior to establishing a tolerance or an exemption from tolerance. Also, the FQPA amendments require EPA to make expedited decisions on antimicrobial pesticides. The Federal Environmental Pesticide Control Act (FEPCA) and The Pesticide Registration Improvement Act (PRIA) had substantial impact on the data requirements and costs association with registration of a pesticidal chemical. These two amendments are worth discussing in depth because they have significant impact on the cost of registration.

The Pesticide Registration Improvement Act changed the relationship between EPA and the regulated community. This legislation authorized EPA to recoup the cost of its review process from the applicants requesting registration. The amount of fee is determined by the nature of the registration activity being requested. Basically, costs increase with the more complex applications (i.e., more supporting data and use of more Agency scientific resources). Associated with the legislation is the fee associated with every category of registration action and the corresponding decision time period allotted to the Agency to reach a regulatory decision on the action. The goal is to create a more predictable evaluation process for affected pesticide decisions. The legislation also promotes shorter decision review periods for reduced risk applications.

Reduced risk initiative was required by the FQPA amendments to FIFRA. *(7)* This initiative mandates expedited reviews for applications for registration and amendments to registration for pesticides that may reasonably be expected to reduce risks from pesticides to human health or nontarget organisms, to reduce potential contamination of groundwater, surface water or other value environmental resources, or to broaden or make available integrated pest management strategies.

To initiate the process, the applicant must demonstrate how the use of the proposed registration or amendment of their current product registration may reasonably be expected to meet the above criteria. The information demonstrating how a product meets the above criteria is called the Reduced-Risk Rationale.

The Agency has defined all the factors that must be addressed in the reduced risk rationale package. If all these factors are not addressed, the request for reduced risk review is considered by the Agency to be incomplete. The documentation must discuss the inherent properties of the new product that leads to the reduced-risk, as well as a comparison of those properties to commonly used alternatives if appropriate. The complexities in drafting a reduced-risk rationale often are best left to those with extensive experience and knowledge of the registration process. The applicant is notified no later than 30 days after receipt if the submission is complete.

The Pesticide Registration Improvement Act was a joint effort that involved the regulated community, concerned citizen groups, and the Agency. In anticipation of PRIA, the regulatory service fees were determined based on the amount of Agency resources utilized for a given type of action. The list of actions and the related fees are published and available from the Office of Pesticide Program's web site for fees. These fees range from $1,000.00 for a product that is identical in all respects to one already registered to $525,000.00 for a new active ingredient requesting experimental use permits and temporary tolerances. For wood preservation products, fees can vary depending if the product is used for control of microbes or for control of insects. The Federal Insecticide, Fungicide and Rodenticide Act contains language that permits those chemical wood preservatives used against microorganisms to be considered as an antimicrobial product reviewed in the Antimicrobial Division of the Office of Pesticide Programs (OPP). Conversely all other chemical wood preservatives would be considered conventional chemicals reviewed in the Registration Division of OPP. From the fee structure in place, a new wood preservation active ingredient for use out doors reviewed in the Registration Division would require a fee of $330,000.00. Whereas a new wood preservation active ingredient for use out doors reviewed by the Antimicrobial Division would only require a fee of $150,000.00. It will not always be easy to discern exactly what registration fee is attached to wood preservation chemicals. The Agency has recognized this problem, telling applicants not to send money until the Agency determines the appropriate category of action and invoices or bills the applicant.

The Pesticide Registration Improvement Act also recognizes that for a small company the registration fee may be so prohibitive as to put the company out of business. In certain cases, the Agency will waive or reduce the fee for service. The Federal Insecticide, Fungicide and Rodenticide Act, Section 4 (i)(5)(E)(ii), defines small business as a legal entity with 500 or fewer employees and 3-year average annual gross global sales from pesticides are $60,000,000.00 or less. Gloss global sales are the total sales of the applicant and all affiliates. The Agency will waive 50% of the required fee for those entities the meet the above criteria. It will waive 100% of the required fee if the 3-year annual average of gross global sales from pesticides is $10,000,000.00 or less. A request for waiver from the fee for service must be appropriately documented.

The Federal Environmental Pesticide Control Act of 1972 drastically changed the way pesticides were being evaluated in the registration process. With the passage of FEPCA, FIFRA was changed to an Act not only concerned with the efficacy and safe use of a pesticidal product by the consumer but also concerned about societal and environmental health and risks. For the Agency to approve a pesticide product registration not only did the product require proper labeling for efficacy and consumer safety, it had to demonstrate that it would not pose an environmental or health related risk greater than the societal benefit achieve by the use of the product. FEPCA required the Agency to define by regulation how it would do business: It required the Agency to delineate its data requirements and the registration, reregistration, and classification procedures among out processes. The expansion of the data requirements and reregistration had substantial impact on the regulated community. "Before pesticides can be marketed and used in the United States, EPA evaluates them thoroughly to ensure that they will meet federal safety standards to protect human health and the environment. The process of registering a pesticide is a scientific, legal, and administrative procedure through which EPA examines the ingredients of the pesticide; the particular site or crop on which it is to be used; the amount, frequency, and timing of its use; and the appropriate storage and disposal practices. Pesticides that meet the requirements are granted a license or "registration" that permits their distribution, sale, and use according to specific use directions and requirements identified on the label.

In evaluating a pesticide registration application, EPA assesses a wide variety of potential human health and environmental effects associated with use of the product. The producer of the pesticide must provide data from tests done according to EPA guidelines and carried out according to the EPA standards of Good Laboratory Practices."*(7)* As required by the FIFRA, the Agency has published the list of data needed to support registration of a pesticide chemical or product according to major categories of use (i.e., in-door non-food, in-door food, terrestrial non-food, etc.). These lists are given published in 40 Code of Federal Regulations (CFR) Part 158. Data lists are given for the conventional chemical pesticide, for the biochemical pesticide and for the microbial pesticide.

The data lists indicate whether the data in required or conditionally required. Within each type of pesticide, the list is further differentiated by scientific areas such mammalian toxicity, fish and wild-life toxicity, plant protection (non-target plants), non-target insects, residue chemistry, environmental fate and transport, spray drift, exposure and product performance. Each scientific area requires data that will allow the Agency to estimate the hazards associated with the pesticide chemical and its uses and the exposure to the hazards from the various uses. The end-result is an Agency calculation of the estimated over-all risks associated with a given pesticide for a given use. Theoretically, the data lists in 40 CFR Part 158 are to assist and enlighten the regulated industry as to the data that would be required to support the registration of a new pesticide chemical or of a pesticidal product. However, this is not the case.

Inherent in developing the database to support the registration of a pesticide is the risk of what the data may or may not demonstrate. Toxicity studies, such as teratogenic, reproduction, chronic feeding, or carcinogenic studies, require several years and many thousands of dollars to run. The results of which may demonstrate a hazard that, because of the exposure, will result in unacceptable levels of risks or margins of safety. Or toxicity to fish, other wildlife, and non-target insects and plants could results in limiting the extent of one's planned uses. These studies could demonstrate effects that are totally unacceptable to EPA. Even though, a chemical may not demonstrate any toxicological significant risk, the product efficacy field testing may not demonstrate the same degree of effectiveness as was demonstrated under laboratory conditions. It is very important to have a research regimen and in place that allows for critical go or no go decisions early in the development process.

Because of the inherent investment risks and costs involved with the development of new pesticidal chemicals, most companies are not equipped financially, physically or technologically to undertake chemical development. Companies most apt to develop new pesticide chemicals are aligned with the agricultural industry and therefore only remotely connected to silvaculture. New pesticide chemicals are developed for the major agricultural markets (i.e., field crops and orchard crops). They are not interested in small specialty markets. Therefore most companies selling to specialty market hope to piggyback on the already registered chemicals.

Even this approach is not without its risks and difficulties. Generally new pesticide chemicals are protected by patents or by the exclusivity provision of FIFRA. The primary producer must want to work with another company to share their data and knowledge about the pesticide. Often, this is not the case. Unless there is a desire to transfer new technology into the smaller market, it is unlikely the small market will receive it until patent and exclusivity rights of the chemical producer have expired. Even then, the company wishing to piggyback must develop the additional data to support the proposed uses. The inherent investment risks in developing new data also becomes an issue as mentioned above.

As stated above 40 CFR Part 158 lays out the data needs for several different broad use scenarios. The list indicates whether the data is required, conditionally required or required for an experimental use permit (EUP). In the list of studies in 40 CFR Part 158 there is an indication what the test substance must be. Some studies are carried out on a technical grade of the active ingredient, the pure active ingredient, the end-use product or a typical end-use product. The studies lists are footnoted to clarified when a conditional study must be carried out. Because the variety of chemical pesticides and uses, 40 CFR Part 158 does not covered every use scenario and is difficult to interpret, with many studies, if a study is really required. The advise of a consultant is often useful in determining a basic test regimen of testing. Discussions with EPA personnel regarding the testing regiment may be warranted. Notwithstanding all the safe guards to ensure the all the correct data are available at time of the submission of the application, EPA may require additional data after reviewing the original application.

In registering a new pesticide chemical, it is necessary to have answers to several preliminary questions. The questions relate to formulation, the application technique, claimed protection and use of the treated wood. First, what formulation or formulations you want to register? Second, are you planning to apply the pesticide chemical by spray, brush, or soaking, by pressure treatment or any other technique? Third, what claims will you be making for the treated wood? And fourth, is the treated wood to be used indoors, outdoors, for food or feed contact surfaces, for soil contact or buried in the soil? These questions determine what kind and how much data are required.

If I as a chemical producer wished to register a pesticide chemical for pressure treatment to protect wood used for bulkhead structures from the attack of marine organisms, I can look in Appendix A of 40 CFR 158, *DATA REQUIREMENTS FOR REGISTRATION: USE PATTERN INDEX,* to determine what general use pattern indicates under item 8, Wood or Wood Structure Protection Treatments fall within the general use pattern a Domestic outdoor or indoor. However, Appendix A also indicates that boat bottoms and other submerged structures fall into the general use pattern of Aquatic Noncrop. Without past experience or input from a knowledgeable person, it is not clear what data are needed based on the general use pattern. Because of the impact on the aquatic environment, EPA would consider this an aquatic use.

If the producer, after searching the Appendix A, sees item 8, makes a determination that the proposed use fall in the general use category as a outdoor noncrop use. Then, using 40 CFR Part 158, the database that EPA expects to be submitted with an application for registration can theoretically be determined. The required data can be divided into those data developed on the pesticide chemical, per se, and those data developed on the formulated product as sold.

The difficulty encountered using 40 CFR Part 158 is it was developed with agricultural pesticides as the model and not with wood preservatives. Hence interpretation is not easy and often requires professional guidance from consultants and EPA.

The chemistry data requirements are discussed first in 40 CFR Part 158. The EPA requires information on the product composition, a description of materials used to product, a description of the production process and the formulation process, a discussion of formation of any impurities, a preliminary analysis of the pesticide chemical. Certified limits of the ingredients in the pesticide product and an enforcement method for determining the active ingredient in the product. The EPA also requires a number of physical and chemical properties of the pesticidal chemical and of the formulated product. Many of the required studies, such as preliminary analyses, storage stability, and chemical stability, are required to be carried out under EPA's standard of good laboratory practice regulation. With a new pesticide ingredient, EPA requires the submission of a chemical standard.

The data requirements listed under the residue chemistry section of 40 CFR 158 are very confusing to the uninitiated, especially to a person not in the mainstream agriculture chemical business. The data lists indicate that the chemical identity, direction for use, residues in potable water and residue in fish are required. The chemical identity applies to the technical grade of the active ingredient and direction for use covers the end use product. The information discussed in the above paragraph is being requested again with labeling. The other two studies are not required if the pesticide is not directly applied to way. That is fairly straightforward. But would the study be required if the treated lumber was used in a fish farm? The answer to this question is it depends on the nature of pesticide chemical. It is not always clear if a study is or is not needed. EPA might require this data or additional information on the nature of the residues in the wood; their capability to leach into the water; or the concentration of residues immediately around the submersed lumber.

In the environment fate data requirements, the difference in data requirements between the use category domestic outdoor and aquatic non-food is significant and more costly to carry out. Both use patterns require a hydrolysis study and a leaching and adsorption/desorption study. However, the domestic door use only requires an aerobic soil metabolism study and a field dissipation study. Whereas the aquatic non-food requires a photodegradation in water study, an aerobic aquatic metabolism study, an aquatic soil (sediment) dissipation study. Should the wrong decision be made concerning the use pattern, studies might be carried out that are not required. Or the use of the treated lumber might require data to support both the domestic outdoor or aquatic non-food.

The data requirements for toxicity studies are generally straightforward regarding a technical product or end use product. The database for these products is generally known as the "six pack." Six toxicological studies needed to support primarily the labeling of the product formulated for distribution and sale. They are the acute oral LD_{50}, the acute dermal LD_{50}, the inhalation LC_{50}, the primary eye irritation, the primary dermal irritation, and skin sensitization. The inhalation study is not required for product that does not produce a respirable vapor or is not sprayed. For products that have a pH less than 2 or greater than 10, the pH may be used as the basis for requesting a waiver of the primary skin and eye irritation studies. Products on the extremes of the pH scale are known to be very corrosive. The skin sensitization does not need to be run if the product is not formulated with any known sensitizers. These studies are also used in the over all risk assessment performed by EPA. The remainder of the toxicity studies listed in 40 CFR Part 158 relate to the active ingredient, per se, and normally the responsibility of the primary producer of the chemical. For a new pesticide chemical used in wood preservation for submerged structures – this would be a chemical never before seen by EPA – the data requirement are extensive and not as straightforward. Normally EPA would expect the six-pack of acute studies a battery of mutagenicity studies (gene mutation, structural chromosomal aberration & other genotoxic effects). And, if the chemical is neurotoxic, EPA requires a delayed neurotoxicity study. The remaining toxicity studies may be required depending on exposure or toxicity findings. Only with expert advice and consultation with EPA is one reasonably certain that a study is not needed.

The exposure issue associated with the treatment and use of treated lumber is critical in determining what conditional studies will also be required to support this use. If wood treaters might be women of child bearing age or the treated lumber might be used in a manner the women of child bearing age could be exposed or used around residences or public facilities, EPA would likely want to see a dermal exposure study, a teratogenicity study in one species and a 12-month chronic feeding study. Depending on the results, other toxicological studies or exposure studies may be needed.

In use exposure studies, per se, are not a direct requirement of FIFRA. These exposure studies are requested or carried out if there is a toxicological end-point for which the margin of exposure (MOE) is approaching the target of 100 or unacceptable to EPA. EPA normally uses a computer model (PHED Version 1.1) for assessing human exposures during pesticide handling activities in the absence of exposure data.*(9)* An applicant with approval from EPA may design a study that reflects actual exposures occurring during various uses of the treated material.

Studies required by FIFRA to enable EPA to evaluate potential effects to fish and non-target organisms (Plants, birds, mammals, etc.) are develop in a three tier progression. Tier one contains the basic data requirements. These data are the same for domestic outdoor and for aquatic non-food. They are an avian acute oral LD_{50} study, two avian dietary LC_{50} studies, two fresh water fish LC_{50}

studies, and a freshwater water invertebrate LC_{50}. However, in our example the treated wood will be used in a marine environment, therefore EPA may require data in species other than those given in 40 CFR part 158 or EPA may require additional studies be carried out using marine organisms. Clearly, early outside consultations are important, if the goal is to have a cohesive and complete database to support the application for registration.

And, lastly, with a new pesticidal active ingredient, EPA requires product performance data. The nature of the performance data required must address the manner in which the chemical is applied (i.e., dip, brush-on, sprays or pressure treatment). Modifications of the Stake method, Ground board method, and the modified ground board method are acceptable. There are no recommended protocols to study the effectiveness of wood treatment of infested lumber or wood treatment for above ground uses. Protocols for these studies should be pre-approved by EPA prior to initiating the study. The data must be obtained from 3 geographically located sites where there is year round pest pressure. The data must show complete resistance to termite attack for at least 5 years. If acceptable data shows effective for at least 2 year but less than 5 years, the product may be registered based on a labeling statement that require annual inspection.*(9)*

It becomes quite obvious that the time, costs and risks active ingredient are very high in bringing a new pesticide chemical to the market place. Strategic planning for the regimen of data is critical. There must be in-place criteria guiding a decision to stop the development or to continue with additional studies. The cost of developing a database to support registration of a new pesticide chemical continues to rise well into the 10's of millions of dollars. To save costs, a decision to stop the development project needs to occur as early as possible. And it needs to be made based on the results from either preliminary investigations or on the results obtained in ongoing studies. Because of the high financial risks involved, this approach is not viable for small market products.

A less timely, costly or risky approach to registering a new pesticide chemical for wood preservation is expanding the use patterns of an already registered pesticide. If a currently registered pesticidal chemical is identified as having those properties necessary for use by the wood preservation industry, working with the manufacturer or registrant of the chemical, a smaller database would have to be developed. Given the extensive database already developed, a much clearer picture of the additional data needs can be derived. Also, the impact of the additional exposure on the chemical's MOE's from the proposed expanded new wood preservation uses can be determined using the available database. This calculation allows the developer to know whether EPA would accept the propose uses or if a safe use issue might arise.

Most companies in the wood preservation field are not prime manufacturers of the pesticidal active ingredient. They desire only to obtain registration for an end use product that they can distribute and sell in their normal channels of distribution. Although this is less daunting than registering a new active ingredient, it can be extremely frustrating for the neophyte. The application has to meet EPA criteria for acceptability. Many times, just getting in the proverbial door can be a challenge. After months of receiving rejection after rejection, the registrant will turn to a consultant to assist in the paper work alone. So let's review a typical end use product registration requirements.

An application for registration must have all the appropriate forms. There are six forms that must be required:

The application form gives EPA pertinent information about the product (such child resistant packaging, the various sizes of packages, the similarity to currently registered products, etc.). The application form also must give the service fee classification and the fee for the purposed registration.

The confidential statement of formulation that gives EPA information about some physical characteristics and identifies each of the ingredients in the product along with the percent by weight for each. The percent by weight is given for the nominal concentration and certifies the upper and lower limits for each ingredient.

The formulator's exemption statement form that states the applicant is purchasing and using a registered manufacturing use product. The purchased product is identified by name and EPA Registration Number.

The certification as to the citation of data form tells EPA how the applicant is supporting the registration. The FIFRA requires that each pesticide product registered by EPA be supported by its own database. There are several ways this can be accomplished. There is the cite-all method of support, the selective method of support, and the selective method with cite-all for some of the data. These methods of support will be discussed more in-depth later in this paper.

The Data Matrix form delineates the data identified in 40 CFR Part 158 as being required to support the registration of the pesticide product. The form requires the applicant to identify who submitted the data and the status of the data. The data may be from the public domain (i.e., scientific journals, government publications, etc.) or old data (data submitted to EPA 15 or more years ago) or exclusive use data (data submitted to support a new active ingredient can only be used with the expressed written consent of the owner for 10 –13 years following the date it was first registered). There is also a form that is releasable to the public with most of the above information blacked out.

The last form is the certification of the physical chemical properties. The applicant indicates to EPA that there exist test reports for each of the end-points given. These report are available and can be sent to EPA for review if required. This form is must be included in the Report B – Chemical/Physical Properties Report.

Let us back up some and discuss the actual database needed to support an end-use product. Based on 40 CFR Part 158, the database that may needed to support registration of an end-use product is shown Table I:

Table I. Registration Data Requirements for an end-use product

OPPTS GUIDELINE NO.	*STUDY TITLE*	*OPPTS GUIDELINE NO.*	*STUDY TITLE*
REPORT A: IDENTITY, COMPOSITION AND ANALYSIS		*GROUP B – PHYSICAL CHEMICAL PROPERTIES (CONT.)*	
830.1550	**PRODUCT IDENTITY AND COMPOSITION**	830.7100	**VISCOSITY**
830.1600	**DESCRIPTION OF MATERIALS USED TO PRODUCE THE PRODUCT**	830.6319	**MISCIBILITY**
830.1650	**DESCRIPTION OF THE FORMULATION PROCESS**	830.6320	**CORROSION CHARACTERISTIC**
830.1670	**DISCUSSION OF FORMATION OF IMPURITIES**	830.6321	**DIELECTRIC BREAKDOWN CONSTANT**
830.1700	**PRELIMINARY ANALYSIS***	*TOXICITY STUDIES*	
830.1750	**CERTIFIED LIMITS**	871.1100	**ACUTE ORAL TOXICITY – RAT***
830.1800	**ENFORCEMENT ANALYTICAL METHOD**	871.1200	**ACUTE DERMAL TOXICITY***
GROUP B – PHYSICAL CHEMICAL PROPERTIES		871.1300	**ACUTE INHALATION TOXICITY – RAT***
830.6302	**COLOR**	871.2400	**PRIMARY EYE IRRITATION – RABBIT***
830.6303	**PHYSICAL STATE**	871.2500	**PRIMARY SKIN IRRITATION***
830.6304	**ODOR**	871.2600	**DERMAL SENSITIZATION***
830.7200	**MELTING POINT**	*PRODUCT PERFORMANCE*	
830.7220	**BOILING POINT**	810.3600	**STRUCTURAL TREATMENTS**
830.7300	**DENSITY, BULK DENSITY, OR SPECIFIC GRAVITY**		
830.7840	**SOLUBILITY***		
830.7950	**VAPOR PRESSURE***		
830.7370	**DISSOCIATION CONSTANT**		

*Study must be carried out in accordance with Good Laboratory Practices (40 CFR Part 160)

In developing the application package, each one of the data points should be address. If the data are not needed, the application should give the reasons why. The footnotes in 40 CFR Part 158 gives the conditions requiring the data. Some data may be waived. In these cases, the basis for requesting a waiver from the data requirement must be presented.

As stated above, there are two ways FIFRA recognizes as acceptable means to meet these requirements: the Selective Method and the Cite-All Method. One can also use a combination of the Selective and Cite-all Method to support an application for registration. Each of the support methods needs to be discussed.

The Selective Method of Support tells EPA what studies you want the Agency to use in support of your product. An applicant can carry out each study on the proposed pesticide product and submit the results to EPA. This is the cleanest way of supporting the registration of a pesticide product. However, there are costs associated with this approach. There are the costs of the studies themselves and the cost associated with the time to obtain the final reports and for EPA to review the results. The second way, an applicant can meet some or all of the requirements, is identifying a study EPA has in its files for each requirement. These must be studies carried out on a product that is substantially similar to your product or a plausible explanation must be given to EPA why the identified can be used to support your product. Also, understand that data submitted to EPA is compensable for 15 years from the date the data were submitted. Consultants familiar with the EPA and its workings can be an enormous assistance to someone not fully versed in FIFRA and pesticide registration. And finally, an applicant can develop some of the data, identify available data and use Cite-All for the remainder.

The Cite-All Method tells EPA to use whatever data it has in its files to support the application. This method is often seen as a blank check approach. As discussed in 40 CFR §152.86, the user of Cite-All must certify that he has make an offer to pay to each and every person on the Data Submitters List regarding the chemical of question. He must indicate a willingness to enter into negotiations over the cost of the data. And the user of the Cite-All Method also acknowledges that the application relies on all data submitted with the application and any previously submitted data. Previously submitted data must be concerned with properties or effects of the applicant's product, or an identical or substantially similar product or on any of the active ingredients in question. This method of support is extremely useful in shortening the time to obtain a registration provided there are identical or substantially similar products already registered. The down side to the Cite-All is the mandated negotiation with a possibility of binding arbitration concerning the cost of using someone's data.

In summary, the data requirements for any pesticide product can be determined from 40 CFR Part 158.

EPA's pesticide laws include provisions to ensure the protection of fish and wildlife. EPA requires and evaluates extensive ecological effects test data before registering a new pesticide or reregistering an existing pesticide. Therefore, EPA already performs much of the scientific analysis that the Services perform in the consultation process.

Under the ESA, and in consultation with FWS and NOAA Fisheries, EPA must ensure that its regulatory actions are not likely to jeopardize threatened and endangered species or destroy or adversely modify their critical habitats. EPA's pesticide risk assessment and regulatory processes ensure that protections are in place for all populations of nontarget species. Because endangered species need specific protection, EPA has developed risk assessment procedures to determine whether individuals of an endangered species have the potential to be harmed by a pesticide, and if so, what specific protections may be appropriate.

References

1. Federal Food Drug and Cosmetic Act, URL http://www.epa.gov/pesticides/regulating/laws.htm.
2. Federal Insecticide Fungicide and Rodenticide Act, URL http://www.epa.gov/pesticides/regulating/laws.htm
3. 86 Stat. 975, Pub. L. 92-516, 7 U. S. C. 136.
4. Federal Register, Vol. 40, No. 1239—Thursday, July 3, 1975.
5. Fees, URL http://www.epa.gov/pesticides/regulating/laws.htm.
6. Food Quality Protection Act, http://www.epa.gov/oppfead1/fqpa.
7. Pesticide Registration Notice 97-3, URL http://www.epa.gov/opppmsd1.
8. HED Science Advisory Council for Exposure, Policy 007, "Use of Values from the Pesticide Programs," January 1999.
9. OPPTS 810.3600 Structural Treatments, Product Performance Harmonized Guidelines URL http://www.epa.gov/docs/OPPTS_Harmonized.

Chapter 31

American Wood-Preservers' Association and Building Code Procedures for New Preservative Systems

Craig R. McIntyre

McIntyre Associates, Inc., 8565 East Grandview Lake Drive, Walls, MS 38680 (www.mcintyre-inc.com)

Once a new preservative system is under development, consideration must be given to obtaining recognition with appropriate bodies to allow commercialization. Both the American Wood-Preservers' Association (AWPA) and the International Code Council (ICC) provide avenues to allow new preservative systems to be added to their respective publications. Both organizations require generally similar testing regimes but there are significant differences in the scope of and the time frames needed for such testing to be accomplished. There are also significant differences in the requirements placed on the testing institutions by the different review bodies. Both procedures and the differences in the two procedures are discussed fully in this chapter.

Why List at All?

Prior to discussing the specifics of the actual process, some consideration should be given to the need for such publications. To be honest, one can sell wood without AWPA or ICC-ES review in the open market of today. Yet almost all organizations pursuing new wood preservative systems choose to obtain publications indicating conformance with either the AWPA or the ICC. Usually conformances are sought from both organizations. One must consider why proponents of new systems pursue such publications if it is not necessary.

The first reason is that preservatives not subjected to AWPA and/or ICC scrutiny have had very limited distribution in the marketplace. The systems may enjoy a small, regional market or perhaps a small niche market but the major retailers typically shy away from these less defined systems as too risky.

Obviously, a key component of the risk is that non-reviewed preservatives would be used in an application where they are not suitable. At the worst, there is the risk that someone would use the preservative treated wood for structural purposes even though structural uses would not be recommended and there would be a failure. Therefore, non-reviewed preservatives are further confined to non-structural uses such as fences, mailbox posts and the like. Some companies risk producing wood for decks built close to the ground but generally this application is considered too hard to control. Avoiding some risk is the second reason.

In today's litigious environment, producing or selling a non-reviewed product leaves one with a very poor defense in this author's opinion. The obvious concern is that the product would be deemed to be "not to industry standards" and hence leave the producer/seller open to a variety of liability issues. In truth, it seems that avoiding some liability issues are the major and third reason that organizations pursue standardization.

The AWPA Process

On the surface, obtaining a listing with the AWPA appears to be a series of simple two-step processes: first, the preservative system itself is listed in the Book of Standards and then the wood treated with that system is listed. In turn, listing the preservative only requires two steps: first, one defines the system and then one tests that system. System definition also only requires two steps: one describes the chemical system and then an analysis method is provided to ensure that the system meets the description. In regard to the actual testing, one need only consider two general uses: above ground and ground contact applications. Within these two general applications, there are a number of sub-categories that must be considered for the AWPA listing.

Above Ground Issues

Recently, the AWPA moved from their traditional commodity specifications to "Use Classifications" wherein the actual use of the commodity is considered the defining principle. The AWPA uses a scale of increasing numbers to designate the increasing severity of the use or exposure. For above ground uses, the appropriate designations are:

UC1—Dry interior uses
UC2—Damp interior uses
UC3A—Coated exterior uses
UC3B—Wet exterior uses

Thus, the least hazardous use, UC1, would be applications such as interior walls where the material would be expected to remain dry in service. A slightly more hazardous environment, UC2, would be where the material may become damp or occasionally wetted. Sill plates or bathroom floors would be examples of UC2 uses. Still more hazardous would be the uses covered by UC3A such as painted millwork where the coating is an integral part of the protective envelope. Lastly, the most extreme of the above ground uses are where the wood is continually exposed to the environment and it can become thoroughly wetted. Uses such as deck boards, fencing and the like fall into UC3B.

In order to obtain qualification for all above ground uses, it would appear that one must simply fulfill the necessary above ground test requirements for the most severe use, UC3B. Unfortunately, as of this writing, there are no procedures in the AWPA solely for above ground uses. Rather than specify certain tests and/or performance, the AWPA has historically used ground contact testing to determine ground contact retention levels and then factored the above ground retention levels at some percentage of the ground contact level. Granted that this is not very scientific but that is the system in place.

A task group within the AWPA has been working to rectify the procedure and specify tests appropriate for above ground uses. This task group has proposed revisions that delineate the various tests as either "Required" or "Recommended" for the various uses. Within that framework, it would appear that in the near future, the AWPA will have a system that lists the testing required for UC3B.

Assuming that the task group modifications are accepted by the AWPA[1], the total number of tests that are required or recommended for above ground uses can be outlined as shown in Table I:

---

[1] Note added in proof: The recommendations were accepted in August 2005.

Table I. Number of Above Ground Tests per Use Classification.

Use Class	Efficacy Tests		Depletion Tests		Mechanical/
	Lab	Field	Lab	Field	Chemical Prop.
UC1	1	1	1	0	5
UC2	2	2	2	1	5-6
UC3A	3	3	2	1	5-6
UC3B	3	3	2	1	6-7

As shown, there is an increasing level of complexity in the testing as one goes from UC1 to UC3B. Complicating the issue is the fact that the costs increase in proportion with the level of complexity and the number of years required to complete the testing also increases (see Table II). Obviously, the thought is that more complex and lengthy test requirements are demanded for products that are used in more critical or more severe exposures.

Table II. Estimated Costs and Time Frames for Above Ground Tests

Use Class	Total Testing Costs	Testing Time, years
UC1	$25-50,000	2
UC2	$30-60,000	2
UC3A	$40-80,000	3-5
UC3B	$50-100,000	3-5

The costs in Table II are out-of-pocket costs paid to the testing organization for conducting the tests. There may be additional costs for sample preparation, travel expenses and so on. As well, it is highly recommended that someone representing the sponsor witness the sample preparation and/or portions of the testing. Some costs can be saved by conducting "in-house" testing since the AWPA traditionally has accepted such testing as being valid, i.e. unbiased.

The testing time cannot be significantly compressed beyond that shown. There is an induction period before any significant attack occurs on the untreated control specimens and failure of the controls is one of the criteria for test validity. The shorter times shown also assume that the sponsor will pay additional costs for conducting the testing in high hazard zones such as Hawaii.

Ground Contact Issues

For ground contact uses (AWPA UC4A/B/C), the total testing costs increase to $150,000 to $300,000 for the additional tests needed to augment the above ground tests. The time frame for completion of the ground contact testing is 3 to

7 years. In actual fact, the 3 years is overly optimistic but it is the exposure time from a high decay zone that the AWPA considers as the minimum. Typically, most preservative systems brought to the AWPA for consideration have 5 years or more of ground contact exposure data.

It is understood that the "exact" formulation may not be present in the data submitted to the AWPA since minor modifications are expected as the development work continues. However, it is expected that the data will be representative of the final formulation as proposed and that the retentions of active ingredients will span the retentions proposed for listing.

Critical Time Issue

The time lag between the "finalization" of the preservative system and the end of the testing makes it critical that test planning be done as soon as reasonable. For our purposes, reasonable can be defined as that time when the system is well understood but perhaps not in its final form, when it is known that the system is robust and has an excellent chance for success and when it is believed that labeling and registration issues with the EPA will be overcome.

It should be understood that there may still be performance or property problems and that further refinement of the formulation is probably going to occur. Sometimes formulation problems are discovered during the scale-up and repetitive treating done for the testing and it is best to find these earlier rather than later. Some additional costs are going to be incurred as well since it will be necessary to test more retentions or possibly formulation variations than if only the final formulation were tested. However, the slight additional costs are relatively small compared to the lost market opportunity that would be incurred during a wait of several months to a year to fully develop a formulation.

AWPA Submittal

When all of the required testing is complete and the data substantiates the performance of the system, it is appropriate to submit the data to the AWPA for their consideration. In keeping with the two step process, there are two general committees in the AWPA: the Preservatives General Committee (P-General) and the Treatments General Committee (T-General). Within those committees, there are a number of subcommittees that have limited scopes of activity. For preservative systems, the important subcommittees are P3, Organic Systems; P4, Inorganic Systems and P5, Analytical Procedures.

Depending on the system characteristics, a "data package" is submitted to either P3 or P4 and the analytical methodology, if new, is submitted to P5. All of these submittals must be sent 45 days before either the Spring (May) or the Fall (September) meeting of the subcommittees.

The data package follows a common format that initially describes the system, gives the chemical formula (if appropriate) and provides chemical and physical property data. The efficacy test data is given next and typically both graphs and tables are used to summarize the testing. At the heart of the data package are the proposed retention levels advocated by the sponsor as the minimum above ground and ground contact levels. Naturally, the recommended retentions are the minimum that can be substantiated by the data.

AWPA Voting Process

Once the data package is submitted, a series of votes occurs in the AWPA. In simplistic form, the voting process can be summarized in Figure 1.

Applicant prepares data package and recommends retentions

↓

P-Subcommittee reviews data and votes to approve retentions (2/3 majority)

↓

P-General Committee votes to substantiate subcommittee vote (1/2 majority)

↓

Executive Committee votes to substantiate procedure (1/2 majority)

↓

Preservative system listed in AWPA Book of Standards

Figure 1. The AWPA Voting Process

The P-Subcommittee vote is perhaps the most crucial and this vote is taken at an open meeting. At that meeting, there is unlimited debate and many of the people on the subcommittee are or will be competitors. If there are any unusual aspects to the data, such as poor performance in one of the tests, then the debate can be very heated as to the allowable degree of safety in the recommended retention levels. Frequently, strong arguments are made that the recommended retentions are too low or that there is some other problem with the system under consideration and, therefore, the retentions should be raised. For this reason, it is imperative that the proponents or their representatives understand and appreciate the subtleties of the testing protocols if they are to successfully counter these arguments. A recorded vote is taken and all negative voters must submit a technical reason substantiating their vote. It is not necessary for the proponent to address the negatives on this ballot.

Typically, there are 30-40 voters present at the meeting and two-thirds of these must approve the retentions. This is the typical majority required by other consensus-based standard writing organizations such as ASTM-International. In effect though, it means that a one-third minority can and sometimes do control the vote and the proponent must resubmit the proposal. When this happens, the proponent can gather additional data or develop a rational explanation for the errant data, or both if needed, and resubmit the data package with the same recommended retentions or raise the retentions to some higher value.

If the subcommittee vote is favorable, the issue moves on to a written ballot of the P-General Committee. This allows those members who did not attend the particular meeting where the issue was discussed a chance to vote. Negatives on this ballot must be accompanied by written reasons and any negatives are returned to the originating subcommittee for a determination of persuasiveness. The ballot for persuasiveness must pass by a one-half majority and, if successful, the issue moves on to the final vote by the Executive Committee.

The Executive Committee reviews the ballot items to ensure that the proper procedures were followed in resolution of the negatives and that no other procedural errors occurred. The Executive Committee does not consider technical issues and is confined solely to judging procedural issues as given in the AWPA Regulations. Once the Executive Committee has approved the ballot item and a valid EPA registration is given, it is immediately listed in the AWPA Book of Standards. To facilitate this, the AWPA maintains a web site (www.awpa.org) with the latest updates to the Book of Standards.

EPA Registration Issue

Although mentioned above, an important point is that an EPA registration is required before a preservative system can be listed in the Book of Standards. In recent years, there have been some preservative systems that have been in limbo for considerable periods awaiting EPA registration. Obviously, detailed discussion of EPA labeling is beyond the scope of this article but the proponent of new systems should begin the EPA process as soon as reasonable as well.

New AWPA Procedures

Two new procedures will soon be adopted by the AWPA to help expedite the processing of new preservative systems. The first procedure is that all data packages will undergo a "data screen" to ensure that the package is substantially complete. No judgment is made during the screen about the quality of the data or the level of the retention recommendations. Rather, it is intended to only flag any data omissions so that the proponent is aware of these and can attempt to supply the missing data.

The second change is institution of the AWPA Preservatives Review Board (PRB) as an expedited means for the proponent of a preservative system to obtain expert review of data supporting a proposal for standardization. The three person PRB will be drawn from a pool consisting of impartial, general interest members. At the end of the PRB review, the proponent receives a report with the consensus opinion of the PRB and any minority opinions regarding the recommended retention levels. This report is not binding on the P-subcommittee but it is believed that some of the debate can be lessened if a review by three independent experts is available. There will be a nominal cost of $4000 for the PRB review and this money is used to reimburse the experts for their time.

AWPA Treatment Voting Process

Once the preservative system is listed by the Preservatives committee of the AWPA, a similar voting process occurs in the Treatments committee to ratify the retention levels recommended. In this case, a data package is submitted to the Treatments subcommittee showing that the preservative system can be routinely used in a commercial manner. The proponents must show that the recommended retentions are achievable and that penetration indicators are available for the system. As well, several lumber species with different treating characteristics should be used in development of the package.

Assuming the treatment data package is complete, the voting process is essentially the same for the Treatments subcommittee and T-General as before. There can be a debate at the subcommittee meeting and then a ballot is taken. If a two-thirds majority is obtained, the proposal goes on to the written ballot of the T-General Committee. If successful, the proposal is then ratified by the Executive Committee and is listed in the Book of Standards in the commodities sections. As before, the negatives on the written ballot must be resolved.

With the thought that the AWPA process can be shortened somewhat, an allowance is made for submittal of the P and T data packages concurrently. During each meeting week, the P subcommittees meet before the T subcommittees so it is possible for a system to undergo both votes in the same week. However, in recent years, this has been relatively rare and it typically takes a meeting or two for the membership to become familiar with and comfortable with a new system. Thus, most new systems are not successful at the first meeting and it takes two to four meetings for complete listing to occur.

AWPA Considerations

With the arduous process in mind, one wonders why the proponents endure. The answer of course is that once listed in the Book of Standards, a preservative system is deemed to be commercially acceptable and "meet" building code provisions. This allows access to all major markets for preservative treated lumber.

A second and important consideration is that once a system is listed, it is essentially listed for as long as some relatively simple "maintenance" is done. For the first few years after listing, it is required that data updates be provided to the AWPA but these update packages need only include recent data. At five year intervals, the appropriate standards must be reaffirmed but generally this is accomplished fairly readily.

A third and also important consideration that there are no listing or continuing fees paid to the AWPA. The costs associated with obtaining an AWPA listing are essentially the testing and data package preparation costs. Many companies use internal resources to minimize these and, obviously, the costs diminish rapidly once the initial testing phase is passed.

The Building Code Process

The International Code Council (ICC) is the dominant national code issuing body in the USA. For preservative treated lumber, the two codes of interest are the International Residential Code (IRC) and the International Building Code (IBC). Both of these have various structural applications where treated wood is required to be used. Local jurisdictions typically adopt these codes in whole or augment the codes with local requirements.

ICC Organization

In addition to the main organization, the ICC has three subsidiaries: the International Accreditation Service (IAS), the International Code Council Evaluation Services (ICC-ES) and the ICC Foundation. The Foundation sponsors research programs related to building products and is of no further interest here. However, the other two subsidiaries play major roles in the recognition of new preservative systems.

ICC-ES

The ICC-ES provides a mechanism for recognizing new products that are then deemed as compliant to the code provisions. Basically, the proponent develops an Acceptance Criteria (AC) that defines the product in a generic manner, specifies the necessary tests used to document the performance of the product and specifies the limitations as needed for the product. The proponent then gathers the necessary test data and submits it to the ICC-ES who review the data and then issue an Evaluation Service Report (ESR). Upon the issuance of the ESR, the product has been evaluated and found to be in compliance with the code.

An important point is that the proponent is able to select the performance tests and hopefully favors those tests that are already underway. For wood preservatives, the range of tests includes numerous efficacy tests as well as strength and fastener corrosion tests. The efficacy and corrosion tests are usually conducted by AWPA standard methods but recently similar test methods from international consensus test organizations have been accepted.

Acceptance Criteria

A draft of the AC is due about three months earlier than one of the three ICC-ES meetings per year held in February, June and October. A submittal fee of $8100 is required at this time as well. The draft undergoes an internal ICC-ES review and then, 30 days before the meeting is published on the ICC-ES web site (www.icc-es.org).

During the 30-day period and at the AC meeting, public comments are welcome on the proposed AC. Usually these comments take the form of "advice on deficient aspects" of the AC, such as more strenuous testing is required or more restrictions need to be placed on the product. Sometimes, organizations supply their own test data on similar products. In many respects, the debate at the public AC meeting is very similar to the debate at the AWPA subcommittee meeting. The AC can be and frequently is amended on the floor to accommodate the various comments.

It is absolutely critical that a representative of the proponent be at the meeting who understands the implications of changing the test or test criteria in terms of additional costs and time. The meeting debate is highly technical and there are many references to either code paragraphs or other ACs. The combination of test, code and AC references can sometimes be confusing to the uninitiated so it is best to have a representative who has weathered several AC meetings.

After the comments are heard, the ES committee votes on approving the AC. Most of the time, the AC is approved with modifications but there are a few that are approved without change. An additional option is to table the AC for review; at one recent meeting over one-third of the proposals were tabled. If the AC is approved, then the proponent sets about conducting the various testing specified in the AC.

Testing and Accreditation

The testing specified in the AC is generally the same as that required by the AWPA. However, an important distinction is that any testing that was initiated after July 2004 and submitted to the ICC-ES must be done at a laboratory accredited by the IAS. Unfortunately, as of this time, there are no laboratories that are accredited by the IAS to perform efficacy testing.

Recognizing this conundrum, the IAS has special provisions for one-time approvals for specific tests. This requires an on-site inspection of the testing facility prior to the test initiation and the proponent is charged about $3000 to cover fees and travel costs.

Evaluation Service Report

After the test data is collected, the various test reports are submitted to the ICC-ES for their review. If there are any anomalies in the data, the proponent is asked for a plausible explanation and if adequately explained, the review is concluded. In some cases though, the test must be repeated. If there are reported field problems with a similar product, the ICC-ES may request additional assurances that such problems will not occur with the current product.

At the conclusion of the review, the ICC-ES prepares and issues an ESR. When this occurs, the product is deemed to meet the requirements of the IBC and IRC. At the end of the ESR, there are typically several restrictions placed on the use of the product to prevent its misuse.

In general, the review takes several months to complete. The shortest review time span that is known was six months (and I am happy to report that I was associated with this ESR). The longest known is over four years and still running (and I am equally happy to report that I am not associated with this ESR). It should be noted that these time spans are after the testing is complete.

The initial ESR is issued with a one year expiration date and costs about $6000. After the one year period, the ESR is reexamined for any deficiencies and then reissued with two year reexamination periods. If there are deficiencies, the proponent must address them at the reexamination. The reexamination fees range from $5 -10,000 and there are additional fees for listing manufacturing plants and other services. In short, there are costs for maintaining an ESR and these costs continue as long as the ESR is valid.

Above Ground and Ground Contact Issues

The ICC-ES makes clear distinctions in the testing required for above ground versus ground contact uses. It is possible within the ICC-ES framework to obtain an ESR that limits the wood to only above ground applications (AWPA UC3B and less). Since accelerated testing is accepted by ICC-ES for above ground uses, the listing process can be expedited and the proponent can enter the market in a timelier manner. Obviously, the necessary tests would be underway for ground contact listings and the ESR would be amended later when the supporting ground contact data is available.

AWPA and ICC Differences

One aspect that remains undiscussed is that there are significant differences between the AWPA and the ICC procedures. The major distinction is that the AWPA requires that the preservative formulation be publicly divulged and listed in its Book of Standards. Certain proprietary ingredients can remain undisclosed if they are minor components but any and all active ingredients must be specifically listed.

The formulation must also be divulged to the ICC-ES but the information is not published or otherwise divulged to the public. (In fact, the ICC-ES is very careful to maintain all information in its files as confidential.) Thus, a "trade-secret" formulation can be recognized by the ICC but not the AWPA.

A second distinction is that, in the past, the ICC-ES did not require evidence of EPA registration prior to the issuance of an ESR and, as discussed, the AWPA requires EPA registration prior to listing in the Book of Standards. Even though it is illegal to sell the product without EPA registration, the issuance of an ESR was exploited by some companies in order to gain a degree of credibility about their products. However, as of this writing, the ICC is beginning to require EPA registration so this distinction may be moot in the future.

Lastly, the costs of maintaining the AWPA listing or the ESR report are very different. Both require similar significant expenses in the beginning phases for testing and data package preparation, but there are no continuing costs for the AWPA listing. There are maintenance costs associated with an ESR and these can be significant if numerous plants are listed.

Summary

Although somewhat formidable at first glance, the procedures for obtaining either AWPA or ICC recognition for a new preservative system are reasonably structured and orderly. There are a number of similarities between the two processes as well.

The AWPA process involves first preparing a data package of appropriate testing information on the proposed system. The testing regime is well defined and involves standardized tests published by the AWPA. This data package is then submitted to the AWPA and undergoes a series of public votes within their committee framework. As discussed, there are a number of pitfalls to avoid during the AWPA process. Also, some recent improvements to the AWPA process should help streamline the standardization process.

Similarly, the ICC involves testing and documenting the properties of the preservative system. With relatively few exceptions, the same test procedures are used as for the AWPA. However, different processes for evaluation are used and these are discussed in detail.

The critical aspects for success are to carefully plan the testing regime, select appropriate tests, select appropriate testing organizations and to start as soon as reasonable. It should be kept in mind that it may be difficult to obtain such listings though. It can also be somewhat expensive. But it can be done.

Chapter 32

Environmental Regulations and the Wood Preserving Industry

H. M. Rollins

H. M. Rollins Company, Inc., P.O. Box 3471, Gulfport, MS 39505

Introduction

The wood preserving industry began operations in the United States in the late 1800's. The age of the industry predates modern environmental laws, and the negative environmental legacy associated with the long operating history resulted in a specific focus on the industry as environmental laws and regulations were developed. The wood preserving industry is now subject to a variety of environmental laws and regulations concerning air, water, wastes, pesticides, community awareness, and remediation of historic contamination.

This chapter will briefly explain the regulatory programs potentially affecting the industry.

Water Regulations

The Federal Water Pollution Control Act of 1972 and subsequent amendments in 1977 and later, now commonly called the Clean Water Act (CWA), are the underlying statutes for the regulations promulgated by the United States Environmental Protection Agency (EPA) that control discharges of pollutants to waters of the United States. The federal regulations are applicable until a state adopts conforming state regulations, and then the state becomes authorized as the lead regulatory entity. The states cannot adopt regulations less stringent than the federal rules, but they can adopt more stringent requirements. Most states' rules are nearly identical to the federal program.

A key component of the CWA is a permitting system that requires industries to obtain permits for the discharge of process wastewater pollutants either directly to surface waters or indirectly by discharging to publicly owned treatment works (POTW). This permitting scheme is referred to as the National Pollutant Discharge Elimination System (NPDES), and the permits are referred to as NPDES permits.

Regulations for the control of pollutants in process wastewaters from the wood preserving industry are found in the Code of Federal Regulations (CFR) in Subparts F, G, and H of 40 CFR Part 429. These regulations specify minimum treatment standards required for existing as well as new sources. The regulations have not changed since the early 1970's. Most NPDES permits issued today contain much more stringent discharge limits than found in the federal regulations, because they are based on site-specific conditions.

For water borne and nonpressure processes, the regulations in Subpart F require zero discharge of process wastewater pollutants for both existing and new sources. For existing plants in the steaming subcategory, Subpart G allows a process wastewater discharge either to surface waters or to a POTW, but the new source performance standard is zero discharge of process wastewater pollutants. Plants in the Boulton subcategory are covered by Subpart H, which allows existing plants to discharge to a POTW but not directly to surface waters, and specifies zero discharge of process wastewater pollutants for new sources.

Process wastewater management is a major environmental issue for the industry today. The wastewaters from creosote and oil borne plants are very difficult to treat because they are chemically complex, containing possibly thousands of compounds derived from the preservatives and wood extractives that are removed during the wood conditioning or treating process. These wastewaters typically will have three phases, a highly contaminated free water phase, an emulsion phase, and a free oil phase. The volume and concentration of contaminants varies significantly over time, depending upon the volume of rainfall in the process area that must be collected and treated. Process wastewaters and wastewater treatment residuals are listed hazardous wastes. This has led many facilities to roof their process areas to minimize the quantity of rainfall that must be managed as hazardous waste and to employ kiln drying on all material to be treated in order to minimize process wastewater generation.

Wood preserving facilities are also subject to the storm water permitting requirements found at 40 CFR Part 122.26. Permitting for storm water is typically in the form of a General Permit applicable to an industry as a whole; however, the regulatory agencies in some states require individual NPDES permits for wood preserving plants. The Federal General Permit contains no discharge limitations for wood preserving plants, but does require periodic monitoring and reporting of certain pollutant indicator parameters. Some states have established limits for preservative constituents in storm water discharges in individual permits issued to wood preserving plants, and it is expected that this trend will spread in the future. Due to the large land areas required for treated wood storage at facilities treating industrial products such as utility poles or crossties, it would be operationally impractical to collect and treat all storm water in areas of the country subject to frequent high intensity rainfall events, such as the southeast. Therefore, if future limitations on storm water discharges are imposed, the industry will be forced to cover its treated product storage areas. This could challenge the financial viability of the industry.

Another regulatory program designed to protect surface waters is found at 40 CFR Part 112. This requires any facility storing more than 1,320 gallons of oil to develop a Spill Prevention, Control and Countermeasure (SPCC) Plan. This plan must be certified by a Registered Professional Engineer and generally requires, among other things, secondary containment for all storage tanks that is adequate to contain the largest volume of any tank plus any potential rainfall accumulation. All oil borne and creosote wood preserving plants must have an SPCC Plan. Similar secondary containment requirements would normally apply to water borne facilities under general provisions of most states' water regulations, which require protection against the potential release of materials that could be harmful to surface waters. These provisions are typically incorporated into the zero discharge NPDES permits issued to wood preserving plants.

Waste Management Regulations

Management of wastes at wood preserving plants is governed by regulations issued to implement the provisions of the Resource Conservation and Recovery Act (RCRA) of 1976. These regulations are normally referred to as the hazardous waste regulations. These rules were amended by the Hazardous and Solid Waste Amendments (HSWA) of 1984. The initial RCRA regulations were effective on November 19, 1980, and these rules listed one specific waste from the wood preserving industry as a hazardous waste. This waste was designated K001, and was defined as "Bottom sediment sludge from the treatment of wastewaters from wood preserving processes that use creosote and/or pentachlorophenol."

This waste listing came about because the impoundments historically used to manage process wastewaters in the industry had accumulated large volumes of sludges. In addition to the listed waste K001, the initial regulations also included toxicity characteristic wastes under which a waste would be classified as a hazardous waste if leachate concentrations of a specific list of compounds in a specified waste leachate test exceeded threshold concentrations. Wastes from wood preserving plants using arsenic or chromium-based preservative systems would normally be classified as hazardous under this procedure. The waste codes assigned to these wastes were: D004, for arsenic containing wastes; and D007, for chromium containing wastes. Generators of hazardous waste must comply with the generator regulations in 40 CFR Part 262 and the applicable portions of 40 CFR Part 265.

Additional wood preserving industry wastes were listed as hazardous wastes in 1990. A separate listing was provided for each of the three major preservative types, with F032 designated for plants with equipment that presently treats, or previously was used to treat, with pentachlorophenol, F034 for plants that treat with creosote, and F035 for plants that treat with inorganic arsenic formulations, such as CCA or ACZA. The listing description included preservative drippage, spent formulations, process wastewaters, and all process residuals. Therefore, all wastes with the potential to be contaminated by constituents of these preservatives are now listed as hazardous wastes. Details on the listings can be found at 40 CFR Part 261.31, and the hazardous constituents cited as the basis for listing these wastes can be found at 40 CFR Part 261, Appendix VII.

In conjunction with the listing of F032, F034, and F035 wastes, the EPA required wood preserving plants to install impervious drip pads to prevent environmental releases of preservative drippage from freshly treated wood. The design and operating requirements of these drip pads are found at Subpart W of 40 CFR Parts 264 and 265. The design of Subpart W drip pads must be certified by a Registered Professional Engineer. Plants are required to keep freshly treated wood on the drip pad until any drippage of preservative ceases, and they must maintain records to verify this practice. Treated wood storage areas must be inspected for any incidental drippage every day the plant is operating and it must be immediately cleaned up and managed as hazardous waste.

The hazardous wastes generated by the wood preserving industry must be shipped using a hazardous waste manifest and must be transported by companies licensed to transport hazardous waste. Generator requirements for record keeping and tracking of shipments are found in 40 CFR Part 262. Wood treaters can only ship wastes to licensed and permitted hazardous waste treatment, storage, and disposal facilities. Hazardous wastes can only be land disposed after they have been treated to meet the standards in the "Land Ban Treatment Standards" found in 40 CFR Part 268.

There is an exemption in the hazardous waste regulations that is very important to the wood preserving industry. This is known as the "wastewater treatment unit exemption" which is found at 40 CFR Parts 264.1 and 265.1. This provision exempts the wastewater treatment units at wood preserving plants from the provisions of the RCRA regulations. The exemption applies to all treatment units, even those having no discharge, such as those at water borne plants that recycle all process wastewaters. Without this exemption, wood treaters would have to obtain full RCRA permits as hazardous waste treatment facilities.

Most older wood treating plants historically used unlined surface impoundments for the management of process wastewaters. These impoundments became land-based hazardous waste management units when the RCRA regulations were issued in 1980. In almost all cases, environmental investigations required as part of these regulations identified groundwater contamination associated with these impoundments. Most of these impoundments were closed in the late 1980's and facilities were required to obtain RCRA Post-Closure Permits that provided for the care, monitoring, and maintenance of the closed impoundments for a minimum period of 30 years. Corrective action for the groundwater contamination was also required as part of the post-closure permitting requirements. Facilities were required to provide financial assurance to ensure that resources were available to complete post-closure care and groundwater remediation.

The Hazardous and Solid Waste Amendments of 1984 expanded coverage of the hazardous waste regulations to all areas of a facility that may be contaminated with "hazardous constituents", regardless of their source. The rules implementing these provisions have required wood treaters to address soil and groundwater contamination existing in areas of the site other than the surface impoundments. The financial burden imposed by compliance with the RCRA regulations forced many facilities to cease operation. A number of these sites became subject to clean up under the "Superfund" program discussed later.

Due to the recent prohibitions placed on the use of arsenic-based preservative systems in residential applications, many plants have changed to copper-based preservative systems. Wastes from the use of these preservatives would not meet the existing hazardous waste listing definition if employed at a new plant. However, plants that change from the use of creosote, pentachlorophenol, or inorganic arsenical preservatives to a system that does not generate hazardous waste must still manage all process wastes as hazardous waste unless the process equipment and drip pad are decontaminated in accordance with a state and/or EPA approved decontamination plan. Facilities changing to these preservative systems may be able to defer closure of the RCRA Subpart W drip pad until the facility ceases to treat wood, at which time a full RCRA closure of the drip pad and any underlying contaminated soils must be completed.

As a result of the RCRA regulations, wood treating facilities operating today have a minimal impact on the environment. Actions to address prior contamination are underway at most plants, although it may not be possible to ever achieve a complete clean up.

Community Awareness Regulations

Title III of the Superfund Amendments and Reauthorization Act (SARA) of 1986, known as the Emergency Planning and Community Right-to-Know Act, requires industry to file several annual reports that contain information concerning the volume of toxic or hazardous chemicals present on the industrial site and to report all environmental releases and off-site transfers of these chemicals in wastes or permitted discharges.

Section 312 of SARA requires all facilities storing over a threshold amount of a listed chemical to file what is commonly called a "Tier II" report by March 1 of each year. The list of chemicals requiring reporting is separated into "hazardous" and "extremely hazardous" categories. The reporting threshold for extremely hazardous substances is typically 500 pounds and that of hazardous substances is 10,000 pounds. Most wood preservatives are classified as hazardous substances for Tier II reporting purposes. The information contained in Tier II reports is entered into a computer database that is made available to the public.

Another annual report is required by Section 313 of SARA. This report is due July 1 of each year and is known as the Toxic Chemical Release Inventory Report (TRI), commonly called the Form R report. The applicability of this report is based upon the annual quantity of the chemical manufactured, processed, or otherwise used, rather than simply stored. Wood preservatives used to treat wood are considered to be "processed" and the reporting threshold would be the annual usage of 25,000 pounds. However, the reporting threshold for chemicals considered to be persistent, bioaccumlative toxic (PBT) chemicals is much lower. The polycyclic aromatic compounds (PACs) found in creosote and the dioxin and dioxin-like compounds (DDLCs) present as microcontaminants in pentachlorophenol are considered to be PBT compounds. For PACs, the threshold is 100 pounds and for DDLCs, the threshold is 0.1 grams. All major wood preservative systems in use today would be subject to the Form R reporting requirement. Facilities with less than the equivalent of 10 full-time employees, or 20,000 annual man-hours, are exempt from the requirement to file a Form R report. In addition to the wood preservatives themselves, there are other sources of Form R reportable chemicals at wood preserving operations. Among these may be wood-waste or other combustion sources, which make TRI reportable compounds as combustion products, and components of oils used as carriers for preservatives such as pentachlorophenol or copper naphthenate, or as diluents in creosote-petroleum systems.

A facility subject to the Form R reporting requirement must account for and report all environmental releases and all transfers of the listed chemicals. This includes releases to air, land, surface waters, on-site disposal, off-site disposal, and transfers to off-site facilities for further treatment, such as incineration, fuel blending, or discharges to POTWs. The preparation of a Form R report is technically complex and requires that all engineering estimates be well documented.

Information contained in Form R reports is compiled in a publicly available national database that is searchable by facility name, location, chemical compound, industry type, or a variety of other terms. This database is used by EPA to track trends in environmental performance and to prioritize industry sectors for additional regulatory effort.

Pesticide Regulations

All wood preservatives must be approved by EPA under the provisions of the Federal Insecticide, Fungicide and Rodenticide Act (FIFRA) of 1975, as subsequently amended. The requirements for pesticide registration are covered in other chapters.

From a wood preserving plant operational perspective, the primary FIFRA requirements are: maintain a licensed pesticide applicator on staff if the preservative is a restricted use pesticide; follow all label directions concerning personal protective equipment and application practices; and place a pesticide label on all tanks storing preservative.

Historic Release Regulations

Many wood preserving plants have significant soil and groundwater contamination associated with historic industry standard operating practices. If this contamination was associated with a distinct operating unit or practice that was closed prior to the effective date of the RCRA regulations in 1980, then this contamination is normally addressed under the provisions of the Comprehensive Environmental Response, Compensation, and Liability Act (CERCLA) of 1980, as amended. These regulations are normally referred to as the "Superfund" regulations. Many closed wood preserving plants are being addressed as Superfund sites.

Normally, areas of wood preserving plants that experienced environmental releases of hazardous constituents after the effective date of the RCRA regulations, but are not hazardous waste management units, are addressed as Solid Waste Management Units (SWMUs) under the RCRA program, not the CERCLA program.

Air Regulations

The Clean Air Act of 1955 and numerous amendments, including the Clean Air Act Amendments of 1990, control the emissions of pollutants from industrial facilities. The primary focus of the air regulations today are industrial facilities classified as "major sources". A facility is a major source of air pollutants if it emits: more than 100 tons annually of any of the criteria pollutants such as particulate matter (PM), nitrogen oxides (NO_X), carbon monoxide (CO), sulfur dioxide (SO_2), or volatile organic compounds (VOC); or, more than 10 tons per year of a single hazardous air pollutant (HAP) or a total of 25 tons of all HAP combined.

The air regulations are highly complex and the requirements vary depending upon whether a source is existing, being modified, or newly constructed, and whether the source is located in an area presently meeting all national air quality standards. In most cases, the regulatory requirements are based on a potential to emit derived as the maximum possible emissions from the process and an assumption that the process operates 24 hours per day, 365 days per year.

A number of wood preservatives or preservative constituents are on the HAP list. Wood waste and other combustion units and dry kilns at wood preserving plants emit both criteria and HAP pollutants. Wood preserving plants must be able to demonstrate through emissions data and engineering estimates whether the facility is a major source requiring a 40 CFR Part 70 permit, commonly known as a Title V permit.

There is significant variability in air regulations from one state to another. Some states require state permits even for facilities that do not meet the major source definition, and the threshold for requiring a permit varies between states.

Other Regulations

There are a number of other environmental regulations at the federal, state, and local level that could potentially impact the wood preserving industry. Most wood preserving facilities are required to develop a written waste minimization plan under the provisions of the Pollution Prevention Act of 1990. An annual report of waste minimization activities is also required as part of this program. There are environmental provisions of the Occupational Safety and Health Act (OSHA) concerning hazardous waste management and training, worker exposures, and the development of Material Safety Data Sheets for treated wood products. Other examples may include local environmental permitting requirements, local emergency response planning requirements, and local public information dissemination requirements.

The wood preserving industry is one of the largest users of pesticidal products, and it can be expected that, as a result, the industry will continue to be highly regulated by all levels of government.

Chapter 33

Disposal Management of Preservative-Treated Wood Products

Timothy Townsend[1], Brajesh Dubey[1], and Helena Solo-Gabriele[2]

[1]Department of Environmental Engineering Sciences, University of Florida, P.O. Box 116450, Gainesville, FL 32611
[2]Department of Civil, Architectural, and Environmental Engineering, University of Miami, Coral Gables, FL 33124

When treated wood structures reach the end of their useful lifespan they must be disposed of. Proper management of the preservative-containing wood is necessary to minimize any potential risk to human health and the environment. Treated wood products enter the waste stream in small amounts as remnants during construction activities, and at a much larger magnitude as a result of demolition. The first step in discarded treated wood management is determination of appropriate regulatory status, i.e. is should the waste be managed as a hazardous waste. Testing finds that creosote and pentachlorophenol treated wood do not typically leach preservative concentrations at sufficient quantities to be characterized as a hazardous waste in the US. Chromated copper arsenate (CCA) treated wood does often leach arsenic (As) and chromium (Cr) above hazardous waste thresholds, but in most of the US, CCA-treated wood is excluded from being defined as a hazardous waste because of a specific exemption under resource conservation and recovery act (RCRA). Other national or state specific regulations may impact disposal-end management requirements. The primary management options for discarded treated wood include landfilling, combustion and recycling. Landfilling is typically the least expensive option, but this practice is accompanied by concerns such as placement and compaction difficulties and the potential for contamination of leachate and groundwater.

Recent research suggests that As leaching may pose a risk to groundwater at unlined C&D debris landfills and leachate management problems at a lined facility. Combustion is commonly practiced to recover energy from creosote and pentachlorophenol treated wood. Arsenic from CCA-treated wood may volatilizes at typical combustion temperatures so combustion facilities should be equipped with appropriate air pollution control devices. Recycling practices and extraction technologies such as bioremediation, chemical extraction, electrodialytic extraction and liquefaction have been developed, but these systems are more costly than traditional disposal options and separation of treated wood from the mixed wood stream can be challenging. As alternative treated wood products are developed and used, manufacturers should consider disposal-end management.

Introduction

Treated wood products are designed to resist decay in harsh environments for many years, but ultimately these products are removed from service and discarded. The preservative chemicals added to treated wood function through their toxicity to organisms that would otherwise attack and deteriorate the wood. The presence of the potentially toxic chemicals raises a natural concern over environmental and human health risks when these materials are disposed. Historically, research on preservative treatment of wood focused on development of alternative preservatives and the ability of treated wood products to perform over their lifespan. In recent years research focusing on environmental and disposal issues has become more prominent. In this chapter, an overview of treated wood disposal issues is presented. Much of the discussion focuses on CCA-treated wood as this preservative system has been investigated most heavily with regard to disposal. Where appropriate, however, the discussion is generalized to include all treated wood products.

Treated Wood Within the Wood Waste Stream

Treated wood products enter the wood waste stream as remnants of new construction activities (a relatively small contribution) and from demolition and renovation of existing structures. Reasons for removal from service include: (1) the product has reached the end of its useful life and no longer possesses desired structural requirements, (2) a structure containing treated wood is demolished, destroyed or otherwise damaged, and (3) the structure is removed because of aesthetic reasons. Cooper and Ung (*1*) estimated that the service life of an untreated southern yellow pine fence post in the environment is between one and three years, whereas a post treated with preservatives can last from twenty to thirty years. In studies focusing on residential decks, McQueen and Stevens (*2*) and Alderman (*3*) estimated a lifespan of 9 and 12.8 years, respectively. The reason for early removal, as determined from surveys, was aesthetics due to color changes and cracking of the wood product as a result of weathering.

The amount of a particular treated wood product discarded in any given time period depends on the amount historically entering the marketplace and the product's lifespan. Records of how much treated wood enters the waste stream are not tracked, but estimates can be made using historical production data and estimate of product lifespan. This process has been applied to estimating the amount of CCA-treated wood discarded in Florida, US. Solo-Gabriele and Townsend (*4*) estimated that approximately 140,000 m^3 of CCA-treated wood products were disposed in Florida during the year 1996. This quantity is forecasted to increase to 900,000 m^3 by 2015 This estimate was further refined in 2005 (*5*) to account for the different lifespans of different products and to factor in the withdrawal of CCA from most residential applications in January 2004 (Figure 1). Currently the newer treated wood products, those that replaced CCA for residential use, are entering the waste stream primarily as construction remnants (*6*).

Treated wood products enter the waste stream both as homogenous loads (i.e., loads containg primarily treated wood) and mixed with other waste components. Some treated wood products such as utility poles and railroad ties are likely to enter the waste stream as a homogenous source, as will demolished structures constructed in near entirety of treated wood (e.g., a deck). Other treated wood products become mixed with other discarded materials when they are removed along with other wastes (e.g., a construction site, demolition of multi material structure). In many cases, treated wood products become mixed with construction and demolition (C&D) debris, but they can also be encountered along with domestic refuse and yard trash. Tolaymat et al. (*7*) sampled C&D debris wood in 1997 and found on average that 5-6% of the C&D debris wood in Florida contained CCA-treated wood. Later studies (*6, 8*) found as much as 22 % CCA-treated wood in Florida C&D debris wood. The proportion of treated wood in a given waste stream will depend on the historical and current use of that material in the vicinity.

The manner in which treated wood is stored and transported for recycling or disposal differs depending on the source. Segregated sources such as utility poles, railroad ties, and demolished treated wood structures are typically managed directly by the contractor performing the work (or a hired hauling company). Discarded treated wood from these sources is transported directly to

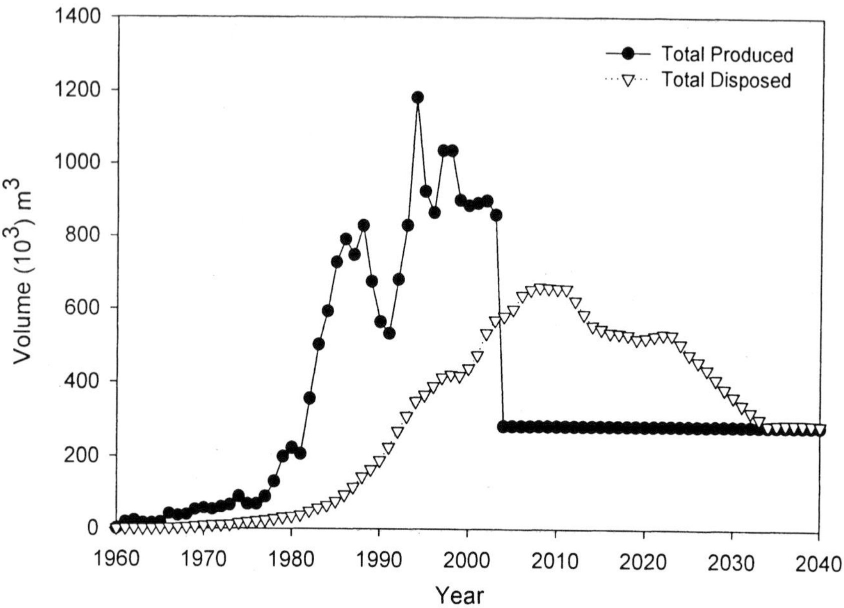

Figure 1. Projected Production and Disposal Quantities for CCA-Treated Wood in United States (Reproduced from reference 5. Copyright 2006 American Chemical Association.)

the disposal facility (in most cases some type of landfill or combustion facility). When treated wood products are produced during C&D activities along with other wastes, the mixed stream is normally stored in large open-top containers (which is periodically hauled away) or placed directly in a transport vehicle. Waste from these operations are often sent to a C&D debris landfill or C&D debris recycling facility; in some cases they might be hauled to a municipal solid waste (MSW) landfill. Treated wood wastes produced at the home are managed as C&D debris, mixed with household refuse and sometimes yard trash.

Solid Waste Regulations Pertaining to Treated Wood

The manner in which discarded treated wood is managed can be greatly influenced by regulatory requirements. Regulations vary from country to country, but in most nations a cornerstone of solid waste regulations is whether the waste is considered "hazardous", and if so, how this hazardous waste must be managed. A hazardous designation would usually require more elaborate containment and long-term tracking of the waste.

In the US, a solid waste is any discarded or abandoned material that is not otherwise exempted from the regulations defined under the Resource Conservation and Recovery Act (RCRA). Any solid waste, unless otherwise exempted, may potentially be a hazardous waste by either being included on a list of defined hazardous wastes (listed hazardous wastes) or by meeting a certain physical or chemical characteristic (characteristic hazardous wastes). Discarded treated wood products are not listed hazardous wastes. The characteristic that would classify discarded treated wood as hazardous is the toxicity characteristic (TC). The TC is determined by performing a leaching method known as the toxicity characteristic leaching procedure (TCLP, *9*). The TCLP utilizes a buffered organic acid solution as an extraction fluid, and was designed to simulate contaminant leaching under acid-forming conditions in a MSW landfill environment (*10, 11*). For details of how this leaching test is performed refer to Townsend et al. (*12*). If the concentration in the TCLP leachate (mg/L) exceeds the defined threshold concentration in the regulation, the waste is characterized as a TC hazardous waste. The TC list only contains a limited number of elements or compounds.

Creosote and pentachlorophenol (PCP) treated wood have been reported to not leach compounds above the TC threshold concentrations (*13*), and are thus not expected to be hazardous waste. As and Cr are on the TC list (each having a TC threshold concentration of 5 mg/L) while copper (Cu) is not. New, as well as weathered CCA-treated wood samples, have been found to exceed the TC limits for As on many occasions, and less frequently for Cr (*12, 14*). CCA-treated wood is, however, excluded from the definition of hazardous waste under RCRA in the US. No other treated wood products are excluded. It is also important to note the exclusion is not necessarily adopted by every state, and that CCA-treated wood remains a solid waste under RCRA and must meet all other applicable requirements.

The regulatory requirements for solid wastes that do not meet the definition of a hazardous waste are less well defined. Non-hazardous solid wastes are typically grouped into a number of different categories, including MSW, C&D debris and industrial wastes. While location and design requirements have been promulgated for MSW landfills at the federal level in the US, industrial waste and C&D debris disposal are regulated at the state level; the regulatory requirements vary dramatically among states (*15*). In Florida, for example, C&D debris can be disposed in unlined landfills, while in New Jersey C&D debris can only be disposed in a lined facility. Non-hazardous solid wastes are often disposed in manners other then landfilling. The beneficial use of waste materials through activities such as land application is becoming more common, and again, regulations governing such activities are governed at the state level. The regulatory requirements that must be addressed when disposing of treated wood products depend on how the materials are managed (e.g., are they disposed in a lined or unlined landfill or they are being reused). A batch leaching test typically used for simulating the C&D landfill environment and the beneficial reuse scenarios is the synthetic precipitation leaching procedure (SPLP, US EPA, 2003). This test simulates the natural precipitation conditions. Details of this test can be found in Townsend et al. (*16*).

Disposal in Landfills

Under most circumstances, disposal in landfills is the least expensive option for managing discarded treated wood. As introduced earlier, several different landfill types may be available to the waste generator (depending on regulatory restrictions). Hazardous waste landfills are most protective, but would likely be used only if regulations classified the waste as hazardous. MSW landfills in most developed countries are lined with a barrier layer (a liner system) to contain and remove leachate so it does not contaminate groundwater. The intercepted leachate is subsequently treated prior to discharge to the environment. C&D debris landfills in many cases are unlined, making them the least expensive uncontrolled landfilling option. Leachate at unlined landfills enters the underlying soil and groundwater; impacts under this scenario may be decreased through natural attenuation and dilution of pollutants.

Potential problems posed by disposing treated wood products in landfills include problems with handling and compaction (which are not a function of the preservative type) and potential impacts on leachate quality. Many lined landfills haul their leachate to off-site treatment facilities, and in many cases these treatment facilities do not accept leachate with concentrations higher than a certain limit (pretreatment standards). Elevated concentrations of pollutants may also result in price surcharges by the treatment facilities. Some wastewater treatment facilities may also regulate metal concentrations because of concerns over impact on biosolids (i.e. sludge) quality.

The impact of treated wood disposal on landfill leachate can be evaluated in several manners. Both batch and column leaching tests have been performed. Since the TCLP was designed to simulate conditions occurring in a MSW landfill when acids are produced from decomposing organic waste, the relatively high level of preservative leaching from CCA (*12, 14*) and Cu-based preservatives (*17*) suggests that pollutant concentration might be elevated in leachate from landfills where this wood is disposed. Most landfills do not, however, contain acids at concentrations equivalent to the TCLP, and if such concentrations were to occur they would only be present for a relatively short time. Based on this observation, one might expect that the TCLP would overestimate pollutant release compared to actual landfill environments. This has been observed in the case of lead (*18*), but others have found that elements such as As and Cr (metalloids that exist as oxy-anions) often leach from waste more in landfill leachates than predicted by TCLP.

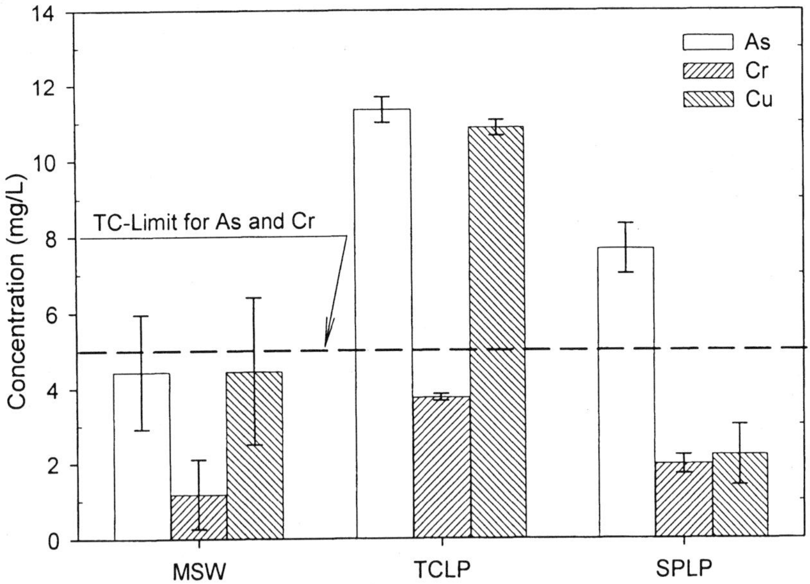

Figure 2. Average Preservative Concentrations from CCA-Treated Wood in Different Batch Leaching Tests (from Dubey 2005)

Townsend et al., (*12, 14*) showed that CCA-treated wood leaches As in most cases at concentrations that would require management as a hazardous waste if it were not otherwise excluded under RCRA. The observation that the preservative elements leach similar As concentrations using the simulated rainwater tests (following the SPLP) indicates that leaching may be a concern regardless of the landfill environment. Dubey (*19*) found statistically similar concentrations of As and Cr in TCLP leachates and in the leachates produced when MSW leachates were used as the leaching fluid (Figure 2). The highest As and Cr leaching was observed with TCLP followed by SPLP and MSW leachate (average leachate concentrations from 26 leachates). Cu leached the most under TCLP followed by MSW leachate and SPLP.

Batch leaching experiments conducted on ammonical copper quaternary (ACQ) and copper boron azole (CBA) indicate that for most of these preservative systems, Cu leaches at greater concentrations than from CCA-treated wood (*17*). The Cu levels in these wood products are of greater magnitude relative to CCA-treated wood and they also leach a greater percentage of the initial Cu. Dubey (*19*) leached ACQ and CBA sawdust using the TCLP and SPLP methods, along with a modification of the method and using leachates from MSW landfills. Results of this study showed that Cu leaching from ACQ and CBA treated wood was similar in magnitude when leached with the standard TCLP and SPLP leaching fluids and with landfill leachate (for details refer to *19*).

The batch leaching results described suggest that the preservative leaching from metal based preservatives such as CCA, ACQ and CBA may impact landfill leachate quality. Although batch leaching tests can provide an indication of leaching potential, subsequent interaction of the leached metals with the surrounding landfill environment may also affect metal concentrations in landfill leachates. Landfills are biologically active and naturally develop into very reducing environments, which can affect the leaching behavior of the waste and subsequent solubility of contaminants in the leachate. Several investigators have constructed simulated landfill columns to evaluate leaching of treated wood in various landfill environments.

Pohland et al. (*20*) co-disposed PCP treated wood in simulated MSW landfills and found that leached PCP was transformed to its daughter products without inhibitory effects on landfill stabilization. Weber et al. (*21*) constructed several lined test-pits containing C&D debris and determined that CCA-treated wood contributed to elevated As concentrations. In a more controlled study, Jambeck (*22*) co-disposed 2% CCA-treated wood in simulated MSW landfills along with other MSW components. The highest As and Cr concentrations in acidic phase were 3.75 and 3.96 mg/L, respectively. In methanogenic phase the maximum concentrations recorded for As and Cr were 0.79 and 0.39 mg/L respectively. The variation of As and Cr concentrations over time are presented in Figure 3.

Jang (*23*), Jambeck (*22*), and Dubey (*19*) constructed simulated C&D debris landfills which contained varying amounts of treated wood. As concentration in the leachate was found to be a major concern in terms of elevated metal concentrations in the leachate. CCA-treated wood represented 5% of the total weight of the waste included in the CCA lysimeter in the study by Dubey (*19*), while Jang (*23*) and Jambeck (*22*) used 0.5% and 10% CCA-treated wood, respectively. In all three studies, elevated As concentrations were observed in the CCA-containing lysimeters as compared to control lysimeters. A comparison of the results from these studies (Figure 4) indicates that higher proportions of CCA-treated wood resulted in higher As concentrations in the corresponding leachate.

Dubey (*19*) examined the leaching of ACQ in the simulated C&D debris landfills and found that Cu concentrations throughout the duration of the experiment were below the detection limit (4 μg/L). Similar observations were made in previous experiments (*22, 21, 23*) when CCA-treated wood was disposed with C&D debris. The low levels of Cu in the C&D debris lysimeters was attributed to the presence of gypsum drywall ($CaSO_4$, $2H_2O$) which is dissociated in aqueous solution to produce sulfate. This sulfate, through the action of sulfate-reducing bacteria, produces hydrogen sulfide which in turn

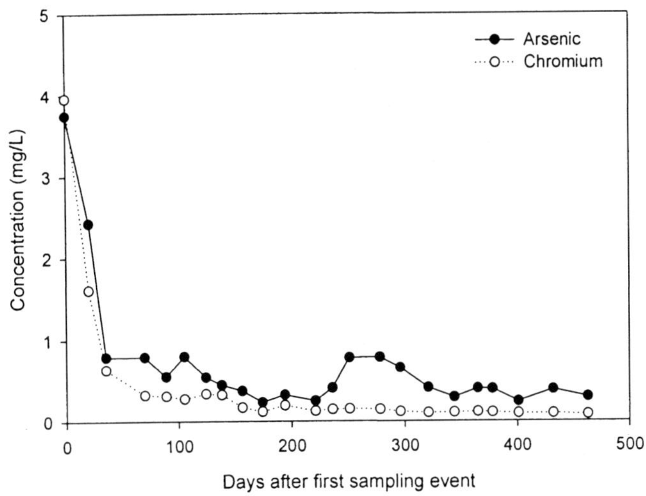

Figure 3. Arsenic and Chromium Leaching from CCA-Treated Wood in Simulated MSW Environment (from Jambeck, 2004)

reacts with Cu to form a stable insoluble complex of CuS. Thus, although batch leaching studies suggest Cu in treated wood might lead to elevated leachate concentrations, simulated landfills indicate that Cu will be attenuated in the landfill.

The As found in CCA-treated wood does appear to pose the greatest concern to lined landfills operations. It is doubtful that a sufficient amount of CCA-treated wood would be disposed to cause leachate to reach hazardous waste levels, but concentrations could certainly become sufficiently elevated to pose leachate disposal problems. Loading of CCA-treated wood in unlined landfills appears to pose a groundwater contamination threat as a result of As. Current data suggest that a few sites in Florida are beginning to see As issues but most do not have these occurrences. The question is whether the As will become attenuated in the soil underneath the landfill or remain mobile and be the source of off-site contamination.

Disposal through Combustion

Combustion constitutes a common form of managing waste wood. This occurs both as combustion of bulk loads of wood (e.g., utility poles, railroad ties) and loads of mixed C&D wood. From an environmental standpoint, the combustion of treated wood poses potential problems with respect to air

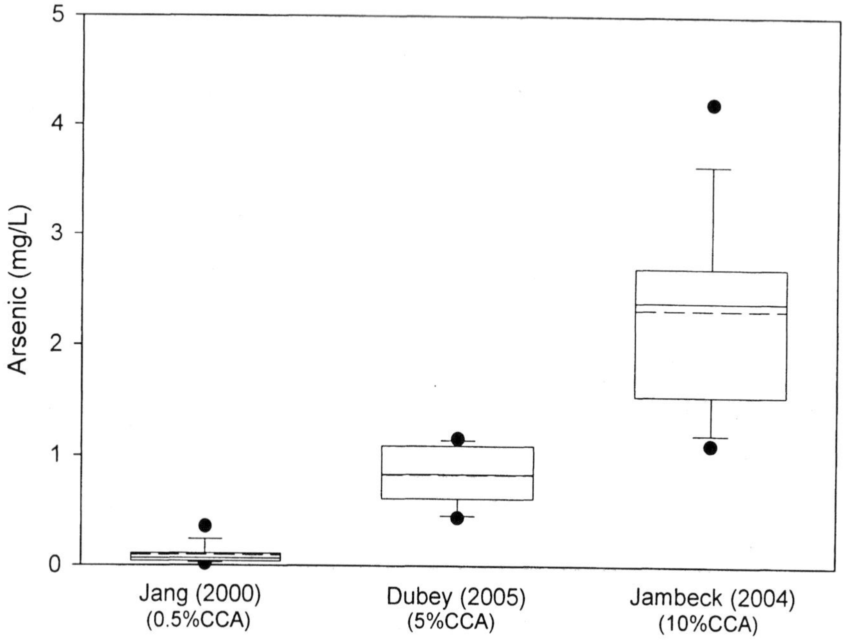

Figure 4. Arsenic concentration in the CCA lysimeter leachates from three C&D lysimeter studies conducted at University of Florida (The black dots on the figure show the 5th and 95th percentile. The box shows the range of data between 25th and 75th percentile, solid straight line correspond the median value and the dotten line shows the mean value of the data set.) The percentage of CCA, by weight, used in the lysimeters in indicated in parenthesis for each corresponding study. (from Dubey, 2005)

emissions and ash generation. Facilities that combust wood will, in nearly all cases, have some form of air pollution control equipment (APC). The type of APC needed depends on the type of emission hazards presented by the preservatives in the wood and the regulatory requirements. Proper combustion techniques can be used to destroy organic preservative compounds used in wood. Several facilities in the US routinely combust pentachlorophenol and creosote treated wood; issues with dioxin and furan emissions are controlled by maintaining proper temperature and gas mixtures.

The control of inorganic elements such as Cr, Cu and As, however, requires additional pollutant removal systems such as bag houses, chemical scrubbing and carbon injection. Some metals will be more prone to air emission problems. As, for example, volatizes and is thus an issue when CCA-treated wood is combusted (*24, 25, 26, 27*). Thermodynamic chemical modeling techniques (*28*) or basic data on the volatility of chemical species from a chemical reference book can be used to assess the potential of inorganic compounds to present an air emission problem. Recently, several sorbents have been studied to control the emission of volatile metals during combustion and leaching from the ash of CCA-treated wood (*28, 29*). Calcium-based sorbents showed great retention capacity of As and Cr during the combustion process and also showed substantial reduction in the leaching of As and Cu. Sodium-based sorbents successfully retained metals in the ash below 750°C, but the products were highly leachable.

The presence of preservative treated wood can also have a major impact on the management requirements for the resulting ash. Organic preservatives should not prove to be a large problem as they will likely be destroyed in well-operated combustion systems, but inorganic preservative compounds will magnify in concentration. Research on the impact of CCA-treated wood on ash properties (*30*) found that at levels of approximately 5% CCA-treated wood in a mixture of wood, the ash would fail the TCLP. The ash, unlike the raw wood, is not exempted from the US regulatory definition of hazardous waste. Combustion of Cu-based treated wood products should be less of a concern as Cu in the ash is relatively immobile (*30*) and is not a TC element. However, even if the ash is not a hazardous waste,elevated metals concentrations might limit reuse options of the ash or dissuade subtitle D MSW landfill operators from accepting the ash.

Co-combustion of treated wood with other fuel such as untreated wood, coal and in cement kilns has also been attempted; the metal concentrations in the ash produced in these combustion will be more dilute. Use of treated wood within cement kilns may have an added advantage in that the wood does not have to be chipped prior to introduction within the kiln and that the inorganics within the wood, including metals, may be stabilized within the cement matrix. However, the Cr in CCA-treated wood may be converted to its hexavalent form in the alkaline cement environment, increasing the possibility of leaching (*31*). Emission from these facilities is also an issue (*32*). Metal concentrations in the cement kiln dust may also limit diposal options.

Recycling Issues of Treated Wood Products

As with all solid waste streams, a potential option for managing treated wood is recycling. When discussing recycling and treated wood, two distinct issues merit further discussion: impacts of treated wood on recycling of mixed wood, and recycling of treated wood that has been separated from rest of the wood waste stream.

The presence of CCA-treated wood in recycled wood from C&D debris recycling facilities is an issue in Florida. These facilities accept debris from a variety of sources, separate the debris into material components including wood, and grind the wood to produce boiler fuel or landscape mulch. The presence of CCA-treated wood was found to greatly impact the recyclibility of recovered C&D debris wood. Leach testing of ground wood from multiple facilities in Florida found that most samples leached As at levels that posed a concern for land application (*33, 34*). In a recent study (*34*), research was performed to determine the extent of contamination of CCA-treated wood in consumer landscaping mulch and to determine whether visual methods or rapid X-ray fluorescence (XRF) technology can be used to identify suspect mulch. Samples were collected throughout Florida and evaluated. Visual identification methods, such as colorant addition or presence of plywood, were found effective to preliminarily screen suspect mulch. XRF analysis was found to be effective for identifying mulch containing higher than 75 mg/kg As. A majority of colored mulch was found to contain between 1 to 5% of CCA-treated wood. The maximum As concentration measured in the mulch samples evaluated was 230 mg/kg, which was 100 times above the Florida residential direct exposure regulatory guideline of 2.1 mg/kg.

Research on evaluating technologies for separating treated wood (particularly CCA) from other wood products has been conducted to solve the problems faced by the mixed C&D debris recycling facilities. Blassino et al. (*35*) described the use of chemical stains for treated wood identification. The PAN indicator, which reacts with the Cu in treated wood to produce a characteristic color on the wood within approximately 10 seconds, was found to be adequate to sort small quantities of wood. If many tons of wood are to be sorted, stains were not considered suitable because of excessive labor costs. A disadvantage associated with the PAN indicator is its lack of specificity to As. As other Cu-based wood preservatives begin to dominate the future wood waste stream in the US., it will be important to develop rapid detection methods specific for As. Efforts are underway to develop an As specific test kit (*36*). Faster and more automated technologies for identifying the metals in treated wood have been investigated recently including near-infrared (NIR) spectroscopy, laser induced breakdown spectroscopy (LIBS), XRF (*8, 36, 37*) and other methods (*38*). Recently, hand held XRF units have been used for research purposes to document their utility to further augment sorting practices (*6*).

Treated wood separated from the mixed waste stream or collected directly as a homogeneous source could possibily be recycled using several different technologies. These technologies focus on either extracting the preservative from the wood or combining the wood into another product. Composite materials can be made from treated wood fibers or in some cases from the wood fibres from which the metals have been extracted (*39*). Wood composites include wood-based composite materials, wood cement composite materials, wood-gypsum composites and wood-plastic composites (*40*). Extraction technologies for treated wood products include bioremediation, chemical extraction, electrodialytic extraction and liquefaction. Bioremediation, chemical extraction and electrodialytic processes generally result in the removal of the metal from the wood fibre without excessive degradation of the wood fibre structure. Liquefaction converts the wood to a viscous liquid containing metals.

Bioremediation technologies have been developed for CCA-treated wood using two different types of microbes, bacteria and fungi. In many cases, chemical extraction of the metals is required before addition of microbes due to toxicity of the CCA at high concentrations. Clausen (*41*) developed a two-step process which includes chemical extraction with oxalic acid followed by bioleaching using metal-tolerant bacteria. This process was capable of removing 70% of the Cu, 81% of the As and 100% of the Cr from CCA-treated wood. Several other studies (*42, 43, 44*) have developed complimentary technologies. Other studies (*45, 46, 47*) have focused on chemical extraction of CCA-treated wood. A considerable amount of research has been conducted looking at electrodialytic extraction of Cu, Cr and As from treated wood products. Lab and pilot studies (*48, 49*) have shown that this technology is capable of removing almost 90% of As from either wood or ash. Research has also been conducted on liquefying treated wood using organic solvents and acids at temperatures of 120 to 250°C (*50, 51*). The process converts treated wood into a viscous liquid which can be diluted to ease further processing. The metals are separated after precipitation and coagulation. The resulting liquefied wood, after removal of the metals, can be used as a feedstock for polymeric materials.

Summary and Conclusion

Although efforts have been made to develop technologies for recycling treated wood products or extracting the preservative from the wood and combining the wood into another product, due to economic reasons the majority of treated wood products are usually landfilled or combusted. Proper disposal management is needed to avoid potential leachate and groundwater contamination issues from landfills and air and ash issues from combustion. Leaching studies indicate that As and Cr leaching may be a major concern when CCA-treated wood is disposed in landfill environments, but Cu leaching should be less of a concern. Several sorting technologies for C&D recycling facilities are being developed to prevent contamination of other wood waste used for mulch or boiler fuels. For CCA-treated wood disposal through combustion, sorbent technologies are being investigated to reduce air emissions of As and to reduce leachability of preservative components from the ash.

Acknowledgement

The authors acknowledge the Hinkley Center (formerly known as Florida Center) for Solid and Hazardous Waste Management and the Florida Center for Environmental Solutions (National Science Foundation Grant No. 0126172) for their support of research on treated wood.

References

1. Cooper, P.; Ung, Y. A simple quantitative measure of CCA fixation, *Forest Prod. J.*, **1993**, *43*(5), 19-20.
2. McQueen J.; Stevens, J. Disposal of CCA treated wood. *Forest Prod. J.*, **1998**, *48*(11/12): 86-90.
3. Alderman, D. An Investigation into Attitudes Towards Recycling CCA Treated Lumber, Ph.D. dissertation, Department of Wood Science and Forest Products, Virginia Tech, Blacksburg, Virginia, 2001.
4. Solo-Gabriele, H.M.; Townsend, T. Disposal Practices and Management Alternatives for CCA-Treated Wood Waste. *Waste Manag. Res.*, **1999**, *17*, 378-389.
5. Khan, B.; Jambeck, J.; Solo-Gabriele, H.; Townsend, T.; Cai, Y. Release of arsenic to the environment from CCA-treated wood: Part II – leaching and speciation during disposal. *Environ. Sci. Technol.*, **2006**, *40*(3), 994-999.
6. Jacobi, G. The extent of CCA-treated wood in consumer mulches and evaluation of technologies to remove treated wood from recycling facilities, Masters thesis, University of Miami, Coral Gables, FL, 2005.
7. Tolaymat, T.; Townsend, T.; Solo-Gabriele, H. Chromated Copper Arsenate-Treated Wood in Recovered Construction and Demolition Debris. *Environ. Eng. Sci.*, **2000**, *17* (1), 19-28.
8. Solo-Gabriele, H.,; Townsend, T.; Hahn, D.; Moskal, T.; Hosein, N.; Jambeck, J.; Jacobi, G. Evaluation of XRF and LIBS technologies for on-line sorting of CCA-treated wood waste, *Waste Manag.*, **2004**, *24*, 413-424.
9. US EPA. Test Methods for Evaluating Solid Waste, SW846, Office of Solid Waste and Emergency Response, Washington D.C, 2003.
10. Francis, C.; Maskarinec, M.; Goyert, J. Mobility of Toxic Compounds from Hazardous Wastes, OENL-6044, U.S. DOE., Oak Ridge, Tennessee, 1984.
11. Francis C.; Maskarinec, M. Field and Laboratory Studies in Support of a Hazardous Waste Extraction Test, DE86 007068, *U.S. Environmental Protection Agency,* Oak Ridge, TN.,1986.
12. Townsend, T.; Tolaymat, T.; Solo-Gabriele, H.; Dubey, B.; Stook, K.; Wadanambi, L. Leaching of CCA treated wood: implications for waste disposal, *J. Hazard Mater.*, **2004**, *B114*, 75-91.

13. EPRI. Pentachlorophenol-treated wood poles and cross arms: Toxicity characteristic leaching procedure results. *Report summary EN7062S Electric Power Research Institute*, Palo Alto, CA, 2001.
14. Townsend, T.; Dubey, B.; Tolaymat, T.; Solo-Gabriele, H. Preservative leaching from weathered CCA-treated wood. *J. Environ. Manag*, **2005**, *75*(2), 105-113.
15. Clark, C.; Jambeck, J.; Townsend, T., A review of construction and demolition debris regulations in the U.S., *Crit. Rev. Environ. Sci. Technol.*, 2006, *36*(2) .
16. Townsend, T.; Dubey, B.; Tolaymat, T. Interpretation of SPLP results for assessing risk to groundwater from land applied granular waste, *J. Environ. Eng. Sci.*, **2006**, *23*(1), 236-248.
17. Stook, K.; Tolaymat, T.; Ward, M.; Dubey, B.; Townsend, T.; Solo-Gabriele, H.; Bitton, G. Relative leaching and aquatic toxicity of pressure-treated wood products using batch leaching tests, *Environ. Sci. Technol.*, **2005**, *39*(01), 155-163.
18. Jang, Y.; Townsend. T. Leaching of Lead from Computer Printed Wire Boards and Cathode Ray Tubes by Municipal Solid Waste Landfill Leachate. *Environ. Sci. Technol.*, **2003**, *37*, 4718-4784.
19. Dubey, B. Comparison of Environmental Impacts of Wood Treated with Chromated Copper Arsenate (CCA) and Three As-Free Preservatives, *PhD Dissertation, University of Florida*, Gainesville, FL, 2005.
20. Pohland, F.; Karadagli, J.; Kim, C.; Battaglia, F. Landfill Codisposal of Pentachlorophenol (PCP)-treated Waste Wood with Municipal Solid Waste. *Water Sci. Technol.*, **1998**, *38*(2), 169-175.
21. Weber, J.; Jang, Y.; Townsend, T.; Laux, S. Leachate from Land Disposed Residential Construction Waste, *J. Environ. Eng.*, **2002**, *128*(3), 237-245.
22. Jambeck, J. The disposal of CCA-treated wood in simulated landfills: potential impacts, *PhD Dissertation, University of Florida*, Gainesville, FL,**2004**.
23. Jang, Y. A study of construction and demolition waste leachate from laboratory landfill-simulators, *PhD dissertation, University of Florida*, Gainesville, FL, 2000.
24. McMahon, C.; Bush, P.; Woolson, E. How Much Arsenic is Released When CCA-Treated Wood is Burned. *Forest Prod. J.*, **1986**, *36*(1912), 45-50.
25. Hirata, T.; Inoue, M.; Fukui, Y. Pyrolysis and Combustion Toxcicity of Wood treated with CCA, *Wood Sci. Technol.*, **1993**, *27*, 35-47.
26. Pasek, E.; McIntyre, C. Treatment and Recycle of CCA Hazardous Waste. *24th Annual Meeting of the International Research Group on Wood Prservation, Stockholm*, Sweden, 1993.
27. Helsen, L.; Van Den Bulck, E. Thermal Treatment Technologies for Treated Wood, book chapter in the book titled "Environmental Impacts of Treated

Wood" edited by Tim Townsend and Helena Solo-Gabriele, CRC Press, Boca Raton, FL, (in-press).

28. Iida, K.; Pierman, J.; Tolaymat, T.; Townsend, T.; Wu, C. Control of Heavy Metal Emissions and Leaching from Incineration of CCA-Treated Woods Using Mineral Sorbents, *J. Environ. Eng.*, **2004**, *130*(2), 184-192
29. Misra, A.; Dubey, B.; Wu, C.; Townsend, T.; Solo-Gabriele, H. Controlling Heavy Metal Emission and Leaching in Combustion Environments by using Sorbent Technology during Disposal of Chromated Copper Arsenate (CCA) Treated Wood, *In the proceedings of 2005 Florida Section A&WMA Annual Conference, St. Pete Beach*, FL, USA, Oct 23-25, 2005.
30. Solo-Gabriele, H.; Townsend, T.; Messick, B.; Calitu, V. Characteristics of CCA- Treated Wood Ash. *J. Hazard. Mater,* **2002**, *89* (2-3): 213-232.
31. Cooper, P.; Ung. T.; Kazi, F.; and Qi, D. 2003. Two approaches to recycling of CCA-treated wood: extraction for recycling and wood cement composites. *Proceedings of Managing the Treated Wood Resource, II, Special Seminar* Sponsored by the American Wood Preservers' Association and the Utility Solid Waste Activities Group held in Boston, MA. American Wood Preservers' Association, Granbury, TX. p. 65-76.
32. Fahy, C.; Glitten, M.; Orcult, M.; Wallace, T. Spent Pressure Treated Wood: An Analysis of Management Options. *Tufts University*, Medford, MA, 1993.
33. Townsend, T.; Solo-Gabriele, H.; Stook, K.; Tolaymat, T., Song, J.; Hosein, N.; Khan, B. New Lines of CCA-Treated Wood Research: In-Service and Disposal Issues. *Florida Center for Solid and Hazardous Waste Management*. Gainesville, FL, Report #01-X, 2001.
34. Jacobi, G.; Solo-Gabriele, H.; Dubey, B.; Townsend, T.; Shiba, T. Evaluation of Commercial Landscaping Mulch for Possible Contamination from CCA-treated Wood. *Waste Manag* (in-press).
35. Blassino, M.; Solo-Gabriele, H.; Townsend, T. Pilot Scale Evaluation of Sorting Technologies for CCA-Treated Wood Waste. *Waste Manag. Res.,*, **2002**, *20*, 290-301.
36. Solo-Gabriele, H.; Townsend, T.; Khan, B.; Song, J.; Jambeck, J.; Dubey, B.; Jang, Y. Arsenic and Chromium Speciation of Leachates from CCA-Treated Wood, *Florida Center for Solid and Hazardous Waste Management*, Gainesville, FL, Report #03-07, 2003.
37. Homan, W.; Miltz, H. Evaluation of Rapid Methods for Detecting Wood Preservatives in Waste Wood, *IRG/WP 94-50024. International Research Group on Wood Preservation*: Stockholm, Sweden, 1994.
38. Peylo, A.; Peek, R. Rapid Analytical Methods for Wood Waste – An Overview. *IRG/WP 98-50104, International Research Group on Wood Preservation, Stockholm*, Sweden, 1998.
39. Clausen, C. Reusing Remediated CCA-Treated Wood, Proceedings of Managing the Treated Wood Resource, II, Special seminar sponsored by the

American Wood Preservers' Association and the Utilities Solid Waste Activities Group held in Boston, MA. American Wood Preservers' Association, Granbury, TX, **2003**, p. 49-56.
40. Kamdem, D.P., 2006. Recycling of Wood Treated With Chromated Copper Arsenate into Composite Construction Materials. In "Environmental Impacts of Preservative Treated Wood", Townsend and Solo-Gabriele (eds.). CRC Press, Boca Raton, FL.
41. Clausen, C. CCA Removal from Treated Wood Using a Dual Remediation Process. *Waste Manag. Res.* **2000**, *18*, 485-488.
42. Humar, M.; Pohleven, F. Bioremediation of Waste Copper/Chromium Treated Wood Using Wood Decay Fungi, *Proceedings of Environmental Impacts of Preservative Treated Wood Conference, Florida Centre for Environmental Solutions*, Gainesville, FL, 2004, p. 245-258.
43. Kartal, S.; Imamura, Y. Fungal Remediation of CCA-Treated Wood, *IRG/WP 04-50210. International Research Group on Wood Preservation: Stockholm*, Sweden, 2004.
44. Illman, B.; Yang, V. Bioremediation and Degradation of CCA-Treated Wood Waste, *Proceedings of the Environmental Impact of Preservative Treated Wood Conference, Florida Centre for Environmental Solutions*, Gainesville, FL, 2004, p. 259-269.
45. Smith, R.; Shiau, R. Steam Processing of Treated Wood Waste for CCA Removal, Technical Report published by the Centre for Forests Products Marketing, Virginia Tech, Blacksburg, VA, 1997.
46. Shiau, R.; Smith, R.; Avellar, B. Effects of Steam Explosion Processing and Organic Acids on CCA Removal from Treated Wood Waste, *Wood Sci. Technol.*, **2000**, *34*, 377-388.
47. Oskoui, K. Recovery and Resue of the Wood and Chromated Copper Arsenate (CCA) from CCA-Treated Wood – A Technical Paper, *Proceedings of the Environmental Impacts of Preservative Treated Wood Conference, FL, Florida Center for Environmental Solutions*, Gainesville, FL, 2004, p. 238-244.
48. Christensen, I.; Pedersen, A.; Ottosen, L.; Riberio, A. Electrodyalytic Remediation of CCA-Treated Wood in Larger Scale, *Proceedings of the Environmental Impacts of Preservative Treated Wood Conference, Florida Center for Environmental Solutions,* Gainesville, FL. 2004, p. 227-237.
49. Ottosen, L.; Pedersen, A.; Christensen, I. Characterization of residues from thermal treatment of CCA impreganated wood, chemical and electrochemical extraction. *Proceedings of the Environmental Impacts of Preservative Treated Wood Conference, Florida Centre for Environmental Solutions, Gainesville, FL*, 2004, p. 295-311.
50. Hse, C.; Shupe, T.; Lin, L. A Complete Closed-Loop CCA-Treated Wood

Preservatives in Waste Wood, *IRG/WP 94-50024, Internatioanl Research Group on Wood Preservation:* Stockholm, Sweden, 2003.

51. Shupe, T.; Hse, C. Recycling of Preservative-Treated Wood Using Chemical Extraction Technologies, book chapter in the book titled "Environmental Impacts of Treated Wood" edited by Tim Townsend and Helena Solo-Gabriele, CRC Press, Boca Raton, FL. .

Worldwide Trends in Wood Preservation

The deck photograph is courtesy of Tor P. Schultz of Mississippi State University. (See page 12 of color inserts.)

Chapter 34

Wood Protection in Europe: Developments Expected up to 2010

Hans Leithoff, Piet Blancquaert, Mark van der Flaas, and Alex Valcke

Janssen Pharmaceutica N.V., Turnhoutseweg 30, B–2340 Beerse, Belgium

Introduction

In the past the wood preservation market in Europe was scattered and characterized by wide differences in national regulations and regional traditions. At this time the main criteria in the development of new wood preservatives were efficacy and economics.

At present, however, all developments of wood preservatives in Europe are preliminary dependant on the European Biocide Directive 98/8/EC (BPD (*1*)). Its main objectives are summarized in the following sentences on the official website of the European Commission "The Biocidal Product Directive aims to harmonize the European market for biocidal products and their active substances. At the same time it aims to provide a high level of protection for humans, animals and the environment." (*2*). In the future, only active substances and biocidal products assessed and authorized according to the BPD regulations can be placed on the European market. Therefore, this article focuses on the impact of the ongoing BPD assessments on wood preservation in Europe.

The history of bans and restrictions on classical all-purpose wood preservatives like PCP, creosote, and CCA, which were the first widely used preservatives, is also considered (*3*). These restrictions/bans accelerated the development of alternative biocides/preservatives, but the efficacy of the newer biocides is limited to certain target organisms compared to the broad efficacy of the earlier biocides. Later biocides consequently have a much better toxicological and ecotoxicological profile (*4*). A further development of this tendency might be a so called "fit for purpose" product targeted to specific uses and/or market segments and the expected service life of the treated end-use product (*5, 6*).

Furthermore, although the BPD will result in harmonization some country specific differences within the European Union (EU) will remain and still have to be considered. It is impossible to address these differences in detail, but, some general background information is given here in the introduction to illustrate the situation:

Prior to the implementation of BPD a tremendous but not complete harmonization of the classification and efficacy testing of wood preservatives was achieved mainly by activities of the European Committee of Standardization (CEN TC 38) and the affiliated national committees. Consequently, within the last 30 years a number of European standards were published (*7*). However, the existing national differences with regard to product registration and/or approval and the related evaluation schemes prevented a single European assessment of wood preservatives and are at present still strong barriers for the free trade of wood preservatives within Europe. The BPD has clearly the objective to pull down these barriers, but, while the approval of active substances is controlled by the European Commission, the wood preservative products are going to be authorized at the member state level. To facilitate and harmonize the authorization of these products, the principle of mutual recognition has been included in the BPD: One member state evaluates and authorizes the biocidal product and based on this first evaluation, other member states will authorize the product for their national market and can limit their authorization due to administrative aspects only.

At the other hand a number of powerful and independent regional or even national associations/agencies established a variety of voluntary approval schemes and provide quality control schemes for the product and its production (e.g. NWPC in Denmark, Finland, Norway and Sweden; CTB P^+ in France; RAL GZ 830 in Germany). Furthermore, approval boards like DIBt (Deutscher Industrieverband für Bautechnik, Germany) grant national approvals required for products employed in a particular area and/or application. Likely such organizations will try to maintain at least some business also after full implementation of the BPD. Some organizations already indicated that they intend to ensure superior quality by selecting those BPD authorized products which fulfill their additional requirements / traditional rules (*8*).

Biocidal Products Directive (BPD)

Concept and Structure of BPD

To achieve harmonization of authorizations and registrations with regard to biocidal products in the EU, the Council and Parliament agreed to unify the assessment of the active substances and biocidal products and so adopted in

1998 the Biocidal Products Directive 98/8/EC (*1*). The BPD defines that the assessment is not done centrally but by the individual agencies (Competent Authorities = CA) of the member states (Rapporteur Member States = RMS). In the first phase the active substance itself is assessed, and in a second phase the biocidal product, the wood preservative, will be assessed and authorized at member state level when it is considered effective and safe for man and the environment.

First stage: Assessment of the active substance

In case the active substance is considered as unacceptable, a non-inclusion decision will be adopted and published. It will impose a deadline for the member states to withdraw the substance and products that employ it from the European market. This phase-out period will be 6 to 12 months.

In case of a positive assessment the biocide will be listed in Annex I of the BPD for the use in a specific product types (e.g. in product type 8 = wood preservation). In that process the list of endpoints is fixed (e.g. Acceptable Operator Exposure Level or AOEL for human risk assessment and the Predicted No Effect Concentration or PNEC for environmental risk assessment), and – if necessary – mitigation measures or restrictions on use may be required. A decision for Annex I listing is published as a Commission Directive addressed to the member states and provides binding rules of the biocidal product authorizations, the second stage.

The applicant, (in most cases the manufacturer of an active substance, pays an application fee covering the work done by the RMS. The fees vary from around 100,000 € to 350,000 €.

To allow a harmonized implementation of the BPD by the CA's of all member states, guidance documents and guidelines were developed and adopted. These include the Technical Notes for Guidance (TNsG, (*9*)) and the Technical Guidance Documents (TGDs (*10*)). The latter are also used for the assessment of new and existing chemicals. Where the BPD and its related Commission Directives and Regulations are legally binding documents, the guidance documents are not legally binding.

The (existing) active substance evaluation is done by the CA of the RMS decided by the Commission (for new active substances the applicant can select the RMS). Before the full evaluation starts, the dossier is checked in a process that can take up to six months. Once considered complete, the scientific evaluation starts and should not last for more than 12 months. Only when data gaps are discovered during the evaluation period the clock can be stopped to allow the generation of the additional data required. The applicant cooperates quite closely with the RMS during the whole process.

Once the CA has finalized its Competent Authority Report (CAR) it is sent to the Commission, circulated to the other CA's and made publicly available on the Commissions website. This public version does not contain the confidential parts and can be found at: http://forum.europa.eu.int/Public/irc/env/bio_reports/library?l=/wood_preservatives_1&vm=detailed&sb=Title

As the experience and expertise in the 27 member states is not yet at the same level, the CARs are discussed during one or more Technical Meetings (TM). These meetings take place three to five times per year and are intended to obtain a high quality of the evaluations, development of further guidance and harmonization of interpretations. Participants at the TMs are the technical and scientific experts of the RMSs.

Following the TM discussions on an active substance, the CAR is transformed into an Assessment Report and a proposal for Annex I inclusion (or non-inclusion) is prepared. The proposal then goes to the CA meeting (gathering four times per year in Brussels) for further discussions and for other considerations like politics and economics. The final decision is taken at the Standing Committee for Biocidal Products (SCBP) by qualified majority voting (see Figure 1).

From the description of the process above, it should be clear that this is a time consuming exercise, and in view of the end of the transitional period on 14 May 2010, the Commission and the member states are under considerable time pressure. For wood preservatives only, dossiers on 38 active substances were submitted and only one decision had been taken before the end of 2006 (sulfuryl fluoride (*11*). A decision on 5 to 10 substances is expected by the end of 2007.

The second stage: Product authorization.

In this second stage, an applicant submits a dossier on an existing wood preservative he wants to keep on the market to a member state before the effective Annex I inclusion of the active substance(s) in his product. This dossier includes a confirmation that the active substance(s) used in the product has (have) the same specifications as the Annex I listed substance(s), a letter of access to the protected data and an application fee (estimated to be around 40,000 to 50,000 €), product data including efficacy, leaching behavior, and a human and environmental exposure and risk assessment.

The dossier is assessed by the RMS (chosen by the applicant but preferably the RMS of the active substance) using comparable timelines as for the active substance (three months for completeness check followed by 12 months evaluation). This first member state can grant an authorization and the applicant can use this first authorization within two months to apply for another EU member state for authorization for the same product by the principle of mutual

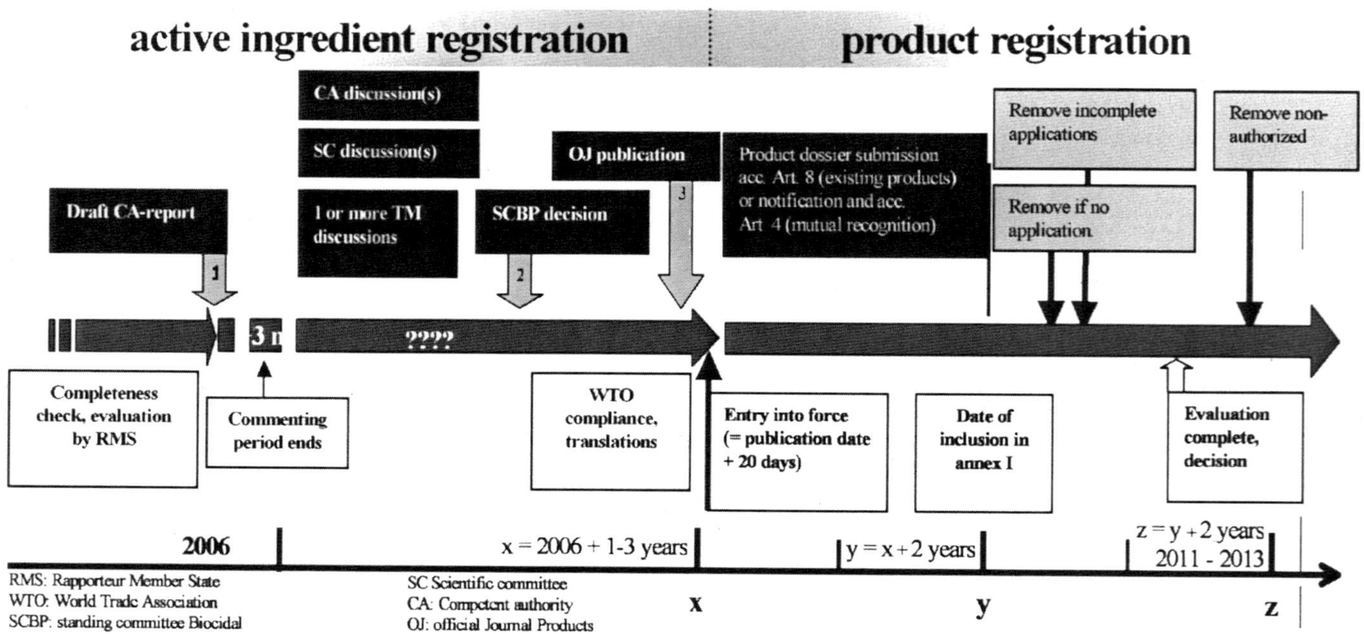

Figure 1. Timeline for BPD decision on Annex I listing of active substances and the subsequent product authorization under the BPD.

recognition. The use of mutual recognition is necessary to allow for the full implementation of the BPD and reduce the workload of all parties involved. Mutual recognition should be mainly an administrative act and refusal needs to be substantiated by the member state.

In case no product dossier has been submitted or it has been considered incomplete, the products concerned will go into a phase-out period and be withdrawn from the market within one year after the Annex I listing of the active substance(s). For no more than 2 years following the Annex I listing of the active substance shall products that remain on the market be re-authorized at national level under the BPD requirements (see Figure 1).

This is a very challenging goal and a comparable approach under the directive 91/414/EEC, the so-called sister directive of the BPD regulating plant protection products, has been shown to be unrealistic.

As the first phase, the evaluation of the existing active substances, is not yet finalized and may be further delayed all parties involved have some years to prepare and to make the system work. The first set of compounds with broader use than sulfuryl fluoride consists of the organic fungicides dichlofluanid, tebuconazole, propiconazole, tolylfluanid and IPBC. Decisions on these compounds are expected by late 2007 with publication early 2008, so that effective Annex I listing of these substances can be foreseen for early 2010. Wood preservative dossiers for existing products based on these fungicides need to be submitted no later than the effective listing of the last active substance in the product entering Annex I. In case an insecticide is also included in a preservative system, the submission will be delayed until that insecticide is also listed on annex I. None of the insecticides submitted under PT 8 has progressed as far as some of the fungicides in the approval process.

Dossiers of new wood preservatives based on annex I listed active substances can be submitted any time after the annex I listing but can only be marketed once national authorization has been granted by a member state.

New Active Substances

Wood preservatives based on new active substances (new active substances are substances not listed in Annex II of Commission Regulation 1048/2005 (*12*), see above) can only be marketed after Annex I inclusion of the new active substance or following provisional approval. This provisional approval can be granted at member state level for a period of three years during the evaluation period of the new active substance. This way new active substances can enter the market more rapidly and gradually as mutual recognition does not apply in this case. However, a formal submission for provisional approval must be submitted in all member states where the applicant intends to market his product during the evaluation period.

The provision for provisional approval did not exist in national biocide schemes and comes from the plant protection products directive 91/414/EEC. It should be noticed that the current proposal to amend this directive, the provisional approval provision, has been deleted. As the preparation of the amendment of the BPD starts in 2007 it can be expected that within 5 to 10 years this provision will also be deleted.

1 September 2006

Commission Regulation 2032/2003 of 4 November 2003 (*13, 14*) giving further direction for the 10-year review program also defines the date when all non-defended active substances and biocidal products including them must be taken off the market: 1st of September 2006. Specifically for wood preservatives this means that only the active substances for which a complete dossier was submitted can stay on the market (see Table 1). Use of all other actives should have been discontinued. Some member states have granted additional phase-out periods, although the EU Commission repeatedly indicated that the date was known sufficiently in advance so that national measures should have been taken to remove the non-defended substances from the market.

Chromium: A Biocide or a Non-biocidal Additive for Fixation Only?

Chromium was identified as an active substance but no Chromium dossier for PT8 was submitted under the BPD. Legally, this requires the withdrawal of all chromium-containing wood preservatives from the EU market by 1st of September 2006. However, reflecting the function of chromium as a fixation agent in wood preservatives as well as the definition of the required "*sufficient efficacy under real-use conditions*", this interpretation was not at all common sense in the CA-meeting (*15*). It was also impossible to agree on chromium as non-biocidal additive. Therefore, the European Commission judged chromium in principle as an active ingredient which, consequently, has to be withdrawn from the EU market. But on the product level the European Commission referred the assessment whether chromium is an active or fixation compound in a particular wood preservative to the national level of each Member States' CA. In case suitable data shows chromium to be "*non efficient under real-use conditions*" against target organisms it can be assessed as a fixation agent and, by this, manufacturers got temporary approval to maintain chromium-containing products in Germany, Sweden, the Nederlands and Austria. Other countries such as United Kingdom, Belgium and France judged data presented as incomplete or not convincing and phased out chromium containing wood preservatives by 1st of

September 2006 (*16*). However, as the BPD also requires human and environmental exposure and risk assessments of "substances of concern" (substances that are classified under the Dangerous Substances Directive, 67/548/EEC and its amendments and Adaptations to Technical Progress (*17*)), a substantial amount of data on chromium will still need to be submitted.

Impact of BPD on Product Development until 2010

Considering the facts related to BPD as explained above, it is obvious that many ongoing development projects are regulatory driven. The most prominent aspect is the identity and number of actives available for wood preservation in near future. Under these circumstances the list of active ingredients under review for PT 8 (wood preservation), listed in Table I, becomes a logical starting point for further discussion. Considering some very common actives previously used but not supported under the BPD, this table allows us to address some changes which have already happened or are to be expected and are caused directly or indirectly by BPD.

Actives to be withdrawn from the market

Only 40 of the active ingredients listed for PT 8 in the Annex 5 of the second review regulation EC 2032/2003 (*13, 14*) are still supported (see Table I, see also *27*). In consequence a number of wood preservatives / active ingredients had to be withdrawn from the market by 1st of September 2006. Only some of them can be discussed here:

Arsenic /CCA

Although the use of arsenic was already restricted to certain applications, before September 2006, CCA could still be used for some important but severe hazard applications like transmission poles or some industrial/professional applications (3). Furthermore, most probably CCA will not be replaced mid term by arsenic-free but chromium-containing wood preservatives (see above). A shift to chromium-free wood preservatives was already seen in 2004 (*18*). In some heavy duty Use Class IV applications the end users judged chromium- and copper-based wood preservatives as very important non-replaceable wood preservatives, mainly due to an excellent record of good performance derived directly from practice. For such applications it is likely that some end customers are going to import the treated wood from non EU countries (which is still

Table I. Biocides in the review process for inclusion in Annex I of the BPD for PT8 (wood protection) by 31st August 2006 (*14, 31, 32*).

type	generic name	CAS number	insecticide*		fungicide**			
			ins.	term.	SS	BS	BAS	SR
organic fungicides	Cyproconazole	94361-06-5					X	
	Dazomet	533-74-4			?			
	DCOIT	64359-81-5			X	X	(X)	
	Dichlofluanid	1085-98-9				X		
	Fenpropimorph	67564-91-4			X	X	(X)	
	Guazatine triacetate	115044-19-4			X	(X)		
	IPBC	55406-53-6			X	X	(X)	
	K-HDO	66603-10-9					X	
	Potassium sorbate	24634-61-5			X			
	Propiconazole	60207-90-1			X	X	X	(X)
	Tebuconazole	107534-96-3					X	(X)
	Thiabendazole	148-79-8			X	X		
	Tolylfluanid	731-27-1				X	(X)	
	Sorbic acid	110-44-1			X			
organic insecticides	Bifenthrin	82657-04-3	X	X				
	Chlorfenapyr	122453-73-0	X	X				
	Clothianidin	210880-92-5	X	X				
	Cypermethrin	52315-07-8	X	X				
	Etofenprox	80844-07-1	X	X				
	Fenoxycarb	72490-01-8	X					
	Flufenoxuron	101463-69-8	X					
	Permethrin	52645-53-1	X	X				
	Thiacloprid	111988-49-9	X	X				
	Thiamethoxam	153719-23-4	X	X				
	Sulphuryl difluoride	2699-79-8	fumigation X					

Legend see Table I. *(Continued)*.

Table I. (Continued).

type	generic name	CAS number	insecticide*		fungicide**			
			ins.	term.	SS	BS	BAS	SR
quaternary ammonium compounds	BKC	-	(X)	(X)	X	X	(X)	(X)
	DDAC	-	(X)	(X)	X	X	(X)	(X)
	Laurylamine dipropylendiamine	2372-82-9	(X)	(X)	X	X	(X)	(X)
	polymeric Betain	214710-34-6	(X)	(X)	X	X	(X)	(X)
	Bardap 26	94667-33-1	(X)	(X)	X	X	(X)	(X)
	TMAC	61789-18-2	(X)	(X)	X	X	(X)	(X)
boron compounds	Boric acid	10043-35-3	**X**	**X**	(X)	(X)	**X**	
	Diboron trioxide	1303-86-2	**X**				**X**	
	Disodium octaborate tetrahydrate	12280-03-4	**X**	**X**	(X)	(X)	**X**	
	Disodium tetraborate, anhydrous	1330-43-4	**X**	**X**	(X)	(X)	**X**	
copper containing compounds	Copper dihydroxide	20427-59-2	X	X	X		X	X
	Copper oxide	1317-38-0	X	X	X		X	X
	Copper(II) carbonate-copper(II) hydroxide (1:1)	12069-69-1	X	X	X		X	X
	CuHDO	312600-89-8	X	X	X		X	X
others	Creosotes	8001-58-9	X	X	X	X	X	X

* Insecticide with efficacy against: ins. = wood boring insects; term. = termites

** Fungicides with efficacy against: SS = sap-stain; BS = blue stain; Bas = wood decaying basidiomycetes, SR = soft rot fungi

X **cross in bold:** Spectrum of the single active derived from BPD dossier / draft evaluation report see: http://forum.europa.eu.int/Public/irc/env/bio_reports/library?l=/wood_preservatives_1&vm=detailed&sb=Title

X cross: Spectrum of the single active derived from literature or registrations

(X) cross in brackets: Either efficacy claimed but the compound is usually not used as single active or: efficacy not claimed, but active is used in practice (e.g. in two way products)

allowed) or switch to completely non wood materials like concrete, steel, or polymeric products.

TBTO

TBTO was used mainly in solvent borne wood preservatives for joinery and cladding. It was recognized quite early that TBTO was not supported under the BPD and was/is replaced mainly by a combination of an azole (propiconazole/ tebuconazole) plus a moldicide (e.g. IPBC, dichlofluanid).

Sodium Pentachlorophenol (Na PCP)

Independent of BPD, Na-PCP was already banned before 2006 in most European countries. In France, Ireland, Portugal, Spain and UK the use was permitted with strong restrictions (*17*). Therefore, Na-PCP has already been replaced by a number of other actives, depending on the specific application.

Fluor salts

These compounds have essentially been discontinued even before implementation of the BPD, mainly due to some disadvantages including corrosivity and high depletion.

Copper sulphate, dicopper oxide

Used mainly in chromium containing wood preservatives but, in many cases, already replaced by other copper compounds, defended under the BPD.

Deltamethrin and Cyfluthrin

Organic insecticides used in various wood preservatives. Both are going to be replaced by other pyrethroids or by neonicotinoids, pyrroles or insect growth regulators already undergoing BPD approval. However, due to the mode of action insect growth regulators are not suitable as a preventive termiticide in wood protection.

Actives Available on the European Market

Table I lists 40 actives being reviewed for wood preservation and gives a general overall view on their spectrum of activity against wood destroying or inhabiting organisms.

However, the number of actives is may be a bit misleading because three entries (BKC, DDAC and TMAC) cover a whole group of similar quaternary ammonium compounds (Quat). On the other hand, some actives listed are mainly used in very special applications, e.g. sulfuryl difluoride (fumigation), potassium sorbate, sorbic acid (sap stain products) or K-HDO (engineered wood). Others are used only in solvent borne formulations like dichlofluanid and tolylfluanid. Most of the actives listed are quite well known in wood preservation and thus their efficacy, spectrum of activity and applications can be obtained in more detail from existing literature or registrations.

There are only a limited number of actives which have a very broad spectrum and can be used to control all target organisms. However, it is remarkable that a broad spectrum of an individual biocide is not automatically linked to its use as single active ingredient in final wood preservative formulations. Approx. 1300 products mentioned in documents available from Austria, France, Germany, the Netherlands and the United Kingdom (*20 - 24*) were used to screen the compositions under this aspect (multiple registration in several countries included). From these products the majority (53%) contain only one active ingredient while 28% contain two actives, 15% contain three actives and only 3% contain four actives, but, none contains more than four actives.

Although it has to be considered that a lot of entries in these lists are only registered but are either not sold or have a very limited market share, these data can be used to characterize the typical design of wood preservatives. For example, it was observed that products which contain only one active ingredient claim in most cases a very limited use spectrum, e.g. pure curative insecticide for UC I and II only or efficacy against blue stain only. Obviously the use of a single active ingredient is most commonly employed in a targeted product. Here the highly effective biocides with a limited but targeted spectrum can be used at a very low dosages and therefore provide strong economical and toxicological benefits.

Two-way combinations are used either to:

a) assemble actives with completely different spectra like an insecticide and a fungicide, e.g. combination of an azole with an insecticide or
b) optimize the efficacy of a broad spectrum biocide by adding an active ingredient with a spectrum limited to target organisms which otherwise could be controlled only with a much higher dosage, e.g. combination of an azole with a strong blue stain agent or
c) optimize the efficacy by combining two broad spectrum biocides with somewhat complementary spectra or synergistic effects, e.g. the combination of copper with Quats or boron.
d) improve formulations, e.g. combination of two boron compounds.

Two way combinations do not necessarily provide protection against fungi and insects. Quite often the combination of two actives is used to only optimize fungicidal efficacy. Interestingly, the combination of two insecticides seems to be very uncommon.

In case of three- and four-way formulations, biocides with similar spectrum of activity as mentioned above are employed, however, often two quite similar actives are combined with a third (and possibly a fourth) complementary biocide (e.g. two fungicides combined with an insecticide), but combinations of three broad spectrum biocides were recorded as well (e.g. combination of a boron compound, a copper compound and a Quat). Three- and four-way formulations usually claim to protect the timber against fungal decay as well as insect attack. Four way combinations usually consist of one insecticide and three fungicides, of which one fungicide usually inhibits predominantly blue stain and/or mold fungi. However, in many cases the combination was used more to optimize the efficacy against fungi or the properties of the formulation than to optimize the insecticidal efficacy.

It is likely that the number of products with more than three active ingredients will increase, although the development of such formulations can be relatively complicated and the registration might require more effort. Two main reasons may explain this statement:

- By combining highly targeted biocides the final formulation can benefit from complementary or synergistic effects which will result in superior efficacy and consequently lower retentions compared to products using fewer actives.
- The strict risk assessments for human health and environment will favor formulations which are effective at low retentions. Even in cases where the total amount of active ingredients is similar to a product with only one active, the combination of actives with different modes of action might result in a considerable better assessment due to the lower amount of the individual actives, their limited spectrum and consequently a lowered effect on non-target organisms.

At the first view it seems that 40 actives should provide sufficient combinations for any future development. But, from 14 organic fungicides listed only 5 are frequently mentioned in the registrations: dichlofluanid, IPBC, propiconazole, tebuconazole and tolylfluanid, with dichlo- and tolylfluanid only used in solvent borne systems. A similar situation counts for the organic insecticides: Here only four (cypermethrin, fenoxycarb, flufenoxuron and permethrin) of 11 actives are frequently listed. Furthermore, only the two pyrethroids are termiticides.

Under the aspect "combination of actives", the four boron compounds listed are so similar in their mode of action and spectrum that they should be summarized as one group where their individual value is mainly related to special formulation issues. The same is valid for the three inorganic copper compounds. This illustrates that due to the special character of the individual actives the number of real options for future development is more limited than would be anticipated simply from the number of active ingredients.

Based on the situation described, some major shifts in products used in UC III & IV can be expected while the influence on products used in UC I and II will likely be much less (*21* - *23*).

At this time, the major wood preservatives used in UC IV are based on a mixture of Cu-HDO, didecylpolyoxyethylammoniumborate (polymeric betain) or Quats with copper carbonate and boron compounds, as well as the combination of azoles with copper compounds plus an insecticide (e.g. a boron compound). Creosote can also be employed, but some legal restrictions limit its use to severe industrial applications like railway ties and utility poles (*17*).

Due to the very limited number of inorganic actives still available, future development will focus on organic biocides. These allow the development of products optimized for the intended use and with an efficacy strongly focused on the target organisms. At the same time water borne systems will replace more and more solvent borne formulations. Therefore, formulation issues in general, especially with respect to long term efficacy and leaching behavior, will become a key issue. It is likely that more effort will be taken to optimize formulations with all types of non biocidal additives.

Role of Biocide-free Treatment Methods and Materials

Only four major types of biocide free treatments and materials shall be mentioned here: Heat treatment, Acetylation, Resin treatments and Wood Polymer Composites (WPCs).

All three chemical methods as well as the WPCs are currently commercially available in Europe. Numerous literature is available on properties and benefit of these materials and describe the different processes used (e.g. *25* - 28). Within this chapter the discussion will be restricted to the three following aspects:

a) Durability of commercially available products;
b) Marketing concept commonly used in European; and
c) Market segment addressed.

Durability of the Commercially Available Biocide-Free Products

Most of the commercially available products do not claim full protection against decay and insects, but use the durability classification as described in EN 350-1/EN 460 (29, 30) for natural timbers. Most products are classified as durability class 2, some even as durability class 1. In many cases full protection could be achieved, but an adjustment of the treatment parameters to maximum durability would cause an unacceptable decrease in mechanical properties or a dramatic and unacceptable price increase. Therefore, the final product reflects a compromise between the different properties and thus marketing emphasized the low maintenance and improved durability as well as the anti shrinking efficiency (ASE), higher dimensional stability, low equilibrium moisture content or water uptake, the improved hardness, the color change, etc.

Marketing Concept for Introduction in the European Market

Most of the recently launched products use the concept of selling a "final product" and avoid focusing too much on the treatment method or the chemicals used. The aim is always to give the treated wood an individual name and characterize it as new material or a kind of new wood species with characteristic and unique properties.

Market Segment

The main market segments are exterior decking, cladding, garden furniture, window frames and doors. All products approach the high quality, high price segment. The use of the material for load bearing structures, e.g. for frames is in many cases not intended (price) or excluded (not compliant with national building regulations national approval so not).

With regard to the given description there is no need for a separate characterization of WPC, although it differs strongly from treated timber with regard to the production processes, many physical properties and appearance.

By creating a new product profile with unique characteristics these new products gained acceptance by the customer and consequently market share. One of the main driving forces behind their developments was the aim to create an improved material with high durability without employing biocides. However, for customers durability is only one of many desired properties and, for example, the appearance or the low maintenance claim might be more important.

These biocide-free products might be seen as direct competitors to preservative-treated wood. However, considering the price and the market segment these materials most likely compete with highly durable and expensive tropical timbers like Bankirai (*Shorea spp.*) than with the classical biocide-treated softwood. But these materials might been considered as a kind of trendsetter. Their performance might become a benchmark and create expectations which are challenging to achieve with traditional wood preservation techniques. Although the efficacy of future preservatives is often assumed to be the dominant performance criterion, other properties shortly might become much more important to consumers for certain applications. The marketing of WPC decking products which claim low maintenance and excellent weathering resistance of the exposed wood surface might illustrate this.

Summary

Due to the ongoing implementation of the BPD, most mid term development of wood preservatives in Europe is regulatory driven. A maximum of 40 active ingredients will be available, some of which are of limited value due to their special characteristics. Organic active ingredients may play a key role in designing targeted and highly efficient wood preservatives which will likely consist of two or more actives. The BPD mandate should not be viewed as a threat but as an opportunity to prove that the biocides used in wood preservation are safe and treated timber can be used without negative impact to human health and the environment. This will help to develop wood preservatives which will reinforce the image of wood as a natural, valuable, high-tech and sustainable resource and could be an important element in competing with new materials like WPC or chemically modified timbers.

References

1. Directive 98/8/EC of the European Parliament and of the Council of 16 February 1998 concerning the placing of biocidal products on the market (30th of August 2006: http://ec.europa.eu/environment/biocides/pdf/dir_98_8_biocides.pdf)
2. Citation: http://ec.europa.eu/environment/biocides/index.htm
3. Council directive of 27 July 1976 on the approximation of the laws, regulations and administrative provisions of the Member States relating to restrictions on the marketing and use of certain dangerous substances and preparations (76/769/EEC) consolidated text: CONSELEG: 1976L0769 – 16/03/2004) (30th of August 2006: http://ec.europa.eu/enterprise/chemicals/legislation/markrestr/consolid_1976L0769_en.pdf)
4. Pallaske, M. Chemical wood protection: Improvement of biocides in use. Presentation given at COST Action E22, Estoril / Portugal March 22nd – 23rd 2004 (30th of August 2006: http://www.bfafh.de/cost37.htm)
5. Van Acker, J. The way forward: Testing and specifying wood protective systems. Presentation given at COST Action E22, Reinbeck / Germany November 8th – 9th 2001 (30th of August 2006: http://www.bfafh.de/inst4/43/ppt/3forward.pdf)
6. Van Acker, J.; Stevens, J.M. Increased biological durability differs for traditional wood preservation and new non-biocidal systems (NBS).

International Research Group on Wood Preservation, Stockholm, 2000 IRG/WP 00-20212, 11 pp.

7. Willeitner, H. Historical perspective of wood preservation and the use of standard test methods to evaluate efficacy. Presentation given at COST Action E37, Oslo / Norway June 20th of 2005 (30th of August 2006: http://www.bfafh.de/inst4/45/ppt/4history.pdf)
8. Peek, R.-D. Prüfbestimmungen des DIBt. oral communication, Sept. 2006.
9. Technical Notes for Guidance. (31st of January 2007: http://ecb.jrc.it/biocides.)
10. Technical Guidance Documents. (31st of January 2007: http://ecb.jrc.it/tgd/)
11. Commission Directive 2006/140/EC of 20 December 2006, OJL: 414/78 of 30.12.2006
12. Commission Regulation (EC) 1048/2005 of 13 June 2005. OJL 178/1 of 09.07.2005
13. Commission Regulation (EC) 2032/2003 of November 2003. OJ L 307 24.11.2003.
14. Commission Regulation (EC) No 2032/2003 as amended by Regulation 1048/2005 ((31st of January 2007: http://ec.europa.eu/environment/biocides/pdf/consol_reg_2032_2003.pdf)
15. Guidance document agreed between the Commission services and the competent authorities of Member States on the role of chromium in wood preservation. Brussels 04.07.2005. (30th of August 2006: http://ec.europa.eu/environment/biocides/pdf/nfg_cr_040705.pdf)
16. Conclusions reached by member Sates competent authorities regarding the efficacy of chromium as a biocidal active substance in wood preservation products – Interim report. CA-Sept06-Doc.8.2
17. Dangerous Substances Directive, 67/548/EEC and its amendments and Adaptations to Technical Progress. (31st of January 2007: http://ecb.jrc.it/assessment-of-chemicals/)
18. Hughes, A. The tools at our disposal. Presentation given at COST E22 Estoril/Portugal 22nd – 23rd March 2004. (18th of September 2006: http://www.bfafh.de/inst4/43/pdf/7tooldis.pdf)
20. Irmschler, H.-J.; Quitt, H. Holzschutzmittelverzeichnis. Erich Schmidt Verlag, Berlin. 2006, 286 pp.
21. Österreichisches Holzschutzmittelverzeichnis2006.editor: Arbeitsgemeinschaft Holzschutzmittel, Wien, Austria128 pp.
22 Pesticide bluebook. Wood preservatives. (18th of September 2006: http://www.hse.gov.uk/pesticides/bluebook/section12.pdf)
23. Pesticide database of college voor de toelating van bestrijdingsmiddelen. (18th of September 2006: http://www.ctb.agro.nl)
24. Produits de Traitement Certifies CTB-P+ (18th of September 2006: http://www.ctba.fr/document_produit/Listes%20P+%20nov03.pdf)

25. Proceedings of the first European Conference on Wood preservation. Ghent, Belgium 3rd – 4th April 2003, 414 pp
26. Proceedings of the second European Conference on Wood preservation. Göttingen, Germany 6th – 7th October 2005
27. Erster Deutscher WPC–Kongress. 8th – 9th November 2005 Köln, Germany
28. Proceedings of Wood Plastic Composites Conference. 14th – 16th of September 2004, Vienna, Austria.
29. EN 350-1 : 1994 Durability of wood and wood-based products - Natural durability of solid wood - Part 1 : Guide to the principles of testing and classification of the natural durability of wood.
30. EN 460 : 1994 Durability of wood and wood-based products - Natural durability of solid wood - Guide to the durability requirements for wood to be used in hazard classes
31. Active substance in PT8 and 14 for which at least one dossier was submitted by 28 March 2004. (30th of August 2006: http://ec.europa.eu/environment/biocides/pdf/040430_received_dossiers.pdf
32. Active substance in PT8 and 14 for which no dossier was submitted by 28 March 2004, but which have been taken over by new participants. (30th of August 2006: http://ec.europa.eu/environment/biocides/pdf/take_over.pdf)

Chapter 35

Wood Preservation Trends in North America

H. M. Barnes

Forest Products Laboratory, Forest and Wildlife Research Center, Mississippi State University, Mississippi State, MS 39762

This paper discusses recent trends in the wood preservation industry in North America. Particular emphasis is placed on the development of copper-rich second generation preservative systems, including problems (disposal, mold, copper leaching, and corrosion) and costs associated with the new systems. The effect of EPA restrictions on CCA is presented. Use and restrictions placed on current organic systems, especially creosote, are reviewed. Near and long term future directions for the wood preservation industry, including low level metallic systems, total organic preservative systems, combination systems, borates, the impact of substitute materials, and non-biocidal treatments such as polymer treatment, heat treatment, and chemical modification are covered. Finally, implications for new treatment processes and products such as engineered wood composites are presented.

Introduction

Wood preservation has changed dramatically since man first attempted to preserve wood. This paper will discuss modern timber preservation and the impacts changes have made in the industry. Following a short glance backwards to see where we have been, the current situation will be discussed and a look into the crystal ball of the future will be presented.

By all accounts, the first reference to treating wood can be found with Noah's building of the ark. The early Egyptians attempted to preserve wood by coating them with natural oils. The Greeks made widespread use of durable woods and construction techniques which kept wood dry. Keeping wood dry remains our best method for preserving timber as attested to by the thousand year old timbers in Japanese shrines and temples and the 28 Scandinavian stave churches built after 1100 (**1**). The Romans continued the preservation techniques of the Greeks and extended it to treating boats with certain salts which reduced susceptibility to fire. Early chemical treatments for wood included treatment with mercuric chloride and blue vitriol (copper sulfate) as early as 1705. Crook (**2**) received an early colonial patent in the Province of South Carolina which incorporated the "Oyle or Spirit of Tarr".

The modern timber preservation industry started in the 1830s. Kyan patented the use of mercuric chloride for treatment of rope, wood, and canvas in 1832 (**3**). The impetus for timber preservation in the UK was the Royal Navy and the need to protect British ships from destruction from shipworms, the major factor in the loss of the Spanish Armada in 1588. Moll and others used creosote to protect wood in 1836, but it was left to Behell (**4**) to devise a process to effectively treat wood and launch the modern timber preservation industry. In that same year, Boucherie was granted a French patent for his sap displacement method of treating with salt solutions (**5**). In the US, the railroads provided the impetus for the timber preservation industry. Burnett (**6**) used zinc chloride to treat wood, and this was a principal treatment for crossties in the US until the 1920s. The first commercial plant for waterborne salt treatments was built in 1848 in Lowell, MA for Kyanizing timbers. In the U.S., the first Bethell-process plan was built in 1865 in Somerset, MA. The advent of modern timber preservation in the US is linked to the production of crossties by the L & N Railroad at a plant in West Pascagoula (now Gautier), MS in 1875. Two other significant developments occurred in the 1800s. The first was the publication of Boulton's (**7**) treatise on wood preservation. The second was the pioneering work done in the pathology arena by Hartig (**8**), considered to be the father of forest pathology. Several excellent historical treatises on wood preservation during this early period can be found in the literature for readers interested in more detailed historical accounts (**3, 9, 10**). The author has also published several reviews of wood preservation, preservatives and fire retardants, and processes which may have merit for the interested reader (**11-16**).

Older Preservative Systems

Table 1 gives a listing of the important commercial preservatives developed prior to the 21st century. Preservative systems are generally classified as oils or oilborne systems, or waterborne systems.

Table 1. Chronological listing of commercially important preservatives in North America

Preservative	Abbreviation	Year
Oils and Oilborne Systems		
Creosote	CR	1831
Copper naphthenate	CuN	1899
Pentachlorophenol	PCP	1931
Waterborne Systems		
Zinc chloride		1838
Fluor-chome-arsenic-phenol	FCAP	1930s
Acid copper chromate	ACC	1928
Chromated copper arsenate	CCA	1938
Ammoniacal copper arsenate	ACA	1939
Ammoniacal copper zinc arsenate	ACZA	1980s
Borates	SBX	1980s (in US)
Ammoniacal copper quaternary ammonium	ACQ	1990s
Copper azole, copper boron azole	CA, CBA	1990s

Creosote, an oil, is by far the oldest commercial system, coming to the fore with the development of Bethell's full-cell treating process. A brief history of creosote can be found in the literature (**17**). *Copper naphthenate* has been used as a wood preservative since 1889. It was first used in Germany and has been in commercial use since 1911, primarily as an amendment to creosote during World War II. It was recognized in the AWPA standards in 1949, but did not gain wide use for pressure treatments until the late 1980s when regulatory activities stimulated interest in the product because of its general use classification. Soon thereafter, copper naphthenate in #2 fuel oil began to be used for cross arms, bridges, utility poles, fence posts and lumber. Copper naphthenate is also used in non-pressure applications, including field-applied preservatives and coatings (**18**). The last major commercial oilborne system is pentachlorophenol (*penta*) which is a crystalline chemical compound (C_6Cl_5OH), formed by the reaction of chlorine on phenol. British patent 296,332, issued to W. Iwanowski and J. Turski (**19**), covers the use of di-, tri-, and polychlorinated phenols for wood-preserving purposes. In the U.S., L.P. Curtin (**20**) patented the use of "chlorine derivatives of coal-tar acids of higher molecular weight than the cresols", expressing a preference for chlorinated phenols. The production of chlorinated phenols in the US for wood preserving experiments did not begin until about 1930 (**3**).

Chromated copper arsenate (CCA) was patented by Kamesam (**21**) in 1938 and, until 2004, was the major waterborne preservative in use in North America. Three forms were standardized by AWPA (**22**), type A in 1953, type B in 1964, and type C in 1969 with type C dominating the marketplace today. The three types differ in their ratio of Cu:Cr:As. The voluntary withdrawal of CCA for most residential uses has greatly impact its consumption. An additional acidic system, *acid copper chromate* (ACC or Celcure®) was patented in 1928 by Gunn (**3**) and was standardized in the 1950s. The other major arsenical preservative, *ammoniacal copper arsenate* (ACA or Chemonite®) was standardized in 1950. It was modified by replacing some of the arsenic with zinc in the 1980s. This formulation is known as *ammoniacal copper zinc arsenate* (ACZA or Chemonite II®). Because ACA and ACZA are alkaline and impart the wood with vivid color, they have generally been used for industrial, rather than residential, products and are used to treat refractory western conifers.

During the early 1930s, Dr. Carl Schmittutz of Bad Kissingen, Germany, invented a process and formula for the preservation of wood. He organized the Osmose Wood Impregnating Company of Leipzig, Germany, and obtained patents for this process in many countries throughout the world, including the United States and Canada. The original Osmose patents described a preservative process using sodium fluoride, potassium bichromate, sodium arsenate, and dinitrophenol. This preservative was known in the industry as *FCAP*. Penetration of preservatives was achieved through the process of diffusion or "osmosis" into green wood or wood of high moisture content. One early commercial use of this preservative in the United States was a timber dipping and stacking process used by coal mines. These mines had a plentiful supply of green timbers that could be treated on-site for use as mine timbers. Another early use was the development of a paste formulation of FCAP preservative for in-place treatment of utility poles in the groundline area. Similar formulations and processes are still in use today for the groundline treatment of utility poles (**23, 24**).

Boron compounds offer some of the most effective and versatile wood preservative systems available today and combine the properties of broad-spectrum efficacy with low acute mammalian toxicity. Oxides of boron, the active ingredients in boron systems, are ubiquitous within the environment, are essential plant micronutrients, and are added regularly to agricultural land as fertilizers. Products treated with borates include the following: lumber, plywood, oriented strand board (OSB), siding, engineered wood, wood fiber-plastic composites, millwork, windows, doors, furniture, telephone poles, railroad ties, and log homes (**25**).

New Wood Preservatives

The major U.S. wood waterborne preservative was chromated copper arsenate (CCA), with oil-borne pentachlorophenol (penta) and creosote used to a lesser extent. CCA, along with smaller amounts of ammoniacal copper zinc arsenate (ACZA) used in western North America and the mid-west, have been used to treat about 80% of all treated wood products in the U.S. (**26**) and have been the only systems used to protect lumber for residential applications. In 2004, about 6.1 billion board feet of the SYP lumber was pressure treated with some type of preservative system (**27**), down 7% from the 2003 level due to residual inventory as treaters shifted to the new copper-based systems. CCA is an inexpensive and highly effective preservative and several EPA and other studies have found that CCA-treated wood poses negligible risk when used in residential construction, gardens, or for items such as playground equipment. However, recent public perceptions on possible arsenic exposure led to a rapid agreement to restrict the use of CCA-treated material to industrial applications by 2004 in the U.S. and in Canada. CCA has already limited use patterns in over 26 other countries. Since the largest market for treated wood is residential, this could reduce production of CCA-treated wood products by about 68%. Micklewright (**26**) showed that of the 581.4 million cubic feet of wood treated with waterborne preservatives, 477.8 million cubic feet were lumber and timbers, of which >98% was CCA. He reported a consumption of 74.4 million pounds of waterborne preservatives of which 98% would be 72.9 million lbs. By comparison, data for 2004 indicates that of the estimated 183.2 million pounds of waterborne preservatives used, only 58.2 million pounds were arsenicals representing 32% of the market for wood treated with waterborne preservatives (**28**). The balance of the waterborne preservative consumption was taken up by non-arsenicals such as borates, ammoniacal copper quats, copper azole, and acid copper chromate. These data can be seen in Table 2. The largest growth was seen in inorganic boron compounds, and the copper-rich ACQ and CA systems.

Most European countries have already limited CCA use with further restrictions being considered, and Japan had quickly changed earlier to preservatives that do not contain arsenic or chromium. Even though many uses of CCA were voluntarily removed after Dec. 31, 2003, many use patterns will continue. Most of these uses are considered industrial uses and include such

Table 2. Estimated iIndustry totals of each waterborne preservative consumed on a dry oxide concentrate basis in 2004 (from Vlosky 2006).

Waterborne Preservatives	Pounds (million, oxide basis, dry)	Percent of Total (All Compounds)
ACC (acid copper chromate)	1.6	0.9
ACQ (ammoniacal copper quaternary ammonium)	85.6	46.7
ACZA (ammoniacal copper zinc arsenate)	0.6	0.3
CA (copper azole)	29.6	16.2
CCA (chromated copper arsenate)	57.6	31.4
SBX (inorganic boron)	8.2	4.5
Total	183.2	

things as poles, pilings, timbers (6- x 6-in or greater), posts, and other large members. CCA-treated stock for the permanent wood foundation is a notable exception to the withdrawal scheme. These may be found in the AWPA Book of Standards (**22**).

Excellent coverage for most of the new generation preservative systems can be found in a soon to be published monograph (**29**). In the ACS Symposium Series (**30**), some thought provoking papers on new preservative systems can be found in the section on New Preservative-Protection Systems. In this series of papers, detailed discussions on ACQ, CA, CBA, and azoles can be found and will not be repeated here. Particularly interesting are the discussions on copper HDO (or copper xyligen/CX), a system recently standardized by AWPA, and PXTS (polymeric xylenol tetrasulfide), a new generation organic system also standardized by AWPA in 2005. An organic 'cocktail' system consisting of tebuconazole, propiconazole, and imidacloprid is also being developed. Table 3 lists many of the newer systems being researched or being brought to standard along with other systems discussed in detail in this series of articles.

Another interesting concept is to improve the performance of oranic preservative systems by adding non-biocidal additives, specifically antioxidants, metal chelators, or water repellent additives, that enhance the efficacy of and/or reduce degradation of organic biocides (**31-33**)

There are several concerns with the second generation ammoniacal/amine copper systems. First, they are more expensive, costing 2-4 times more than CCA. The second major concern is corrosion. The increase in corrosion may be linked to the dissolution of zinc leading to increased attack on iron at the oxygen-wood interface. Increased mold growth is attributed to the newer systems, supposedly because of the enriched nitrogen environment. Part of this problem is perception since the molds growing on this treated substrate are whitish and hence more visible that the darker molds which were always present on CCA-treated wood. Because of concerns with the leaching of copper in these copper-rich systems, aquatic toxicity is a concern for marine uses. Disposal may also present some challenges. From a treating perspective, these newer systems are less forgiving and formulation is trickier than with CCA.

Wood Preserving Processes

Little has changed commercially since the advent of pressure treatment of wood in the 1830s and the major treating processes have changed little since their inception. Bethell (**4**) patented the full-cell process for treating wood with creosote for the Royal Navy. In the early 1900s, two empty-cell processes were patented by Rüping (**34**) and Lowry (**35**). A detailed discussion of treating technology can be found in the literature (**36**) including discussions of the Oscillating Pressure Method, Alternating Pressure Method, Pulsation, Royal,

Table 3. New generation biocides for protection of wood.

Borates	Zirconium compounds
Uncomplexed Copper Systems	Quaternary Ammonium Compounds
• ACQ	• DDAC
• CA/CBA	• ABAC
• Iron quats	• BAC
Complexed Metal-based Systems	• ADBAC
• Copper-8-quinolinolate	• IPBC
• Copper HDO (copper xyligen)	Synthetic Pyrethroids
• CuN (waterborne)	• Permethrin
• ZnN	• Bifenthrin
• CDDC	• Cypermethrin
• TBTO	• Cyfluthrin
Zinc Systems	• Deltamethrin
• Zinc/boron versatate	Organic Agrochemicals
• Zinc 8-hydroxyquinoline	• TCMTB
• Zinc naphthenate	• Chlorothalonil
• Zinc + dicarboxylic acid hydrazide	• Dichlofluanid
	• Isothiazolone
PXTS (polymeric xylenol tetrasulfide)	• Fipronil
	• Imidachloprid
PTI (propiconazole, tebuconazole, imidacloprid)	• Methylene bis-thiocyanate
Polymeric betaine	
Copper betaine	
Azoles	
• Cyproconazole	
• Propiconazole	
• Tebuconazole	

HPP, sap displacement (PresCap, SlurrySeal, Gewecke), Cellon, Dow, MSU, MCI, and Multi-Phase Pressure processes. A modern day refinement of the Bethell process, the modified full-cell process, has found widespread use in North America. In this process, the initial vacuum is reduced to 12-16 in Hg for a shorter time period than the conventional full-cell process. The final vacuum may be extended thus yielding treated wood with lower solution absorption of 22-32 pcf compared to 40 pcf for the conventional full-cell process. This results in much lower shipping and drying costs for wood, especially those treated with waterborne systems.

There has been considerable interest in recent years in the development of vapor phase treatments for wood and wood composites. Treatment with gas phase components would eliminate the problems which exist with the liquid tension interface in current treatment practices (**37**). Trimethyl borate (TMB) has been successfully used to treat a wide range of wood composites (**38-42**).

Even more fascinating is the potential for treating wood using supercritical CO_2 ($ScCO_2$) as a carrier (**43-45**). In this case, there are no problems with the high surface tension associated with liquid treatment. Evans (**46**) reports that a plant for $ScCO_2$ treatment has been commissioned in Denmark. The use of $ScCO_2$ in composites is particularly appealing (**47**). Successful treatment of composites with a IPBC + silafluofen mixture has been achieved (**48**). $ScCO_2$ treatment on a wide range of composites showed minimal loss in mechanical properties for most composites (**49**). The notable exception was a large loss of bending strength in OSB. Previously, Kim *et al.* (**50**) had shown some loss in bending strength when treating southern pine with TCMTB using $ScCO_2$ treatment. For above-ground exposure in Hilo, Hawaii, Morrell *et al.* (**51**) showed excellent performance of plywood, MDF, particleboard, and OSB treated with tebuconazole using $ScCO_2$ treatment so long as retention was high enough. Kang *et al.* (**52**) noted that the movement of cyproconazole in the $ScCO_2$ treatment of ponderosa pine was influenced more by diffusion than by bulk flow.

The use of $ScCO_2$ was originally developed to extract flavors or decaffeinate coffee. Its use to improve treatability of Douglas-fir by extracting fatty acids has been investigated (**53**). $ScCO_2$ extraction has been used to extract PAHs and organo-chlorine compounds from wood, demonstrating its potential in waste recycling (**54, 55**).

Wood Modification Treatments

As an alternative to biocide treatment, considerable research has been undertaken on wood modification treatments to improve the durability of wood and wood-based materials. Modification techniques have long been the ‘holy grail' of wood science and have been described quite elegantly by Stamm (**56**) in his classic treatise on wood science. A number of treatments including formaldehyde crosslinking, resin impregnation, and acetylation have led to increased durability. Much of the North American work in recent years has centered at the US Forest Products Lab under the direction of Roger Rowell (**57, 58**). Most of this work has shown that acetylation and treatment with alkylene oxides enhance the durability of lignocellulosic materials. Graft copolymerizaton has been shown to have some value in improving the water repellency of wood (**59**), and polymer/boron systems have shown some efficacy (**60**). No commercial wood modification plants have been initiated in North America.

International work on modification has centered in Europe which has developed a European Thematic Network for Wood Modification (**61**). Successful European Conferences on Wood Modification were held in 2003 and 2005. Silicone compounds and their derivatives have shown promise as water repellent agents (**62, 63**) and some derivatives, such as silafluofen, have shown promise as termiticides (**48, 64**). Acetylation and furfurylation (**65**) continue to be studied and commercialized in Europe and Japan. Dizman *et al.* (**66**) have reported good results with the modification of alder and spruce particleboards using acetic, maleic, succinic, and phthalic anhydrides.

The major emphasis in Europe has been on the commercialization of heat-treated wood (**67, 68**). Commercial operations (15-20 companies) in Finland, France, Germany, Austria, Switzerland, and Holland are supplying the marketplace with 200-300,000 m^3 per year. Welzbacher and Rapp (**69**) indicate that commercially produced material from four processes increase the ground contact durability of wood, but not to the extent needed for ground contact applications. Evans (**46**) also presents an excellent discussion of wood modification, and a section on wood modification was presented at a recent conference (**70**). An excellent monograph on wood modification treatments is available (**71**).

Concerns and Challenges

Any discussion of trends must include some soul searching and crystal ball gazing. Some of the problems and concerns with second generation preservative systems were presented earlier. In all likelihood, these systems will remain viable for the next 10-15 years, but pressure on the use of heavy metals in preservative formulations, including copper, is worldwide. Several European countries have moved to eliminate systems containing copper. We are likely to see a move to all organic systems similar to the 'cocktail' system described earlier. These systems are already making headway in Europe. Organic systems have some shortcomings which will need to be overcome. They are generally expensive and have limited bioactivity. They are either oilborne or formulated as emlsifiable concentrates or mirco-emulsions. This means either a more expensive carrier system in the case of oilborne systems or a trickier formulating process in the case of emulsions. The trick is to understand how to "solubilize" the insoluble actives and then add a mixture of anionic and cationic surfactants which enable dilution with water and dramatically improve reduction of leaching. Appearance of the treated wood may be a problem, and if they are non-fixed systems, leaching could be problematic. The future may yield nano-particle systems and micro-emulsions. In all likelihood, we will be faced with a shorter service life for treated wood. Lower retentions mean less environmental impact. Envelope treatments may become more commonplace and barrier wraps may be employed to boost service life. Society is becoming chemophobic which makes non-biocidal treatments more attractive. The use of antagonistic microbes may lead to improved durability. The use of emerging technologies to marry wood and other materials could lead to products with increased durability.

Challenges abound for wood preservation and treated wood products. Four critical challenges face the industry: the mold issue, Formosan termite, engineered wood composites, and public education. The mold issue is an emotional one (**72**), and the CDC (**73**) has stated that no scientific proof exists that *Stachybotrys* has caused health problems. "*The mold issue has only become a problem because the public now perceives it as a health threat and . . . attorneys are bringing the issue before juries to seek large judgments.*" (**16**). The devastating hurricanes of 2005 have exacerbated the problem. Amburgey (**74**) has presented some common sense solutions to the clean-up of the mold problem. In all instances the first rule should be – cure the moisture problem!

The Formosan termite has become a \$2 billion+ problem in coastal areas south of the 35° parallel. In the affected areas, there is a high demand for treated wood. Borate-treated stock for above ground applications would seem a good start on a solution. Research in Hawaii, Mississippi, and Louisiana is aimed at elimination of this threat.

Engineered wood composites (EWC) are the wave of the future. Increasing the durability of these building materials is essential. Considerable research is being focused on EWC (**75**) as new EWC products, biocide addition methods, and preservative systems are brought forward. The extensive work being done on EWC durability modeling should prove a boon to our understanding (**76-78**).

Lastly, we simply must do a better job of public education. The public perceives a health risk with treated wood, whether real or imagined. Few are familiar with consumer information sheets which accompany treated wood products. It is incumbent on the industry to improve technology transfer so the public can deal with the real facts about wood preservation and treated wood products.

References

1. Jjensenius, J. H. URL http://www.stavechurch.org/index.html.
2. Crook, William. *Province of South Carolina Patent* No. 19, 1716.
3. Hunt, G. M.; Garratt, G. A. 1967. *Wood Preservation*. McGraw-Hill. New York, 1967.
4. Bethell, J. *British Patent* No. 7731, 1838.
5. Burt, H. P. *Minutes of Proceedings, Institution of Civil Engineers (London)* **1853**, *12*, 206-43.
6. Burnett, W. *British Patent* 7747, 1838.
7. Boulton, S. B. *Minutes of Proceedings, Institution of Civil Engineers (London)* **1884**, *78*, 97-211.
8. Hartig, R. *Textbook of the Diseases of Trees*. MacMillan and Co. New York, 1894. (translation from the German (*Lehrbuch der Baumkrankheiten*) by W. Sommerville).
9. Graham, R. D. *Wood Deterioration and Its Prevention by Preservative Treatments. Vol. I. Degradation and Protection of Wood*, Syracuse Univ. Press. Syracuse, N.Y, 1973, Chapter 1.
10. Wilkinson, J. G. *Industrial Timber Preservation*. Associated Business Press, London, England, UK, 1979.
11. Barnes, H. M. *Forest Prod. J.* **1985**, *35* (1), 13-22
12. Barnes, H. M. Proceedings, The First International Symposium on the Development of Natural Resources and Environmental Preservation, New Horizons in Agricultural Science, Institute of Natural Resources and Environment, Korea University, Seoul, Korea, 1992, 161-198.
13. Barnes, H. M. Proceedings, 1993 Eurowood Oxford Fire Conference, Timber Research & Development Association, Mansfield College, Oxford University, Oxford, England, UK, 1993.
14. Barnes, H. M. Int. Res. Grp. on Wood Pres., Doc. No. IRG/WP/93-30018, 1993.
15. Barnes, H. M.; Murphy, R. J. *Forest Prod. J.* **1995**, *45*(9), 16-26.
16. Freeman, M. H.; Shupe, T. F.; Vlosky, R. P.; Barnes, H. M. *Forest Prod. J.* **2003**, *53*(10), 8-15.
17. Barnes, H. M.; Ingram, L. L. *Proceedings, American Wood-Preservers' Association* **1995**, *91*, 108-117.
18. The Pacific Wood Preserving Companies. URL http:// www.pacificwood.com/ faq2.cfm?FaqID=70
19. Iwanowski, W.; Turski, J. *British Patent* No. 296,332, 1928
20. Curtin, L. P. *US Patent* No. 1,722,323, 1929.
21. Kamesan, S. *U.S. Patent* No. 2,106,978, 1938
22. American Wood-Preservers' Association. *Book of Standards*. AWPA, Selma, AL, 2005.
23. McNamara, W.S. Proceedings, First International Conference on Wood Protection with Diffusible Preservatives, Proceedings 47355, Forest Products Research Society, Madison, WI, 1990, 19-21.

23. McNamara, W.S. Proceedings, First International Conference on Wood Protection with Diffusible Preservatives, Proceedings 47355, Forest Products Research Society, Madison, WI, 1990, 19-21.
24. Osmose. URL http://www.osmose.com/about/history/
25. US Borax. URL http://www.borax.com/borates2g.html (see also http://www.borax.com/wood/)
26. Micklewright, J. T. Report to the Wood Preserving Industry in the United States, American Wood-Preservers' Association, Granbury, TX, 1998.
27. Southern Forest Products Association. URL http://www.sfpa.org/industry_statistics/SP_treated.htm
28. Vlosky, R. P. Report to the Southern Forest Products Association, Kenner, LA, 2006.
29. Freeman, M. H.; Nicholas, D. D.; Schultz, T. P. *Environmental Impacts of Treated Wood*, CRC Press, Boca Raton, FL (in press).
30. Goodell, B.; Nicholas, D. D.; Schultz, T. P. (eds.). *Wood Deterioration and Preservation: Advances in Our Changing World*, ACS Symposium Series 845, Am. Chem. Society, Washington, DC, 2003.
31. Schultz, T.P., D.D. Nicholas. *Phytochemistry* **2000**, *54*, 47-52.
32. Schultz, T.P.; Nicholas, D. D. *Phytochemistry* **2002**, *61*, 555-560.
33. Schultz, T.P.; Nicholas, D. D; Henry, W. P.; Pittman, C. U.; Wipf, D. O.; Goodell, B. *Wood & Fiber Science* **2005**, *37*(1), 175-184.
34. Rüping, M. *U. S. Patent* No. 709,799, 1902
35. Lowry, C. B. *U. S. Patent* No. 831,450, 1906.
36. Barnes, H. M. Proceedings, Wood Protection 2006, Forest Products Society, Madison, WI (in press).
37. Scheurch, C. 1968. *Forest Prod. J.* **1968**, *18*(3), 47-53.
38. Hashim, R; Dickinson, D. J.; Murphy, R. J.; Dinwoodie, J. International Res. Group on Wood Preservation, Doc. No. IRGIWP/3727-92, 1992.
39. Hashim, R; Dickinson, D. J.; Murphy, R. J.; Dinwoodie, J. *Forest Prod. J.* **1994**, *44*(10), 73-79.
40. Barnes, H. M.; Murphy, R. J. *Wood & Fiber Science* **2005**, *37*(3), 379-383.
41 Barnes, H. M.; Murphy, R. J. *Composites Part A: Applied Science and Manufacturing* (in press).
42. Murphy, Richard J; Barnes, H. Michael; Dickinson, David J. Proceedings, Enhancing the Durability of Lumber & Engineered Wood Products, Publication No. 7249, Forest Products Society, Madison, WI, 2002, 251-255.
43. Junsophonsri, S. *Solubility of biocides in pure and modified supercritical carbon dioxide*. Unpublished Master's thesis, Oregon State Univ., Corvallis, OR, 1994.
44. Morrell, J.J.; Levien, K. L.; Demessie, E. S.; Kumar, S.; Smith, S.;Barnes, H. M. 1993. *Proc., Canadian Wood Preservation Assoc.* **1993**, *14*, 6-25.

45. Morrell, J.J.; Levien, K. L.; Demessie, E. S.; Barnes, H. M. Proc., Inter. Conf. on Wood Poles & Piles, Colorado State Univ., Ft. Collins, CO, 1994, 325-337.
46. Evans, P. *Forest Prod. J.* **2003**, *53*(1), 14-22.
47. Oberdorfer, G.; Humphrey, P. E.; Leichti, R. J.; Morrell, J. J. International Research Group on Wood Preservation, Doc. No. IRG/WP 00-40175, 2000.
48. Tsunoda, K.; Muin, M. International Research Group on Wood Preservation, Doc. No. IRG/WP 03-40251, 2003.
49. Muin, M.; Adachi, A.; Tsunoda, K. International Research Group on Wood Preservation, Doc. No. IRG/WP 01-40199, 2001.
50. Kim, G. –H; Kumar, S.; Demessie, E. S.; Levien, K. L.; Morrell, J. J. International Research Group on Wood Preservation, Doc. No. IRG/WP 97-40080, 1997.
51. Morrell, J. J.; Acda, M. N.; Zahora, A. R. International Research Group on Wood Preservation, Doc. No. IRG/WP 05-30364, 2005.
52. Kang, S-M.; Levien, K. L.; Morrell, J. J. *Wood & Fiber Science* **2006**, *38*(1), 64-73.
53. Kumar, S.; Morrell, J. J. International Research Group on Wood Preservation, Doc. No. IRG/WP 93-40008, 1993.
54. Schrive, L.; Perre, C.; Labat, G. International Research Group on Wood Preservation, Doc. No. IRG/WP 98-50101-18a, 1998.
55. Legay, S.; Marchal, P.; Labat, G. International Research Group on Wood Preservation, Doc. No. IRG/WP 98-50101-18b, 1998.
56. Stamm, A. J. *Wood and Cellulose Science.* Ronald Press. New York, 1964.
57. Rowell, R. M. *Handbook on Wood and Cellulosic Materials.* Marcel Dekker, Inc. New York, 1991
58. Rowell, R. M. 1999. Proceedings, International Workshop of Frontiers of Surface Modification and Characterization of Lignocellulosic Fibers. Fiskebackskil, Sweden, 1999, 31-47.
59. Morshed, M. M.; Nicholas, D. D.; Pittman, Jr., C. U.; Schultz, T. P. International Research Group on Wood Preservation, Doc. No. IRG/WP 03-40257, 2003.
60. Murphy, R. J.; Barnes, H. M.; Gray, S. M. *Forest Prod. J.* **1995**, *45*(9), 77-81.
61. Jones, D.; Homan, W.; Van Acker, J. International Research Group on Wood Preservation, Doc. No. IRG/WP 03-40268, 2003.
62. Rowell, R.M.; Banks, W. B. Water repellency and dimensional stability of wood. USDA Forest Service, Forest Products Laboratory, General Technical Report FPL-50, Madison, WI, 1985, 24 pp.
63. Hager, R. International Research Group on Wood Preservation, Doc. No. IRG/WP 95-30062, 1995.
64. Adams, A. J.; Jermannaud, A.; Serment, M. –M. 1995. International Research Group on Wood Preservation, Doc. No. IRG/WP 95-30069, 1995.

65. Balfas, J.; Evans, P. D. International Research Group on Wood Preservation, Doc. No. IRG/WP 94-40017, 1994.
66. Dizman, E.; Yildiz, Ü.; Kalaycioglu, H.; Yildiz, S; Temiz, A.; Gezer, E. D. International Res. Group on Wood Preservation, Doc. No. IRGIWP/05-40300, 2005.
67. Tjeerdsma, B. F.; Boonstra, M.; Pizzi, A.; Tekely, P.; Militz, H. 1998. *Holz als Roh- und Werkstoff* **1998**, *56*(3), 149-153.
68. Welzbacher, C. R.; Rapp, A. O. International Research Group on Wood Preservation, Doc. No. IRG/WP 02-40229, 2002.
69. Welzbacher, C. R.; Rapp, A. O. International Research Group on Wood Preservation, Doc. No. IRG/WP 04-40312, 2004.
70. Barnes, H. M. (ed). Proceedings, Wood Protection 2006, Forest Products Society, Madison, WI (in press).
71. Hill, Callum A. S. *Wood Modification: Chemical, Thermal and Other Processes*. Wiley. New York, 2006.
72. Vlosky, R. P.; Shupe, T. F. *Forest Product Journal* **2004**, *54*(12), 289-295.
73. Centers for Disease Control and Prevention. URL http://www.cdc.gov/health/mold.html
74. Amburgey, T.L. MSU Extension Forestry Technical Note, MTN-E14, Mississippi State University, 2005, 2 pp.
75. Kirkpatrick, J. W.; Barnes, H. M. International Research Group on Wood Protection, Document No. IRG/WP 06-(in press), 2006.
76. Curling, S.F.; Carll, C. C.; Micales, J. A.; TenWolde, A.; Winandy, J. E.. *Holzforschung* **2002**, *57*(1), 8-12.
77. Curling, S. F.; Clausen, C. A.; Winandy, J. E. *International Biodeterioration & Biodegradation*, **2002**, *49*, 13-19.
78. Curling, S. F.; Clausen, C. A.; Winandy, J. E. *Forest Prod. J.* **2002**, *52*(7/8), 34-39.

Chapter 36

Trends in Wood Protection in Asia and Oceania: A Perspective

Alan Preston and Lehong Jin

Chemical Specialties, Inc., 200 East Woodburn Road, Suite 350, Charlotte, NC 28217

Introduction

The availability of plentiful native wood resources in many countries in the western Pacific established a long tradition of the use of wood for structural applications in utility, construction and agricultural applications. In the tropical regions of Asia and Oceania, this native resource has been characterized by a wide variety of hardwoods, many of which had excellent natural durability. In the case of Japan and New Zealand, the dominant native species were softwood species with good durability in construction applications but with few species having good natural durability in ground contact. These resource realities shaped the growth and applications of wood protection technologies in the region. Other key drivers have been the presence or otherwise of termites and climate and traditions with the use of other materials in place of wood.

As a consequence of these various factors, both New Zealand and Japan have long established wood preservation industries and technology bases for the treatment of plantation softwoods, while Australia and the continental countries of Asia traditionally relied on their durable hardwood resources for many applications until quite late in the 20th Century. However, the establishment and harvesting of softwood plantations in Australia saw a move towards wood preservation that earlier was restricted to lyctid control of susceptible hardwoods as well as treatments of some less durable Eucalyptus species. In Malaysia, Thailand and tropical China the establishment and eventual rotation harvesting of large rubberwood plantations gave rise over the last quarter century to a thriving treatment industry based around borate treatments of rubberwood lumber for use in furniture applications.

While it is convenient in this survey to group Asia and Oceania together as a geographically based package, in reality most activity is driven by national rather than regional factors, and these are resource, population, wealth, tradition and application driven.

Asia

Japan

Like many northern hemisphere countries with a temperate climate, Japan is a society with a long tradition of using wood for many applications. This relates to the historic availability of a natural softwood resource of species with an appropriate level of natural durability. As the population has increased the pressures on the natural resource increased and the availability of the naturally durable old growth trees diminished. Plantation forestry of these species, especially *Cryptomeria japonica* (Sugi) and *Pinus densiflora* (Japanese pine) flourished, but the utilization of these plantation forests necessitated the development and growth of the wood preservation industry. This, coupled with the importation of large volumes of Western hemlock (*Tsuga heterophylla*) from Canada and also Douglas fir, gave rise to the primary structural building materials used in Japan today, all of which require treatment to control either termite infestation, especially by *Coptotermes formosanus* (Formosan termite), which is now endemic in regions from Tokyo south and west, or from decay. The wood treating industry is thus well developed with many pressure treatment plants, both large and small. Many of larger plants are associated with construction companies, but independent treaters are also common. Japan has well-established standards and procedures in the wood treating area, although third party inspection of treatments is rather uncommon.

Japan is a country of considerable natural beauty and the population lives in a relatively small proportion of the land area. These two factors, coupled with the high overall wealth of the country may be key drivers in leading to a society that is quite environmentally sensitive and in the mid-1990s Japan was one of the first countries to switch away from CCA because of ecotoxicological concerns with arsenic and chromium as used at wood treating plants. Today ACQ and copper azole are the dominant wood preservatives in commercial use. In addition, a number of other preservatives, quats and borates among them, are used for above ground-only applications.

In the area of research in wood preservation, Japan has a strong and flourishing research community based around several prominent university programs, the government research organization, as well as several industrial laboratories. There is a history of innovative research, which continues to this day. Japanese society is very aware of product quality and appearance, and this is reflected also in the utilization and research on protected wood products. Perhaps because markets for many products in Japan tend to be far more fragmented than is the case in North America, this has the positive effect of allowing relatively niche product developments to flourish, and this has been the case in wood protection markets, too, in Japan.

Korea

The situation in Korea is somewhat similar to that in Taiwan but recent growth has been more robust in this market. In Korea there has been no historic widespread use of treated wood, and it appears that some of the activity in the last ten years has been focused on export to Japan has the main target. Trends in the types of wood preservatives used in Korea have generally followed that of Japan, in part because of the export sector needing to parallel that of Japan.

Korea has an active academic research base in wood protection. This is focused around research at several universities as well as the Forest Research Institute of Korea.

China Region

Taiwan

Taiwan's involvement in wood protection is small although in the last few years there has been increased activity as commercial contacts with China have increased. The market in Taiwan is small, with only a small domestic use of treated wood, due to a weak tradition in wood utilization. The Taiwan market has had an export based focus, especially towards Japan, but the increasing economic wealth of Taiwan, coupled with the recent growth of the industry in China, will presumably have an impact on such export outlets.

Within Taiwan there is a small research base, but it would seem unlikely that this will grow unless there are significant cultural or industry changes to provide further impetus.

Mainland China

As in many other areas of commerce, the wood preservation industry has undergone rapid change in the last few years. Traditionally, the industry in China was dominated by creosote treatments of railroad ties and poles, but this has been in decline with the advent of alternative substrates for these applications. In the tropical regions, especially Hainan Island, the development and harvesting of extensive rubberwood plantations has given rise to the establishment of treating plants that specialize in borate and mold prevention based treatments for rubberwood for use in furniture and other interior applications. There is also some use of pressure treated fire retardants in China.

However, it is in the area of water-based preservatives for lumber applications that the wood treating industry is seeing rapid growth and change. The driving force for the popularity of using treated wood has come from the architects, designers and building specifiers. The uses of treated wood are mainly in the following applications:

- Park and lakeside walkway projects controlled by government agencies
- Residential decking and decorative garden applications
- Structural house cladding, wooden bridges, fences, etc.

The number of chemical treating plants has increased dramatically during last two years. More than 150 treating plants are operational in China with a capacity of up to 300 to 500 m^3 treated wood production per month. The Shanghai area has the biggest share of the treating plants with 40 to 50, Beijing has 15, and then the other three most developed provinces are Guangdong, Jiangsu, and Zhejiang with 10 to 15 plants each. The rest of the country has around 30 to 35 plants. The demand for treated wood is on a steep upward growth curve, and as a consequence more and/or larger treating plants can be expected to come on line in the next several years. It appears increasingly likely that the treated wood market in China will become significant within Asia and perhaps eventually also as a global influence.

The major preservative used in China is still CCA and that is anticipated to remain so for the next 5 to 10 years since the major players in the industry are speaking for the continuing safe use of CCA and the consumers do not know the difference since their main criteria is treated wood with a green color. CCA is used at 65-70% of treating plants and while ACQ is being used in 30-35% of treating plants for domestic applications. Some usage of copper azole is also reported.

The main driver in preservative choice is cost. There are eight known domestic preservative manufacturers in China and these are concentrated in three areas, namely Guangzhou, Shanghai and in Jiangxi province. Imported wood preservative concentrates are also used in some plant. In 2005 it is anticipated that the total treated wood used could reach 400,000 m^3.

While a considerable amount of the wood being treated in China is exported to Japan, especially with ACQ formulations, the greatest change in the last few years has been the rise of a significant domestic market for treated lumber. Unlike in some markets around the world, this newly arisen China treated lumber market puts an emphasis on appearance and lumber condition in service, although in-plant quality control let alone a third party system remains dormant. The emphasis is to make sure the wood looks uniformly green and well penetrated and major effort is taken to ensure good looking wood surfaces. After treatment and drying, the treated lumber surfaces are often re-planed or sanded if necessary.

For government projects appearance and preservative type are often carefully specified and inspected. Both domestically sourced as well as imported treated wood products are used, and this is often southern yellow pine lumber sourced from the U.S. When southern pine is used, surface checking of the treated product is considered undesirable and hence the use of surface stain coatings, kerfing or profiling, or the use of pressure treated water repellents has started to gain more influence than is the case in other countries.

The China market for wood preservatives has been reliant on a new standards system for wood treatment that is based around the AWPA and the European standards. However, as noted above, quantitative quality control or third party inspection enforcement is lacking. It is also noted that most preservative standards listed by the AWPA, such as for CCA and various types of ACQ, have been followed, but because the AWPA standards do not list specific formulation information, some of the preservative products made in China are distinctly different from those used in the U.S., Japan or other countries. These differences have the potential to unwittingly impact the long-term efficacy of some of these domestically produced preservatives. Moreover, the absence of a well established third party quality inspection scheme for either the preservatives or treated wood, may serve to exacerbate such potential problems over time.

China has a significant research base in wood products protection, which is centered in the forest product based Universities such as the Beijing, Nanjing and Northeast Forestry universities, and research institutes such as the Forestry Academy of Science in Beijing and Guangdong Forestry Research Institute. However, there has also been a tendency for some key researchers involved in such research to embark on associated entrepreneurial activity in the industry, which clearly has the potential to weaken the development of innovative and fundamental knowledge but rather put more emphasis on application research. This may also create potential conflicts of interest as well as potential concerns in regards to intellectual property rights.

Malaysia

Perhaps because of its historic political association with Britain, Malaysia has long had a comparatively well-developed market, albeit small and static, for wood preservatives. CCA has been, and remains, the dominant preservative in Malaysia, although LOSP treatments enjoy a small niche and rubberwood treatments based around borate systems have proliferated over the years. The Malaysian market is marked by the presence of many small wood treatment plants, which are unlikely to change from their current preservatives unless some outside pressure necessitate such a change.

The research base in wood protection was centered around the Forest Research Institute Malaysia in Kuala Lumpur, but in recent years this has declined somewhat while universities in Sarawak and Penang have become more active than in the past.

Oceania

Polynesia, Melanesia and Micronesia

The islands of the Pacific that one could include under this section cover a vast area of the world's surface but the habitable land mass, with the exception of New Guinea is very small indeed. Leaving out Hawaii, as being the predominant Polynesian part of the U.S., and New Zealand for convenience even though it is part of Polynesia too, the use of treated wood in this diffuse area tends to be dominated by former colonial ties, or the relative proximity of larger supplier countries such as New Zealand. CCA remains the dominant preservative used throughout the region. Small treatment plants dot the islands, but maintaining stable manufacturing staff at these facilities is a major challenge for continuity of supply from them. Some regions of Melanesia had a significant natural hardwood resource, but these have diminished with extraction harvesting and much of the treatment carried out today is with imported softwoods from plantations. An exception to this is the pine plantations of Fiji, which provide a valuable domestic wood resource.

Australia

Australia is a large, relatively under-populated country, with a strong tradition in the use of wood products, but with a wood treatment market that in many sectors has developed relatively recently. The compelling factor that has influenced the use of wood and the growth of various sectors of the wood treatment market is the presence of large areas of native hardwood forests, many of whose species had excellent natural durability that made them attractive for utilization without necessitating treatment. Indeed, some native Australian species such as Turpentine (*Syncarpia*) have sufficient natural durability that they served well untreated as marine piling. The use of durable hardwoods continues for both roundwood and lumber applications, in some instances requiring treatment or removal of the thin sapwood zone and in other case not. Some seventy years ago, the harvesting and utilization of lyctid susceptible hardwoods gave rise to a dip treatment borate market for the protection of such hardwoods from powder post beetle infestation. At the time, this was one of the first successful applications of borate technology in wood protection, a practice that continues to this day.

As the softwood plantation of slash, radiata and hoop pine matured, so to did the utilization of these species in applications that required protection from biodeterioration. For these applications CCA has been the dominant preservative. Initially, many of these applications were agricultural, but over the last quarter century there has been a steady increase in the use of CCA treated lumber for construction applications. As the resource of durable hardwood pole stock diminished, there was a commensurate increase in the use of treated poles, especially with CCA. This has been generally successful but Australia has included two unique situations with this. One was the occurrence in the late 1970s of soft rot in Spotted Gum (*Eucalyptus maculata*) poles treated with CCA. Research revealed that this was caused by the greater susceptibility of lower lignin hardwood species, even when treated, to soft rot in the treated zone as compared with softwoods and high lignin hardwoods with comparable treatment retentions. The issue was overcome by raising the CCA retention requirements for such poles. The other factor that water-based metallic preservatives such as CCA have faced in some applications in Australia relates to the environment. In many regions Australia has long period of hot dry weather that creates the potential for rapidly moving wildfires. With metallic water-based preservatives this has been known to lead to several instances of spectacular destruction of the treated material by after-glow of the wood following fire. While CCA has remained the most used preservative, it is anticipated that in 2006 residential applications of treated wood will be mandated to be non-arsenical, although this is a relatively small segment of the treated wood market in Australia.

Besides the use of CCA for many applications, over the last twenty years light organic solvent preservatives (LOSP) have gained favor for some specialized products in Australia. LOSP treatments have moved in some cases into exposed applications such as timber decking, with mixed success, and more recently into insecticidal treatments to protect framing lumber from termite attack. This latter market has also recently seen the rise of surface applied insecticide treatments using water-based formulations for the protection of framing lumber from termite attack. At this time there are lingering questions in regards to the effectiveness of both the organic solvent and water based low penetration treatments, especially if cut ends of the wood product are not retreated to reinstate a protective complete envelope around the untreated inner zone.

New Zealand

For many years New Zealand has had the distinction of being the highest per capita user of treated wood in the world, and this remains so today, albeit at a somewhat lower level than in the past. While the native New Zealand forests were predominately Podocarpus softwoods excellent properties for many applications, these were essentially cleared or consumed many years ago and replaced, after much highly innovative research dating to the nineteenth century, with northern hemisphere pine species, especially *Pinus radiata*, and Douglas fir, while more recently Eucalypt plantations have become common as additional fiber feedstock resource to supplement the fiber predominantly obtained from radiate pine. New Zealand is a largely agricultural country, with a low population that is concentrated in several large and hilly cities, and a damp temperate coastal climate. These factors have long fueled the use of treated wood in agricultural, utility and construction applications. There has been limited use of creosote for railroad sleepers (crossties) over the years, and also some use of pentachlorophenol for pole treatments. The vast majority of the treated wood used in exterior applications has been CCA and this dates back some 50 years now. It has to be said that the use of CCA has been extremely successful in New Zealand and this remains so to this day. The use of borates for the protection of internal framing lumber was pioneered in New Zealand in the 1950s. While there has been no requirement for termite protection in New Zealand during this time (there have been no native or introduced termites that attack forest products in the country) the borate treatment served to protect the plantation pine framing from *Anobiid* attack. This also proved to be highly successful and cost effective. In the mid-1990s changes in the New Zealand building code allowed the use of pine framing without borate or any other treatment. Within a few years a substantial rise in internal decay in residential construction was observed, and subsequently borate treatments are now common as a means to protect the framing lumber from such decay, be it incidental or building practice related.

The use of LOSP treatments has grown steadily in New Zealand over the last quarter century, and remains a preferred protectant for joinery (millwork) applications and plywood. Whether oil-price rises or VOC considerations eventually impact such uses in New Zealand remains to be seen.

A major driver for the treated wood industries of both Australia in New Zealand over the last half-century has been the presence of strong research bases in both countries. In New Zealand the erstwhile government-run Forest Research Institute was the predominant force, while in Australia the Forest Products division of CSIRO, as well as state laboratories in Queensland and New South Wales, were the key resources for many years. As in so many other countries, government funding of such laboratories has evolved rapidly in the last 15 years, and the overall total scientific resource in the two countries has been drastically reduced. An interesting outcome of this has been the appearance of ensis, which was launched in 2004 as a combination of CSIRO Forest Products and Forest Research (the previously privatized Forest Research Institute). Such trans-national research cooperatives or entities are highly unusual, and it remains to be seen whether the combined entity will be able to stem the seemingly inexorable decline in funding for forest products research.

Conclusion

As we conclude this perspective on the wood preservative markets in Asia and Oceania, it is possibly worthwhile to consider the current relative global market values for this area of commerce. The value of preservative markets around the world in 2004 showed that greater than 50% of the value was concentrated in North America, with near 20% in Europe and the rest of the world taking up the remaining part. While Latin America is a small, but potentially significant growth market, it would be fair to say that Australasia remains relatively static in size, although export opportunities for specialty wood treatments offer growth opportunities beyond GDP growth. Asia is essentially many distinct markets, not all of which have been addressed in this perspective, and while Japan remains the largest market for treated wood in Asia with the strongest tradition of such use, it is clear that China and its rapidly growing middle class offers the greatest growth opportunity for the foreseeable future, albeit from a small base.

Indexes

Author Index

Subject Index

B

C

E

F

I

J

K

L

P

Q

R

S

T

U

V

W

X

Y

Z